1997
MASONRY
Codes
and
Specifications

Masonry
Institute of
America

CRC Press
Boca Raton   New York

Front Cover

Performing Arts Center
San Luis Obispo, CA

Architect Firm
DMJM/Warnecke
Los Angeles, CA

Structural Engineering Firm
Nabih Youssef & Associates
Los Angeles, CA

Photo by Thomas Escobar, MIA

General Contractor
Centex Golden Construction Co.
San Luis Obispo, CA

Masonry Contractor
R & R Masonry, Inc.
North Hollywood, CA

Block Manufacturer
Truestone, Inc.
Oxnard, CA

Library of Congress Cataloging-in-Publication Division

Catalog record available from the Library of Congress

Published by Masonry Institute of America and CRC Press LLC. Direct all inquiries to CRC Press LLC, 2000 Corporate Blvd., N.W., Boca Raton, FL 33431.

MASONRY ORGANIZATIONS

The organizations listed below provide information on the material, design and construction of masonry systems.

BRICK INSTITUTE OF CALIFORNIA   800/924-2742
2550 Beverly Boulevard   Fax: 213/388-6958
Los Angeles, CA. 90057-1085

CONCRETE MASONRY ASSOCIATION   916/722-1700
OF CALIFORNIA AND NEVADA   Fax: 916/722-1819
6060 Sunrise Drive, #1875
Citrus heights, CA 95610

MASONRY INSTITUTE OF SACRAMENTO   916/966-7666
7844 Madison Avenue   Fax: 916/966-1411
Fair Oaks, CA 95628-3587

MASONRY INSTITUTE   510/253-8537
62 Hacienda Circle   Fax: 510/254-4963
Orinda, CA 94563-1735

MASONRY INSTITUTE OF AMERICA   213/388-0472
2550 Beverly Boulevard   Fax: 213/388-6958
Los Angeles, Ca 90057-1085

MASONRY INSTITUTE OF FRESNO c/o KSM   209/435-3533
375 W. Bedford #103
Fresno, CA 93711

MASONRY RESOURCE OF SOUTHERN CALIFORNIA   800/339-4977
2550 Beverly Boulevard
Los Angeles, CA 90057-1085

SANTA CLARA-SAN BENITO JOINT MASONRY INSTITUTE   510/831-9090
62 Hacienda Circle
Orinda, CA 94563

VENTURA MASONRY INSTITUTE   805/487-3722
Kincaid Masonry
1910 Lincoln Court
Oxnard, CA 93033

WESTERN STATES CLAY PRODUCTS ASSOCIATION
2550 Beverly Boulevard
Los Angeles, CA 90057-1085

INTRODUCTION

The Masonry Institute of America believes that the best way to extend and improve the use of masonry is through education and the dissemination of information. Accordingly, the masonry promotion groups of California have joined together uniting their efforts to provide more information to the design professional and builder.

This manual, published by the Masonry Institute of America and CRC Press is intended for use as a ready reference which furnishes, in one document, the various code requirements for masonry from the Uniform Building Code and Standards, the California State Building Code and the American Society for Testing and Materials (ASTM) Standards that govern the specification of quality and testing of materials.

Included in this manual are Guide Specifications for masonry construction set forth in the CSI format with notes to the specifier.

This "1997 Masonry Codes and Specifications" manual was prepared by the Masonry Institute of America, Los Angeles, California. Every effort has been taken to ensure that all data and information furnished is as accurate as possible. The editors and publishers cannot assume or accept any responsibility or liability, including liability for negligence, for errors or oversights in this data and information, in the use of such information.

Edited by:

John Chrysler, P.E.   and   **Thomas Escobar, Assoc. AIA**
Executive Director   *Architectural Representative*

Also making significant contributions to this manual were:

Petra Stadtfeld   **James E. Amrhein, S.E.**
Type Setting   *Consultant*

Masonry Institute of America
Los Angeles, California
http://www.masonryinstitute.org

TABLE OF CONTENTS

EXCERPTS FROM

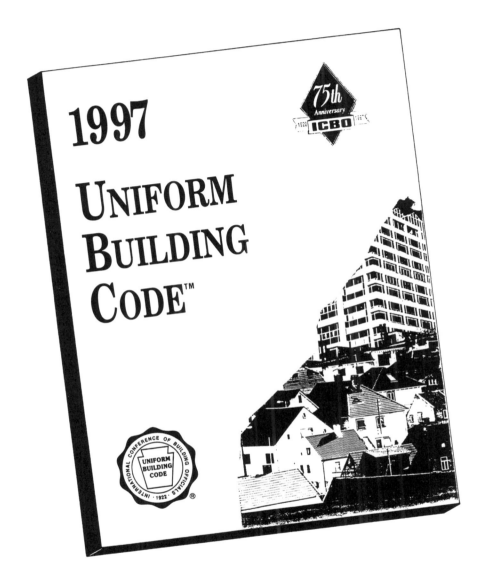

"PORTIONS OF THE 1997 UNIFORM BUILDING CODE™, COPYRIGHT © 1997, AND THE 1997 UNIFORM BUILDING CODE STANDARDS, COPYRIGHT © 1997, ARE REPRODUCED, HEREIN, WITH PERMISSION OF THE PUBLISHER, THE INTERNATIONAL CONFERENCE OF BUILDING OFFICIALS."

International Conference of Building Officials
5360 WORKMAN MILL ROAD
WHITTIER, CALIFORNIA 90601-2298
(800) 284-4406 • (562) 699-0541

Preface

The *Uniform Building Code*™ is dedicated to the development of better building construction and greater safety to the public by uniformity in building laws. The code is founded on broad-based principles that make possible the use of new materials and new construction systems.

The *Uniform Building Code* was first enacted by the International Conference of Building Officials at the Sixth Annual Business Meeting held in Phoenix, Arizona, October 18-21, 1927. Revised editions of this code have been published since that time at approximate three-year intervals. New editions incorporate changes approved since the last edition.

The *Uniform Building Code* is designed to be compatible with related publications to provide a complete set of documents for regulatory use. See the publications list following this preface for a listing of the complete family of Uniform Codes and related publications.

Code Changes. The ICBO code development process has been suspended by the Board of Directors and, because of this action, changes to the *Uniform Building Code* will not be processed. For more information, write to the International Conference of Building Officials, 5360 Workman Mill Road, Whittier, California 90601-2298. An analysis of changes between editions is published in the *Analysis of Revisions to the Uniform Codes*.

Marginal Markings. Solid vertical lines in the margins within the body of the code indicate a change from the requirements of the 1994 edition except where an entire chapter was revised, a new chapter was added or a change was minor. Where an entire chapter was revised or a new chapter was added, a notation appears at the beginning of that chapter. The letter **F** repeating in line vertically in the margin indicates that the provision is maintained under the code change procedures of the International Fire Code Institute. Deletion indicators (◆) are provided in the margin where a paragraph or item listing has been deleted if the deletion resulted in a change of requirements.

Three-Volume Set. Provisions of the *Uniform Building Code* have been divided into a three-volume set. Volume 1 accommodates administrative, fire- and life-safety, and field inspection provisions. Chapters 1 through 15 and Chapters 24 through 35 are printed in Volume 1 in their entirety. Any appendix chapters associated with these chapters are printed in their entirety at the end of Volume 1. Excerpts of certain chapters from Volume 2 are reprinted in Volume 1 to provide greater usability.

Volume 2 accommodates structural engineering design provisions, and specifically contains Chapters 16 through 23 printed in their entirety. Included in this volume are design standards that have been added to their respective chapters as divisions of the chapters. Any appendix chapters associated with these chapters are printed in their entirety at the end of Volume 2. Excerpts of certain chapters from Volume 1 are reprinted in Volume 2 to provide greater usability.

Volume 3 contains material, testing and installation standards.

Metrication. The *Uniform Building Code* was metricated in the 1994 edition. The metric conversions are provided in parenthesis following the English units. Where industry has made metric conversions available, the conversions conform to current industry standards.

Formulas are also provided with metric equivalents. Metric equivalent formulas immediately follow the English formula and are denoted by "For **SI:**" preceding the metric equivalent. Some formulas do not use dimensions and, thus, are not provided with a metric equivalent. Multiplying conversion factors have been provided for formulas where metric forms were unavailable. Tables are provided with multiplying conversion factors in subheadings for each tabulated unit of measurement.

SECTION 104 — ORGANIZATION AND ENFORCEMENT

104.2.7 Modifications. When there are practical difficulties involved in carrying out the provisions of this code, the building official may grant modifications for individual cases. The building official shall first find that a special individual reason makes the strict letter of this code impractical and that the modification is in conformance with the intent and purpose of this code and that such modification does not lessen any fire-protection requirements or any degree of structural integrity. The details of any action granting modifications shall be recorded and entered in the files of the code enforcement agency.

104.2.8 Alternate materials, alternate design and methods of construction. The provisions of this code are not intended to prevent the use of any material, alternate design or method of construction not specifically prescribed by this code, provided any alternate has been approved and its use authorized by the building official.

The building official may approve any such alternate, provided the building official finds that the proposed design is satisfactory and complies with the provisions of this code and that the material, method or work offered is, for the purpose intended, at least the equivalent of that prescribed in this code in suitability, strength, effectiveness, fire resistance, durability, safety and sanitation.

The building official shall require that sufficient evidence or proof be submitted to substantiate any claims that may be made regarding its use. The details of any action granting approval of an alternate shall be recorded and entered in the files of the code enforcement agency.

104.2.9 Tests. Whenever there is insufficient evidence of compliance with any of the provisions of this code or evidence that any material or construction does not conform to the requirements of this code, the building official may require tests as proof of compliance to be made at no expense to this jurisdiction.

Test methods shall be as specified by this code or by other recognized test standards. If there are no recognized and accepted test methods for the proposed alternate, the building official shall determine test procedures.

All tests shall be made by an approved agency. Reports of such tests shall be retained by the building official for the period required for the retention of public records.

*Only portions of this section are shown which are particularly applicable to masonry construction. For additional information see the UBC.

Chapter 16

STRUCTURAL DESIGN REQUIREMENTS

NOTE: This chapter has been revised in its entirety.

Division I—GENERAL DESIGN REQUIREMENTS

SECTION 1611 — OTHER MINIMUM LOADS

1611.1 General. In addition to the other design loads specified in this chapter, structures shall be designed to resist the loads specified in this section and the special loads set forth in Table 16-B.

ANCHORAGE OF CONCRETE OR MASONRY WALLS

1611.4 Anchorage of Concrete and Masonry Walls. Concrete and masonry walls shall be anchored as required by Section 1605.2.3. Such anchorage shall be capable of resisting the load combinations of Section 1612.2 or 1612.3 using the greater of the wind or earthquake loads required by this chapter or a minimum horizontal force of 280 pounds per linear foot (4.09 kN/m) of wall, substituted for E.

WALLS AND STRUCTURAL FRAMING

1611.5 Interior Wall Loads. Interior walls, permanent partitions and temporary partitions that exceed 6 feet (1829 mm) in height shall be designed to resist all loads to which they are subjected but not less than a load, L, of 5 psf (0.24 kN/m^2) applied perpendicular to the walls. The 5 psf (0.24 kN/m^2) load need not be applied simultaneously with wind or seismic loads. The deflection of such walls under a load of 5 psf (0.24 kN/m^2) shall not exceed $1/240$ of the span for walls with brittle finishes and $1/120$ of the span for walls with flexible finishes. See Table 16-O for earthquake design requirements where such requirements are more restrictive.

> **EXCEPTION:** Flexible, folding or portable partitions are not required to meet the load and deflection criteria but must be anchored to the supporting structure to meet the provisions of this code.

RETAINING WALLS

1611.6 Retaining Walls. Retaining walls shall be designed to resist loads due to the lateral pressure of retained material in accordance with accepted engineering practice. Walls retaining drained soil, where the surface of the retained soil is level, shall be designed for a load, H, equivalent to that exerted by a fluid weighing not less than 30 psf per foot of depth (4.71 kN/m^2/m) and having a depth equal to that of the retained soil. Any surcharge shall be in addition to the equivalent fluid pressure.

Retaining walls shall be designed to resist sliding by at least 1.5 times the lateral force and overturning by at least 1.5 times the overturning moment, using allowable stress design loads.

---

*Only portions of this section are shown which are particularly applicable to masonry construction. For additional information see the UBC.

Division III—WIND DESIGN

SECTION 1615 — GENERAL

Every building or structure and every portion thereof shall be designed and constructed to resist the wind effects determined in accordance with the requirements of this division. Wind shall be assumed to come from any horizontal direction. No reduction in wind pressure shall be taken for the shielding effect of adjacent structures.

Structures sensitive to dynamic effects, such as buildings with a height-to-width ratio greater than five, structures sensitive to wind-excited oscillations, such as vortex shedding or icing, and buildings over 400 feet (121.9 m) in height, shall be, and any structure may be, designed in accordance with approved national standards.

The provisions of this section do not apply to building and foundation systems in those areas subject to scour and water pressure by wind and wave action. Buildings and foundations subject to such loads shall be designed in accordance with approved national standards.

Division IV—EARTHQUAKE DESIGN

SECTION 1626 — GENERAL

1626.1 Purpose. The purpose of the earthquake provisions herein is primarily to safeguard against major structural failures and loss of life, not to limit damage or maintain function.

1626.2 Minimum Seismic Design. Structures and portions thereof shall, as a minimum, be designed and constructed to resist the effects of seismic ground motions as provided in this division.

1626.3 Seismic and Wind Design. When the code-prescribed wind design produces greater effects, the wind design shall govern, but detailing requirements and limitations prescribed in this section and referenced sections shall be followed.

SECTION 1633 — DETAILED SYSTEMS DESIGN REQUIREMENTS

1633.1 General. All structural framing systems shall comply with the requirements of Section 1629. Only the elements of the designated seismic-force-resisting system shall be used to resist design forces. The individual components shall be designed to resist the prescribed design seismic forces acting on them. The components shall also comply with the specific requirements for the material contained in Chapters 19 through 23. In addition, such framing systems and components shall comply with the detailed system design requirements contained in Section 1633.

All building components in Seismic Zones 2, 3 and 4 shall be designed to resist the effects of the seismic forces prescribed herein and the effects of gravity loadings from dead, floor live and snow loads.

Consideration shall be given to design for uplift effects caused by seismic loads.

In Seismic Zones 2, 3 and 4, provision shall be made for the effects of earthquake forces acting in a direction other than the principal axes in each of the following circumstances:

The structure has plan irregularity Type 5 as given in Table 16-M.

The structure has plan irregularity Type 1 as given in Table 16-M for both major axes.

A column of a structure forms part of two or more intersecting lateral-force-resisting systems.

> **EXCEPTION:** If the axial load in the column due to seismic forces acting in either direction is less than 20 percent of the column axial load capacity.

The requirement that orthogonal effects be considered may be satisfied by designing such elements for 100 percent of the prescribed design seismic forces in one direction plus 30 percent of the prescribed design seismic forces in the perpendicular direction. The combination requiring the greater component strength shall be used for design. Alternatively, the effects of the two orthogonal directions may be combined on a square root of the sum of the squares (SRSS) basis. When the SRSS method of combining directional effects is used, each term computed shall be assigned the sign that will result in the most conservative result.

1633.2 Structural Framing Systems.

1633.2.1 General. Four types of general building framing systems defined in Section 1629.6 are recognized in these provisions and shown in Table 16-N. Each type is subdivided by the types of vertical elements used to resist lateral seismic forces. Special framing requirements are given in this section and in Chapters 19 through 23.

1633.2.2 Detailing for combinations of systems. For components common to different structural systems, the more restrictive detailing requirements shall be used.

1633.2.3 Connections. Connections that resist design seismic forces shall be designed and detailed on the drawings.

1633.2.4 Deformation compatibility. All structural framing elements and their connections, not required by design to be part of the lateral-force-resisting system, shall be designed and/or detailed to be adequate to maintain support of design dead plus live loads when subjected to the expected deformations caused by seismic forces. $P\Delta$ effects on such elements shall be considered. Expected deformations shall be determined as the greater of the Maximum Inelastic Response Displacement, Δ_M, considering $P\Delta$ effects determined in accordance with Section 1630.9.2 or the deformation induced by a story drift of 0.0025 times the story height. When computing expected deformations, the stiffening effect of those elements not part of the lateral-force-resisting system shall be neglected.

For elements not part of the lateral-force-resisting system, the forces inducted by the expected deformation may be considered as ultimate or factored forces. When computing the forces induced by expected deformations, the restraining effect of adjoining rigid structures and nonstructural elements shall be considered and a rational value of member and restraint stiffness shall be used. Inelastic deformations of members and connections may be considered in the evaluation, provided the assumed calculated capacities are consistent with member and connection design and detailing.

For concrete and masonry elements that are part of the lateral-force-resisting system, the assumed flexural and shear stiffness properties shall not exceed one half of the gross section properties unless a rational cracked-section analysis is performed. Additional deformations that may result from foundation flexibility and diaphragm deflections shall be considered. For concrete elements not part of the lateral-force-resisting system, see Section 1921.7.

1633.2.4.1 Adjoining rigid elements. Moment-resisting frames and shear walls may be enclosed by or adjoined by more rigid elements, provided it can be shown that the participation or failure of the more rigid elements will not impair the vertical and lateral-load-resisting ability of the gravity load and lateral-force-resisting systems. The effects of adjoining rigid elements shall be considered when assessing whether a structure shall be designated regular or irregular in Section 1629.5.1.

1633.2.4.2 Exterior elements. Exterior nonbearing, nonshear wall panels or elements that are attached to or enclose the exterior shall be designed to resist the forces per Formula (32-1) or (32–2) and shall accommodate movements of the structure based on Δ_M and temperature changes. Such elements shall be supported by means of cast-in-place concrete or by mechanical connections and fasteners in accordance with the following provisions:

1. Connections and panel joints shall allow for a relative movement between stories of not less than two times story drift caused by wind, the calculated story drift based on Δ_M or $^1/_2$ inch (12.7 mm), whichever is greater.

2. Connections to permit movement in the plane of the panel for story drift shall be sliding connections using slotted or oversize holes, connections that permit movement by bending of steel, or other connections providing equivalent sliding and ductility capacity.

3. Bodies of connections shall have sufficient ductility and rotation capacity to preclude fracture of the concrete or brittle failures at or near welds.

4. The body of the connection shall be designed for the force determined by Formula (32-2), where $R_p = 3.0$ and $a_p = 1.0$.

5. All fasteners in the connecting system, such as bolts, inserts, welds and dowels, shall be designed for the forces determined by Formula (32-2), where $R_p = 1.0$ and $a_p = 1.0$.

6. Fasteners embedded in concrete shall be attached to, or hooked around, reinforcing steel or otherwise terminated to effectively transfer forces to the reinforcing steel.

1633.2.5 Ties and continuity. All parts of a structure shall be interconnected and the connections shall be capable of transmitting the seismic force induced by the parts being connected. As a minimum, any smaller portion of the building shall be tied to the remainder of the building with elements having at least a strength to resist 0.5 $C_a I$ times the weight of the smaller portion.

A positive connection for resisting a horizontal force acting parallel to the member shall be provided for each beam, girder or truss. This force shall not be less than 0.5 $C_a I$ times the dead plus live load.

1633.2.6 Collector elements. Collector elements shall be provided that are capable of transferring the seismic forces originating in other portions of the structure to the element providing the resistance to those forces.

Collector elements, splices and their connections to resisting elements shall resist the forces determined in accordance with Formula (33-1). In addition, collector elements, splices, and their connections to resisting elements shall have the design strength to resist the combined loads resulting from the special seismic load of Section 1612.4.

> **EXCEPTION:** In structures, or portions thereof, braced entirely by light-frame wood shear walls or light-frame steel and wood structural panel shear wall systems, collector elements, splices and connections to resisting elements need only be designed to resist forces in accordance with Formula (33-1).

The quantity E_M need not exceed the maximum force that can be transferred to the collector by the diaphragm and other elements of the lateral-force-resisting system. For Allowable Stress Design, the design strength may be determined using an allowable stress increase of 1.7 and a resistance factor, ϕ, of 1.0. This increase shall not be combined with the one-third stress increase permitted by Section 1612.3, but may be combined with the duration of load increase permitted in Division III of Chapter 23.

1633.2.7 Concrete frames. Concrete frames required by design to be part of the lateral-force-resisting system shall conform to the following:

1. In Seismic Zones 3 and 4 they shall be special moment-resisting frames.

2. In Seismic Zone 2 they shall, as a minimum, be intermediate moment-resisting frames.

1633.2.8 Anchorage of concrete or masonry walls. Concrete or masonry walls shall be anchored to all floors and roofs that provide out-of-plane lateral support of the wall. The anchorage shall provide a positive direct connection between the wall and floor or roof construction capable of resisting the larger of the horizontal forces specified in this section and Sections 1611.4 and 1632. In addition, in Seismic Zones 3 and 4, diaphragm to wall anchorage using embedded straps shall have the straps attached to or hooked around the reinforcing steel or otherwise terminated to effectively transfer forces to the reinforcing steel. Requirements for developing anchorage forces in diaphragms are given in Section 1633.2.9. Diaphragm deformation shall be considered in the design of the supported walls.

1633.2.8.1 Out-of-plane wall anchorage to flexible diaphragms. This section shall apply in Seismic Zones 3 and 4 where flexible diaphragms, as defined in Section 1630.6, provide lateral support for walls.

1. Elements of the wall anchorage system shall be designed for the forces specified in Section 1632 where $R_p = 3.0$ and $a_p = 1.5$.

In Seismic Zone 4, the value of F_p used for the design of the elements of the wall anchorage system shall not be less than 420 pounds per lineal foot (6.1 kN per lineal meter) of wall substituted for E.

See Section 1611.4 for minimum design forces in other seismic zones.

2. When elements of the wall anchorage system are not loaded concentrically or are not perpendicular to the wall, the system shall be designed to resist all components of the forces induced by the eccentricity.

3. When pilasters are present in the wall, the anchorage force at the pilasters shall be calculated considering the additional load transferred from the wall panels to the pilasters. However, the minimum anchorage force at a floor or roof shall be that specified in Section 1633.2.8.1, Item 1.

4. The strength design forces for steel elements of the wall anchorage system shall be 1.4 times the forces otherwise required by this section.

5. The strength design forces for wood elements of the wall anchorage system shall be 0.85 times the force otherwise required by this section and these wood elements shall have a minimum actual net thickness of $2^1/_2$ inches (63.5 mm).

1633.2.9 Diaphragms.

1. The deflection in the plane of the diaphragm shall not exceed the permissible deflection of the attached elements. Permissible deflection shall be that deflection that will permit the attached ele-

ment to maintain its structural integrity under the individual loading and continue to support the prescribed loads.

2. Floor and roof diaphragms shall be designed to resist the forces determined in accordance with the following formula:

$$F_{px} = \frac{F_t + \displaystyle\sum_{i=x}^{n} F_i}{\displaystyle\sum_{i=x}^{n} w_i} w_{px} \qquad (33\text{-}1)$$

The force F_{px} determined from Formula (33-1) need not exceed $1.0C_a I w_{px}$, but shall not be less than $0.5C_a I w_{px}$.

When the diaphragm is required to transfer design seismic forces from the vertical-resisting elements above the diaphragm to other vertical-resisting elements below the diaphragm due to offset in the placement of the elements or to changes in stiffness in the vertical elements, these forces shall be added to those determined from Formula (33-1).

3. Design seismic forces for flexible diaphragms providing lateral supports for walls or frames of masonry or concrete shall be determined using Formula (33-1) based on the load determined in accordance with Section 1630.2 using a R not exceeding 4.

4. Diaphragms supporting concrete or masonry walls shall have continuous ties or struts between diaphragm chords to distribute the anchorage forces specified in Section 1633.2.8. Added chords of subdiaphragms may be used to form subdiaphragms to transmit the anchorage forces to the main continuous crossties. The maximum length-to-width ratio of the wood structural subdiaphragm shall be $2^1/_2$:1.

5. Where wood diaphragms are used to laterally support concrete or masonry walls, the anchorage shall conform to Section 1633.2.8. In Seismic Zones 2, 3 and 4, anchorage shall not be accomplished by use of toenails or nails subject to withdrawal, wood ledgers or framing shall not be used in cross-grain bending or cross-grain tension, and the continuous ties required by Item 4 shall be in addition to the diaphragm sheathing.

6. Connections of diaphragms to the vertical elements in structures in Seismic Zones 3 and 4, having a plan irregularity of Type 1, 2, 3 or 4 in Table 16-M, shall be designed without considering the one-third increase usually permitted in allowable stresses for elements resisting earthquake forces.

7. In structures in Seismic Zones 3 and 4 having a plan irregularity of Type 2 in Table 16-M, diaphragm chords and drag members shall be designed considering independent movement of the projecting wings of the structure. Each of these diaphragm elements shall be designed for the more severe of the following two assumptions:

Motion of the projecting wings in the same direction.

Motion of the projecting wings in opposing directions.

EXCEPTION: This requirement may be deemed satisfied if the procedures of Section 1631 in conjunction with a three-dimensional model have been used to determine the lateral seismic forces for design.

Chapter 17 *
STRUCTURAL TESTS AND INSPECTIONS

SECTION 1701 — SPECIAL INSPECTIONS

1701.1 General. In addition to the inspections required by Section 108, the owner or the engineer or architect of record acting as the owner's agent shall employ one or more special inspectors who shall provide inspections during construction on the types of work listed under Section 1701.5.

> **EXCEPTION:** The building official may waive the requirement for the employment of a special inspector if the construction is of a minor nature.

1701.2 Special Inspector. The special inspector shall be a qualified person who shall demonstrate competence, to the satisfaction of the building official, for inspection of the particular type of construction or operation requiring special inspection.

1701.3 Duties and Responsibilities of the Special Inspector. The special inspector shall observe the work assigned for conformance to the approved design drawings and specifications.

The special inspector shall furnish inspection reports to the building official, the engineer or architect of record, and other designated persons. All discrepancies shall be brought to the immediate attention of the contractor for correction, then, if uncorrected, to the proper design authority and to the building official.

The special inspector shall submit a final signed report stating whether the work requiring special inspection was, to the best of the inspector's knowledge, in conformance to the approved plans and specifications and the applicable workmanship provisions of this code.

1701.5 Types of Work. Except as provided in Section 1701.1, the types of work listed below shall be inspected by a special inspector.

7. **Structural masonry.**

   7.1 For masonry, other than fully grouted open-end hollow-unit masonry, during preparation and taking of any required prisms or test specimens, placing of all masonry units, placement of reinforcement, inspection of grout space, immediately prior to closing of cleanouts, and during all grouting operations.

> **EXCEPTION:** For hollow-unit masonry where the f'_m is no more than 1,500 psi (10.34 MPa) for concrete units or 2,600 psi (17.93 MPa) for clay units, special inspection may be performed as required for fully grouted open-end hollow-unit masonry specified in Item 7.2.

   7.2 For fully grouted open-end hollow-unit masonry during preparation and taking of any required prisms or test specimens, at the start of laying units, after the placement of reinforcing steel, grout space prior to each grouting operation, and during all grouting operations.

> **EXCEPTION:** Special inspection as required in Items 7.1 and 7.2 need not be provided when design stresses have been adjusted as specified in Chapter 21 to permit noncontinuous inspection.

SECTION 1702 — STRUCTURAL OBSERVATION

Structural observation shall be provided in Seismic Zone 3 or 4 when one of the following conditions exists:

1. The structure is defined in Table 16-K as Occupancy Category I, II or III,

2. The structure is required to comply with Section 403,

3. The structure is in Seismic Zone 4, N_a as set forth in Table 16-S is greater than one, and a lateral design is required for the entire structure,

> **EXCEPTION:** One- and two-story Group R, Division 3 and Group U Occupancies and one- and two-story Groups B, F, M and S Occupancies.

4. When so designated by the architect or engineer of record, or

5. When such observation is specifically required by the building official.

The owner shall employ the engineer or architect responsible for the structural design, or another engineer or architect designated by the engineer or architect responsible for the structural design, to perform structural observation as defined in Section 220. Observed deficiencies shall be reported in writing to the owner's representative, special inspector, contractor and the building official. The structural observer shall submit to the building official a written statement that the site visits have been made and identifying any reported deficiencies that, to the best of the structural observer's knowledge, have not been resolved.

---

*Only portions of this section are shown which are particularly applicable to masonry construction. For additional information see the UBC.

Chapter 21
MASONRY

SECTION 2101 — GENERAL

2101.1 Scope. The materials, design, construction and quality assurance of masonry shall be in accordance with this chapter.

2101.2 Design Methods. Masonry shall comply with the provisions of one of the following design methods in this chapter as well as the requirements of Sections 2101 through 2105.

2101.2.1 Working stress design. Masonry designed by the working stress design method shall comply with the provisions of Sections 2106 and 2107.

2101.2.2 Strength design. Masonry designed by the strength design method shall comply with the provisions of Sections 2106 and 2108.

2101.2.3 Empirical design. Masonry designed by the empirical design method shall comply with the provisions of Sections 2106.1 and 2109.

2101.2.4 Glass masonry. Glass masonry shall comply with the provisions of Section 2110.

2101.3 Definitions. For the purpose of this chapter, certain terms are defined as follows:

AREAS:

Bedded Area is the area of the surface of a masonry unit which is in contact with mortar in the plane of the joint.

Effective Area of Reinforcement is the cross-sectional area of reinforcement multiplied by the cosine of the angle between the reinforcement and the direction for which effective area is to be determined.

Gross Area is the total cross-sectional area of a specified section.

Net Area is the gross cross-sectional area minus the area of ungrouted cores, notches, cells and unbedded areas. Net area is the actual surface area of a cross section of masonry.

Transformed Area is the equivalent area of one material to a second based on the ratio of moduli of elasticity of the first material to the second.

BOND:

Adhesion Bond is the adhesion between masonry units and mortar or grout.

Reinforcing Bond is the adhesion between steel reinforcement and mortar or grout.

BOND BEAM is a horizontal grouted element within masonry in which reinforcement is embedded.

CELL is a void space having a gross cross-sectional area greater than $1^1/_2$ square inches (967 mm^2).

CLEANOUT is an opening to the bottom of a grout space of sufficient size and spacing to allow the removal of debris.

COLLAR JOINT is the mortared or grouted space between wythes of masonry.

COLUMN, REINFORCED, is a vertical structural member in which both the reinforcement and masonry resist compression.

COLUMN, UNREINFORCED, is a vertical structural member whose horizontal dimension measured at right angles to the thickness does not exceed three times the thickness.

DIMENSIONS:

Actual Dimensions are the measured dimensions of a designated item. The actual dimension shall not vary from the specified dimension by more than the amount allowed in the appropriate standard of quality in Section 2102.

Nominal Dimensions of masonry units are equal to its specified dimensions plus the thickness of the joint with which the unit is laid.

Specified Dimensions are the dimensions specified for the manufacture or construction of masonry, masonry units, joints or any other component of a structure.

GROUT LIFT is an increment of grout height within the total grout pour.

GROUT POUR is the total height of masonry wall to be grouted prior to the erection of additional masonry. A grout pour will consist of one or more grout lifts.

GROUTED MASONRY:

Grouted Hollow-unit Masonry is that form of grouted masonry construction in which certain designated cells of hollow units are continuously filled with grout.

Grouted Multiwythe Masonry is that form of grouted masonry construction in which the space between the wythes is solidly or periodically filled with grout.

JOINTS:

Bed Joint is the mortar joint that is horizontal at the time the masonry units are placed.

Head Joint is the mortar joint having a vertical transverse plane.

MASONRY UNIT is brick, tile, stone, glass block or concrete block conforming to the requirements specified in Section 2102.

Hollow-masonry Unit is a masonry unit whose net cross-sectional areas (solid area) in any plane parallel to the surface containing cores, cells or deep frogs is less than 75 percent of its gross cross-sectional area measured in the same plane.

Solid-masonry Unit is a masonry unit whose net cross-sectional area in any plane parallel to the surface containing the cores or cells is at least 75 percent of the gross cross-sectional area measured in the same plane.

PRISM is an assemblage of masonry units and mortar with or without grout used as a test specimen for determining properties of the masonry.

REINFORCED MASONRY is that form of masonry construction in which reinforcement acting in conjunction with the masonry is used to resist forces.

SHELL is the outer portion of a hollow masonry unit as placed in masonry.

WALLS

Bonded Wall is a masonry wall in which two or more wythes are bonded to act as a structural unit.

Cavity Wall is a wall containing continuous air space with a minimum width of 2 inches (51 mm) and a maximum width of

$4^{1}/_{2}$ inches (114 mm) between wythes which are tied with metal ties.

WALL TIE is a mechanical metal fastener which connects wythes of masonry to each other or to other materials.

WEB is an interior solid portion of a hollow-masonry unit as placed in masonry.

WYTHE is the portion of a wall which is one masonry unit in thickness. A collar joint is not considered a wythe.

2101.4 Notations.

A_b = cross-sectional area of anchor bolt, square inches (mm²).

A_e = effective area of masonry, square inches (mm²).

A_g = gross area of wall, square inches (mm²).

A_{jh} = total area of special horizontal reinforcement through wall frame joint, square inches (mm²).

A_{mv} = net area of masonry section bounded by wall thickness and length of section in direction of shear force considered, square inches (mm²).

A_p = area of tension (pullout) cone of embedded anchor bolt projected onto surface of masonry, square inches (mm²).

A_s = effective cross-sectional area of reinforcement in column or flexural member, square inches (mm²).

A_{se} = effective area of reinforcement, square inches (mm²).

A_{sh} = total cross-sectional area of rectangular tie reinforcement for confined core, square inches (mm²).

A_v = area of reinforcement required for shear reinforcement perpendicular to longitudinal reinforcement, square inches (mm²).

A'_s = effective cross-sectional area of compression reinforcement in flexural member, square inches (mm²).

a = depth of equivalent rectangular stress block, inches (mm).

B_{sn} = nominal shear strength of anchor bolt, pounds (N).

B_t = allowable tensile force on anchor bolt, pounds (N).

B_{tn} = nominal tensile strength of anchor bolt, pounds (N).

B_v = allowable shear force on anchor bolt, pounds (N).

b = effective width of rectangular member or width of flange for T and I sections, inches (mm).

b_{su} = factored shear force supported by anchor bolt, pounds (N).

b_t = computed tensile force on anchor bolt, pounds (N).

b_{tu} = factored tensile force supported by anchor bolt, pounds (N).

b_v = computed shear force on anchor bolt, pounds (N).

b' = width of web in T or I section, inches (mm).

C_d = nominal shear strength coefficient as obtained from Table 21-K.

c = distance from neutral axis to extreme fiber, inches (mm).

D = dead loads, or related internal moments and forces.

d = distance from compression face of flexural member to centroid of longitudinal tensile reinforcement, inches (mm).

d_b = diameter of reinforcing bar, inches (mm).

d_{bb} = diameter of largest beam longitudinal reinforcing bar passing through, or anchored in, a joint, inches (mm).

d_{bp} = diameter of largest pier longitudinal reinforcing bar passing through a joint, inches (mm).

E = load effects of earthquake, or related internal moments and forces.

E_m = modulus of elasticity of masonry, pounds per square inch (MPa).

e = eccentricity of P_{uf}, inches (mm).

e_{mu} = maximum usable compressive strain of masonry.

F = loads due to weight and pressure of fluids or related moments and forces.

F_a = allowable average axial compressive stress in columns for centroidally applied axial load only, pounds per square inch (MPa).

F_b = allowable flexural compressive stress in members subjected to bending load only, pounds per square inch (MPa).

F_{br} = allowable bearing stress in masonry, pounds per square inch (MPa).

F_s = allowable stress in reinforcement, pounds per square inch (MPa).

F_{sc} = allowable compressive stress in column reinforcement, pounds per square inch (MPa).

F_t = allowable flexural tensile stress in masonry, pounds per square inch (MPa).

F_v = allowable shear stress in masonry, pounds per square inch (MPa).

f_a = computed axial compressive stress due to design axial load, pounds per square inch (MPa).

f_b = computed flexural stress in extreme fiber due to design bending loads only, pounds per square inch (MPa).

f_{md} = computed compressive stress due to dead load only, pounds per square inch (MPa).

f_r = modulus of rupture, pounds per square inch (MPa).

f_s = computed stress in reinforcement due to design loads, pounds per square inch (MPa).

f_v = computed shear stress due to design load, pounds per square inch (MPa).

f_y = tensile yield stress of reinforcement, pounds per square inch (MPa).

fyh = tensile yield stress of horizontal reinforcement, pounds per square inch (MPa).

f'_g = specified compressive strength of grout at age of 28 days, pounds per square inch (MPa).

f'_m = specified compressive strength of masonry at age of 28 days, pounds per square inch (MPa).

G = shear modulus of masonry, pounds per square inch (MPa).

H = loads due to weight and pressure of soil, water in soil or related internal moments and forces.

h = height of wall between points of support, inches (mm).

h_b = beam depth, inches (mm).

h_c = cross-sectional dimension of grouted core measured center to center of confining reinforcement, inches (mm).

h_p = pier depth in plane of wall frame, inches (mm).

h' = effective height of wall or column, inches (mm).

I = moment of inertia about neutral axis of cross-sectional area, inches⁴ (mm⁴).

I_e = effective moment of inertia, inches⁴ (mm⁴).

I_g, I_{cr} = gross, cracked moment of inertia of wall cross section, inches⁴ (mm⁴).

j = ratio or distance between centroid of flexural compressive forces and centroid of tensile forces of depth, d.

K = reinforcement cover or clear spacing, whichever is less, inches (mm).

k = ratio of depth of compressive stress in flexural member to depth, d.

L = live loads, or related internal moments and forces.

L_w = length of wall, inches (mm).

l = length of wall or segment, inches (mm).

l_b = embedment depth of anchor bolt, inches (mm).

l_{be} = anchor bolt edge distance, the least distance measured from edge of masonry to surface of anchor bolt, inches (mm).

l_d = required development length of reinforcement, inches (mm).

M = design moment, inch-pounds (N·mm).

M_a = maximum moment in member at stage deflection is computed, inch-pounds (N·mm).

M_c = moment capacity of compression reinforcement in flexural member about centroid of tensile force, inch-pounds (N·mm).

M_{cr} = nominal cracking moment strength in masonry, inch-pounds (N·mm).

M_m = moment of compressive force in masonry about centroid of tensile force in reinforcement, inch-pounds (N·mm).

M_n = nominal moment strength, inch-pounds (N·mm).

M_s = moment of tensile force in reinforcement about centroid of compressive force in masonry, inch-pounds (N·mm).

M_{ser} = service moment at midheight of panel, including $P\Delta$ effects, inch-pounds (N·mm).

M_u = factored moment, inch-pounds (N·mm).

n = modular ratio.

  = E_s / E_m.

P = design axial load, pounds (N).

P_a = allowable centroidal axial load for reinforced masonry columns, pounds (N).

P_b = nominal balanced design axial strength, pounds (N).

P_f = load from tributary floor or roof area, pounds (N).

P_n = nominal axial strength in masonry, pounds (N).

P_o = nominal axial load strength in masonry without flexure, pounds (N).

P_u = factored axial load, pounds (N).

P_{uf} = factored load from tributary floor or roof loads, pounds (N).

P_{uw} = factored weight of wall tributary to section under consideration, pounds (N).

P_w = weight of wall tributary to section under consideration, pounds (N).

r = radius of gyration (based on specified unit dimensions or Tables 21-H-1, 21-H-2 and 21-H-3), inches (mm).

r_b = ratio of area of reinforcing bars cut off to total area of reinforcing bars at the section.

S = section modulus, inches^3 (mm^3).

s = spacing of stirrups or of bent bars in direction parallel to that of main reinforcement, inches (mm).

T = effects of temperature, creep, shrinkage and differential settlement.

t = effective thickness of wythe, wall or column, inches (mm).

U = required strength to resist factored loads, or related internal moments and forces.

u = bond stress per unit of surface area of reinforcing bar, pounds per square inch (MPa).

V = total design shear force, pounds (N).

V_{jh} = total horizontal joint shear, pounds (N).

V_m = nominal shear strength of masonry, pounds (N).

V_n = nominal shear strength, pounds (N).

V_s = nominal shear strength of shear reinforcement, pounds (N).

V_u = required shear strength in masonry, pounds (N).

W = wind load, or related internal moments in forces.

w_u = factored distributed lateral load.

Δ_s = horizontal deflection at midheight under factored load, inches (mm).

Δ_u = deflection due to factored loads, inches (mm).

ρ = ratio of area of flexural tensile reinforcement, A_s, to area bd.

ρ_b = reinforcement ratio producing balanced strain conditions.

ρ_n = ratio of distributed shear reinforcement on plane perpendicular to plane of A_{mv}.

Σ_o = sum of perimeters of all longitudinal reinforcement, inches (mm).

$\sqrt{f'_m}$ = square root of specified strength of masonry at the age of 28 days, pounds per square inch (MPa).

ϕ = strength-reduction factor.

SECTION 2102 — MATERIAL STANDARDS

2102.1 Quality. Materials used in masonry shall conform to the requirements stated herein. If no requirements are specified in this section for a material, quality shall be based on generally accepted good practice, subject to the approval of the building official.

Reclaimed or previously used masonry units shall meet the applicable requirements as for new masonry units of the same material for their intended use.

2102.2 Standards of Quality. The standards listed below labeled a "UBC Standard" are also listed in Chapter 35, Part II, and are part of this code. The other standards listed below are recognized standards. See Sections 3503 and 3504.

1. **Aggregates.**

    1.1 ASTM C 144, Aggregates for Masonry Mortar

    1.2 ASTM C 404, Aggregates for Grout

2. **Cement.**

    2.1 UBC Standard 21-11, Cement, Masonry. (Plastic cement conforming to the requirements of UBC Standard 25-1 may be used in lieu of masonry cement when it also conforms to UBC Standard 21-11.)

    2.2 ASTM C 150, Portland Cement

    2.3 UBC Standard 21-14, Mortar Cement

3. **Lime.**

    3.1 UBC Standard 21-12, Quicklime for Structural Purposes

    3.2 UBC Standard 21-13, Hydrated Lime for Masonry Purposes. When Types N and NA hydrated lime are used in masonry mortar, they shall comply with the provisions of UBC Standard 21-15, Section 21.1506.7, excluding the plasticity requirement.

4. **Masonry units of clay or shale.**

   4.1 ASTM C 34, Structural Clay Load-bearing Wall Tile

   4.2 ASTM C 56, Structural Clay Nonload-bearing Tile

   4.3 UBC Standard 21-1, Section 21.101, Building Brick (solid units)

   4.4 ASTM C 126, Ceramic Glazed Structural Clay Facing Tile, Facing Brick and Solid Masonry Units. Load-bearing glazed brick shall conform to the weathering and structural requirements of UBC Standard 21-1, Section 21.106, Facing Brick

   4.5 UBC Standard 21-1, Section 21.106, Facing Brick (solid units)

   4.6 UBC Standard 21-1, Section 21.107, Hollow Brick

   4.7 ASTM C 67, Sampling and Testing Brick and Structural Clay Tile

   4.8 ASTM C 212, Structural Clay Facing Tile

   4.9 ASTM C 530, Structural Clay Non-Loadbearing Screen Tile

5. **Masonry units of concrete.**

   5.1 UBC Standard 21-3, Concrete Building Brick

   5.2 UBC Standard 21-4, Hollow and Solid Load-bearing Concrete Masonry Units

   5.3 UBC Standard 21-5, Nonload-bearing Concrete Masonry Units

   5.4 ASTM C 140, Sampling and Testing Concrete Masonry Units

   5.5 ASTM C 426, Standard Test Method for Drying Shrinkage of Concrete Block

6. **Masonry units of other materials.**

   6.1 **Calcium silicate.**

   UBC Standard 21-2, Calcium Silicate Face Brick (Sand-lime Brick)

   6.2 UBC Standard 21-9, Unburned Clay Masonry Units and Standard Methods of Sampling and Testing Unburned Clay Masonry Units

   6.3 ACI-704, Cast Stone

   6.4 UBC Standard 21-17, Test Method for Compressive Strength of Masonry Prisms

7. **Connectors.**

   7.1 Wall ties and anchors made from steel wire shall conform to UBC Standard 21-10, Part II, and other steel wall ties and anchors shall conform to A 36 in accordance with UBC Standard 22-1. Wall ties and anchors made from copper, brass or other nonferrous metal shall have a minimum tensile yield strength of 30,000 psi (207 MPa).

   7.2 All such items not fully embedded in mortar or grout shall either be corrosion resistant or shall be coated after fabrication with copper, zinc or a metal having at least equivalent corrosion-resistant properties.

8. **Mortar.**

   8.1 UBC Standard 21-15, Mortar for Unit Masonry and Reinforced Masonry other than Gypsum

   8.2 UBC Standard 21-16, Field Tests Specimens for Mortar

   8.3 UBC Standard 21-20, Standard Test Method for Flexural Bond Strength of Mortar Cement

9. **Grout.**

   9.1 UBC Standard 21-18, Method of Sampling and Testing Grout

   9.2 UBC Standard 21-19, Grout for Masonry

10. **Reinforcement.**

   10.1 UBC Standard 21-10, Part I, Joint Reinforcement for Masonry

   10.2 ASTM A 615, A 616, A 617, A 706, A 767, and A 775, Deformed and Plain Billet-steel Bars, Rail-steel Deformed and Plain Bars, Axle-steel Deformed and Plain Bars, and Deformed Low-alloy Bars for Concrete Reinforcement

   10.3 UBC Standard 21-10, Part II, Cold-drawn Steel Wire for Concrete Reinforcement

SECTION 2103 — MORTAR AND GROUT

2103.1 General. Mortar and grout shall comply with the provisions of this section. Special mortars, grouts or bonding systems may be used, subject to satisfactory evidence of their capabilities when approved by the building official.

2103.2 Materials. Materials used as ingredients in mortar and grout shall conform to the applicable requirements in Section 2102. Cementitious materials for grout shall be one or both of the following: lime and portland cement. Cementitious materials for mortar shall be one or more of the following: lime, masonry cement, portland cement and mortar cement. Cementitious materials or additives shall not contain epoxy resins and derivatives, phenols, asbestos fibers or fireclays.

Water used in mortar or grout shall be clean and free of deleterious amounts of acid, alkalies or organic material or other harmful substances.

2103.3 Mortar.

2103.3.1 General. Mortar shall consist of a mixture of cementitious materials and aggregate to which sufficient water and approved additives, if any, have been added to achieve a workable, plastic consistency.

2103.3.2 Selecting proportions. Mortar with specified proportions of ingredients that differ from the mortar proportions of Table 21-A may be approved for use when it is demonstrated by laboratory or field experience that this mortar with the specified proportions of ingredients, when combined with the masonry units to be used in the structure, will achieve the specified compressive strength f'_m. Water content shall be adjusted to provide proper workability under existing field conditions. When the proportion of ingredients is not specified, the proportions by mortar type shall be used as given in Table 21-A.

2103.4 Grout.

2103.4.1 General. Grout shall consist of a mixture of cementitious materials and aggregate to which water has been added such that the mixture will flow without segregation of the constituents. The specified compressive strength of grout, f'_g, shall not be less than 2,000 psi (13.8 MPa).

2103.4.2 Selecting proportions. Water content shall be adjusted to provide proper workability and to enable proper placement under existing field conditions, without segregation. Grout shall be specified by one of the following methods:

1. Proportions of ingredients and any additives shall be based on laboratory or field experience with the grout ingredients and the masonry units to be used. The grout shall be specified by the proportion of its constituents in terms of parts by volume, or

2. Minimum compressive strength which will produce the required prism strength, or

3. Proportions by grout type shall be used as given in Table 21-B.

2103.5 Additives and Admixtures.

2103.5.1 General. Additives and admixtures to mortar or grout shall not be used unless approved by the building official.

2103.5.2 Antifreeze compounds. Antifreeze liquids, chloride salts or other such substances shall not be used in mortar or grout.

2103.5.3 Air entrainment. Air-entraining substances shall not be used in mortar or grout unless tests are conducted to determine compliance with the requirements of this code.

2103.5.4 Colors. Only pure mineral oxide, carbon black or synthetic colors may be used. Carbon black shall be limited to a maximum of 3 percent of the weight of the cement.

SECTION 2104 — CONSTRUCTION

2104.1 General. Masonry shall be constructed according to the provisions of this section.

2104.2 Materials: Handling, Storage and Preparation. All materials shall comply with applicable requirements of Section 2102. Storage, handling and preparation at the site shall conform also to the following:

1. Masonry materials shall be stored so that at the time of use the materials are clean and structurally suitable for the intended use.

2. All metal reinforcement shall be free from loose rust and other coatings that would inhibit reinforcing bond.

3. At the time of laying, burned clay units and sand lime units shall have an initial rate of absorption not exceeding 0.035 ounce per square inch (1.6 L/m^2) during a period of one minute. In the absorption test, the surface of the unit shall be held $^1/_8$ inch (3 mm) below the surface of the water.

4. Concrete masonry units shall not be wetted unless otherwise approved.

5. Materials shall be stored in a manner such that deterioration or intrusion of foreign materials is prevented and that the material will be capable of meeting applicable requirements at the time of mixing or placement.

6. The method of measuring materials for mortar and grout shall be such that proportions of the materials can be controlled.

7. Mortar or grout mixed at the jobsite shall be mixed for a period of time not less than three minutes or more than 10 minutes in a mechanical mixer with the amount of water required to provide the desired workability. Hand mixing of small amounts of mortar is permitted. Mortar may be retempered. Mortar or grout which has hardened or stiffened due to hydration of the cement shall not be used. In no case shall mortar be used two and one-half hours, nor grout used one and one-half hours, after the initial mixing water has been added to the dry ingredients at the jobsite.

EXCEPTION: Dry mixes for mortar and grout which are blended in the factory and mixed at the jobsite shall be mixed in mechanical mixers until workable, but not to exceed 10 minute

2104.3 Cold-weather Construction.

2104.3.1 General. All materials shall be delivered in a usable condition and stored to prevent wetting by capillary action, rain and snow.

The tops of all walls not enclosed or sheltered shall be covered with a strong weather-resistive material at the end of each day or shutdown.

Partially completed walls shall be covered at all times when work is not in progress. Covers shall be draped over the wall and extend a minimum of 2 feet (600 mm) down both sides and shall be securely held in place, except when additional protection is required in Section 2104.3.4.

2104.3.2 Preparation. If ice or snow has inadvertently formed on a masonry bed, it shall be thawed by application of heat carefully applied until top surface of the masonry is dry to the touch.

A section of masonry deemed frozen and damaged shall be removed before continuing construction of that section.

2104.3.3 Construction. Masonry units shall be dry at time of placement. Wet or frozen masonry units shall not be laid.

Special requirements for various temperature ranges are as follows:

1. Air temperature 40°F to 32°F (4.5°C. to 0°C): Sand or mixing water shall be heated to produce mortar temperatures between 40°F and 120°F (4.5°C. and 49°C).

2. Air temperature 32°F to 25°F (0°C. to –4°C): Sand and mixing water shall be heated to produce mortar temperatures between 40°F and 120°F (4.5°C. and 49°C). Maintain temperatures of mortar on boards above freezing.

3. Air temperature 25°F to 20°F (–4°C. to –7°C): Sand and mixing water shall be heated to produce mortar temperatures between 40°F and 120°F (4.5°C. and 49°C). Maintain mortar temperatures on boards above freezing. Salamanders or other sources of heat shall be used on both sides of walls under construction. Windbreaks shall be employed when wind is in excess of 15 miles per hour (24 km/h).

4. Air temperature 20°F (–7°C) and below: Sand and mixing water shall be heated to produce mortar temperatures between 40°F and 120°F (4.5°C. and 49°C). Enclosure and auxiliary heat shall be provided to maintain air temperature above freezing. Temperature of units when laid shall not be less than 20°F (–7°C).

2104.3.4 Protection. When the mean daily air temperature is 40°F to 32°F (4.5°C. to 0°C), masonry shall be protected from rain or snow for 24 hours by covering with a weather-resistive membrane.

When the mean daily air temperature is 32°F to 25°F (0°C. to –4°C), masonry shall be completely covered with a weather-resistive membrane for 24 hours.

When the mean daily air temperature is 25°F to 20°F (–4°C. to –7°C), masonry shall be completely covered with insulating blankets or equally protected for 24 hours.

When the mean daily air temperature is 20°F (–7°C) or below, masonry temperature shall be maintained above freezing for 24 hours by enclosure and supplementary heat, by electric heating blankets, infrared heat lamps or other approved methods.

2104.3.5 Placing grout and protection of grouted masonry. When air temperatures fall below 40°F (4.5°C), grout mixing water and aggregate shall be heated to produce grout temperatures between 40°F and 120°F (4.5°C. and 49°C).

Masonry to be grouted shall be maintained above freezing during grout placement and for at least 24 hours after placement.

When atmospheric temperatures fall below 20°F (–7°C), enclosures shall be provided around the masonry during grout placement and for at least 24 hours after placement.

2104.4 Placing Masonry Units.

2104.4.1 Mortar. The mortar shall be sufficiently plastic and units shall be placed with sufficient pressure to extrude mortar from the joint and produce a tight joint. Deep furrowing which produces voids shall not be used.

The initial bed joint thickness shall not be less than $1/4$ inch (6 mm) or more than 1 inch (25 mm); subsequent bed joints shall not be less than $1/4$ inch (6 mm) or more than $5/8$ inch (16 mm) in thickness.

2104.4.2 Surfaces. Surfaces to be in contact with mortar or grout shall be clean and free of deleterious materials.

2104.4.3 Solid masonry units. Solid masonry units shall have full head and bed joints.

2104.4.4 Hollow-masonry units. All head and bed joints shall be filled solidly with mortar for a distance in from the face of the unit not less than the thickness of the shell.

Head joints of open-end units with beveled ends that are to be fully grouted need not be mortared. The beveled ends shall form a grout key which permits grout within $5/8$ inch (16 mm) of the face of the unit. The units shall be tightly butted to prevent leakage of grout.

2104.5 Reinforcement Placing. Reinforcement details shall conform to the requirements of this chapter. Metal reinforcement shall be located in accordance with the plans and specifications. Reinforcement shall be secured against displacement prior to grouting by wire positioners or other suitable devices at intervals not exceeding 200 bar diameters.

Tolerances for the placement of reinforcement in walls and flexural elements shall be plus or minus $1/2$ inch (12.7 mm) for d equal to 8 inches (200 mm) or less, $\pm$ 1 inch ($\pm$ 25 mm) for d equal to 24 inches (600 mm) or less but greater than 8 inches (200 mm), and $\pm$ $1^1/4$ inches (32 mm) for d greater than 24 inches (600 mm).

Tolerance for longitudinal location of reinforcement shall be $\pm$ 2 inches (51 mm).

2104.6 Grouted Masonry.

2104.6.1 General conditions. Grouted masonry shall be constructed in such a manner that all elements of the masonry act together as a structural element.

Prior to grouting, the grout space shall be clean so that all spaces to be filled with grout do not contain mortar projections greater than $1/2$ inch (12.7 mm), mortar droppings or other foreign material. Grout shall be placed so that all spaces designated to be grouted shall be filled with grout and the grout shall be confined to those specific spaces.

Grout materials and water content shall be controlled to provide adequate fluidity for placement without segregation of the constituents, and shall be mixed thoroughly.

The grouting of any section of wall shall be completed in one day with no interruptions greater than one hour.

Between grout pours, a horizontal construction joint shall be formed by stopping all wythes at the same elevation and with the grout stopping a minimum of $1^1/2$ inches (38 mm) below a mortar joint, except at the top of the wall. Where bond beams occur, the grout pour shall be stopped a minimum of $1/2$ inch (12.7 mm) below the top of the masonry.

Size and height limitations of the grout space or cell shall not be less than shown in Table 21-C. Higher grout pours or smaller cavity widths or cell size than shown in Table 21-C may be used when approved, if it is demonstrated that grout spaces will be properly filled.

Cleanouts shall be provided for all grout pours over 5 feet (1524 mm) in height.

Where required, cleanouts shall be provided in the bottom course at every vertical bar but shall not be spaced more than 32 inches (813 mm) on center for solidly grouted masonry. When cleanouts are required, they shall be sealed after inspection and before grouting.

Where cleanouts are not provided, special provisions must be made to keep the bottom and sides of the grout spaces, as well as the minimum total clear area as required by Table 21-C, clean and clear prior to grouting.

Units may be laid to the full height of the grout pour and grout shall be placed in a continuous pour in grout lifts not exceeding 6 feet (1830 mm). When approved, grout lifts may be greater than 6 feet (1830 mm) if it can be demonstrated the grout spaces can be properly filled.

All cells and spaces containing reinforcement shall be filled with grout.

2104.6.2 Construction requirements. Reinforcement shall be placed prior to grouting. Bolts shall be accurately set with templates or by approved equivalent means and held in place to prevent dislocation during grouting.

Segregation of the grout materials and damage to the masonry shall be avoided during the grouting process.

Grout shall be consolidated by mechanical vibration during placement before loss of plasticity in a manner to fill the grout space. Grout pours greater than 12 inches (300 mm) in height shall be reconsolidated by mechanical vibration to minimize voids due to water loss. Grout pours 12 inches (300 mm) or less in height shall be mechanically vibrated or puddled.

In one-story buildings having wood-frame exterior walls, foundations not over 24 inches (600 mm) high measured from the top of the footing may be constructed of hollow-masonry units laid in running bond without mortared head joints. Any standard shape unit may be used, provided the masonry units permit horizontal flow of grout to adjacent units. Grout shall be solidly poured to the full height in one lift and shall be puddled or mechanically vibrated.

In nonstructural elements which do not exceed 8 feet (2440 mm) in height above the highest point of lateral support, including fireplaces and residential chimneys, mortar of pouring consistency may be substituted for grout when the masonry is constructed and grouted in pours of 12 inches (300 mm) or less in height.

In multiwythe grouted masonry, vertical barriers of masonry shall be built across the grout space the entire height of the grout pour and spaced not more than 30 feet (9144 mm) horizontally. The grouting of any section of wall between barriers shall be completed in one day with no interruption longer than one hour.

2104.7 Aluminum Equipment. Grout shall not be handled nor pumped utilizing aluminum equipment unless it can be demonstrated with the materials and equipment to be used that there will be no deleterious effect on the strength of the grout.

2104.8 Joint Reinforcement. Wire joint reinforcement used in the design as principal reinforcement in hollow-unit construction shall be continuous between supports unless splices are made by lapping:

1. Fifty-four wire diameters in a grouted cell, or

2. Seventy-five wire diameters in the mortared bed joint, or

3. In alternate bed joints of running bond masonry a distance not less than 54 diameters plus twice the spacing of the bed joints, or

4. As required by calculation and specific location in areas of minimum stress, such as points of inflection.

Side wires shall be deformed and shall conform to UBC Standard 21-10, Part I, Joint Reinforcement for Masonry.

SECTION 2105 — QUALITY ASSURANCE

2105.1 General. Quality assurance shall be provided to ensure that materials, construction and workmanship are in compliance with the plans and specifications, and the applicable requirements of this chapter. When required, inspection records shall be maintained and made available to the building official.

2105.2 Scope. Quality assurance shall include, but is not limited to, assurance that:

1. Masonry units, reinforcement, cement, lime, aggregate and all other materials meet the requirements of the applicable standards of quality and that they are properly stored and prepared for use.

2. Mortar and grout are properly mixed using specified proportions of ingredients. The method of measuring materials for mortar and grout shall be such that proportions of materials are controlled.

3. Construction details, procedures and workmanship are in accordance with the plans and specifications.

4. Placement, splices and reinforcement sizes are in accordance with the provisions of this chapter and the plans and specifications.

2105.3 Compliance with f'_m.

2105.3.1 General. Compliance with the requirements for the specified compressive strength of masonry f'_m shall be in accordance with one of the sections in this subsection.

2105.3.2 Masonry prism testing. The compressive strength of masonry determined in accordance with UBC Standard 21-17 for each set of prisms shall equal or exceed f'_m. Compressive strength of prisms shall be based on tests at 28 days. Compressive strength at seven days or three days may be used provided a relationship between seven-day and three-day and 28-day strength has been established for the project prior to the start of construction. Verification by masonry prism testing shall meet the following:

1. A set of five masonry prisms shall be built and tested in accordance with UBC Standard 21-17 prior to the start of construction. Materials used for the construction of the prisms shall be taken from those specified to be used in the project. Prisms shall be constructed under the observation of the engineer or special inspector or an approved agency and tested by an approved agency.

2. When full allowable stresses are used in design, a set of three prisms shall be built and tested during construction in accordance with UBC Standard 21-17 for each 5,000 square feet (465 m^2) of wall area, but not less than one set of three masonry prisms for the project.

3. When one half the allowable masonry stresses are used in design, testing during construction is not required. A letter of certification from the supplier of the materials used to verify the f'_m in accordance with Section 2105.3.2, Item 1, shall be provided at the time of, or prior to, delivery of the materials to the jobsite to ensure the materials used in construction are representative of the materials used to construct the prisms prior to construction.

2105.3.3 Masonry prism test record. Compressive strength verification by masonry prism test records shall meet the following:

1. A masonry prism test record approved by the building official of at least 30 masonry prisms which were built and tested in accordance with UBC Standard 21-17. Prisms shall have been constructed under the observation of an engineer or special inspector or an approved agency and shall have been tested by an approved agency.

2. Masonry prisms shall be representative of the corresponding construction.

3. The average compressive strength of the test record shall equal or exceed 1.33 f'_m.

4. When full allowable stresses are used in design, a set of three masonry prisms shall be built during construction in accordance with UBC Standard 21-17 for each 5,000 square feet (465 m^2) of wall area, but not less than one set of three prisms for the project.

5. When one half the allowable masonry stresses are used in design, field testing during construction is not required. A letter of certification from the supplier of the materials to the jobsite shall be provided at the time of, or prior to, delivery of the materials to assure the materials used in construction are representative of the materials used to develop the prism test record in accordance with Section 2105.3.3, Item 1.

2105.3.4 Unit strength method. Verification by the unit strength method shall meet the following:

1. When full allowable stresses are used in design, units shall be tested prior to construction and test units during construction for each 5,000 square feet (465 m^2) of wall area for compressive strength to show compliance with the compressive strength required in Table 21-D; and

> **EXCEPTION:** Prior to the start of construction, prism testing may be used in lieu of testing the unit strength. During construction, prism testing may also be used in lieu of testing the unit strength and the grout as required by Section 2105.3.4, Item 4.

2. When one half the allowable masonry stresses are used in design, testing is not required for the units. A letter of certification from the manufacturer of the units shall be provided at the time of, or prior to, delivery of the units to the jobsite to assure the units comply with the compressive strength required in Table 21-D; and

3. Mortar shall comply with the mortar type required in Table 21-D; and

4. When full stresses are used in design for concrete masonry, grout shall be tested for each 5,000 square feet (465 m^2) of wall area, but not less than one test per project, to show compliance with the compressive strength required in Table 21-D, Footnote 4.

5. When one half the allowable stresses are used in design for concrete masonry, testing is not required for the grout. A letter of certification from the supplier of the grout shall be provided at the time of, or prior to, delivery of the grout to the jobsite to assure the grout complies with the compressive strength required in Table 21-D, Footnote 4; or

6. When full allowable stresses are used in design for clay masonry, grout proportions shall be verified by the engineer or special inspector or an approved agency to conform with Table 21-B.

7. When one half the allowable masonry stresses are used in design for clay masonry, a letter of certification from the supplier of the grout shall be provided at the time of, or prior to, delivery of the grout to the jobsite to assure the grout conforms to the proportions of Table 21-B.

2105.3.5 Testing prisms from constructed masonry. When approved by the building official, acceptance of masonry which does not meet the requirements of Section 2105.3.2, 2105.3.3 or 2105.3.4 shall be permitted to be based on tests of prisms cut from the masonry construction in accordance with the following:

1. A set of three masonry prisms that are at least 28 days old shall be saw cut from the masonry for each 5,000 square feet (465 m^2) of the wall area that is in question but not less than one set of three masonry prisms for the project. The length, width and height dimensions of the prisms shall comply with the requirements of UBC Standard 21-17. Transporting, preparation and testing of prisms shall be in accordance with UBC Standard 21-17.

2. The compressive strength of prisms shall be the value calculated in accordance with UBC Standard 21-17, Section 21.1707.2, except that the net cross-sectional area of the prism shall be based on the net mortar bedded area.

3. Compliance with the requirement for the specified compressive strength of masonry, f'_m, shall be considered satisfied provided the modified compressive strength equals or exceeds the specified f'_m. Additional testing of specimens cut from locations in question shall be permitted.

2105.4 Mortar Testing. When required, mortar shall be tested in accordance with UBC Standard 21-16.

2105.5 Grout Testing. When required, grout shall be tested in accordance with UBC Standard 21-18.

SECTION 2106 — GENERAL DESIGN REQUIREMENTS

2106.1 General.

2106.1.1 Scope. The design of masonry structures shall comply with the working stress design provisions of Section 2107, or the strength design provisions of Section 2108 or the empirical design provisions of Section 2109, and with the provisions of this section. Unless otherwise stated, all calculations shall be made using or based on specified dimensions.

2106.1.2 Plans. Plans submitted for approval shall describe the required design strengths of masonry materials and inspection requirements for which all parts of the structure were designed, and any load test requirements.

2106.1.3 Design loads. See Chapter 16 for design loads.

2106.1.4 Stack bond. In bearing and nonbearing walls, except veneer walls, if less than 75 percent of the units in any transverse vertical plane lap the ends of the units below a distance less than one half the height of the unit, or less than one fourth the length of the unit, the wall shall be considered laid in stack bond.

2106.1.5 Multiwythe walls.

2106.1.5.1 General. All wythes of multiwythe walls shall be bonded by grout or tied together by corrosion-resistant wall ties or joint reinforcement conforming to the requirements of Section 2102, and as set forth in this section.

2106.1.5.2 Wall ties in cavity wall construction. Wall ties shall be of sufficient length to engage all wythes. The portion of the wall ties within the wythe shall be completely embedded in mortar or

grout. The ends of the wall ties shall be bent to 90-degree angles with an extension not less than 2 inches (51 mm) long. Wall ties not completely embedded in mortar or grout between wythes shall be a single piece with each end engaged in each wythe.

There shall be at least one $^3/_{16}$-inch-diameter (9.5 mm) wall tie for each $4^1/_2$ square feet (0.42 m^2) of wall area. For cavity walls in which the width of the cavity is greater than 3 inches (75 mm), but not more than $4^1/_2$ inches (115 mm), at least one $^3/_{16}$-inch-diameter (9.5 mm) wall tie for each 3 square feet (0.28 m^2) of wall area shall be provided.

Ties in alternate courses shall be staggered. The maximum vertical distance between ties shall not exceed 24 inches (610 mm) and the maximum horizontal distance between ties shall not exceed 36 inches (914 mm).

Additional ties spaced not more than 36 inches (914 mm) apart shall be provided around openings within a distance of 12 inches (305 mm) from the edge of the opening.

Adjustable wall ties shall meet the following requirements:

1. One tie shall be provided for each 1.77 square feet (0.16 m^2) of wall area. Horizontal and vertical spacing shall not exceed 16 inches (406 mm). Maximum misalignment of bed joints from one wythe to the other shall be $1^1/_4$ inches (32 mm).

2. Maximum clearance between the connecting parts of the tie shall be $^1/_{16}$ inch (1.6 mm). When used, pintle ties shall have at least two $^3/_{16}$-inch-diameter (4.8 mm) pintle legs.

Wall ties of different size and spacing that provide equivalent strength between wythes may be used.

2106.1.5.3 Wall ties for grouted multiwythe construction. Wythes of multiwythe walls shall be bonded together with at least $^3/_{16}$-inch-diameter (4.8 mm) steel wall tie for each 2 square feet (0.19 m^2) of area. Wall ties of different size and spacing that provide equivalent strength between wythes may be used.

2106.1.5.4 Joint reinforcement. Prefabricated joint reinforcement for masonry walls shall have at least one cross wire of at least No. 9 gage steel for each 2 square feet (0.19 m^2) of wall area. The vertical spacing of the joint reinforcement shall not exceed 16 inches (406 mm). The longitudinal wires shall be thoroughly embedded in the bed joint mortar. The joint reinforcement shall engage all wythes.

Where the space between tied wythes is solidly filled with grout or mortar, the allowable stresses and other provisions for masonry bonded walls shall apply. Where the space is not filled, tied walls shall conform to the allowable stress, lateral support, thickness (excluding cavity), height and tie requirements for cavity walls.

2106.1.6 Vertical support. Structural members providing vertical support of masonry shall provide a bearing surface on which the initial bed joint shall not be less than $^1/_4$ inch (6 mm) or more than 1 inch (25 mm) in thickness and shall be of noncombustible material, except where masonry is a nonstructural decorative feature or wearing surface.

2106.1.7 Lateral support. Lateral support of masonry may be provided by cross walls, columns, pilasters, counterforts or buttresses where spanning horizontally or by floors, beams, girts or roofs where spanning vertically.

The clear distance between lateral supports of a beam shall not exceed 32 times the least width of the compression area.

2106.1.8 Protection of ties and joint reinforcement. A minimum of $^5/_8$-inch (16 mm) mortar cover shall be provided between ties or joint reinforcement and any exposed face. The thickness of grout or mortar between masonry units and joint reinforcement shall not be less than $^1/_4$ inch (6 mm), except that $^1/_4$ inch (6 mm) or smaller diameter reinforcement or bolts may be placed in bed joints which are at least twice the thickness of the reinforcement or bolts.

2106.1.9 Pipes and conduits embedded in masonry. Pipes or conduit shall not be embedded in any masonry in a manner that will reduce the capacity of the masonry to less than that necessary for required strength or required fire protection.

Placement of pipes or conduits in unfilled cores of hollow-unit masonry shall not be considered as embedment.

> **EXCEPTIONS:** 1. Rigid electric conduits may be embedded in structural masonry when their locations have been detailed on the approved plan.
>
> 2. Any pipe or conduit may pass vertically or horizontally through any masonry by means of a sleeve at least large enough to pass any hub or coupling on the pipeline. Such sleeves shall not be placed closer than three diameters, center to center, nor shall they unduly impair the strength of construction.

2106.1.10 Load tests. When a load test is required, the member or portion of the structure under consideration shall be subjected to a superimposed load equal to twice the design live load plus one half of the dead load. This load shall be left in position for a period of 24 hours before removal. If, during the test or upon removal of the load, the member or portion of the structure shows evidence of failure, such changes or modifications as are necessary to make the structure adequate for the rated capacity shall be made; or where approved, a lower rating shall be established. A flexural member shall be considered to have passed the test if the maximum deflection D at the end of the 24-hour period does not exceed the value of Formulas (6-1) or (6-2) and the beams and slabs show a recovery of at least 75 percent of the observed deflection within 24 hours after removal of the load.

$$D = \frac{l}{200} \qquad (6\text{-}1)$$

$$D = \frac{l^2}{4,000t} \qquad (6\text{-}2)$$

2106.1.11 Reuse of masonry units. Masonry units may be reused when clean, whole and conforming to the other requirements of this section. All structural properties of masonry of reclaimed units shall be determined by approved test.

2106.1.12 Special provisions in areas of seismic risk.

2106.1.12.1 General. Masonry structures constructed in the seismic zones shown in Figure 16-2 shall be designed in accordance with the design requirements of this chapter and the special provisions for each seismic zone given in this section.

2106.1.12.2 Special provisions for Seismic Zones 0 and 1. There are no special design and construction provisions in this section for structures built in Seismic Zones 0 and 1.

2106.1.12.3 Special provisions for Seismic Zone 2. Masonry structures in Seismic Zone 2 shall comply with the following special provisions:

1. Columns shall be reinforced as specified in Sections 2106.3.6, 2106.3.7 and 2107.2.13.

2. Vertical wall reinforcement of at least 0.20 square inch (130 mm^2) in cross-sectional area shall be provided continuously from support to support at each corner, at each side of each opening, at the ends of walls and at maximum spacing of 4 feet (1219 mm) apart horizontally throughout walls.

3. Horizontal wall reinforcement not less than 0.2 square inch (130 mm^2) in cross-sectional area shall be provided (1) at the bottom and top of wall openings and shall extend not less than 24 inches (610 mm) or less than 40 bar diameters past the opening, (2) continuously at structurally connected roof and floor levels and at the top of walls, (3) at the bottom of walls or in the top of foundations when doweled in walls, and (4) at maximum spacing of 10 feet (3048 mm) unless uniformly distributed joint reinforcement is provided. Reinforcement at the top and bottom of openings when continuous in walls may be used in determining the maximum spacing specified in Item 1 of this paragraph.

4. Where stack bond is used, the minimum horizontal reinforcement ratio shall be 0.0007bt. This ratio shall be satisfied by uniformly distributed joint reinforcement or by horizontal reinforcement spaced not over 4 feet (1219 mm) and fully embedded in grout or mortar.

5. The following materials shall not be used as part of the vertical or lateral load-resisting systems: Type O mortar, masonry cement, plastic cement, nonloadbearing masonry units and glass block.

2106.1.12.4 Special provisions for Seismic Zones 3 and 4. All masonry structures built in Seismic Zones 3 and 4 shall be designed and constructed in accordance with requirements for Seismic Zone 2 and with the following additional requirements and limitations:

> **EXCEPTION:** One- and two-story masonry buildings of Group R, Division 3 and Group U Occupancies located in Seismic Zone 3 having masonry wall h'/t ratios not greater than 27 and using running bond construction when provisions of Section 2106.1.12.3 are met.

1. **Column reinforcement ties.** In columns that are stressed by tensile or compressive axial overturning forces from seismic loading, the spacing of column ties shall not exceed 8 inches (203 mm) for the full height of such columns. In all other columns, ties shall be spaced a maximum of 8 inches (203 mm) in the tops and bottoms of the columns for a distance of one sixth of the clear column height, 18 inches (457 mm), or the maximum column cross-sectional dimension, whichever is greater. Tie spacing for the remaining column height shall not exceed the lessor of 16 bar diameters, 48 tie diameters, the least column cross-sectional dimension, or 18 inches (457 mm).

Column ties shall terminate with a minimum 135-degree hook with extensions not less than six bar diameters or 4 inches (102 mm). Such extensions shall engage the longitudinal column reinforcement and project into the interior of the column. Hooks shall comply with Section 2107.2.2.5, Item 3.

> **EXCEPTION:** Where ties are placed in horizontal bed joints, hooks shall consist of a 90-degree bend having an inside radius of not less than four tie diameters plus an extension of 32 tie diameters.

2. **Shear Walls.**

   2.1 **Reinforcement.** The portion of the reinforcement required to resist shear shall be uniformly distributed and shall be joint reinforcement, deformed bars or a combination thereof. The spacing of reinforcement in each direction shall not exceed one half the length of the element, nor one half the height of the element, nor 48 inches (1219 mm).

   Joint reinforcement used in exterior walls and considered in the determination of the shear strength of the member shall be hot-dipped galvanized in accordance with UBC Standard 21-10.

   Reinforcement required to resist in-plane shear shall be terminated with a standard hook as defined in Section 2107.2.2.5 or with an extension of proper embedment length beyond the reinforcement at the end of the wall section. The hook or extension may be turned up, down or horizontally. Provisions shall be made not to obstruct grout placement. Wall reinforcement terminating in columns or beams shall be fully anchored into these elements.

2.2 **Bond.** Multiwythe grouted masonry shear walls shall be designed with consideration of the adhesion bond strength between the grout and masonry units. When bond strengths are not known from previous tests, the bond strength shall be determined by tests.

2.3 **Wall reinforcement.** All walls shall be reinforced with both vertical and horizontal reinforcement. The sum of the areas of horizontal and vertical reinforcement shall be at least 0.002 times the gross cross-sectional area of the wall, and the minimum area of reinforcement in either direction shall not be less than 0.0007 times the gross cross-sectional area of the wall. The minimum steel requirements for Seismic Zone 2 in Section 2106.1.12.3, Items 2 and 3, may be included in the sum. The spacing of reinforcement shall not exceed 4 feet (1219 mm). The diameter of reinforcement shall not be less than $^3/_8$ inch (9.5 mm) except that joint reinforcement may be considered as a part or all of the requirement for minimum reinforcement. Reinforcement shall be continuous around wall corners and through intersections. Only reinforcement which is continuous in the wall or element shall be considered in computing the minimum area of reinforcement. Reinforcement with splices conforming to Section 2107.2.2.6 shall be considered as continuous reinforcement.

2.4 **Stack bond.** Where stack bond is used, the minimum horizontal reinforcement ratio shall be 0.0015bt. Where open-end units are used and grouted solid, the minimum horizontal reinforcement ratio shall be 0.0007bt.

Reinforced hollow-unit stacked bond construction which is part of the seismic-resisting system shall use open-end units so that all head joints are made solid, shall use bond beam units to facilitate the flow of grout and shall be grouted solid.

3. **Type N mortar.** Type N mortar shall not be used as part of the vertical- or lateral-load-resisting system.

4. **Concrete abutting structural masonry.** Concrete abutting structural masonry, such as at starter courses or at wall intersections not designed as true separation joints, shall be roughened to a full amplitude of $^1/_{16}$ inch (1.6 mm) and shall be bonded to the masonry in accordance with the requirements of this chapter as if it were masonry. Unless keys or proper reinforcement is provided, vertical joints as specified in Section 2106.1.4 shall be considered to be stack bond and the reinforcement as required for stack bond shall extend through the joint and be anchored into the concrete.

2106.2 Working Stress Design and Strength Design Requirements for Unreinforced and Reinforced Masonry.

2106.2.1 General. In addition to the requirements of Section 2106.1, the design of masonry structures by the working stress design method and strength design method shall comply with the requirements of this section. Additionally, the design of reinforced masonry structures by these design methods shall comply with the requirements of Section 2106.3.

2106.2.2 Specified compressive strength of masonry. The allowable stresses for the design of masonry shall be based on a value of f'_m selected for the construction.

Verification of the value of f'_m shall be based on compliance with Section 2105.3. Unless otherwise specified, f'_m shall be based on 28-day tests. If other than a 28-day test age is used, the value of f'_m shall be as indicated in design drawings or specifications. Design drawings shall show the value of f'_m for which each part of the structure is designed.

2106.2.3 Effective thickness.

2106.2.3.1 Single-wythe walls. The effective thickness of single-wythe walls of either solid or hollow units is the specified thickness of the wall.

2106.2.3.2 Multiwythe walls. The effective thickness of multiwythe walls is the specified thickness of the wall if the space between wythes is filled with mortar or grout. For walls with an open space between wythes, the effective thickness shall be determined as for cavity walls.

2106.2.3.3 Cavity walls. Where both wythes of a cavity wall are axially loaded, each wythe shall be considered to act independently and the effective thickness of each wythe is as defined in Section 2106.2.3.1. Where only one wythe is axially loaded, the effective thickness of the cavity wall is taken as the square root of the sum of the squares of the specified thicknesses of the wythes.

Where a cavity wall is composed of a single wythe and a multiwythe, and both sides are axially loaded, each side of the cavity wall shall be considered to act independently and the effective thickness of each side is as defined in Sections 2106.2.3.1 and 2106.2.3.2. Where only one side is axially loaded, the effective thickness of the cavity wall is the square root of the sum of the squares of the specified thicknesses of the sides.

2106.2.3.4 Columns. The effective thickness for rectangular columns in the direction considered is the specified thickness. The effective thickness for nonrectangular columns is the thickness of the square column with the same moment of inertia about its axis as that about the axis considered in the actual column.

2106.2.4 Effective height. The effective height of columns and walls shall be taken as the clear height of members laterally supported at the top and bottom in a direction normal to the member axis considered. For members not supported at the top normal to the axis considered, the effective height is twice the height of the member above the support. Effective height less than clear height may be used if justified.

2106.2.5 Effective area. The effective cross-sectional area shall be based on the minimum bedded area of hollow units, or the gross area of solid units plus any grouted area. Where hollow units are used with cells perpendicular to the direction of stress, the effective area shall be the lesser of the minimum bedded area or the minimum cross-sectional area. Where bed joints are raked, the effective area shall be correspondingly reduced. Effective areas for cavity walls shall be that of the loaded wythes.

2106.2.6 Effective width of intersecting walls. Where a shear wall is anchored to an intersecting wall or walls, the width of the overhanging flange formed by the intersected wall on either side of the shear wall, which may be assumed working with the shear wall for purposes of flexural stiffness calculations, shall not exceed six times the thickness of the intersected wall. Limits of the effective flange may be waived if justified. Only the effective area of the wall parallel to the shear forces may be assumed to carry horizontal shear.

2106.2.7 Distribution of concentrated vertical loads in walls. The length of wall laid up in running bond which may be considered capable of working at the maximum allowable compressive stress to resist vertical concentrated loads shall not exceed the center-to-center distance between such loads, nor the width of bearing area plus four times the wall thickness. Concentrated vertical loads shall not be assumed to be distributed across continuous vertical mortar or control joints unless elements designed to distribute the concentrated vertical loads are employed.

2106.2.8 Loads on nonbearing walls. Masonry walls used as interior partitions or as exterior surfaces of a building which do not carry vertical loads imposed by other elements of the building shall be designed to carry their own weight plus any superimposed finish and lateral forces. Bonding or anchorage of nonbearing walls shall be adequate to support the walls and to transfer lateral forces to the supporting elements.

2106.2.9 Vertical deflection. Elements supporting masonry shall be designed so that their vertical deflection will not exceed $^1/_{600}$ of the clear span under total loads. Lintels shall bear on supporting masonry on each end such that allowable stresses in the supporting masonry are not exceeded. A minimum bearing length of 4 inches (102 mm) shall be provided for lintels bearing on masonry.

2106.2.10 Structural continuity. Intersecting structural elements intended to act as a unit shall be anchored together to resist the design forces.

2106.2.11 Walls intersecting with floors and roofs. Walls shall be anchored to all floors, roofs or other elements which provide lateral support for the wall. Where floors or roofs are designed to transmit horizontal forces to walls, the anchorage to such walls shall be designed to resist the horizontal force.

2106.2.12 Modulus of elasticity of materials.

2106.2.12.1 Modulus of elasticity of masonry. The moduli for masonry may be estimated as provided below. Actual values, where required, shall be established by test. The modulus of elasticity of masonry shall be determined by the secant method in which the slope of the line for the modulus of elasticity is taken from $0.05 f'_m$ to a point on the curve at $0.33 f'_m$. These values are not to be reduced by one half as set forth in Section 2107.1.2.

Modulus of elasticity of clay or shale unit masonry.

$$E_m = 750 f'_m, \text{ 3,000,000 psi (20.5 GPa) maximum} \qquad (6\text{-}3)$$

Modulus of elasticity of concrete unit masonry.

$$E_m = 750 f'_m, \text{ 3,000,000 psi (20.5 GPa) maximum} \qquad (6\text{-}4)$$

2106.2.12.2 Modulus of elasticity of steel.

$$E_s = 29,000,000 \text{ psi (200 GPa)} \qquad (6\text{-}5)$$

2106.2.13 Shear modulus of masonry.

$$G = 0.4 E_m \qquad (6\text{-}6)$$

2106.2.14 Placement of embedded anchor bolts.

2106.2.14.1 General. Placement requirements for plate anchor bolts, headed anchor bolts and bent bar anchor bolts shall be determined in accordance with this subsection. Bent bar anchor bolts shall have a hook with a 90-degree bend with an inside diameter of three bolt diameters, plus an extension of one and one half bolt diameters at the free end. Plate anchor bolts shall have a plate welded to the shank to provide anchorage equivalent to headed anchor bolts.

The effective embedment depth l_b for plate or headed anchor bolts shall be the length of embedment measured perpendicular from the surface of the masonry to the bearing surface of the plate

or head of the anchorage, and l_b for bent bar anchors shall be the length of embedment measured perpendicular from the surface of the masonry to the bearing surface of the bent end minus one anchor bolt diameter. All bolts shall be grouted in place with at least 1 inch (25 mm) of grout between the bolt and the masonry, except that $^1/_4$-inch-diameter (6.4 mm) bolts may be placed in bed joints which are at least $^1/_2$ inch (12.7 mm) in thickness.

2106.2.14.2 Minimum edge distance. The minimum anchor bolt edge distance l_{be} measured from the edge of the masonry parallel with the anchor bolt to the surface of the anchor bolt shall be $1^1/_2$ inches (38 mm).

2106.2.14.3 Minimum embedment depth. The minimum embedment depth of anchor bolts l_b shall be four bolt diameters but not less than 2 inches (51 mm).

2106.2.14.4 Minimum spacing between bolts. The minimum center-to-center distance between anchor bolts shall be four bolt diameters.

2106.2.15 Flexural resistance of cavity walls. For computing the flexural resistance of cavity walls, lateral loads perpendicular to the plane of the wall shall be distributed to the wythes according to their respective flexural rigidities.

2106.3 Working Stress Design and Strength Design Requirements for Reinforced Masonry.

2106.3.1 General. In addition to the requirements of Sections 2106.1 and 2106.2, the design of reinforced masonry structures by the working stress design method or the strength design method shall comply with the requirements of this section.

2106.3.2 Plain bars. The use of plain bars larger than $^1/_4$ inch (6.4 mm) in diameter is not permitted.

2106.3.3 Spacing of longitudinal reinforcement. The clear distance between parallel bars, except in columns, shall not be less than the nominal diameter of the bars or 1 inch (25 mm), except that bars in a splice may be in contact. This clear distance requirement applies to the clear distance between a contact splice and adjacent splices or bars.

The clear distance between the surface of a bar and any surface of a masonry unit shall not be less than $^1/_4$ inch (6.4 mm) for fine grout and $^1/_2$ inch (12.7 mm) for coarse grout. Cross webs of hollow units may be used as support for horizontal reinforcement.

2106.3.4 Anchorage of flexural reinforcement. The tension or compression in any bar at any section shall be developed on each side of that section by the required development length. The development length of the bar may be achieved by a combination of an embedment length, anchorage or, for tension only, hooks.

Except at supports or at the free end of cantilevers, every reinforcing bar shall be extended beyond the point at which it is no longer needed to resist tensile stress for a distance equal to 12 bar diameters or the depth of the beam, whichever is greater. No flexural bar shall be terminated in a tensile zone unless at least one of the following conditions is satisfied:

1. The shear is not over one half that permitted, including allowance for shear reinforcement where provided.

2. Additional shear reinforcement in excess of that required is provided each way from the cutoff a distance equal to the depth of the beam. The shear reinforcement spacing shall not exceed $d/8r_b$.

3. The continuing bars provide double the area required for flexure at that point or double the perimeter required for reinforcing bond.

At least one third of the total reinforcement provided for negative moment at the support shall be extended beyond the extreme position of the point of inflection a distance sufficient to develop one half the allowable stress in the bar, not less than $^1/_{16}$ of the clear span, or the depth d of the member, whichever is greater.

Tensile reinforcement for negative moment in any span of a continuous restrained or cantilever beam, or in any member of a rigid frame, shall be adequately anchored by reinforcement bond, hooks or mechanical anchors in or through the supporting member.

At least one third of the required positive moment reinforcement in simple beams or at the freely supported end of continuous beams shall extend along the same face of the beam into the support at least 6 inches (153 mm). At least one fourth of the required positive moment reinforcement at the continuous end of continuous beams shall extend along the same face of the beam into the support at least 6 inches (153 mm).

Compression reinforcement in flexural members shall be anchored by ties or stirrups not less than $^1/_4$ inch (6.4 mm) in diameter, spaced not farther apart than 16 bar diameters or 48 tie diameters, whichever is less. Such ties or stirrups shall be used throughout the distance where compression reinforcement is required.

2106.3.5 Anchorage of shear reinforcement. Single, separate bars used as shear reinforcement shall be anchored at each end by one of the following methods:

1. Hooking tightly around the longitudinal reinforcement through 180 degrees.

2. Embedment above or below the mid-depth of the beam on the compression side a distance sufficient to develop the stress in the bar for plain or deformed bars.

3. By a standard hook, as defined in Section 2107.2.2.5, considered as developing 7,500 psi (52 MPa), plus embedment sufficient to develop the remainder of the stress to which the bar is subjected. The effective embedded length shall not be assumed to exceed the distance between the mid-depth of the beam and the tangent of the hook.

The ends of bars forming a single U or multiple U stirrup shall be anchored by one of the methods set forth in Items 1 through 3 above or shall be bent through an angle of at least 90 degrees tightly around a longitudinal reinforcing bar not less in diameter than the stirrup bar, and shall project beyond the bend at least 12 stirrup diameters.

The loops or closed ends of simple U or multiple U stirrups shall be anchored by bending around the longitudinal reinforcement through an angle of at least 90 degrees and project beyond the end of the bend at least 12 stirrup diameters.

2106.3.6 Lateral ties. All longitudinal bars for columns shall be enclosed by lateral ties. Lateral support shall be provided to the longitudinal bars by the corner of a complete tie having an included angle of not more than 135 degrees or by a standard hook at the end of a tie. The corner bars shall have such support provided by a complete tie enclosing the longitudinal bars. Alternate longitudinal bars shall have such lateral support provided by ties and no bar shall be farther than 6 inches (152 mm) from such laterally supported bar.

Lateral ties and longitudinal bars shall be placed not less than $1^1/_2$ inches (38 mm) and not more than 5 inches (127 mm) from the surface of the column. Lateral ties may be placed against the longitudinal bars or placed in the horizontal bed joints where the requirements of Section 2106.1.8 are met. Spacing of ties shall not exceed 16 longitudinal bar diameters, 48 tie diameters or the least dimension of the column but not more than 18 inches (457 mm).

Ties shall be at least $^1/_4$ inch (6.4 mm) in diameter for No. 7 or smaller longitudinal bars and at least No. 3 for longitudinal bars larger than No. 7. Ties smaller than No. 3 may be used for longitudinal bars larger than No. 7, provided the total cross-sectional area of such smaller ties crossing a longitudinal plane is equal to that of the larger ties at their required spacing.

2106.3.7 Column anchor bolt ties. Additional ties shall be provided around anchor bolts which are set in the top of columns. Such ties shall engage at least four bolts or, alternately, at least four vertical column bars or a combination of bolts and bars totaling at least four. Such ties shall be located within the top 5 inches (127 mm) of the column and shall provide a total of 0.4 square inch (260 mm^2) or more in cross-sectional area. The uppermost tie shall be within 2 inches (51 mm) of the top of the column.

2106.3.8 Effective width b of compression area. In computing flexural stresses in walls where reinforcement occurs, the effective width assumed for running bond masonry shall not exceed six times the nominal wall thickness or the center-to-center distance between reinforcement. Where stack bond is used, the effective width shall not exceed three times the nominal wall thickness or the center-to-center distance between reinforcement or the length of one unit, unless solid grouted open-end units are used.

SECTION 2107 — WORKING STRESS DESIGN OF MASONRY

2107.1 General.

2107.1.1 Scope. The design of masonry structures using working stress design shall comply with the provisions of Section 2106 and this section. Stresses in clay or concrete masonry under service loads shall not exceed the values given in this section.

2107.1.2 Allowable masonry stresses. When quality assurance provisions do not include requirements for special inspection as prescribed in Section 1701, the allowable stresses for masonry in Section 2107 shall be reduced by one half.

When one half allowable masonry stresses are used in Seismic Zones 3 and 4, the value of f'_m from Table 21-D shall be limited to a maximum of 1,500 psi (10 MPa) for concrete masonry and 2,600 psi (18 MPa) for clay masonry unless the value of f'_m is verified by tests in accordance with Section 2105.3.4, Items 1 and 4 or 6. A letter of certification is not required.

When one half allowable masonry stresses are used for design in Seismic Zones 3 and 4, the value of f'_m shall be limited to 1,500 psi (10 MPa) for concrete masonry and 2,600 psi (18 MPa) for clay masonry for Section 2105.3.2, Item 3, and Section 2105.3.3, Item 5, unless the value of f'_m is verified during construction by the testing requirements of Section 2105.3.2, Item 2. A letter of certification is not required.

2107.1.3 Minimum dimensions for masonry structures located in Seismic Zones 3 and 4. Elements of masonry structures located in Seismic Zones 3 and 4 shall be in accordance with this section.

2107.1.3.1 Bearing walls. The nominal thickness of reinforced masonry bearing walls shall not be less than 6 inches (152 mm) except that nominal 4-inch-thick (102 mm) load-bearing reinforced hollow-clay unit masonry walls may be used, provided net area unit strength exceeds 8,000 psi (55 MPa), units are laid in running bond, bar sizes do not exceed $^1/_2$ inch (12.7 mm) with no more than two bars or one splice in a cell, and joints are flush cut, concave or a protruding V section.

2107.1.3.2 Columns. The least nominal dimension of a reinforced masonry column shall be 12 inches (305 mm) except that, for working stress design, if the allowable stresses are reduced by one half, the minimum nominal dimension shall be 8 inches (203 mm).

2107.1.4 Design assumptions. The working stress design procedure is based on working stresses and linear stress-strain distribution assumptions with all stresses in the elastic range as follows:

1. Plane sections before bending remain plane after bending.

2. Stress is proportional to strain.

3. Masonry elements combine to form a homogenous member.

2107.1.5 Embedded anchor bolts.

2107.1.5.1 General. Allowable loads for plate anchor bolts, headed anchor bolts and bent bar anchor bolts shall be determined in accordance with this section.

2107.1.5.2 Tension. Allowable loads in tension shall be the lesser value selected from Tables 21-E-1 and 21-E-2 or shall be determined from the lesser of Formula (7-1) or Formula (7-2).

$$B_t = 0.5 A_p \sqrt{f'_m} \qquad (7\text{-}1)$$

For **SI:** $\qquad B_t = 0.042 A_p \sqrt{f'_m}$

$$B_t = 0.2 A_b f_y \qquad (7\text{-}2)$$

The area A_p shall be the lesser of Formula (7-3) or Formula (7-4) and where the projected areas of adjacent anchor bolts overlap, A_p of each anchor bolt shall be reduced by one half of the overlapping area.

$$A_p = \pi l_b{}^2 \qquad (7\text{-}3)$$

$$A_p = \pi l_{b_e}{}^2 \qquad (7\text{-}4)$$

2107.1.5.3 Shear. Allowable loads in shear shall be the value selected from Table 21-F or shall be determined from the lesser of Formula (7-5) or Formula (7-6).

$$B_v = 350 \sqrt[4]{f'_m A_b} \qquad (7\text{-}5)$$

For **SI:** $\qquad B_v = 1070 \sqrt[4]{f'_m A_b}$

$$B_v = 0.12 A_b f_y \qquad (7\text{-}6)$$

Where the anchor bolt edge distance l_{be} in the direction of load is less than 12 bolt diameters, the value of B_v in Formula (7-5) shall be reduced by linear interpolation to zero at an l_{be} distance of $1^1/_2$ inches (38 mm). Where adjacent anchors are spaced closer than $8d_b$, the allowable shear of the adjacent anchors determined by Formula (7-5) shall be reduced by linear interpolation to 0.75 times the allowable shear value at a center-to-center spacing of four bolt diameters.

2107.1.5.4 Combined shear and tension. Anchor bolts subjected to combined shear and tension shall be designed in accordance with Formula (7-7).

$$\frac{b_t}{B_t} + \frac{b_v}{B_v} \leq 1.0 \qquad (7\text{-}7)$$

2107.1.6 Compression in walls and columns.

2107.1.6.1 Walls, axial loads. Stresses due to compressive forces applied at the centroid of wall may be computed by Formula (7-8) assuming uniform distribution over the effective area.

$$f_a = P / A_e \qquad (7\text{-}8)$$

2107.1.6.2 Columns, axial loads. Stresses due to compressive forces applied at the centroid of columns may be computed by Formula (7-8) assuming uniform distribution over the effective area.

2107.1.6.3 Columns, bending or combined bending and axial loads. Stresses in columns due to combined bending and axial loads shall satisfy the requirements of Section 2107.2.7 where f_a/F_a is replaced by P/P_a. Columns subjected to bending shall meet all applicable requirements for flexural design.

2107.1.7 Shear walls, design loads. When calculating shear or diagonal tension stresses, shear walls which resist seismic forces in Seismic Zones 3 and 4 shall be designed to resist 1.5 times the forces required by Section 1630.

2107.1.8 Design, composite construction.

2107.1.8.1 General. The requirements of this section govern multiwythe masonry in which at least one wythe has strength or composition characteristics different from the other wythe or wythes and is adequately bonded to act as a single structural element.

The following assumptions shall apply to the design of composite masonry:

1. Analysis shall be based on elastic transformed section of the net area.

2. The maximum computed stress in any portion of composite masonry shall not exceed the allowable stress for the material of that portion.

2107.1.8.2 Determination of moduli of elasticity. The modulus of elasticity of each type of masonry in composite construction shall be measured by tests if the modular ratio of the respective types of masonry exceeds 2 to 1 as determined by Section 2106.2.12.

2107.1.8.3 Structural continuity.

2107.1.8.3.1 Bonding of wythes. All wythes of composite masonry elements shall be tied together as specified in Section 2106.1.5.2 as a minimum requirement. Additional ties or the combination of grout and metal ties shall be provided to transfer the calculated stress.

2107.1.8.3.2 Material properties. The effect of dimensional changes of the various materials and different boundary conditions of various wythes shall be included in the design.

2107.1.8.4 Design procedure, transformed sections. In the design of transformed sections, one material is chosen as the reference material, and the other materials are transformed to an equivalent area of the reference material by multiplying the areas of the other materials by the respective ratios of the moduli of elasticity of the other materials to that of the reference material. Thickness of the transformed area and its distance perpendicular to a given bending axis remain unchanged. Effective height or length of the element remains unchanged.

2107.1.9 Reuse of masonry units. The allowable working stresses for reused masonry units shall not exceed 50 percent of those permitted for new masonry units of the same properties.

2107.2 Design of Reinforced Masonry.

2107.2.1 Scope. The requirements of this section are in addition to the requirements of Sections 2106 and 2107.1, and govern masonry in which reinforcement is used to resist forces.

Walls with openings used to resist lateral loads whose pier and beam elements are within the dimensional limits of Section 2108.2.6.1.2 may be designed in accordance with Section 2108.2.6. Walls used to resist lateral loads not meeting the dimensional limits of Section 2108.2.6.1.2 may be designed as walls in accordance with this section or Section 2108.2.5.

2107.2.2 Reinforcement.

2107.2.2.1 Maximum reinforcement size. The maximum size of reinforcement shall be No. 11 bars. Maximum reinforcement area in cells shall be 6 percent of the cell area without splices and 12 percent of the cell area with splices.

2107.2.2.2 Cover. All reinforcing bars, except joint reinforcement, shall be completely embedded in mortar or grout and have a minimum cover, including the masonry unit, of at least $^3/_4$ inch (19 mm), $1^1/_2$ inches (38 mm) of cover when the masonry is exposed to weather and 2 inches (51 mm) of cover when the masonry is exposed to soil.

2107.2.2.3 Development length. The required development length l_d for deformed bars or deformed wire shall be calculated by:

$$l_d = 0.002\, d_b\, f_s \text{ for bars in tension} \qquad (7\text{-}9)$$

For **SI:** $\qquad l_d = 0.29\, d_b\, f_s \text{ for bars in tension}$

$$l_d = 0.0015\, d_b\, f_s \text{ for bars in compression} \qquad (7\text{-}10)$$

For **SI:** $\qquad l_d = 0.22\, d_b\, f_s \text{ for bars in compression}$

Development length for smooth bars shall be twice the length determined by Formula (7-9).

2107.2.2.4 Reinforcement bond stress. Bond stress u in reinforcing bars shall not exceed the following:

Plain Bars	60 psi (413 kPa)
Deformed Bars	200 psi (1378 kPa)
Deformed Bars without Special Inspection	100 psi (689 kPa)

2107.2.2.5 Hooks.

1. The term "standard hook" shall mean one of the following:

    1.1 A 180-degree turn plus extension of at least four bar diameters, but not less than $2^1/_2$ inches (63 mm) at free end of bar.

    1.2 A 90-degree turn plus extension of at least 12 bar diameters at free end of bar.

    1.3 For stirrup and tie anchorage only, either a 90-degree or a 135-degree turn, plus an extension of at least six bar diameters, but not less than $2^1/_2$ inches (63 mm) at the free end of the bar.

2. Inside diameter of bend of the bars, other than for stirrups and ties, shall not be less than that set forth in Table 21-G.

3. Inside diameter of bend for No. 5 or smaller stirrups and ties shall not be less than four bar diameters. Inside diameter of bend for No. 5 or larger stirrups and ties shall not be less than that set forth in Table 21-G.

4. Hooks shall not be permitted in the tension portion of any beam, except at the ends of simple or cantilever beams or at the freely supported end of continuous or restrained beams.

5. Hooks shall not be assumed to carry a load which would produce a tensile stress in the bar greater than 7,500 psi (52 MPa).

6. Hooks shall not be considered effective in adding to the compressive resistance of bars.

7. Any mechanical device capable of developing the strength of the bar without damage to the masonry may be used in lieu of a hook. Data must be presented to show the adequacy of such devices.

2107.2.2.6 Splices. The amount of lap of lapped splices shall be sufficient to transfer the allowable stress of the reinforcement as specified in Sections 2106.3.4, 2107.2.2.3 and 2107.2.12. In no case shall the length of the lapped splice be less than 30 bar diameters for compression or 40 bar diameters for tension.

Welded or mechanical connections shall develop 125 percent of the specified yield strength of the bar in tension.

> **EXCEPTION:** For compression bars in columns that are not part of the seismic-resisting system and are not subject to flexure, only the compressive strength need be developed.

When adjacent splices in grouted masonry are separated by 3 inches (76 mm) or less, the required lap length shall be increased 30 percent.

> **EXCEPTION:** Where lap splices are staggered at least 24 bar diameters, no increase in lap length is required.

See Section 2107.2.12 for lap splice increases.

2107.2.3 Design assumptions. The following assumptions are in addition to those stated in Section 2107.1.4:

1. Masonry carries no tensile stress.

2. Reinforcement is completely surrounded by and bonded to masonry material so that they work together as a homogenous material within the range of allowable working stresses.

2107.2.4 Nonrectangular flexural elements. Flexural elements of nonrectangular cross section shall be designed in accordance with the assumptions given in Sections 2107.1.4 and 2107.2.3.

2107.2.5 Allowable axial compressive stress and force. For members other than reinforced masonry columns, the allowable axial compressive stress F_a shall be determined as follows:

$$F_a = 0.25f'_m\left[1 - \left(\frac{h'}{140r}\right)^2\right] \text{ for } h'/r \le 99 \quad (7\text{-}11)$$

$$F_a = 0.25f'_m\left(\frac{70r}{h'}\right)^2 \text{ for } h'/r > 99 \qquad (7\text{-}12)$$

For reinforced masonry columns, the allowable axial compressive force P_a shall be determined as follows:

$$P_a = [0.25f'_m A_e + 0.65A_s F_{sc}]\left[1 - \left(\frac{h'}{140r}\right)^2\right]$$
$$\text{for } h'/r \le 99 \qquad (7\text{-}13)$$

$$P_a = [0.25f'_m A_e + 0.65A_s F_{sc}]\left(\frac{70r}{h'}\right)^2$$
$$\text{for } h'/r > 99 \qquad (7\text{-}14)$$

2107.2.6 Allowable flexural compressive stress. The allowable flexural compressive stress F_b is:

$$F_b = 0.33\, f'_m, \ 2{,}000 \text{ psi (13.8 MPa) maximum} \qquad (7\text{-}15)$$

2107.2.7 Combined compressive stresses, unity formula. Elements subjected to combined axial and flexural stresses shall be designed in accordance with accepted principles of mechanics or in accordance with Formula (7-16):

$$\frac{f_a}{F_a} + \frac{f_b}{F_b} \le 1 \qquad (7\text{-}16)$$

2107.2.8 Allowable shear stress in flexural members. Where no shear reinforcement is provided, the allowable shear stress F_v in flexural members is:

$$F_v = 1.0\ \sqrt{f'_m}, \ 50 \text{ psi maximum} \qquad (7\text{-}17)$$

For **SI:** $F_v = 0.083 \sqrt{f'_m}$, 345 kPa maximum

> **EXCEPTION:** For a distance of $^1/_{16}$ the clear span beyond the point of inflection, the maximum stress shall be 20 psi (140 kPa).

Where shear reinforcement designed to take entire shear force is provided, the allowable shear stress F_v in flexural members is:

$$F_v = 3.0 \sqrt{f'_m}, \text{ 150 psi maximum} \qquad (7\text{-}18)$$

For **SI:** $F_v = 0.25 \sqrt{f'_m}$, 1.0 MPa maximum

2107.2.9 Allowable shear stress in shear walls. Where in-plane flexural reinforcement is provided and masonry is used to resist all shear, the allowable shear stress F_v in shear walls is:

For $M/Vd < 1$,

$$F_v = {}^1/_3 \left(4 - \frac{M}{Vd}\right)\sqrt{f'_m}, \ \left(80 - 45\frac{M}{Vd}\right) \text{ maximum} \quad (7\text{-}19)$$

For **SI:** $F_v = {}^1/_{36} \left(4 - \frac{M}{Vd}\right)\sqrt{f'_m}, \ \left(80 - 45\frac{M}{Vd}\right)$ maximum

For $M/Vd \geq 1$, $F_v = 1.0 \sqrt{f'_m}$, 35 psi maximum (7-20)

For **SI:** $F_v = {}^1/_{12} \sqrt{f'_m}$, 240 kPa maximum

Where shear reinforcement designed to take all the shear is provided, the allowable shear stress F_v in shear walls is:

For $M/Vd < 1$,

$$F_v = {}^1/_2 \left(4 - \frac{M}{Vd}\right)\sqrt{f'_m}, \ \left(120 - 45\frac{M}{Vd}\right) \text{ maximum} \quad (7\text{-}21)$$

For **SI:** For $M/Vd < 1$,

$$F_v = {}^1/_{24} \left(4 - \frac{M}{Vd}\right)\sqrt{f'_m}, \ \left(120 - 45\frac{M}{Vd}\right) \text{ maximum}$$

For $M/Vd \geq 1$, $F_v = 1.5 \sqrt{f'_m}$, 75 psi maximum (7-22)

For **SI:** For $M/Vd \geq 1$, $F_v = 0.12 \sqrt{f'_m}$, 520 kPa maximum

2107.2.10 Allowable bearing stress. When a member bears on the full area of a masonry element, the allowable bearing stress F_{br} is:

$$F_{br} = 0.26 f'_m \qquad (7\text{-}23)$$

When a member bears on one third or less of a masonry element, the allowable bearing stress F_{br} is:

$$F_{br} = 0.38 f'_m \qquad (7\text{-}24)$$

Formula (7-24) applies only when the least dimension between the edges of the loaded and unloaded areas is a minimum of one fourth of the parallel side dimension of the loaded area. The allowable bearing stress on a reasonably concentric area greater than one third but less than the full area shall be interpolated between the values of Formulas (7-23) and (7-24).

2107.2.11 Allowable stresses in reinforcement. The allowable stresses in reinforcement shall be as follows:

1. **Tensile stress.**

    1.1 Deformed bars,

    $F_s = 0.5 f_y$, 24,000 psi (165 MPa) maximum (7-25)

    1.2 Wire reinforcement,

    $F_s = 0.5 f_y$, 30,000 psi (207 MPa) maximum (7-26)

    1.3 Ties, anchors and smooth bars,

    $F_s = 0.4 f_y$, 20,000 psi (138 MPa) maximum (7-27)

2. **Compressive stress.**

    2.1 Deformed bars in columns,

    $F_{sc} = 0.4 f_y$, 24,000 psi (165 MPa) maximum (7-28)

    2.2 Deformed bars in flexural members,

    $F_s = 0.5 f_y$, 24,000 psi (165 MPa) maximum (7-29)

    2.3 Deformed bars in shear walls which are confined by lateral ties throughout the distance where compression reinforcement is required and where such lateral ties are not less than $^1/_4$ inch in diameter and spaced not farther apart than 16 bar diameters or 48 tie diameters,

    $F_{sc} = 0.4 f_y$, 24,000 psi (165 MPa) maximum (7-30)

2107.2.12 Lap splice increases. In regions of moment where the design tensile stresses in the reinforcement are greater than 80 percent of the allowable steel tensile stress F_s, the lap length of splices shall be increased not less than 50 percent of the minimum required length. Other equivalent means of stress transfer to accomplish the same 50 percent increase may be used.

2107.2.13 Reinforcement for columns. Columns shall be provided with reinforcement as specified in this section.

2107.2.13.1 Vertical reinforcement. The area of vertical reinforcement shall not be less than $0.005 A_e$ and not more than $0.04 A_e$. At least four No. 3 bars shall be provided. The minimum clear distance between parallel bars in columns shall be two and one half times the bar diameter.

2107.2.14 Compression in walls and columns.

2107.2.14.1 General. Stresses due to compressive forces in walls and columns shall be calculated in accordance with Section 2107.2.5.

2107.2.14.2 Walls, bending or combined bending and axial loads. Stresses in walls due to combined bending and axial loads shall satisfy the requirements of Section 2107.2.7 where f_a is given by Formula (7-8). Walls subjected to bending with or without axial loads shall meet all applicable requirements for flexural design.

The design of walls with an h'/t ratio larger than 30 shall be based on forces and moments determined from an analysis of the structure. Such analysis shall consider the influence of axial loads and variable moment of inertia on member stiffness and fixed-end moments, effect of deflections on moments and forces and the effects of duration of loads.

2107.2.15 Flexural design, rectangular flexural elements. Rectangular flexural elements shall be designed in accordance with the following formulas or other methods based on the assumptions given in Sections 2107.1.4, 2107.2.3 and this section.

1. Compressive stress in the masonry:

$$f_b = \frac{M}{bd^2}\left(\frac{2}{jk}\right) \qquad (7\text{-}31)$$

2. Tensile stress in the longitudinal reinforcement:

$$f_s = \frac{M}{A_s jd} \qquad (7\text{-}32)$$

3. Design coefficients:

$$k = \sqrt{(n\rho)^2 + 2n\rho} - n\rho \qquad (7\text{-}33)$$

or

$$k = \frac{1}{1 + \dfrac{f_s}{nf_b}} \qquad (7\text{-}34)$$

$$j = 1 - \frac{k}{3} \qquad (7\text{-}35)$$

2107.2.16 Bond of flexural reinforcement. In flexural members in which tensile reinforcement is parallel to the compressive face, the bond stress shall be computed by the formula:

$$u = \frac{V}{\Sigma_o jd} \qquad (7\text{-}36)$$

2107.2.17 Shear in flexural members and shear walls. The shear stress in flexural members and shear walls shall be computed by:

$$f_v = \frac{V}{bjd} \qquad (7\text{-}37)$$

For members of T or I section, b' shall be substituted for b. Where f_v as computed by Formula (7-37) exceeds the allowable shear stress in masonry, F_v, web reinforcement shall be provided and designed to carry the total shear force. Both vertical and horizontal shear stresses shall be considered.

The area required for shear reinforcement placed perpendicular to the longitudinal reinforcement shall be computed by:

$$A_v = \frac{sV}{F_s d} \qquad (7\text{-}38)$$

Where web reinforcement is required, it shall be so spaced that every 45-degree line extending from a point at $d/2$ of the beam to the longitudinal tension bars shall be crossed by at least one line of web reinforcement.

2107.3 Design of Unreinforced Masonry.

2107.3.1 General. The requirements of this section govern masonry in which reinforcement is not used to resist design forces and are in addition to the requirements of Sections 2106 and 2107.1.

2107.3.2 Allowable axial compressive stress. The allowable axial compressive stress F_a is:

$$F_a = 0.25 f'_m \left[1 - \left(\frac{h'}{140r} \right)^2 \right] \text{ for } h'/r \le 99 \qquad (7\text{-}39)$$

$$F_a = 0.25 f'_m \left(\frac{70r}{h'} \right)^2 \text{ for } h'/r > 99 \qquad (7\text{-}40)$$

2107.3.3 Allowable flexural compressive stress. The allowable flexural compressive stress F_b is:

$$F_b = 0.33 f'_m, \text{ 2,000 psi (13.8 MPa) maximum} \qquad (7\text{-}41)$$

2107.3.4 Combined compressive stresses, unity formula. Elements subjected to combined axial and flexural stresses shall be designed in accordance with accepted principles of mechanics or in accordance with the Formula (7-42):

$$\frac{f_a}{F_a} + \frac{f_b}{F_b} \le 1 \qquad (7\text{-}42)$$

2107.3.5 Allowable tensile stress. Resultant tensile stress due to combined bending and axial load shall not exceed the allowable flexural tensile stress, F_t.

The allowable tensile stress for walls in flexure without tensile reinforcement using portland cement and hydrated lime, or using mortar cement Type M or S mortar, shall not exceed the values in Table 21-I.

Values in Table 21-I for tension normal to head joints are for running bond; no tension is allowed across head joints in stack bond masonry. These values shall not be used for horizontal flexural members.

2107.3.6 Allowable shear stress in flexural members. The allowable shear stress F_v in flexural members is:

$$F_v = 1.0 \sqrt{f'_m}, \text{ 50 psi maximum} \qquad (7\text{-}43)$$

For **SI:** $\qquad F_v = 0.083 \sqrt{f'_m}, \text{ 345 kPa maximum}$

EXCEPTION: For a distance of $^1/_{16}$th the clear span beyond the point of inflection, the maximum stress shall be 20 psi (138 kPa).

2107.3.7 Allowable shear stress in shear walls. The allowable shear stress F_v in shear walls is as follows:

1. Clay units $\qquad F_v = 0.3 \sqrt{f'_m}, \text{ 80 psi maximum} \qquad (7\text{-}44)$

For **SI:** $\qquad F_v = 0.025 \sqrt{f'_m}, \text{ 551 kPa maximum}$

2. Concrete units with Type M or S mortar, $F_v = 34$ psi (234 kPa) maximum.

3. Concrete units with Type N mortar, $F_v = 23$ psi (158 kPa) maximum.

4. The allowable shear stress in unreinforced masonry may be increased by $0.2 f_{md}$.

2107.3.8 Allowable bearing stress. When a member bears on the full area of a masonry element, the allowable bearing stress F_{br} shall be:

$$F_{br} = 0.26 f'_m \qquad (7\text{-}45)$$

When a member bears on one-third or less of a masonry element, the allowable bearing stress F_{br} shall be:

$$F_{br} = 0.38 f'_m \qquad (7\text{-}46)$$

Formula (7-46) applies only when the least dimension between the edges of the loaded and unloaded areas is a minimum of one fourth of the parallel side dimension of the loaded area. The allowable bearing stress on a reasonably concentric area greater than one third but less than the full area shall be interpolated between the values of Formulas (7-45) and (7-46).

2107.3.9 Combined bending and axial loads, compressive stresses. Compressive stresses due to combined bending and axial loads shall satisfy the requirements of Section 2107.3.4.

2107.3.10 Compression in walls and columns. Stresses due to compressive forces in walls and columns shall be calculated in accordance with Section 2107.2.5.

2107.3.11 Flexural design. Stresses due to flexure shall not exceed the values given in Sections 2107.1.2, 2107.3.3 and 2107.3.5, where:

$$f_b = Mc/I \qquad (7\text{-}47)$$

2107.3.12 Shear in flexural members and shear walls. Shear calculations for flexural members and shear walls shall be based on Formula (7-48).

$$f_v = V / A_e \qquad (7\text{-}48)$$

2107.3.13 Corbels. The slope of corbelling (angle measured from the horizontal to the face of the corbelled surface) of unreinforced masonry shall not be less than 60 degrees.

The maximum horizontal projection of corbelling from the plane of the wall shall be such that allowable stresses are not exceeded.

2107.3.14 Stack bond. Masonry units laid in stack bond shall have longitudinal reinforcement of at least 0.00027 times the vertical cross-sectional area of the wall placed horizontally in the bed joints or in bond beams spaced vertically not more than 48 inches (1219 mm) apart.

SECTION 2108 — STRENGTH DESIGN OF MASONRY

2108.1 General.

2108.1.1 General provisions. The design of hollow-unit clay and concrete masonry structures using strength design shall comply with the provisions of Section 2106 and this section.

> **EXCEPTION:** Two-wythe solid-unit masonry may be used under Sections 2108.2.1 and 2108.2.4.

2108.1.2 Quality assurance provisions. Special inspection during construction shall be provided as set forth in Section 1701.5, Item 7.

2108.1.3 Required strength. The required strength shall be determined in accordance with the factored load combinations of Section 1612.2.

2108.1.4 Design strength. Design strength is the nominal strength, multiplied by the strength-reduction factor, ϕ, as specified in this section. Masonry members shall be proportioned such that the design strength exceeds the required strength.

2108.1.4.1 Beams, piers and columns.

2108.1.4.1.1 Flexure. Flexure with or without axial load, the value of ϕ shall be determined from Formula (8-1):

$$\phi = 0.8 - \frac{P_u}{A_e f'_m} \qquad (8\text{-}1)$$
$$\text{and } 0.60 \le \phi \le 0.80$$

2108.1.4.1.2 Shear. Shear: $\phi = 0.60$

2108.1.4.2 Wall design for out-of-plane loads.

2108.1.4.2.1 Walls with factored axial load of 0.04 f'_m or less. Flexure: $\phi = 0.80$.

2108.1.4.2.2 Walls with factored axial load greater than 0.04 f'_m. Axial load and axial load with flexure: $\phi = 0.80$. Shear: $\phi = 0.60$.

2108.1.4.3 Wall design for in-plane loads.

2108.1.4.3.1 Axial load. Axial load and axial load with flexure: $\phi = 0.65$.

For walls with symmetrical reinforcement in which f_y does not exceed 60,000 psi (413 MPa), the value of ϕ may be increased linearly to 0.85 as the value of ϕP_n decreases from $0.10 f'_m A_e$ or $0.25 P_b$ to zero.

For solid grouted walls, the value of P_b may be calculated by Formula (8-2)

$$P_b = 0.85 f'_m b a_b \qquad (8\text{-}2)$$

WHERE:

$$a_b = 0.85d \{e_{mu} / [e_{mu} + (f_y / E_s)]\} \qquad (8\text{-}3)$$

2108.1.4.3.2 Shear. Shear: $\phi = 0.60$.

The value of ϕ may be 0.80 for any shear wall when its nominal shear strength exceeds the shear corresponding to development of its nominal flexural strength for the factored-load combination.

2108.1.4.4 Moment-resisting wall frames.

2108.1.4.4.1 Flexure with or without axial load. The value of ϕ shall be as determined from Formula (8-4); however, the value of ϕ shall not be less than 0.65 nor greater than 0.85.

$$\phi = 0.85 - 2\left(\frac{P_u}{A_n f'_m}\right) \qquad (8\text{-}4)$$

2108.1.4.4.2 Shear. Shear: $\phi = 0.80$.

2108.1.4.5 Anchor. Anchor bolts: $\phi = 0.80$.

2108.1.4.6 Reinforcement.

2108.1.4.6.1 Development. Development: $\phi = 0.80$.

2108.1.4.6.2 Splices. Splices: $\phi = 0.80$.

2108.1.5 Anchor bolts.

2108.1.5.1 Required strength. The required strength of embedded anchor bolts shall be determined from factored loads as specified in Section 2108.1.3.

2108.1.5.2 Nominal anchor bolt strength. The nominal strength of anchor bolts times the strength-reduction factor shall equal or exceed the required strength.

The nominal tensile capacity of anchor bolts shall be determined from the lesser of Formula (8-5) or (8-6).

$$B_{tn} = 1.0 A_p \sqrt{f'_m} \qquad (8\text{-}5)$$

For **SI:** $\qquad B_{tn} = 0.084 A_p \sqrt{f'_m}$

$$B_{tn} = 0.4 A_b f_y \qquad (8\text{-}6)$$

The area A_p shall be the lesser of Formula (8-7) or (8-8) and where the projected areas of adjacent anchor bolts overlap, the value of A_p of each anchor bolt shall be reduced by one half of the overlapping area.

$$A_p = \pi l_b^2 \qquad (8\text{-}7)$$
$$A_p = \pi l_{be}^2 \qquad (8\text{-}8)$$

The nominal shear capacity of anchor bolts shall be determined from the lesser of Formula (8-9) or (8-10).

$$B_{sn} = 900 \sqrt[4]{f'_m A_b} \qquad (8\text{-}9)$$

For **SI:** $\qquad B_{sn} = 2750 \sqrt[4]{f'_m A_b}$

$$B_{sn} = 0.25 A_b f_y \qquad (8\text{-}10)$$

Where the anchor bolt edge distance, l_{be}, in the direction of load is less than 12 bolt diameters, the value of B_{tn} in Formula (8-9) shall be reduced by linear interpolation to zero at an l_{be} distance of $1^1/_2$ inches (38 mm). Where adjacent anchor bolts are spaced closer than $8d_b$, the nominal shear strength of the adjacent anchors determined by Formula (8-9) shall be reduced by linear interpolation to 0.75 times the nominal shear strength at a center-to-center spacing of four bolt diameters.

Anchor bolts subjected to combined shear and tension shall be designed in accordance with Formula (8-11).

$$\frac{b_{tu}}{\phi B_{tn}} + \frac{b_{su}}{\phi B_{sn}} \le 1.0 \qquad (8\text{-}11)$$

2108.1.5.3 Anchor bolt placement. Anchor bolts shall be placed so as to meet the edge distance, embedment depth and spacing requirements of Sections 2106.2.14.2, 2106.2.14.3 and 2106.2.14.4.

2108.2 Reinforced Masonry.

2108.2.1 General.

2108.2.1.1 Scope. The requirements of this section are in addition to the requirements of Sections 2106 and 2108.1 and govern masonry in which reinforcement is used to resist forces.

2108.2.1.2 Design assumptions. The following assumptions apply:

Masonry carries no tensile stress greater than the modulus of rupture.

Reinforcement is completely surrounded by and bonded to masonry material so that they work together as a homogeneous material.

Nominal strength of singly reinforced masonry wall cross sections for combined flexure and axial load shall be based on applicable conditions of equilibrium and compatibility of strains. Strain in reinforcement and masonry walls shall be assumed to be directly proportional to the distance from the neutral axis.

Maximum usable strain, e_{mu}, at the extreme masonry compression fiber shall:

1. Be 0.003 for the design of beams, piers, columns and walls.

2. Not exceed 0.003 for moment-resisting wall frames, unless lateral reinforcement as defined in Section 2108.2.6.2.6 is utilized.

Strain in reinforcement and masonry shall be assumed to be directly proportional to the distance from the neutral axis.

Stress in reinforcement below specified yield strength f_y for grade of reinforcement used shall be taken as E_s times steel strain. For strains greater than that corresponding to f_y, stress in reinforcement shall be considered independent of strain and equal to f_y.

Tensile strength of masonry walls shall be neglected in flexural calculations of strength, except when computing requirements for deflection.

Relationship between masonry compressive stress and masonry strain may be assumed to be rectangular as defined by the following:

Masonry stress of 0.85 f'_m shall be assumed uniformly distributed over an equivalent compression zone bounded by edges of the cross section and a straight line located parallel to the neutral axis at a distance $a = 0.85c$ from the fiber of maximum compressive strain. Distance c from fiber of maximum strain to the neutral axis shall be measured in a direction perpendicular to that axis.

2108.2.2 Reinforcement requirements and details.

2108.2.2.1 Maximum reinforcement. The maximum size of reinforcement shall be No. 9. The diameter of a bar shall not exceed one fourth the least dimension of a cell. No more than two bars shall be placed in a cell of a wall or a wall frame.

2108.2.2.2 Placement. The placement of reinforcement shall comply with the following:

In columns and piers, the clear distance between vertical reinforcing bars shall not be less than one and one-half times the nominal bar diameter, nor less than $1^1/_2$ inches (38 mm).

2108.2.2.3 Cover. All reinforcing bars shall be completely embedded in mortar or grout and shall have a cover of not less than $1^1/_2$ inches (38 mm) nor less than 2.5 d_b.

2108.2.2.4 Standard hooks. A standard hook shall be one of the following:

1. A 180-degree turn plus an extension of at least four bar diameters, but not less than $2^1/_2$ inches (63 mm) at the free end of the bar.

2. A 135-degree turn plus an extension of at least six bar diameters at the free end of the bar.

3. A 90-degree turn plus an extension of at least 12 bar diameters at the free end of the bar.

2108.2.2.5 Minimum bend diameter for reinforcing bars. Diameter of bend measured on the inside of a bar other than for stirrups and ties in sizes No. 3 through No. 5 shall not be less than the values in Table 21-G.

Inside diameter of bends for stirrups and ties shall not be less than $4d_b$ for No. 5 bars and smaller. For bars larger than No. 5, diameter of bend shall be in accordance with Table 21-G.

2108.2.2.6 Development. The calculated tension or compression reinforcement shall be developed in accordance with the following provisions:

The embedment length of reinforcement shall be determined by Formula (8-12).

$$l_d = l_{de} / \phi \qquad (8\text{-}12)$$

WHERE:

$$l_{de} = \frac{0.15d_b^2 f_y}{K\sqrt{f'_m}} \le 52d_b \qquad (8\text{-}13)$$

For **SI:** $\qquad l_{de} = \frac{1.8d_b^2 f_y}{K\sqrt{f'_m}} \le 52d_b$

K shall not exceed $3d_b$.

The minimum embedment length of reinforcement shall be 12 inches (305 mm).

2108.2.2.7 Splices. Reinforcement splices shall comply with one of the following:

1. The minimum length of lap for bars shall be 12 inches (305 mm) or the length determined by Formula (8-14).

$$l_d = l_{de} / \phi \qquad (8\text{-}14)$$

Bars spliced by noncontact lap splices shall be spaced transversely a distance not greater than one fifth the required length of lap or more than 8 inches (203 mm).

2. A welded splice shall have the bars butted and welded to develop in tension 125 percent of the yield strength of the bar, f_y.

3. Mechanical splices shall have the bars connected to develop in tension or compression, as required, at least 125 percent of the yield strength of the bar, f_y.

2108.2.3 Design of beams, piers and columns.

2108.2.3.1 General. The requirements of this section are for the design of masonry beams, piers and columns.

The value of f'_m shall not be less than 1,500 psi (10.3 MPa). For computational purposes, the value of f'_m shall not exceed 4,000 psi (27.6 MPa).

2108.2.3.2 Design assumptions.

Member design forces shall be based on an analysis which considers the relative stiffness of structural members. The calculation of lateral stiffness shall include the contribution of all beams, piers and columns.

The effects of cracking on member stiffness shall be considered.

The drift ratio of piers and columns shall satisfy the limits specified in Chapter 16.

2108.2.3.3 Balanced reinforcement ratio for compression limit state. Calculation of the balanced reinforcement ratio, ρ_b, shall be based on the following assumptions:

1. The distribution of strain across the section shall be assumed to vary linearly from the maximum usable strain, e_{mu}, at the extreme compression fiber of the element, to a yield strain of f_y/E_s at the extreme tension fiber of the element.

2. Compression forces shall be in equilibrium with the sum of tension forces in the reinforcement and the maximum axial load associated with a loading combination $1.0D + 1.0L + (1.4E$ or $1.3W)$.

3. The reinforcement shall be assumed to be uniformly distributed over the depth of the element and the balanced reinforcement ratio shall be calculated as the area of this reinforcement divided by the net area of the element.

4. All longitudinal reinforcement shall be included in calculating the balanced reinforcement ratio except that the contribution of compression reinforcement to resistance of compressive loads shall not be considered.

2108.2.3.4 Required strength. Except as required by Sections 2108.2.3.6 through 2108.2.3.12, the required strength shall be determined in accordance with Section 2108.1.3.

2108.2.3.5 Design strength. Design strength provided by beam, pier or column cross sections in terms of axial force, sheer and moment shall be computed as the nominal strength multiplied by the applicable strength-reduction factor, ϕ, specified in Section 2108.1.4.

2108.2.3.6 Nominal strength.

2108.2.3.6.1 Nominal axial and flexural strength. The nominal axial strength, P_n, and the nominal flexural strength, M_n, of a cross section shall be determined in accordance with the design assumptions of Section 2108.2.1.2 and 2108.2.3.2.

The maximum nominal axial compressive strength shall be determined in accordance with Formula (8-15).

$$P_n = 0.80[0.85f'_m(A_e - A_s) + f_y A_s] \qquad (8\text{-}15)$$

2108.2.3.6.2 Nominal shear strength. The nominal shear strength shall be determined in accordance with Formula (8-16).

$$V_n = V_m + V_s \qquad (8\text{-}16)$$

WHERE:

$$V_m = C_d A_e \sqrt{f'_m}, \; 63C_d A_e \text{ maximum} \qquad (8\text{-}17)$$

For **SI:** $\quad V_m = 0.083 \, C_d A_e \sqrt{f'_m}, \; 63C_d A_e \text{ maximum}$

and

$$V_s = A_e \rho_n f_y \qquad (8\text{-}18)$$

1. The nominal shear strength shall not exceed the value given in Table 21-J.

2. The value of V_m shall be assumed to be zero within any region subjected to net tension factored loads.

3. The value of V_m shall be assumed to be 25 psi (172 kPa) where M_u is greater than $0.7 M_n$. The required moment, M_u, for seismic design for comparison with the $0.7 M_n$ value of this section shall be based on an R_w of 3.

2108.2.3.7 Reinforcement.

1. Where transverse reinforcement is required, the maximum spacing shall not exceed one half the depth of the member nor 48 inches (1219 mm).

2. Flexural reinforcement shall be uniformly distributed throughout the depth of the element.

3. Flexural elements subjected to load reversals shall be symmetrically reinforced.

4. The nominal moment strength at any section along a member shall not be less than one fourth of the maximum moment strength.

5. The flexural reinforcement ratio, ρ, shall not exceed $0.5 \, \rho_b$.

6. Lap splices shall comply with the provisions of Section 2108.2.2.7.

7. Welded splices and mechanical splices which develop at least 125 percent of the specified yield strength of a bar may be used for splicing the reinforcement. Not more than two longitudinal bars shall be spliced at a section. The distance between splices of adjacent bars shall be at least 30 inches (762 mm) along the longitudinal axis.

8. Specified yield strength of reinforcement shall not exceed 60,000 psi (413 MPa). The actual yield strength based on mill tests shall not exceed 1.3 times the specified yield strength.

2108.2.3.8 Seismic design provisions. The lateral seismic load resistance in any line or story level shall be provided by shear walls or wall frames, or a combination of shear walls and wall frames. Shear walls and wall frames shall provide at least 80 percent of the lateral stiffness in any line or story level.

> **EXCEPTION:** Where seismic loads are determined based on R_w not greater than three and where all joints satisfy the provisions of Section 2108.2.6.2.9, the piers may be used to provide seismic load resistance.

2108.2.3.9 Dimensional limits. Dimensions shall be in accordance with the following:

1. **Beams.**

    1.1 The nominal width of a beam shall not be less than 6 inches (153 mm).

    1.2 The clear distance between locations of lateral bracing of the compression side of the beam shall not exceed 32 times the least width of the compression area.

    1.3 The nominal depth of a beam shall not be less than 8 inches (203 mm).

2. **Piers.**

    2.1 The nominal width of a pier shall not be less than 6 inches (153 mm) and shall not exceed 16 inches (406 mm).

    2.2 The distance between lateral supports of a pier shall not exceed 30 times the nominal width of the piers except as provided for in Section 2108.2.3.9, Item 2.3.

    2.3 When the distance between lateral supports of a pier exceeds 30 times the nominal width of the pier, the provisions of Section 2108.2.4 shall be used for design.

    2.4 The nominal length of a pier shall not be less than three times the nominal width of the pier. The nominal length of a pier shall not be greater than six times the nominal width of the pier. The clear height of a pier shall not exceed five times the nominal length of the pier.

    > **EXCEPTION:** The length of a pier may be equal to the width of the pier when the axial force at the location of maximum moment is less than $0.04 f'_m A_g$.

3. **Columns.**

    3.1 The nominal width of a column shall not be less than 12 inches (305 mm).

    3.2 The distance between lateral supports of a column shall not exceed 30 times the nominal width of the column.

    3.3 The nominal length of a column shall not be less than 12 inches (305 mm) and not greater than three times the nominal width of the column.

2108.2.3.10 Beams.

2108.2.3.10.1 Scope. Members designed primarily to resist flexure shall comply with the requirements of this section. The factored axial compressive force on a beam shall not exceed $0.05 A_e f'_m$.

2108.2.3.10.2 Longitudinal reinforcement.

1. The variation in the longitudinal reinforcing bars shall not be greater than one bar size. Not more than two bar sizes shall be used in a beam.

2. The nominal flexural strength of a beam shall not be less than 1.3 times the nominal cracking moment strength of the beam. The modulus of rupture, f_r, for this calculation shall be assumed to be 235 psi (1.6 MPa).

2108.2.3.10.3 Transverse reinforcement.
Transverse reinforcement shall be provided where V_u exceeds V_m. Required shear, V_u, shall include the effects of drift. The value of V_u shall be based on Δ_M. When transverse shear reinforcement is required, the following provisions shall apply:

1. Shear reinforcement shall be a single bar with a 180-degree hook at each end.

2. Shear reinforcement shall be hooked around the longitudinal reinforcement.

3. The minimum transverse shear reinforcement ratio shall be 0.0007.

4. The first transverse bar shall not be more than one fourth of the beam depth from the end of the beam.

2108.2.3.10.4 Construction.
Beams shall be solid grouted.

2108.2.3.11 Piers.

2108.2.3.11.1 Scope. Piers proportioned to resist flexure and shear in conjunction with axial load shall comply with the requirements of this section. The factored axial compression on the piers shall not exceed $0.3 A_e f'_m$.

2108.2.3.11.2 Longitudinal reinforcement. A pier subjected to in-plane stress reversals shall be longitudinally reinforced symmetrically on both sides of the neutral axis of the pier.

1. One bar shall be provided in the end cells.

2. The minimum longitudinal reinforcement ratio shall be 0.0007.

2108.2.3.11.3 Transverse reinforcement. Transverse reinforcement shall be provided where V_u exceeds V_m. Required shear, V_u, shall include the effects of drift. The value of V_u shall be based on Δ_M. When transverse shear reinforcement is required, the following provisions shall apply:

1. Shear reinforcement shall be hooked around the extreme longitudinal bars with a 180-degree hook. Alternatively, at wall intersections, transverse reinforcement with a 90-degree standard hook around a vertical bar in the intersecting wall shall be permitted.

2. The minimum transverse reinforcement ratio shall be 0.0015.

2108.2.3.12 Columns.

2108.2.3.12.1 Scope. Columns shall comply with the requirements of this section.

2108.2.3.12.2 Longitudinal reinforcement. Longitudinal reinforcement shall be a minimum of four bars, one in each corner of the column.

1. Maximum reinforcement area shall be $0.03 A_e$.

2. Minimum reinforcement area shall be $0.005 A_e$.

2108.2.3.12.3 Lateral ties.

1. Lateral ties shall be provided in accordance with Section 2106.3.6.

2. Minimum lateral reinforcement area shall be $0.0018 A_g$.

2108.2.3.12.4 Construction. Columns shall be solid grouted.

2108.2.4 Wall design for out-of-plane loads.

2108.2.4.1 General. The requirements of this section are for the design of walls for out-of-plane loads.

2108.2.4.2 Maximum reinforcement. The reinforcement ratio shall not exceed $0.5\rho_b$.

2108.2.4.3 Moment and deflection calculations. All moment and deflection calculations in Section 2108.2.4 are based on simple support conditions top and bottom. Other support and fixity conditions, moments and deflections shall be calculated using established principles of mechanics.

2108.2.4.4 Walls with axial load of $0.04f'_m$ or less. The procedures set forth in this section, which consider the slenderness of walls by representing effects of axial forces and deflection in calculation of moments, shall be used when the vertical load stress at the location of maximum moment does not exceed $0.04f'_m$ as computed by Formula (8-19). The value of f'_m shall not exceed 6,000 psi (41.3 MPa).

$$\frac{P_w + P_f}{A_g} \leq 0.04 f'_m \qquad (8\text{-}19)$$

Walls shall have a minimum nominal thickness of 6 inches (153 mm).

Required moment and axial force shall be determined at the midheight of the wall and shall be used for design. The factored moment, M_u, at the midheight of the wall shall be determined by Formula (8-20).

$$M_u = \frac{w_u h^2}{8} + P_{uf} \frac{e}{2} + P_u \Delta_u \qquad (8\text{-}20)$$

WHERE:

Δ_u = deflection at midheight of wall due to factored loads

$$P_u = P_{uw} + P_{uf} \qquad (8\text{-}21)$$

The design strength for out-of-plane wall loading shall be determined by Formula (8-22).

$$M_u \leq \phi M_n \qquad (8\text{-}22)$$

WHERE:

$$M_n = A_{se} f_y (d - a/2) \qquad (8\text{-}23)$$

$$A_{se} = (A_s f_y + P_u) / f_y, \text{ effective area of steel} \qquad (8\text{-}24)$$

$$a = (P_u + A_s f_y) / 0.85 f'_m b, \text{ depth of stress block due to factored loads} \qquad (8\text{-}25)$$

2108.2.4.5 Wall with axial load greater than $0.04f'_m$. The procedures set forth in this section shall be used for the design of masonry walls when the vertical load stresses at the location of maximum moment exceed $0.04f'_m$ but are less than $0.2f'_m$ and the slenderness ratio h'/t does not exceed 30.

Design strength provided by the wall cross section in terms of axial force, shear and moment shall be computed as the nominal strength multiplied by the applicable strength-reduction factor, ϕ, specified in Section 2108.1.4. Walls shall be proportioned such that the design strength exceeds the required strength.

The nominal shear strength shall be determined by Formula (8-26).

$$V_n = 2A_{mv} \sqrt{f'_m} \qquad (8\text{-}26)$$

For **SI:**
$$V_n = 0.166 A_{mv} \sqrt{f'_m}$$

2108.2.4.6 Deflection design. The midheight deflection, Δ_s, under service lateral and vertical loads (without load factors) shall be limited by the relation:

$$\Delta_s = 0.007h \qquad (8\text{-}27)$$

$P\Delta$ effects shall be included in deflection calculation. The midheight deflection shall be computed with the following formula:

$$\Delta_s = \frac{5 M_s h^2}{48 E_m I_g} \text{ for } M_{ser} \leq M_{cr} \qquad (8\text{-}28)$$

$$\Delta_s = \frac{5 M_{cr} h^2}{48 E_m I_g} + \frac{5 (M_{ser} - M_{cr})h^2}{48 E_m I_{cr}} \text{ for } M_{cr} < M_{ser} < M_n \qquad (8\text{-}29)$$

The cracking moment strength of the wall shall be determined from the formula:

$$M_{cr} = S f_r \qquad (8\text{-}30)$$

The modulus of rupture, f_r, shall be as follows:

1. For fully grouted hollow-unit masonry,

$$f_r = 4.0 \sqrt{f'_m}, \text{ 235 psi maximum} \qquad (8\text{--}31)$$

For **SI:** $\qquad f_r = 0.33 \sqrt{f'_m}, \text{ 1.6 MPa maximum}$

2. For partially grouted hollow-unit masonry,

$$f_r = 2.5 \sqrt{f'_m}, \text{ 125 psi maximum} \qquad (8\text{-}32)$$

For **SI:** $\qquad f_r = 0.21 \sqrt{f'_m}, \text{ 861 kPa maximum}$

3. For two-wythe brick masonry,

$$f_r = 2.0 \sqrt{f'_m}, \text{ 125 psi maximum} \qquad (8\text{-}33)$$

For **SI:** $\qquad f_r = 0.166 \sqrt{f'_m}, \text{ 861 kPa maximum}$

2108.2.5 Wall design for in-plane loads.

2108.2.5.1 General. The requirements of this section are for the design of walls for in-plane loads.

The value of f'_m shall not be less than 1,500 psi (10.3 MPa) nor greater than 4,000 psi (27.6 MPa).

2108.2.5.2 Reinforcement. Reinforcement shall be in accordance with the following:

1. Minimum reinforcement shall be provided in accordance with Section 2106.1.12.4, Item 2.3, for all seismic areas using this method of analysis.

2. When the shear wall failure mode is in flexure, the nominal flexural strength of the shear wall shall be at least 1.8 times the cracking moment strength of a fully grouted wall or 3.0 times the cracking moment strength of a partially grouted wall from Formula (8-30).

3. The amount of vertical reinforcement shall not be less than one half the horizontal reinforcement.

4. Spacing of horizontal reinforcement within the region defined in Section 2108.2.5.5, Item 3, shall not exceed three times the nominal wall thickness nor 24 inches (610 mm).

2108.2.5.3 Design strength. Design strength provided by the shear wall cross section in terms of axial force, shear and moment shall be computed as the nominal strength multiplied by the applicable strength-reduction factor, ϕ, specified in Section 2108.1.4.3.

2108.2.5.4 Axial strength. The nominal axial strength of the shear wall supporting axial loads only shall be calculated by Formula (8-34).

$$P_o = 0.85 f'_m (A_e - A_s) + f_y A_s \qquad (8\text{-}34)$$

Axial design strength provided by the shear wall cross section shall satisfy Formula (8-35).

$$P_u \leq 0.80 \, \phi \, P_o \qquad (8\text{-}35)$$

2108.2.5.5 Shear strength. Shear strength shall be as follows:

1. The nominal shear strength shall be determined using either Item 2 or 3 below. Maximum nominal shear strength values are determined from Table 21-J.

2. The nominal shear strength of the shear wall shall be determined from Formula (8-36), except as provided in Item 3 below

$$V_n = V_m + V_s \qquad (8\text{-}36)$$

WHERE:

$$V_m = C_d A_{mv} \sqrt{f'_m} \qquad (8\text{-}37)$$

For **SI:** $\qquad V_m = 0.083 \, C_d A_{mv} \sqrt{f'_m}$

and

$$V_s = A_{mv} \, \rho_n f_y \qquad (8\text{-}38)$$

3. For a shear wall whose nominal shear strength exceeds the shear corresponding to development of its nominal flexural strength, two shear regions exist.

For all cross sections within the region defined by the base of the shear wall and a plane at a distance L_w above the base of the shear wall, the nominal shear strength shall be determined from Formula (8-39).

$$V_n = A_{mv} \, \rho_n f_y \qquad (8\text{-}39)$$

The required shear strength for this region shall be calculated at a distance $L_w/2$ above the base of the shear wall, but not to exceed one half story height.

For the other region, the nominal shear strength of the shear wall shall be determined from Formula (8-36).

2108.2.5.6 Boundary members. Boundary members shall be as follows:

1. Boundary members shall be provided at the boundaries of shear walls when the compressive strains in the wall exceed 0.0015. The strain shall be determined using factored forces and R_w equal to 1.5.

2. The minimum length of the boundary member shall be three times the thickness of the wall, but shall include all areas where the compressive strain per Section 2108.2.6.2.7 is greater than 0.0015.

3. Lateral reinforcement shall be provided for the boundary elements. The lateral reinforcement shall be a minimum of No. 3 bars at a maximum of 8-inch (203 mm) spacing within the grouted core or equivalent confinement which can develop an ultimate compressive masonry strain of at least 0.006.

2108.2.6 Design of moment-resisting wall frames.

2108.2.6.1 General requirements.

2108.2.6.1.1 Scope. The requirements of this section are for the design of fully grouted moment-resisting wall frames constructed of reinforced open-end hollow-unit concrete or hollow-unit clay masonry.

2108.2.6.1.2 Dimensional limits. Dimensions shall be in accordance with the following.

Beams. Clear span for the beam shall not be less than two times its depth.

The nominal depth of the beam shall not be less than two units or 16 inches (406 mm), whichever is greater. The nominal beam depth to nominal beam width ratio shall not exceed 6.

The nominal width of the beam shall be the greater of 8 inches (203 mm) or $1/26$ of the clear span between pier faces.

Piers. The nominal depth of piers shall not exceed 96 inches (2438 mm). Nominal depth shall not be less than two full units or 32 inches (813 mm), whichever is greater.

The nominal width of piers shall not be less than the nominal width of the beam, nor less than 8 inches (203 mm) or $1/14$ of the clear height between beam faces, whichever is greater.

The clear height-to-depth ratio of piers shall not exceed 5.

2108.2.6.1.3 Analysis. Member design forces shall be based on an analysis which considers the relative stiffness of pier and beam members, including the stiffening influence of joints.

The calculation of beam moment capacity for the determination of pier design shall include any contribution of floor slab reinforcement.

The out-of-plane drift ratio of all piers shall satisfy the drift-ratio limits specified in Section 1630.10.2.

2108.2.6.2 Design procedure.

2108.2.6.2.1 Required strength. Except as required by Sections 2108.2.6.2.7 and 2108.2.6.2.8, the required strength shall be determined in accordance with Section 2108.1.3.

2108.2.6.2.2 Design strength. Design strength provided by frame member cross sections in terms of axial force, shear and moment shall be computed as the nominal strength multiplied by the applicable strength-reduction factor, ϕ, specified in Section 2108.1.4.4.

Members shall be proportioned such that the design strength exceeds the required strength.

2108.2.6.2.3 Design assumptions for nominal strength. The nominal strength of member cross sections shall be based on assumptions prescribed in Section 2108.2.1.2.

The value of f'_m shall not be less than 1,500 psi (10.3 MPa) or greater than 4,000 psi (27.6 MPa).

2108.2.6.2.4 Reinforcement. The nominal moment strength at any section along a member shall not be less than one fourth of the higher moment strength provided at the two ends of the member.

Lap splices shall be as defined in Section 2108.2.2.7. The center of the lap splice shall be at the center of the member clear length.

Welded splices and mechanical connections conforming to Section 1912.14.3, Items 1 through 4, may be used for splicing the reinforcement at any section provided not more than alternate longitudinal bars are spliced at a section, and the distance between splices of alternate bars is at least 24 inches (610 mm) along the longitudinal axis.

Reinforcement shall not have a specified yield strength greater than 60,000 psi (413 MPa). The actual yield strength based on mill tests shall not exceed the specified yield strength times 1.3.

2108.2.6.2.5 Flexural members (beams). Requirements of this section apply to beams proportioned primarily to resist flexure as follows:

The axial compressive force on beams due to factored loads shall not exceed $0.10 A_n f'_m$.

1. **Longitudinal reinforcement.** At any section of a beam, each masonry unit through the beam depth shall contain longitudinal reinforcement.

The variation in the longitudinal reinforcement area between units at any section shall be not greater than 50 percent, except multiple No. 4 bars shall not be greater than 100 percent of the minimum area of longitudinal reinforcement contained by any one unit, except where splices occur.

Minimum reinforcement ratio calculated over the gross cross section shall be 0.002.

Maximum reinforcement ratio calculated over the gross cross section shall be $0.15 f'_m / f_y$.

2. **Transverse reinforcement.** Transverse reinforcement shall be hooked around top and bottom longitudinal bars with a standard 180-degree hook, as defined in Section 2108.2.2.4, and shall be single pieces.

Within an end region extending one beam depth from pier faces and at any region at which beam flexural yielding may occur during seismic or wind loading, maximum spacing of transverse reinforcement shall not exceed one fourth the nominal depth of the beam.

The maximum spacing of transverse reinforcement shall not exceed one half the nominal depth of the beam.

Minimum reinforcement ratio shall be 0.0015.

The first transverse bar shall not be more than 4 inches (102 mm) from the face of the pier.

2108.2.6.2.6 Members subjected to axial force and flexure.

The requirements set forth in this subsection apply to piers proportioned to resist flexure in conjunction with axial loads.

1. **Longitudinal reinforcement.** A minimum of four longitudinal bars shall be provided at all sections of every pier.

Flexural reinforcement shall be distributed across the member depth. Variation in reinforcement area between reinforced cells shall not exceed 50 percent.

Minimum reinforcement ratio calculated over the gross cross section shall be 0.002.

Maximum reinforcement ratio calculated over the gross cross section shall be $0.15 f'_m / f_y$.

Maximum bar diameter shall be one eighth nominal width of the pier.

2. **Transverse reinforcement.** Transverse reinforcement shall be hooked around the extreme longitudinal bars with standard 180-degree hook as defined in Section 2108.2.2.4.

Within an end region extending one pier depth from the end of the beam, and at any region at which flexural yielding may occur during seismic or wind loading, the maximum spacing of transverse reinforcement shall not exceed one fourth the nominal depth of the pier.

The maximum spacing of transverse reinforcement shall not exceed one half the nominal depth of the pier.

The minimum transverse reinforcement ratio shall be 0.0015.

3. **Lateral reinforcement.** Lateral reinforcement shall be provided to confine the grouted core when compressive strains due to axial and bending forces exceed 0.0015, corresponding to factored forces with R_w equal to 1.5. The unconfined portion of the cross section with strain exceeding 0.0015 shall be neglected in computing the nominal strength of the section.

The total cross-sectional area of rectangular tie reinforcement for the confined core shall not be less than:

$$A_{sh} = 0.09 s h_c f'_m / f_{yh} \qquad (8\text{-}40)$$

Alternatively, equivalent confinement which can develop an ultimate compressive strain of at least 0.006 may be substituted for rectangular tie reinforcement.

2108.2.6.2.7 Pier design forces.
Pier nominal moment strength shall not be less than 1.6 times the pier moment corresponding to the development of beam plastic hinges, except at the foundation level.

Pier axial load based on the development of beam plastic hinges in accordance with the paragraph above and including factored dead and live loads shall not exceed $0.15 A_n f'_m$.

The drift ratio of piers shall satisfy the limits specified in Chapter 16.

The effects of cracking on member stiffness shall be considered.

The base plastic hinge of the pier must form immediately adjacent to the level of lateral support provided at the base or foundation.

2108.2.6.2.8 Shear design.

1. **General.** Beam and pier nominal shear strength shall not be less than 1.4 times the shears corresponding to the development of beam flexural yielding.

It shall be assumed in the calculation of member shear force that moments of opposite sign act at the joint faces and that the member is loaded with the tributary gravity load along its span.

2. **Vertical member shear strength.** The nominal shear strength shall be determined from Formula (8-41):

$$V_n = V_m + V_s \qquad (8\text{-}41)$$

WHERE:

$$V_m = C_d A_{mv} \sqrt{f'_m} \qquad (8\text{-}42)$$

For **SI:** $\qquad V_m = 0.083 C_d A_{mv} \sqrt{f'_m}$

and

$$V_s = A_{mv} \rho_n f_y \qquad (8\text{-}43)$$

The value of V_m shall be zero within an end region extending one pier depth from beam faces and at any region where pier flexural yielding may occur during seismic loading, and at piers subjected to net tension factored loads.

The nominal pier shear strength, V_n, shall not exceed the value determined from Table 21-J.

3. **Beam shear strength.** The nominal shear strength shall be determined from Formula (8-44),

WHERE:

$$V_m = 1.2 A_{mv} \sqrt{f'_m} \qquad (8\text{-}44)$$

For **SI:** $\qquad V_m = 0.01 A_{mv} \sqrt{f'_m}$

The value of V_m shall be zero within an end region extending one beam depth from pier faces and at any region at which beam flexural yielding may occur during seismic loading.

The nominal beam shear strength, V_n, shall be determined from Formula (8-45).

$$V_n \leq 4 A_{mv} \sqrt{f'_m} \qquad (8\text{-}45)$$

For **SI:** $\qquad V_n \leq 0.33 A_{mv} \sqrt{f'_m}$

2108.2.6.2.9 Joints.

1. **General requirements.** Where reinforcing bars extend through a joint, the joint dimensions shall be proportioned such that

$$h_p > 4800 d_{bb} / \sqrt{f'_g} \qquad (8\text{-}46)$$

For **SI:** $\qquad h_p > 57\,827 d_{bb} / \sqrt{f'_g}$

and

$$h_b > 1800 d_{bp} / \sqrt{f'_g} \qquad (8\text{-}47)$$

For **SI:** $\qquad h_b > 21\,685 d_{bp} / \sqrt{f'_g}$

The grout strength shall not exceed 5,000 psi (34.4 MPa) for the purposes of Formulas (8-46) and (8-47).

Joint shear forces shall be calculated on the assumption that the stress in all flexural tension reinforcement of the beams at the pier faces is $1.4 f_y$.

Strength of joint shall be governed by the appropriate strength-reduction factors specified in Section 2108.1.4.4.

Beam longitudinal reinforcement terminating in a pier shall be extended to the far face of the pier and anchored by a standard 90- or 180-degree hook, as defined in Section 2108.2.2.4, bent back to the beam.

Pier longitudinal reinforcement terminating in a beam shall be extended to the far face of the beam and anchored by a standard 90- or 180-degree hook, as defined in Section 2108.2.2.4, bent back to the beam.

2. **Transverse reinforcement.** Special horizontal joint shear reinforcement crossing a potential corner-to-corner diagonal joint shear crack, and anchored by standard hooks, as defined in Section 2108.2.2.4, around the extreme pier reinforcing bars shall be provided such that

$$A_{jh} = 0.5 V_{jh} / f_y \qquad (8\text{-}48)$$

Vertical shear forces may be considered to be carried by a combination of masonry shear-resisting mechanisms and truss mechanisms involving intermediate pier reinforcing bars.

3. **Shear strength.** The nominal horizontal shear strength of the joint shall not exceed $7 \sqrt{f'_m}$ (For **SI:** $0.58 \sqrt{f'_m}$) or 350 psi (2.4 MPa), whichever is less.

SECTION 2109 — EMPIRICAL DESIGN OF MASONRY

2109.1 General. The design of masonry structures using empirical design located in those portions of Seismic Zones 0 and 1 as defined in Part III of Chapter 16 where the basic wind speed is less than 80 miles per hour as defined in Part II of Chapter 16 shall comply with the provisions of Section 2106 and this section, subject to approval of the building official.

2109.2 Height. Buildings relying on masonry walls for lateral load resistance shall not exceed 35 feet (10 668 mm) in height.

2109.3 Lateral Stability. Where the structure depends on masonry walls for lateral stability, shear walls shall be provided parallel to the direction of the lateral forces resisted.

Minimum nominal thickness of masonry shear walls shall be 8 inches (203 mm).

In each direction in which shear walls are required for lateral stability, the minimum cumulative length of shear walls provided shall be 0.4 times the long dimension of the building. The cumulative length of shear walls shall not include openings.

The maximum spacing of shear walls shall not exceed the ratio listed in Table 21-L.

2109.4 Compressive Stresses.

2109.4.1 General. Compressive stresses in masonry due to vertical dead loads plus live loads, excluding wind or seismic loads, shall be determined in accordance with Section 2109.4.3. Dead and live loads shall be in accordance with this code with permitted live load reductions.

2109.4.2 Allowable stresses. The compressive stresses in masonry shall not exceed the values set forth in Table 21-M. The allowable stresses given in Table 21-M for the weakest combination of the units and mortar used in any load wythe shall be used for all loaded wythes of multiwythe walls.

2109.4.3 Stress calculations. Stresses shall be calculated based on specified rather than nominal dimensions. Calculated compressive stresses shall be determined by dividing the design load by the gross cross-sectional area of the member. The area of openings, chases or recesses in walls shall not be included in the gross cross-sectional area of the wall.

2109.4.4 Anchor bolts. Bolt values shall not exceed those set forth in Table 21-N.

2109.5 Lateral Support.
Masonry walls shall be laterally supported in either the horizontal or vertical direction not exceeding the intervals set forth in Table 21-O.

Lateral support shall be provided by cross walls, pilasters, buttresses or structural framing members horizontally or by floors, roof or structural framing members vertically.

Except for parapet walls, the ratio of height to nominal thickness for cantilever walls shall not exceed 6 for solid masonry or 4 for hollow masonry.

In computing the ratio for cavity walls, the value of thickness shall be the sums of the nominal thickness of the inner and outer wythes of the masonry. In walls composed of different classes of units and mortars, the ratio of height or length to thickness shall not exceed that allowed for the weakest of the combinations of units and mortar of which the member is composed.

2109.6 Minimum Thickness.

2109.6.1 General. The nominal thickness of masonry bearing walls in buildings more than one story in height shall not be less than 8 inches (203 mm). Solid masonry walls in one-story buildings may be of 6-inch nominal thickness when not over 9 feet (2743 mm) in height, provided that when gable construction is used, an additional 6 feet (1829 mm) is permitted to the peak of the gable.

> **EXCEPTION:** The thickness of unreinforced grouted brick masonry walls may be 2 inches (51 mm) less than required by this section, but in no case less than 6 inches (152 mm).

2109.6.2 Variation in thickness. Where a change in thickness due to minimum thickness occurs between floor levels, the greater thickness shall be carried up to the higher floor level.

2109.6.3 Decrease in thickness. Where walls of masonry of hollow units or masonry-bonded hollow walls are decreased in thickness, a course or courses of solid masonry shall be constructed between the walls below and the thinner wall above, or special units or construction shall be used to transmit the loads from face shells or wythes to the walls below.

2109.6.4 Parapets. Parapet walls shall be at least 8 inches (203 mm) in thickness and their height shall not exceed three times their thickness. The parapet wall shall not be thinner than the wall below.

2109.6.5 Foundation walls. Mortar used in masonry foundation walls shall be either Type M or S.

Where the height of unbalanced fill (height of finished grade above basement floor or inside grade) and the height of the wall between lateral support does not exceed 8 feet (2438 mm), and when the equivalent fluid weight of unbalanced fill does not exceed 30 pounds per cubic foot (480 kg/m^2), the minimum thickness of foundation walls shall be as set forth in Table 21-P. Maximum depths of unbalanced fill permitted in Table 21-P may be increased with the approval of the building official when local soil conditions warrant such an increase.

Where the height of unbalanced fill, height between lateral supports or equivalent fluid weight of unbalanced fill exceeds that set forth above, foundation walls shall be designed in accordance with Chapter 18.

2109.7 Bond.

2109.7.1 General. The facing and backing of multiwythe masonry walls shall be bonded in accordance with this section.

2109.7.2 Masonry headers. Where the facing and backing of solid masonry construction are bonded by masonry headers, not less than 4 percent of the wall surface of each face shall be composed of headers extending not less than 3 inches (76 mm) into the backing. The distance between adjacent full-length headers shall not exceed 24 inches (610 mm) either vertically or horizontally. In walls in which a single header does not extend through the wall, headers from opposite sides shall overlap at least 3 inches (76 mm), or headers from opposite sides shall be covered with another header course overlapping the header below at least 3 inches (76 mm).

Where two or more hollow units are used to make up the thickness of the wall, the stretcher courses shall be bonded at vertical intervals not exceeding 34 inches (864 mm) by lapping at least 3 inches (76 mm) over the unit below, or by lapping at vertical intervals not exceeding 17 inches (432 mm) with units which are at least 50 percent greater in thickness than the units below.

2109.7.3 Wall ties. Where the facing and backing of masonry walls are bonded with $^3/_{16}$-inch-diameter (4.8 mm) wall ties or metal ties of equivalent stiffness embedded in the horizontal mortar joints, there shall be at least one metal tie for each $4^1/_2$ square feet (0.42 m^2) of wall area. Ties in alternate courses shall be staggered, the maximum vertical distance between ties shall not exceed 24 inches (610 mm), and the maximum horizontal distance shall not exceed 36 inches (914 mm). Rods bent to rectangular shape shall be used with hollow-masonry units laid with the cells vertical. In other walls, the ends of ties shall be bent to 90-degree angles to provide hooks not less than 2 inches (51 mm) long. Additional ties shall be provided at all openings, spaced not more than 3 feet (914 mm) apart around the perimeter and within 12 inches (305 mm) of the opening.

The facing and backing of masonry walls may be bonded with prefabricated joint reinforcement. There shall be at least one cross wire serving as a tie for each $2^2/_3$ square feet (0.25 m^2) of wall area. The vertical spacing of the joint reinforcement shall not exceed 16 inches (406 mm). Cross wires of prefabricated joint reinforcement shall be at least No. 9 gage wire. The longitudinal wire shall be embedded in mortar.

2109.7.4 Longitudinal bond. In each wythe of masonry, head joints in successive courses shall be offset at least one fourth of the unit length or the walls shall be reinforced longitudinally as required in Section 2106.1.12.3, Item 4.

2109.8 Anchorage.

2109.8.1 Intersecting walls. Masonry walls depending on one another for lateral support shall be anchored or bonded at locations where they meet or intersect by one of the following methods:

1. Fifty percent of the units at the intersection shall be laid in an overlapping pattern, with alternating units having a bearing of not less than 3 inches (76 mm) on the unit below.

2. Walls shall be anchored by steel connectors having a minimum section of $^1/_4$ inch by $1^1/_2$ inches (6.4 mm by 38 mm) with ends bent up at least 2 inches (51 mm), or with cross pins to form anchorage. Such anchors shall be at least 24 inches (610 mm) long and the maximum spacing shall be 4 feet (1219 mm) vertically.

3. Walls shall be anchored by joint reinforcement spaced at a maximum distance of 8 inches (203 mm) vertically. Longitudinal rods of such reinforcement shall be at least No. 9 gage and shall extend at least 30 inches (762 mm) in each direction at the intersection.

4. Interior nonbearing walls may be anchored at their intersection, at vertical spacing of not more than 16 inches (406 mm) with joint reinforcement or $^1/_4$-inch (6.4 mm) mesh galvanized hardware cloth.

5. Other metal ties, joint reinforcement or anchors may be used, provided they are spaced to provide equivalent area of anchorage to that required by this section.

2109.8.2 Floor and roof anchorage. Floor and roof diaphragms providing lateral support to masonry walls shall be connected to the masonry walls by one of the following methods:

1. Wood floor joists bearing on masonry walls shall be anchored to the wall by approved metal strap anchors at intervals not exceeding 6 feet (1829 mm). Joists parallel to the wall shall be anchored with metal straps spaced not more than 6 feet (1829 mm) on center extending over and under and secured to at least three joists. Blocking shall be provided between joists at each strap anchor.

2. Steel floor joists shall be anchored to masonry walls with No. 3 bars, or their equivalent, spaced not more than 6 feet (1829 mm) on center. Where joists are parallel to the wall, anchors shall be located at joist cross bridging.

3. Roof structures shall be anchored to masonry walls with $^1/_2$-inch-diameter (12.7 mm) bolts at 6 feet (1829 mm) on center or their equivalent. Bolts shall extend and be embedded at least 15 inches (381 mm) into the masonry, or be hooked or welded to not less than 0.2 square inch (129 mm^2) of bond beam reinforcement placed not less than 6 inches (152 mm) from the top of the wall.

2109.8.3 Walls adjoining structural framing. Where walls are dependent on the structural frame for lateral support, they shall be anchored to the structural members with metal anchors or keyed to the structural members. Metal anchors shall consist of $^1/_2$-inch-diameter (12.7 mm) bolts spaced at a maximum of 4 feet (1219 mm) on center and embedded at least 4 inches (102 mm) into the masonry, or their equivalent area.

2109.9 Unburned Clay Masonry.

2109.9.1 General. Masonry of stabilized unburned clay units shall not be used in any building more than one story in height. The unsupported height of every wall of unburned clay units shall not be more than 10 times the thickness of such walls. Bearing walls shall in no case be less than 16 inches (406 mm) in thickness. All footing walls which support masonry of unburned clay units shall extend to an elevation not less than 6 inches (152 mm) above the adjacent ground at all points.

2109.9.2 Bolts. Bolt values shall not exceed those set forth in Table 21-Q.

2109.10 Stone Masonry.

2109.10.1 General. Stone masonry is that form of construction made with natural or cast stone in which the units are laid and set in mortar with all joints filled.

2109.10.2 Construction. In ashlar masonry, bond stones uniformly distributed shall be provided to the extent of not less than 10 percent of the area of exposed facets. Rubble stone masonry 24 inches (610 mm) or less in thickness shall have bond stones with a maximum spacing of 3 feet (914 mm) vertically and 3 feet (914 mm) horizontally and, if the masonry is of greater thickness than 24 inches (610 mm), shall have one bond stone for each 6 square feet (0.56 m^2) of wall surface on both sides.

2109.10.3 Minimum thickness. The thickness of stone masonry bearing walls shall not be less than 16 inches (406 mm).

SECTION 2110 — GLASS MASONRY

2110.1 General. Masonry of glass blocks may be used in non-load-bearing exterior or interior walls and in openings which might otherwise be filled with windows, either isolated or in continuous bands, provided the glass block panels have a minimum thickness of 3 inches (76 mm) at the mortar joint and the mortared surfaces of the blocks are treated for mortar bonding. Glass block may be solid or hollow and may contain inserts.

2110.2 Mortar Joints. Glass block shall be laid in Type S or N mortar. Both vertical and horizontal mortar joints shall be at least $^1/_4$ inch (6 mm) and not more than $^3/_8$ inch (9.5 mm) thick and shall be completely filled. All mortar contact surfaces shall be treated to ensure adhesion between mortar and glass.

2110.3 Lateral Support. Glass panels shall be laterally supported along each end of the panel.

Lateral support shall be provided by panel anchors spaced not more than 16 inches (406 mm) on center or by channels. The lateral support shall be capable of resisting the horizontal design forces determined in Chapter 16 or a minimum of 200 pounds per lineal foot (2920 N per linear meter) of wall, whichever is greater. The connection shall accommodate movement requirements of Section 2110.6.

2110.4 Reinforcement. Glass block panels shall have joint reinforcement spaced not more than 16 inches (406 mm) on center and located in the mortar bed joint extending the entire length of the panel. A lapping of longitudinal wires for a minimum of 6 inches (152 mm) is required for joint reinforcement splices. Joint reinforcement shall also be placed in the bed joint immediately below and above openings in the panel. Joint reinforcement shall conform to UBC Standard 21-10, Part I. Joint reinforcement in exterior panels shall be hot-dip galvanized in accordance with UBC Standard 21-10, Part I.

2110.5 Size of Panels. Glass block panels for exterior walls shall not exceed 144 square feet (13.4 m^2) of unsupported wall surface or 15 feet (4572 mm) in any dimension. For interior walls, glass block panels shall not exceed 250 square feet (23.2 m^2) of unsupported area or 25 feet (7620 mm) in any dimension.

2110.6 Expansion Joints. Glass block shall be provided with expansion joints along the sides and top, and these joints shall have sufficient thickness to accommodate displacements of the supporting structure, but not less than $^3/_8$ inch (9.5 mm). Expansion joints shall be entirely free of mortar and shall be filled with resilient material.

2110.7 Reuse of Units. Glass block units shall not be reused after being removed from an existing panel.

SECTION 2111 — CHIMNEYS, FIREPLACES AND BARBECUES

Chimneys, flues, fireplaces and barbecues and their connections carrying products of combustion shall be designed, anchored, supported and reinforced as set forth in Chapter 31 and any applicable provisions of this chapter.

TABLE 21-A—MORTAR PROPORTIONS FOR UNIT MASONRY

MORTAR	TYPE	Portland Cement or Blended Cement	Masonry Cement[1]			Mortar Cement[2]			Hydrated Lime or Lime Putty	AGGREGATE MEASURED IN A DAMP, LOOSE CONDITION
			M	S	N	M	S	N		
Cement-lime	M	1	—	—	—	—	—	—	1/4	Not less than 2 1/4 and not more than 3 times the sum of the separate volumes of cementitious materials.
	S	1	—	—	—	—	—	—	over 1/4 to 1/2	
	N	1	—	—	—	—	—	—	over 1/2 to 1 1/4	
	O	1	—	—	—	—	—	—	over 1 1/4 to 2 1/2	
Mortar cement	M	1	—	—	—	—	—	1	—	
	M	—	—	—	—	1	—	—	—	
	S	1/2	—	—	—	—	—	1	—	
	S	—	—	—	—	—	1	—	—	
	N	—	—	—	—	—	—	1	—	
Masonry cement	M	1	—	—	1	—	—	—	—	
	M	—	1	—	—	—	—	—	—	
	S	1/2	—	—	1	—	—	—	—	
	S	—	—	1	—	—	—	—	—	
	N	—	—	—	1	—	—	—	—	
	O	—	—	—	1	—	—	—	—	

[1]Masonry cement conforming to the requirements of UBC Standard 21-11.
[2]Mortar cement conforming to the requirements of UBC Standard 21-14.

TABLE 21-B—GROUT PROPORTIONS BY VOLUME[1]

TYPE	PARTS BY VOLUME OF PORTLAND CEMENT OR BLENDED CEMENT	PARTS BY VOLUME OF HYDRATED LIME OR LIME PUTTY	AGGREGATE MEASURED IN A DAMP, LOOSE CONDITION	
			Fine	Coarse
Fine grout	1	0 to 1/10	2 1/4 to 3 times the sum of the volumes of the cementitious materials	
Coarse grout	1	0 to 1/10	2 1/4 to 3 times the sum of the volumes of the cementitious materials	1 to 2 times the sum of the volumes of the cementitious materials

[1]Grout shall attain a minimum compressive strength at 28 days of 2,000 psi (13.8 MPa). The building official may require a compressive field strength test of grout made in accordance with UBC Standard 21-18.

TABLE 21-C—GROUTING LIMITATIONS

GROUT TYPE	GROUT POUR MAXIMUM HEIGHT (feet)[1]	MINIMUM DIMENSIONS OF THE TOTAL CLEAR AREAS WITHIN GROUT SPACES AND CELLS[2,3]	
		× 25.4 for mm	
	× 304.8 for mm	Multiwythe Masonry	Hollow-unit Masonry
Fine	1	3/4	1 1/2 × 2
Fine	5	1 1/2	1 1/2 × 2
Fine	8	1 1/2	1 1/2 × 3
Fine	12	1 1/2	1 3/4 × 3
Fine	24	2	3 × 3
Coarse	1	1 1/2	1 1/2 × 3
Coarse	5	2	2 1/2 × 3
Coarse	8	2	3 × 3
Coarse	12	2 1/2	3 × 3
Coarse	24	3	3 × 4

[1]See also Section 2104.6.
[2]The actual grout space or grout cell dimensions must be larger than the sum of the following items: (1) The required minimum dimensions of total clear areas in Table 21-C; (2) The width of any mortar projections within the space; and (3) The horizontal projections of the diameters of the horizontal reinforcing bars within a cross section of the grout space or cell.
[3]The minimum dimensions of the total clear areas shall be made up of one or more open areas, with at least one area being 3/4 inch (19 mm) or greater in width.

TABLE 21-A - TABLE 21-C
UBC 2-229

1997 UBC - MASONRY

35

TABLE 21-D—SPECIFIED COMPRESSIVE STRENGTH OF MASONRY, f'_m (psi) BASED ON SPECIFYING THE COMPRESSIVE STRENGTH OF MASONRY UNITS

COMPRESSIVE STRENGTH OF CLAY MASONRY UNITS[1,2] (psi)	SPECIFIED COMPRESSIVE STRENGTH OF MASONRY, f'_m	
	Type M or S Mortar[3] (psi)	Type N Mortar[3] (psi)
	× 6.89 for kPa	
14,000 or more	5,300	4,400
12,000	4,700	3,800
10,000	4,000	3,300
8,000	3,350	2,700
6,000	2,700	2,200
4,000	2,000	1,600

COMPRESSIVE STRENGTH OF CONCRETE MASONRY UNITS[2,4] (psi)	SPECIFIED COMPRESSIVE STRENGTH OF MASONRY, f'_m	
	Type M or S Mortar[3] (psi)	Type N Mortar[3] (psi)
	× 6.89 for kPa	
4,800 or more	3,000	2,800
3,750	2,500	2,350
2,800	2,000	1,850
1,900	1,500	1,350
1,250	1,000	950

[1]Compressive strength of solid clay masonry units is based on gross area. Compressive strength of hollow clay masonry units is based on minimum net area. Values may be interpolated. When hollow clay masonry units are grouted, the grout shall conform to the proportions in Table 21-B.

[2]Assumed assemblage. The specified compressive strength of masonry f'_m is based on gross area strength when using solid units or solid grouted masonry and net area strength when using ungrouted hollow units.

[3]Mortar for unit masonry, proportion specification, as specified in Table 21-A. These values apply to portland cement-lime mortars without added air-entraining materials.

[4]Values may be interpolated. In grouted concrete masonry, the compressive strength of grout shall be equal to or greater than the compressive strength of the concrete masonry units.

TABLE 21-E-1—ALLOWABLE TENSION, B_t, FOR EMBEDDED ANCHOR BOLTS FOR CLAY AND CONCRETE MASONRY, pounds[1,2,3]

f'_m (psi)	EMBEDMENT LENGTH, l_b, or EDGE DISTANCE, l_{be} (inches)						
	2	3	4	5	6	8	10
× 6.89 for kPa	× 25.4 for mm × 4.45 for N						
1,500	240	550	970	1,520	2,190	3,890	6,080
1,800	270	600	1,070	1,670	2,400	4,260	6,660
2,000	280	630	1,120	1,760	2,520	4,500	7,020
2,500	310	710	1,260	1,960	2,830	5,030	7,850
3,000	340	770	1,380	2,150	3,100	5,510	8,600
4,000	400	890	1,590	2,480	3,580	6,360	9,930
5,000	440	1,000	1,780	2,780	4,000	7,110	11,100
6,000	480	1,090	1,950	3,040	4,380	7,790	12,200

[1]The allowable tension values in Table 21-E-1 are based on compressive strength of masonry assemblages. Where yield strength of anchor bolt steel governs, the allowable tension in pounds is given in Table 21-E-2.

[2]Values are for bolts of at least A 307 quality. Bolts shall be those specified in Section 2106.2.14.1.

[3]Values shown are for work with or without special inspection.

TABLE 21-E-2—ALLOWABLE TENSION, B_t, FOR EMBEDDED ANCHOR BOLTS FOR CLAY AND CONCRETE MASONRY, pounds[1,2]

ANCHOR BOLT DIAMETER (inches)							
× 25.4 for mm							
$1/4$	$3/8$	$1/2$	$5/8$	$3/4$	$7/8$	1	$1 1/8$
× 4.45 for N							
350	790	1,410	2,210	3,180	4,330	5,650	7,160

[1]Values are for bolts of at least A 307 quality. Bolts shall be those specified in Section 2106.2.14.1.
[2]Values shown are for work with or without special inspection.

TABLE 21-F—ALLOWABLE SHEAR, B_v, FOR EMBEDDED ANCHOR BOLTS FOR CLAY AND CONCRETE MASONRY, pounds[1,2]

f'_m (psi)	ANCHOR BOLT DIAMETER (inches)						
	× 25.4 for mm						
	$3/8$	$1/2$	$5/8$	$3/4$	$7/8$	1	$1 1/8$
	× 4.45 for N						
1,500	480	850	1,330	1,780	1,920	2,050	2,170
1,800	480	850	1,330	1,860	2,010	2,150	2,280
2,000	480	850	1,330	1,900	2,060	2,200	2,340
2,500	480	850	1,330	1,900	2,180	2,330	2,470
3,000	480	850	1,330	1,900	2,280	2,440	2,590
4,000	480	850	1,330	1,900	2,450	2,620	2,780
5,000	480	850	1,330	1,900	2,590	2,770	2,940
6,000	480	850	1,330	1,900	2,600	2,900	3,080

[1]Values are for bolts of at least A 307 quality. Bolts shall be those specified in Section 2106.2.14.1.
[2]Values shown are for work with or without special inspection.

TABLE 21-G—MINIMUM DIAMETERS OF BEND

BAR SIZE	MINIMUM DIAMETER
No. 3 through No. 8	6 bar diameters
No. 9 through No. 11	8 bar diameters

TABLE 21-H-1—RADIUS OF GYRATION[1] FOR CONCRETE MASONRY UNITS[2]

GROUT SPACING (inches)	NOMINAL WIDTH OF WALL (inches)				
× 25.4 for mm	× 25.4 for mm				
	4	6	8	10	12
Solid grouted	1.04	1.62	2.19	2.77	3.34
16	1.16	1.79	2.43	3.04	3.67
24	1.21	1.87	2.53	3.17	3.82
32	1.24	1.91	2.59	3.25	3.91
40	1.26	1.94	2.63	3.30	3.97
48	1.27	1.96	2.66	3.33	4.02
56	1.28	1.98	2.68	3.36	4.05
64	1.29	1.99	2.70	3.38	4.08
72	1.30	2.00	2.71	3.40	4.10
No grout	1.35	2.08	2.84	3.55	4.29

[1]For single-wythe masonry or for an individual wythe of a cavity wall.

$$r = \sqrt{I/A_e}$$

[2]The radius of gyration shall be based on the specified dimensions of the masonry units or shall be in accordance with the values shown which are based on the minimum dimensions of hollow concrete masonry unit face shells and webs in accordance with UBC Standard 21-4 for two cell units.

TABLE 21-E-2 - TABLE 21-H-1
UBC 2-231

1997 UBC - MASONRY

37

TABLE 21-H-2—RADIUS OF GYRATION[1] FOR CLAY MASONRY UNIT LENGTH, 16 INCHES[2]

GROUT SPACING (inches) × 25.4 for mm	NOMINAL WIDTH OF WALL (inches) × 25.4 for mm				
	4	6	8	10	12
Solid grouted	1.06	1.64	2.23	2.81	3.39
16	1.16	1.78	2.42	3.03	3.65
24	1.20	1.85	2.51	3.13	3.77
32	1.23	1.88	2.56	3.19	3.85
40	1.25	1.91	2.59	3.23	3.90
48	1.26	1.93	2.61	3.26	3.93
56	1.27	1.94	2.63	3.28	3.95
64	1.27	1.95	2.64	3.30	3.97
72	1.28	1.95	2.65	3.31	3.99
No grout	1.32	2.02	2.75	3.42	4.13

[1]For single-wythe masonry or for an individual wythe of a cavity wall.

$$r = \sqrt{I/A_e}$$

[2]The radius of gyration shall be based on the specified dimensions of the masonry units or shall be in accordance with the values shown which are based on the minimum dimensions of hollow clay masonry face shells and webs in accordance with UBC Standard 21-1 for two cell units.

TABLE 21-H-3—RADIUS OF GYRATION[1] FOR CLAY MASONRY UNIT LENGTH, 12 INCHES[2]

GROUT SPACING (inches) × 25.4 for mm	NOMINAL WIDTH OF WALL (inches) × 25.4 for mm				
	4	6	8	10	12
Solid grouted	1.06	1.65	2.24	2.82	3.41
12	1.15	1.77	2.40	3.00	3.61
18	1.19	1.82	2.47	3.08	3.71
24	1.21	1.85	2.51	3.12	3.76
30	1.23	1.87	2.53	3.15	3.80
36	1.24	1.88	2.55	3.17	3.82
42	1.24	1.89	2.56	3.19	3.84
48	1.25	1.90	2.57	3.20	3.85
54	1.25	1.90	2.58	3.21	3.86
60	1.26	1.91	2.59	3.21	3.87
66	1.26	1.91	2.59	3.22	3.88
72	1.26	1.91	2.59	3.22	3.88
No grout	1.29	1.95	2.65	3.28	3.95

[1]For single-wythe masonry or for an individual wythe of a cavity wall.

$$r = \sqrt{I/A_e}$$

[2]The radius of gyration shall be based on the specified dimensions of the masonry units or shall be in accordance with the values shown which are based on the minimum dimensions of hollow clay masonry face shells and webs in accordance with UBC Standard 21-1 for two cell units.

TABLE 21-I—ALLOWABLE FLEXURAL TENSION (psi)

UNIT TYPE	MORTAR TYPE			
	Cement-lime and Mortar Cement		Masonry Cement	
	M or S	N	M or S	N
	× 6.89 for kPa			
Normal to bed joints				
Solid	40	30	24	15
Hollow	25	19	15	9
Normal to head joints				
Solid	80	60	48	30
Hollow	50	38	30	18

TABLE 21-J—MAXIMUM NOMINAL SHEAR STRENGTH VALUES[1,2]

M/Vd	V_n MAXIMUM
≤ 0.25	$6.0 A_e \sqrt{f'_m} \le 380 A_e \ (322 A_e \sqrt{f'_m} \le 1691 A_e)$
≥ 1.00	$4.0 A_e \sqrt{f'_m} \le 250 A_e \ (214 A_e \sqrt{f'_m} \le 1113 A_e)$

[1]M is the maximum bending moment that occurs simultaneously with the shear load V at the section under consideration. Interpolation may be by straight line for M/Vd values between 0.25 and 1.00.

[2]V_n is in pounds (N), and f'_m is in pounds per square inches (kPa).

TABLE 21-K—NOMINAL SHEAR STRENGTH COEFFICIENT

M/Vd[1]	C_d
≤ 0.25	2.4
≥ 1.00	1.2

[1]M is the maximum bending moment that occurs simultaneously with the shear load V at the section under consideration. Interpolation may be by straight line for M/Vd values between 0.25 and 1.00.

TABLE 21-L—SHEAR WALL SPACING REQUIREMENTS FOR EMPIRICAL DESIGN OF MASONRY

FLOOR OR ROOF CONSTRUCTION	MAXIMUM RATIO Shear Wall Spacing to Shear Wall Length
Cast-in-place concrete	5:1
Precast concrete	4:1
Metal deck with concrete fill	3:1
Metal deck with no fill	2:1
Wood diaphragm	2:1

TABLE 21-J - TABLE 21-L
UBC 2-233

1997 UBC - MASONRY

39

TABLE 21-M—ALLOWABLE COMPRESSIVE STRESSES FOR EMPIRICAL DESIGN OF MASONRY

CONSTRUCTION: COMPRESSIVE STRENGTH OF UNIT, GROSS AREA	ALLOWABLE COMPRESSIVE STRESSES[1] GROSS CROSS-SECTIONAL AREA (psi)	
× 6.89 for kPa	× 6.89 for kPa	
	Type M or S Mortar	Type N Mortar
Solid masonry of brick and other solid units of clay or shale; sand-lime or concrete brick:		
8,000 plus, psi	350	300
4,500 psi	225	200
2,500 psi	160	140
1,500 psi	115	100
Grouted masonry, of clay or shale; sand-lime or concrete:		
4,500 plus, psi	275	200
2,500 psi	215	140
1,500 psi	175	100
Solid masonry of solid concrete masonry units:		
3,000 plus, psi	225	200
2,000 psi	160	140
1,200 psi	115	100
Masonry of hollow load-bearing units:		
2,000 plus, psi	140	120
1,500 psi	115	100
1,000 psi	75	70
700 psi	60	55
Hollow walls (cavity or masonry bonded)[2] solid units:		
2,500 plus, psi	160	140
1,500 psi	115	100
Hollow units	75	70
Stone ashlar masonry:		
Granite	720	640
Limestone or marble	450	400
Sandstone or cast stone	360	320
Rubble stone masonry Coarse, rough or random	120	100
Unburned clay masonry	30	—

[1]Linear interpolation may be used for determining allowable stresses for masonry units having compressive strengths which are intermediate between those given in the table.
[2]Where floor and roof loads are carried upon one wythe, the gross cross-sectional area is that of the wythe under load. If both wythes are loaded, the gross cross-sectional area is that of the wall minus the area of the cavity between the wythes.

TABLE 21-N—ALLOWABLE SHEAR ON BOLTS FOR EMPIRICALLY DESIGNED MASONRY EXCEPT UNBURNED CLAY UNITS

DIAMETER BOLT (inches)	EMBEDMENT[1] (inches)	SOLID MASONRY (shear in pounds)	GROUTED MASONRY (shear in pounds)
× 25.4 for mm		× 4.45 for N	
$^1/_2$	4	350	550
$^5/_8$	4	500	750
$^3/_4$	5	750	1,100
$^7/_8$	6	1,000	1,500
1	7	1,250	1,850[2]
$1^1/_8$	8	1,500	2,250[2]

[1]An additional 2 inches of embedment shall be provided for anchor bolts located in the top of columns for buildings located in Seismic Zones 2, 3 and 4.
[2]Permitted only with not less than 2,500 pounds per square inch (17.24 MPa) units.

TABLE 21-O—WALL LATERAL SUPPORT REQUIREMENTS FOR EMPIRICAL DESIGN OF MASONRY

CONSTRUCTION	MAXIMUM *l*/*t* or *h*/*t*
Bearing walls Solid or solid grouted All other	20 18
Nonbearing walls Exterior Interior	18 36

TABLE 21-P—THICKNESS OF FOUNDATION WALLS FOR EMPIRICAL DESIGN OF MASONRY

FOUNDATION WALL CONSTRUCTION	NOMINAL THICKNESS (inches) × 25.4 for mm	MAXIMUM DEPTH OF UNBALANCED FILL (feet) × 304.8 for mm
Masonry of hollow units, ungrouted	8 10 12	4 5 6
Masonry of solid units	8 10 12	5 6 7
Masonry of hollow or solid units, fully grouted	8 10 12	7 8 8
Masonry of hollow units reinforced vertically with No. 4 bars and grout at 24″ o.c. Bars located not less than 4¹/₂″ from pressure side of wall.	8	7

TABLE 21-Q—ALLOWABLE SHEAR ON BOLTS FOR MASONRY OF UNBURNED CLAY UNITS

DIAMETER OF BOLTS (inches)	EMBEDMENTS (inches)	SHEAR (pounds)
× 25.4 for mm		× 4.45 for N
¹/₂	—	—
⁵/₈	12	200
³/₄	15	300
⁷/₈	18	400
1	21	500
1¹/₈	24	600

TABLE 21-O - TABLE 21-Q
UBC 2-235

1997 UBC - MASONRY

41

Chapter 14
EXTERIOR WALL COVERINGS

SECTION 1403 — VENEER

1403.1 Scope.

1403.1.1 General. All veneer and its application shall conform to the requirements of this code. Wainscots not exceeding 4 feet (1219 mm) in height measured above the adjacent ground elevation for exterior veneer or the finish floor elevation for interior veneer may be exempted from the provisions of this chapter if approved by the building official.

1403.1.2 Limitations. Exterior veneer shall not be attached to wood-frame construction at a point more than 30 feet (9144 mm) in height above the noncombustible foundation, except the 30-foot (9144 mm) limit may be increased when special construction is designed to provide for differential movement and when approved by the building official.

1403.2 Definitions. For the purpose of this chapter, certain terms are defined as follows:

BACKING as used in this chapter is the surface or assembly to which veneer is attached.

VENEER is nonstructural facing of brick, concrete, stone, tile, metal, plastic or other similar approved material attached to a backing for the purpose of ornamentation, protection or insulation.

Adhered Veneer is veneer secured and supported through adhesion to an approved bonding material applied over an approved backing.

Anchored Veneer is veneer secured to and supported by approved connectors attached to an approved backing.

Exterior Veneer is veneer applied to weather-exposed surfaces as defined in Section 224.

Interior Veneer is veneer applied to surfaces other than weather-exposed surfaces as defined in Section 224.

1403.3 Materials. Materials used in the application of veneer shall conform to the applicable requirements for such materials as set forth elsewhere in this code.

For masonry units and mortar, see Chapter 21.

For precast concrete units, see Chapter 19.

For portland cement plaster, see Chapter 25.

Anchors, supports and ties shall be noncombustible and corrosion resistant.

When the terms "corrosion resistant" or "noncorrosive" are used in this chapter, they shall mean having a corrosion resistance equal to or greater than a hot-dipped galvanized coating of 1.5 ounces of zinc per square foot (458 g/m²) of surface area. When an element is required to be corrosion resistant or noncorrosive, all of its parts, such as screws, nails, wire, dowels, bolts, nuts, washers, shims, anchors, ties and attachments, shall be corrosion resistant.

1403.4 Design.

1403.4.1 General. The design of all veneer shall comply with the requirements of Chapter 16 and this section.

Veneer shall support no load other than its own weight and the vertical dead load of veneer above.

Surfaces to which veneer is attached shall be designed to support the additional vertical and lateral loads imposed by the veneer.

Consideration shall be given for differential movement of supports, including that caused by temperature changes, shrinkage, creep and deflection.

1403.4.2 Adhered veneer. With the exception of ceramic tile, adhered veneer and its backing shall be designed to have a bond to the supporting element sufficient to withstand a shearing stress of 50 psi (345 kPa).

1403.4.3 Anchored veneer. Anchored veneer and its attachments shall be designed to resist a horizontal force equal to at least twice the weight of the veneer.

1403.5 Adhered Veneer.

1403.5.1 Permitted backing. Backing shall be continuous and may be of any material permitted by this code. It shall have surfaces prepared to secure and support the imposed loads of veneer.

Exterior veneer, including its backing, shall provide a weatherproof covering.

For additional backing requirements, see Section 1402.

1403.5.2 Area limitations. The height and length of veneered areas shall be unlimited except as required to control expansion and contraction and as limited by Section 1403.1.2.

1403.5.3 Unit size limitations. Veneer units shall not exceed 36 inches (914 mm) in the greatest dimension or more than 720 square inches (0.46 m²) in total area and shall not weigh more than 15 pounds per square foot (psf) (73.2 kg/m²) unless approved by the building official.

> **EXCEPTION:** Veneer units weighing less than 3 psf (14.6 kg/m²) shall not be limited in dimension or area.

1403.5.4 Application. In lieu of the design required by Sections 1403.4.1 and 1403.4.2, adhered veneer may be applied by one of the following application methods:

1. A paste of neat portland cement shall be brushed on the backing and the back of the veneer unit. Type S mortar then shall be applied to the backing and the veneer unit. Sufficient mortar shall be used to create a slight excess to be forced out the edges of the units. The units shall be tapped into place so as to completely fill the space between the units and the backing. The resulting thickness of mortar in back of the units shall be not less than $^1/_2$ inch (12.7 mm) or more than $1^1/_4$ inches (32 mm).

2. Units of tile, masonry, stone or terra cotta, not over 1 inch (25 mm) in thickness, shall be restricted to 81 square inches (52 258 mm²) in area unless the back side of each unit is ground or box screeded to true up any deviations from plane. These units and glass mosaic units of tile not over 2 inches by 2 inches by $^3/_8$ inch (51 mm by 51 mm by 9.5 mm) in size may be adhered by means of portland cement. Backing may be of masonry, concrete or portland cement plaster on metal lath. Metal lath shall be fastened to the supports in accordance with the requirements of Chapter 25. Mortar as described in Table 14-A shall be applied to the backing as a setting bed. The setting bed shall be a minimum of $^3/_8$ inch (10 mm) thick and a maximum of $^3/_4$ inch (19 mm) thick. A paste of neat portland cement or one half portland cement and one half

graded sand shall be applied to the back of the exterior veneer units and to the setting bed and the veneer pressed and tapped into place to provide complete coverage between the mortar bed and veneer unit. A cement mortar shall be used to point the veneer.

1403.5.5 Ceramic tile. Portland cement mortars for installing ceramic tile on walls, floors and ceilings shall be as set forth in Table 14-A.

1403.6 Anchored Veneer.

1403.6.1 Permitted backing. Backing may be of any material permitted by this code. Exterior veneer including its backing shall provide a weatherproof covering.

1403.6.2 Height and support limitations. Anchored veneers shall be supported on footings, foundations or other noncombustible support except as provided under Section 2316.

In Seismic Zones 2, 3 and 4, the weight of all anchored veneers installed on structures more than 30 feet (9144 mm) in height above the noncombustible foundation or support shall be supported by noncombustible, corrosion-resistant structural framing. The structural framing shall have horizontal supports spaced not more than 12 feet (3658 mm) vertically above the initial 30-foot (9144 mm) height. The vertical spacing between horizontal supports may be increased when special design techniques, approved by the building official, are used in the construction.

Noncombustible, noncorrosive lintels and noncombustible supports shall be provided over all openings where the veneer unit is not self-spanning. The deflections of all structural lintels and horizontal supports required by this subsection shall not exceed $^1/_{600}$ of the span under full load of the veneer.

1403.6.3 Area limitations. The area and length of anchored veneer walls shall be unlimited, except as required to control expansion and contraction and by Section 1403.1.2.

1403.6.4 Application.

1403.6.4.1 General. In lieu of the design required by Sections 1403.4.1 and 1403.4.3, anchored veneer may be applied in accordance with this section.

1403.6.4.2 Masonry and stone units [5 inches (127 mm) maximum in thickness]. Masonry and stone veneer not exceeding 5 inches (127 mm) in thickness may be anchored directly to structural masonry, concrete or studs in one of the following manners:

1. Wall ties shall be corrosion resistant, and if made of sheet metal, shall have a minimum thickness of 0.030 inch (0.76 mm) (No. 22 galvanized sheet gage) by $^3/_4$ inch (19.1 mm) or, if of wire, shall have a minimum diameter of 0.148 inch (3.76 mm) (No. 9 B.W. gage). Wall ties shall be spaced so as to support not more than 2 square feet (0.19 m^2) of wall area but shall not be more than 24 inches (610 mm) on center horizontally. In Seismic Zones 3 and 4, wall ties shall have a lip or hook on the extended leg that will engage or enclose a horizontal joint reinforcement wire having a diameter of 0.148 inch (3.76 mm) (No. 9 B.W. gage) or equivalent. The joint reinforcement shall be continuous with butt splices between ties permitted.

When applied over stud construction, 2-inch-by-4-inch (51 mm by 102 mm) stud spacing shall not exceed 16 inches (406 mm) on center and 2-inch-by-6-inch (51 mm by 152 mm) stud spacing shall not exceed 24 inches (610 mm) on center. Approved paper shall first be applied over the sheathing or wires between studs except as otherwise provided in Section 1402, and mortar shall be slushed into the 1-inch (25 mm) space between facing and paper.

As an alternate to approved paper with slush fill, an air space of at least 1 inch (25 mm) may be maintained between the backing and the veneer in which case spot bedding at all ties shall be of cement mortar.

2. Veneer may be applied with 1-inch-minimum (25 mm) grouted backing space reinforced by not less than 2-inch-by-2-inch (51 mm by 51 mm) 0.065-inch (1.65 mm) (No. 16 B.W. gage) galvanized wire mesh placed over waterproof paper backing and anchored directly to stud construction.

Two-inch-by-4-inch (51 mm by 102 mm) stud spacing shall not exceed 16 inches (406 mm) on center and 2-inch-by-6-inch (51 mm by 152 mm) stud spacing shall not exceed 24 inches (610 mm) on center. The galvanized wire mesh shall be anchored to wood studs by galvanized steel wire furring nails at 4 inches (102 mm) on center or by barbed galvanized nails at 6 inches (152 mm) on center with a $1^1/_8$-inch-minimum (29 mm) penetration. The galvanized wire mesh may be attached to steel studs by equivalent wire ties. If this method is applied over solid sheathing, the mesh must be furred for embedment in grout. The wire mesh must be attached at the top and bottom with not less than 8-penny (64 mm) common wire nails. The grout fill shall be placed to fill the space intimately around the mesh and veneer facing.

1403.6.4.3 Stone units [10 inches (254 mm) maximum in thickness]. Stone veneer units not exceeding 10 inches (254 mm) in thickness may be anchored directly to structural masonry, concrete or to studs:

1. **With concrete or masonry backing.** Anchor ties shall not be less than 0.109-inch (2.77 mm) (No. 12 B.W. gage) galvanized wire, or approved equal, formed as an exposed eye and extending not less than $^1/_2$ inch (12.7 mm) beyond the face of the backing. The legs of the loops shall not be less than 6 inches (152 mm) in length bent at right angles and laid in the masonry mortar joint and spaced so that the eyes or loops are 12 inches (254 mm) maximum on center in both directions. There shall be provided not less than a 0.109-inch (2.77 mm) (No. 12 B.W. gage) galvanized wire tie, or approved equal, threaded through the exposed loops for every 2 square feet (0.19 m^2) of stone veneer. This tie shall be a loop having legs not less than 15 inches (381 mm) in length bent so that it will lie in the stone veneer mortar joint. The last 2 inches (51 mm) of each wire leg shall have a right angle bend. One inch (25 mm) of cement grout shall be placed between the backing and the stone veneer.

2. **With stud backing.** A 2-inch-by-2-inch (51 mm by 51 mm) 0.065-inch (1.65 mm) (No. 16 B.W. gage) galvanized wire mesh with two layers of waterproof paper backing shall be applied directly to 2-inch-by-4-inch (51 mm by 102 mm) wood studs spaced a maximum of 16 inches (406 mm) on center or 2-inch-by-6-inch (51 mm by 152 mm) wood studs spaced a maximum of 24 inches (610 mm) on center. On studs, the mesh shall be attached with 2-inch-long (51 mm) galvanized steel wire furring nails at 4 inches (102 mm) on center providing a minimum $1^1/_8$-inch (29 mm) penetration into each stud and with 8-penny (64 mm) common nails at 8 inches (203 mm) on center into top and bottom plates. The galvanized wire mesh may be attached to steel studs with equivalent wire ties. There shall not be less than 0.109-inch (2.77 mm) (No. 12 B.W. gage) galvanized wire, or approved equal, looped through the mesh for every 2 square feet (0.19 m^2) of stone veneer. This tie shall be a loop having legs not less than 15 inches (381 mm) in length, bent so that it will lie in the stone veneer mortar joint.

The last 2 inches (51 mm) of each wire leg shall have a right angle bend. One-inch-minimum (25 mm) thickness of cement grout shall be placed between the backing and the stone veneer.

1403.6.4.4 Slab-type units [2 inches (51 mm) maximum in thickness]. For veneer units of marble, travertine, granite or other stone units of slab form, ties of corrosion-resistant dowels shall engage drilled holes located in the middle third of the edge of the units spaced a maximum of 24 inches (610 mm) apart around the periphery of each unit with not less than four ties per veneer unit. Units shall not exceed 20 square feet (1.9 m^2) in area.

If the dowels are not tight fitting, the holes may be drilled not more than $^1/_{16}$ inch (1.6 mm) larger in diameter than the dowel with the hole countersunk to a diameter and depth equal to twice the diameter of the dowel in order to provide a tightfitting key of cement mortar at the dowel locations when the mortar in the joint has set.

All veneer ties shall be corrosion-resistant metal capable of resisting in tension or compression a force equal to two times the weight of the attached veneer.

If made of sheet metal, veneer ties shall not be smaller in area than 0.030 inch (0.76 mm) (No. 22 galvanized sheet gage) by 1 inch (25 mm) or, if made of wire, not smaller in diameter than 0.148-inch (3.76 mm) (No. 9 B.W. gage) wire.

1403.6.4.5 Terra cotta or ceramic units. Tied terra cotta or ceramic veneer units shall not be less than $1^1/_4$ inches (32 mm) in thickness with projecting dovetail webs on the back surface spaced approximately 8 inches (203 mm) on centers. The facing shall be tied to the backing wall with noncorrosive metal anchors of not less than 0.165-inch (4.19 mm) (No. 8 B.W. gage) wire installed at the top of each piece in horizontal bed joints not less than 12 inches (305 mm) or more than 18 inches (457 mm) on centers; these anchors shall be secured to $^1/_4$-inch (6.4 mm) galvanized pencil rods that pass through the vertical aligned loop anchors in the backing wall. The veneer ties shall have sufficient strength to support the full weight of the veneer in tension. The facing shall be set with not less than a 2-inch (51 mm) space from the backing wall and the space shall be filled solidly with portland cement grout and pea gravel. Immediately prior to setting, the backing wall and the facing shall be drenched with clean water and shall be distinctly damp when the grout is poured.

SECTION 1404 — VINYL SIDING

1404.1 General. Vinyl siding conforming to the requirements of this section and complying with UBC Standard 14-2 may be installed on exterior walls of buildings of Type V construction located in areas where the wind speed specified in Figure 16-1 does not exceed 80 miles per hour (129 km/h) and the building height is less than 40 feet (12 192 mm) in Exposure C. If construction is located in areas where wind speed exceeds 80 miles per hour (129 km/h) or building heights are in excess of 40 feet (12 192 mm), data indicating compliance with Chapter 16 must be submitted. Vinyl siding shall be secured to the building to provide weather protection for the exterior walls of the building.

1404.2 Application. The siding shall be applied over sheathing or materials listed in Section 2310. Siding shall be applied to conform with the weather-resistive barrier requirements in Section 1402.1. Siding and accessories shall be installed in accordance with approved manufacturer's instructions.

Nails used to fasten the siding and accessories shall have a minimum $^3/_8$-inch (9.5 mm) head diameter and 0.120-inch (3.05 mm) shank diameter. The nails shall be corrosion resistant and shall be long enough to penetrate the studs or nailing strip at least $^3/_4$ inch (19 mm). Where the siding is installed horizontally, the fastener spacing shall not exceed 16 inches (406 mm) horizontally and 12 inches (305 mm) vertically. Where the siding is installed vertically, the fastener spacing shall not exceed 12 inches (305 mm) horizontally and 12 inches (305 mm) vertically.

TABLE 14-A—CERAMIC TILE SETTING MORTARS

COAT		VOLUME TYPE 1 PORTLAND CEMENT	VOLUME TYPE S HYDRATED LIME	VOLUME SAND Dry	VOLUME SAND Damp	MAXIMUM THICKNESS OF COAT (inches) × 25.4 for mm	MINIMUM INTERVAL BETWEEN COATS (hours)
1. Walls and ceilings over 10 square feet (0.93 m^2)	Scratch	1	$^1/_2$	4	5	$^3/_8$	24
		1	0	3	4	$^3/_8$	24
		1	$^1/_2$	4	5	$^3/_4$	24
	Float or leveling	1	1	6	7	$^3/_4$	24
2. Walls and ceilings 10 square feet (0.93 m^2) or less	Scratch and float	1	$^1/_2$	$2^1/_2$	3	$^3/_8$ $^3/_4$	24
3. Floors	Setting bed	1	0	5	6	$1^1/_4$	—
		1	$^1/_{10}$	5	6	$1^1/_4$	—

Chapter 31
SPECIAL CONSTRUCTION

SECTION 3102 — CHIMNEYS, FIREPLACES AND BARBECUES

3102.1 Scope. Chimneys, flues, fireplaces and barbecues, and their connections, carrying products of combustion shall conform to the requirements of this section.

3102.2 Definitions.

BARBECUE is a stationary open hearth or brazier, either fuel fired or electric, used for food preparation.

CHIMNEY is a hollow shaft containing one or more passageways, vertical or nearly so, for conveying products of combustion to the outside atmosphere.

CHIMNEY CLASSIFICATIONS:

Chimney, High-heat Industrial Appliance-type, is a factory-built, masonry or metal chimney suitable for removing the products of combustion from fuel-burning high-heat appliances producing combustion gases in excess of 2,000°F (1093°C) measured at the appliance flue outlet.

Chimney, Low-heat Industrial Appliance-type, is a factory-built, masonry or metal chimney suitable for removing the products of combustion from fuel-burning low-heat appliances producing combustion gases not in excess of 1,000°F (538°C) under normal operating conditions but capable of producing combustion gases of 1,400°F (760°C) during intermittent forced firing for periods up to one hour. All temperatures are measured at the appliance flue outlet.

Chimney, Medium-heat Industrial Appliance-type, is a factory-built, masonry or metal chimney suitable for removing the products of combustion from fuel-burning medium-heat appliances producing combustion gases not in excess of 2,000°F (1093°C) measured at the appliance flue outlet.

Chimney, Residential Appliance-type, is a factory-built or masonry chimney suitable for removing products of combustion from residential-type appliances producing combustion gases not in excess of 1,000°F (538°C) measured at the appliance flue outlet.

CHIMNEY CONNECTOR is the pipe or breeching that connects a fuel-burning appliance to a chimney. (See Mechanical Code, Chapter 9.)

CHIMNEY, FACTORY-BUILT, is a chimney manufactured at a location other than the building site and composed of listed factory-built components assembled in accordance with the terms of the listing to form the completed chimney.

CHIMNEY LINER is a lining material of fireclay or approved refractory brick. For recognized standards on fireclay refractory brick see Sections 3503 and 3504; ASTM C 27, Fireclay and High-Alumina Refractory Brick; or ASTM C 1261, Firebox Brick for Residential Fireplaces.

FIREBRICK is a refractory brick.

FIREPLACE is a hearth and fire chamber or similar prepared place in which a fire may be made and which is built in conjunction with a chimney.

Factory-built Fireplace is a listed assembly of a fire chamber, its chimney and related factory-made parts designed for unit assembly without requiring field construction. Factory-built fireplaces are not dependent on mortar-filled joints for continued safe use.

Masonry Fireplace is a hearth and fire chamber of solid masonry units such as bricks, stones, masonry units or reinforced concrete provided with a suitable chimney.

MASONRY CHIMNEY is a chimney of masonry units, bricks, stones or listed masonry chimney units lined with approved flue liners. For the purpose of this chapter, masonry chimneys shall include reinforced concrete chimneys.

3102.3 Chimneys, General.

3102.3.1 Chimney support. Chimneys shall be designed, anchored, supported and reinforced as required in this chapter and applicable provisions of Chapters 16, 18, 19, 21 and 22 of this code. A chimney shall not support any structural load other than its own weight unless designed as a supporting member.

3102.3.2 Construction. Each chimney shall be so constructed as to safely convey flue gases not exceeding the maximum temperatures for the type of construction as set forth in Table 31-B and shall be capable of producing a draft at the appliance not less than that required for safe operation.

3102.3.3 Clearance. Clearance to combustible material shall be as required by Table 31-B.

3102.3.4 Lining. When required by Table 31-B, chimneys shall be lined with clay flue tile, firebrick, molded refractory units or other approved lining not less than $5/8$ inch (15.9 mm) thick as set forth in Table 31-B. Chimney liners shall be carefully bedded in approved medium-duty refractory mortar with close-fitting joints left smooth on the inside. Medium-duty fractory motor shall be in accordance with Sections 3503, 3504 and ASTM C 199.

3102.3.5 Area. The minimum net cross-sectional area of the chimney flue for fireplaces shall be determined in accordance with Figure 31-1. The minimum cross-sectional area shown or a flue size providing equivalent net cross-sectional area shall be used. The height of the chimney shall be measured from the firebox floor to the top of the last chimney flue tile. Chimney passageways for low-heat chimneys and incinerators shall not be smaller in area than the vent connection on the appliance attached thereto or not less than that set forth in Table 31-A.

> **EXCEPTION:** Chimney passageways designed by engineering methods approved by the building official.

3102.3.6 Height and termination. Every chimney shall extend above the roof and the highest elevation of any part of a building as shown in Table 31-B. For altitudes over 2,000 feet (610 m), the building official shall be consulted in determining the height of the chimney.

3102.3.7 Cleanouts. Cleanout openings shall be provided within 6 inches (152 mm) of the base of every masonry chimney.

3102.3.8 Spark arrester. Where determined necessary by the building official due to local climatic conditions or where sparks escaping from the chimney would create a hazard, chimneys attached to any appliance or fireplace that burns solid fuel shall be equipped with an approved spark arrester. The net free area of the spark arrester shall not be less than four times the net free area of the outlet of the chimney. The spark arrester screen shall have heat and corrosion resistance equivalent to 0.109-inch (2.77 mm) (No. 12 B.W. gage) wire, 0.042-inch (1.07 mm) (No. 19 B.W. gage) galvanized wire or 0.022-inch (0.56 mm) (No. 24 B.W. gage) stainless steel. Openings shall not permit the passage of spheres having a diameter larger than $1/2$ inch (12.7 mm) and shall not block the passage of spheres having a diameter of less than $3/8$ inch (9.5 mm).

Chimneys used with fireplaces or heating appliances in which solid or liquid fuel is used shall be provided with a spark arrester as required in the Fire Code.

> **EXCEPTION:** Chimneys that are located more than 200 feet (60 960 mm) from any mountainous, brush-covered or forest-covered land or land covered with flammable material and that are not attached to a structure having less than a Class C roof covering, as set forth in Chapter 15.

3102.4 Masonry Chimneys.

3102.4.1 Design. Masonry chimneys shall be designed and constructed to comply with Sections 3102.3.2 and 3102.4.2.

3102.4.2 Walls. Walls of masonry chimneys shall be constructed as set forth in Table 31-B.

3102.4.3 Reinforcing and seismic anchorage. Unless a specific design is provided, every masonry or concrete chimney in Seismic Zones 2, 3 and 4 shall be reinforced with not less than four No. 4 steel reinforcing bars conforming to the provisions of Chapter 19 or 21 of this code. The bars shall extend the full height of the chimney and shall be spliced in accordance with the applicable requirements of Chapter 19 or 21. In masonry chimneys, the vertical bars shall have a minimum cover of $1/2$ inch (12.7 mm) of grout or mortar tempered to a pouring consistency. The bars shall be tied horizontally at 18-inch (457 mm) intervals with not less than $1/4$-inch-diameter (6.4 mm) steel ties. The slope of the inclined portion of the offset in vertical bars shall not exceed 2 units vertical in 1 unit horizontal (200% slope). Two ties shall also be placed at each bend in vertical bars. Where the width of the chimney exceeds 40 inches (1016 mm), two additional No. 4 vertical bars shall be provided for each additional flue incorporated in the chimney or for each additional 40 inches (1016 mm) in width or fraction thereof.

In Seismic Zones 2, 3 and 4, all masonry and concrete chimneys shall be anchored at each floor or ceiling line more than 6 feet (1829 mm) above grade, except when constructed completely within the exterior walls of the building. Anchorage shall consist of two $3/16$-inch-by-1-inch (4.8 mm by 25 mm) steel straps cast at least 12 inches (305 mm) into the chimney with a 180-degree bend with a 6-inch (152 mm) extension around the vertical reinforcing bars in the outer face of the chimney.

Each strap shall be fastened to the structural framework of the building with two $1/2$-inch-diameter (12.7 mm) bolts per strap. Where the joists do not head into the chimney, the anchor strap shall be connected to 2-inch-by-4-inch (51 mm by 102 mm) ties crossing a minimum of four joists. The ties shall be connected to each joist with two 16d nails. As an alternative to the 2-inch-by-4-inch (51 mm by 102 mm) ties, each anchor strap shall be connected to the structural framework by two $1/2$-inch-diameter (12.7 mm) bolts in an approved manner.

3102.4.4 Chimney offset. Masonry chimneys may be offset at a slope of not more than 4 units vertical in 24 units horizontal (16.7% slope), but not more than one third of the dimension of the chimney, in the direction of the offset. The slope of the transition

from the fireplace to the chimney shall not exceed 2 units vertical in 1 unit horizontal (200% slope).

3102.4.5 Change in size or shape. Masonry chimneys shall not change in size or shape within 6 inches (152 mm) above or below any combustible floor, ceiling or roof component penetrated by the chimney.

3102.4.6 Separation of masonry chimney passageways. Two or more flues in a chimney shall be separated by masonry not less than 4 inches (102 mm) thick bonded into the masonry wall of the chimney.

3102.4.7 Inlets. Every inlet to any masonry chimney shall enter the side thereof and shall not be of less than $1/8$-inch-thick (3.2 mm) metal or $5/8$-inch-thick (15.9 mm) refractory material. Where there is no other opening below the inlet other than the cleanout, a masonry plug shall be constructed in the chimney not more than 16 inches (406 mm) below the inlet and the cleanout shall be located where it is accessible above the plug. If the plug is located less than 6 inches (152 mm) below the inlet, the inlet may serve as the cleanout.

3102.5 Factory-built Chimneys and Fireplaces.

3102.5.1 General. Factory-built chimneys and factory-built fireplaces shall be listed and shall be installed in accordance with the terms of their listings and the manufacturer's instructions as specified in the Mechanical Code.

3102.5.2 Hearth extensions. Hearth extensions of listed factory-built fireplaces shall conform to the conditions of listing and the manufacturer's installation instructions.

3102.5.3 Multiple venting in vertical shafts. Factory-built chimneys utilized with listed factory-built fireplaces may be used in a common vertical shaft having the required fire-resistance rating.

3102.6 Metal Chimneys. Metal chimneys shall be constructed and installed to meet the requirements of the Mechanical Code.

Metal chimneys shall be anchored at each floor and roof with two $1^1/2$-inch-by-$1/8$-inch (38 mm by 3.2 mm) metal straps looped around the outside of the chimney installation and nailed with not less than six 8d nails per strap at each joist.

3102.7 Masonry and Concrete Fireplaces and Barbecues.

3102.7.1 General. Masonry fireplaces, barbecues, smoke chambers and fireplace chimneys shall be of masonry or reinforced concrete and shall conform to the requirements of this section.

3102.7.2 Support. Masonry fireplaces shall be supported on foundations designed as specified in Chapters 16, 18 and 21.

When an approved design is not provided, foundations for masonry and concrete fireplaces shall not be less than 12 inches (305 mm) thick, extend not less than 6 inches (152 mm) outside the fireplace wall and project below the natural ground surface in accordance with the depth of foundations set forth in Table 18-I-C.

3102.7.3 Fireplace walls. Masonry walls of fireplaces shall not be less than 8 inches (203 mm) in thickness. Walls of fireboxes shall not be less than 10 inches (254 mm) in thickness, except that where a lining of firebrick is used, such walls shall not be less than a total of 8 inches (203 mm) in thickness. The firebox shall not be less than 20 inches (508 mm) in depth. Joints in firebrick shall not exceed $1/4$ inch (6.4 mm).

> **EXCEPTION:** For Rumford fireplaces, the depth may be reduced to 12 inches (305 mm) when
>
> 1. The depth is at least one third the width of the fireplace opening.

2. The throat is at least 12 inches (305 mm) above the lintel and is at least $^1/_{20}$ of the cross-sectional area of the fireplace opening.

3102.7.4 Hoods. Metal hoods used as part of a fireplace or barbecue shall not be less than 0.036-inch (0.92 mm) (No. 19 carbon sheet steel gage) copper, galvanized steel or other equivalent corrosion-resistant ferrous metal with all seams and connections of smokeproof unsoldered constructions. The hoods shall be sloped at an angle of 45 degrees or less from the vertical and shall extend horizontally at least 6 inches (152 mm) beyond the limits of the firebox. Metal hoods shall be kept a minimum of 18 inches (457 mm) from combustible materials unless approved for reduced clearances.

3102.7.5 Metal heat circulators. Approved metal heat circulators may be installed in fireplaces.

3102.7.6 Smoke chamber. Front and side walls shall not be less than 8 inches (203 mm) in thickness. Smoke chamber back walls shall not be less than 6 inches (152 mm) in thickness. A minimum $^5/_8$-inch-thick (16 mm) clay flue lining, complying with Sections 3503, 3504 and ASTM C 315, shall be permitted to form the inside surface of the 8-inch (203 mm) and 6-inch (152 mm) smoke chamber walls.

3102.7.7 Chimneys. Chimneys for fireplaces shall be constructed as specified in Sections 3102.3, 3102.4 and 3102.5 for residential-type appliances.

3102.7.8 Clearance to combustible material. Combustible materials shall not be placed within 2 inches (51 mm) of fireplace, smoke chamber or chimney walls. Combustible material shall not be placed within 6 inches (152 mm) of the fireplace opening. No such combustible material within 12 inches (305 mm) of the fireplace opening shall project more than $^1/_8$ inch (3.2 mm) for each 1-inch (25 mm) clearance from such opening.

No part of metal hoods used as part of a fireplace or barbecue shall be less than 18 inches (457 mm) from combustible material. This clearance may be reduced to the minimum requirements specified in the Mechanical Code.

3102.7.9 Areas of flues, throats and dampers. The throat shall be at least 8 inches (203 mm) above the fireplace opening and shall be at least 4 inches (102 mm) in depth. The net cross-sectional area of the flue and of the throat between the firebox and the smoke chamber of a fireplace shall not be less than that set forth in Figure 31-1 or Table 31-A. Metal dampers equivalent to not less than 0.097-inch (2.46 mm) (No. 12 carbon sheet metal gage) steel shall be installed. When fully opened, damper openings shall not be less than 90 percent of the required flue area.

3102.7.10 Lintel. Masonry over the fireplace opening shall be supported by a noncombustible lintel unless the masonry is self-supporting.

3102.7.11 Hearth. Masonry fireplaces shall be provided with a brick, concrete, stone or other approved noncombustible hearth slab. This slab shall not be less than 4 inches (102 mm) thick and shall be supported by noncombustible materials or reinforced to carry its own weight and all imposed loads. Combustible forms and centering shall be removed.

3102.7.12 Hearth extensions. Hearths shall extend at least 16 inches (406 mm) from the front of, and at least 8 inches (203 mm) beyond each side of, the fireplace opening. Where the fireplace opening is 6 square feet (0.56 m^2) or larger, the hearth extension shall extend at least 20 inches (508 mm) in front of, and at least 12 inches (305 mm) beyond each side of, the fireplace opening.

Except for fireplaces that open to the exterior of the building, the hearth slab shall be readily distinguishable from the surrounding or adjacent floor.

3102.7.13 Fire blocking. Fire blocking between chimneys and combustible construction shall meet the requirements specified in Section 708.

TABLE 31-A—MINIMUM PASSAGEWAY AREAS FOR MASONRY CHIMNEYS[1]

Type of Masonry Chimney	MINIMUM CROSS-SECTIONAL AREA		Lined with Firebrick or Unlined
	Tile Lined		
	Round	Square or Rectangle	
	$\times$ 645 for mm^2		
1. Residential	50 square inches	50 square inches	85 square inches
2. Fireplace	See Figure 31-1	See Figure 31-1	$^1/_8$ of opening minimum 100 square inches
3. Low heat	50 square inches	57 square inches	135 square inches
4. Incinerator Apartment type 1 opening 2 to 6 openings 7 to 14 openings 15 or more openings	196 square inches 324 square inches 484 square inches 484 square inches plus 10 square inches for each additional opening		Not applicable

NOTE: For altitudes over 2,000 feet (610 m) above sea level, the building official shall be consulted in determining the area of the passageway.
[1]Areas for medium- and high-heat chimneys shall be determined using accepted engineering methods and as approved by the building official.

TABLE 31-B—CONSTRUCTION, CLEARANCE AND TERMINATION REQUIREMENTS FOR MASONRY AND CONCRETE CHIMNEYS

CHIMNEYS SERVING	THICKNESS (min. inches) ×25.4 for mm — Walls	THICKNESS — Lining	HEIGHT ABOVE ROOF OPENING (feet) ×304.8 for mm	HEIGHT ABOVE ANY PART OF BUILDING WITHIN (feet) ×304.8 for mm — 10	25	50	CLEARANCE TO COMBUSTIBLE CONSTRUCTION (inches) ×25.4 for mm — Int. Inst.	Ext. Inst.
1. **RESIDENTIAL-TYPE APPLIANCES**[1,2] (Low Btu input) Clay, shale or concrete brick Reinforced concrete Hollow masonry units Stone	4[3] 4[3] 4[4] 12	$\frac{5}{8}$ fire-clay tile or 2 firebrick	2	2			2	1 or $\frac{1}{2}$ gypsum[5]
Unburned clay units	8	$4\frac{1}{2}$ firebrick						
2. **BUILDING HEATING AND INDUSTRIAL-TYPE LOW-HEAT APPLIANCES**[1,2] [1,000°F (538°C) operating temp.—1,400°F (760°C) maximum] Clay, shale or concrete brick Hollow masonry units Reinforced concrete Stone	8 8[4] 8 12	$\frac{5}{8}$ fire-clay tile or 2 firebrick	3	2			2	2
3. **MEDIUM-HEAT INDUSTRIAL-TYPE APPLIANCES**[1,6] [2,000°F (1093°C) maximum] Clay, shale or concrete brick Hollow masonry units (Grouted solid) Reinforced concrete Stone	8 8 8 12	$4\frac{1}{2}$ medium-duty firebrick	10		10		4	4
4. **HIGH-HEAT INDUSTRIAL-TYPE APPLIANCES**[1,6] [Over 2,000°F (1093°C)] Clay, shale or concrete brick Hollow masonry units (Grouted solid) Reinforced concrete	16[7] 16[7] 16[7]	$4\frac{1}{2}$ high-duty firebrick	20			20	8[8]	8[8]
5. **RESIDENTIAL-TYPE INCINERATORS**	Same as for residential-type appliances as shown above.							
6. **CHUTE-FED AND FLUE-FED INCINERATORS WITH COMBINED HEARTH AND GRATE AREA 7 SQ. FT. (0.65 m²) OR LESS** Clay, shale or concrete brick or hollow units Portion extending to 10 ft. (3048 mm) above combustion chamber roof Portion more than 10 ft. (3048 mm) above combustion chamber roof	4 8	$4\frac{1}{2}$ medium-duty firebrick $\frac{5}{8}$ fire-clay tile liner	3	2			2	2
7. **CHUTE-FED AND FLUE-FED INCINERATORS—COMBINED HEARTH AND GRATE AREAS LARGER THAN 7 SQ. FT.** (0.65 m²) Clay, shale or concrete brick or hollow units grouted solid or reinforced concrete Portion extending to 40 ft. (12 192 mm) above combustion chamber roof Portion more than 40 ft. (12 192 mm) above combustion chamber roof Reinforced concrete	4 8 8	$4\frac{1}{2}$ medium-duty firebrick $\frac{5}{8}$ fire-clay tile liner $4\frac{1}{2}$ medium-duty firebrick laid in medium-duty refract mortar			10		2	2
8. **COMMERCIAL OR INDUSTRIAL-TYPE INCINERATORS**[2] Clay or shale solid brick Reinforced concrete	8 8	$4\frac{1}{2}$ medium-duty firebrick laid in medium-duty refract mortar			10		4	4

[1]See Table 9-A of the Mechanical Code for types of appliances allowed with each type of chimney.
[2]Lining shall extend from bottom to top of chimney.
[3]Chimneys having walls 8 inches (203 mm) or more in thickness may be unlined.
[4]Equivalent thickness including grouted cells when grouted solid. The equivalent thickness may also include the grout thickness between the liner and masonry unit.
[5]Chimneys for residential-type appliances installed entirely on the exterior of the building. For fireplace and barbecue chimneys, see Section 3102.7.8.
[6]Lining to extend from 24 inches (610 mm) below connector to 25 feet (7620 mm) above.
[7]Two 8-inch (203 mm) walls with 2-inch (51 mm) airspace between walls. Outer and inner walls may be of solid masonry units or reinforced concrete or any combination thereof.
[8]Clearance shall be approved by the building official and shall be such that the temperature of combustible materials will not exceed 160°F (710°C).

TABLE 31-B
UBC 1-293

1997 UBC - FIREPLACES

49

MINIMUM CROSS-SECTIONAL
FLUE AREA (SQ. IN.)

ROUND FLUES	SQUARE OR RECTANGULAR FLUES
NOMINAL FLUE SIZE DIAMETER, IN.	NOMINAL FLUE SIZE, IN.

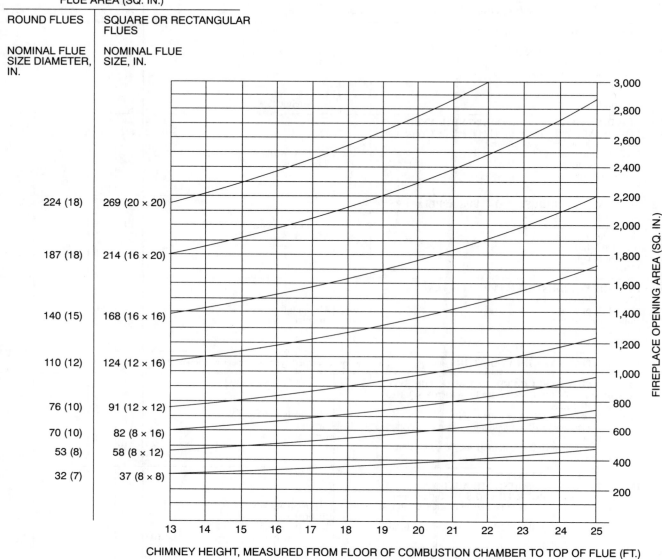

CHIMNEY HEIGHT, MEASURED FROM FLOOR OF COMBUSTION CHAMBER TO TOP OF FLUE (FT.)

For **SI:** 1 inch = 25.4 mm, 1 square inch = 654.16 mm^2, 1 foot = 304.8 mm.

FIGURE 31-1—FLUE SIZES FOR MASONRY CHIMNEYS[1]

[1]The smaller flue area shall be utilized where the fireplace opening area and the chimney height selected intersect between flue area curves.

Chapter 7
FIRE-RESISTANT MATERIALS AND CONSTRUCTION

SECTION 701 — SCOPE

This chapter applies to materials and systems used in the design and construction of a building to safeguard against the spread of fire and smoke within a building and the spread of fire to or from buildings.

SECTION 702 — DEFINITIONS

For the purposes of this chapter, the terms, phrases and words listed in this section and their derivatives shall have the indicated meanings.

ANNULAR SPACE is the opening around the penetrating item.

CONCRETE, CARBONATE AGGREGATE, is concrete made with aggregates consisting mainly of calcium or magnesium carbonate, e.g., limestone or dolomite, and containing 40 percent or less quartz, chert or flint.

CONCRETE, LIGHTWEIGHT AGGREGATE, is concrete made with aggregates of expanded clay, shale, slag or slate or sintered fly ash or any natural lightweight aggregate meeting ASTM C 330 and possessing equivalent fire-resistive properties and weighing 85 to 115 pounds per cubic foot (pcf) (1360 to 1840 kg/m^3).

CONCRETE, SAND-LIGHTWEIGHT, is concrete made with a combination of expanded clay, shale, slag or slate or sintered fly ash or any natural lightweight aggregate meeting ASTM C 330 and possessing equivalent fire-resistive properties and natural sand. Its unit weight is generally between 105 and 120 pcf (1680 and 1920 kg/m^3).

CONCRETE, SILICEOUS AGGREGATE, is concrete made with normal-weight aggregates consisting mainly of silica or compounds other than calcium or magnesium carbonate, and may contain more than 40 percent quartz, chert or flint.

F RATING is the time period the penetration firestop system limits the passage of fire through the penetration when tested in accordance with UBC Standard 7-5.

FIREBLOCKING is building material installed to resist the free passage of flame and gases to other areas of the building through small concealed spaces.

FIRE-RESISTIVE JOINT SYSTEM is an assemblage of specific materials or products that are designed, tested and fire resistive in accordance with UBC Standard 7-1 to resist, for a prescribed period of time, the passage of fire through joints.

JOINT is the linear opening between adjacent fire-resistive assemblies. A joint is a division of a building that allows independent movement of the building, in any plane, which may be caused by thermal, seismic, wind loading or any other loading.

MEMBRANE PENETRATION is an opening made through one side (wall, floor or ceiling membrane) of an assembly.

PENETRATION is an opening created in a membrane or assembly to accommodate penetrating items for electrical, mechanical, plumbing, environmental and communication systems.

EXCEPTION: Ducts.

PENETRATION FIRESTOP SYSTEM is an assemblage of specific materials or products that are designed, tested and fire re-sistive in accordance with UBC Standard 7-5 to resist, for a prescribed period of time, the passage of fire through penetrations.

SPLICE is the result of a factory or field method of joining or connecting two or more lengths of a fire-resistive joint system into a continuous entity.

T RATING is the time period that the penetration firestop system including the penetrating item, limits the maximum temperature rise to 325°F (163°C) above its initial temperature through the penetration on the nonfire side, when tested in accordance with UBC Standard 7-5.

THROUGH-PENETRATION is an opening that passes through both sides of an assembly.

SECTION 703 — FIRE-RESISTIVE MATERIALS AND SYSTEMS

703.1 General. Materials and systems used for fire-resistive purposes shall be limited to those specified in this chapter, unless accepted under the procedure given in Section 703.2 or 703.3.

The materials and details of construction for the fire-resistive systems described in this chapter shall be in accordance with all other provisions of this code except as modified herein.

For the purpose of determining the degree of fire resistance afforded, the materials of construction listed in this chapter shall be assumed to have the fire-resistance rating indicated in Table 7-A, 7-B or 7-C.

As an alternate to Table 7-A, 7-B or 7-C, fire-resistive construction may be approved by the building official on the basis of evidence submitted showing that the construction meets the required fire-resistive classification.

703.2 Qualification by Testing. Material or assembly of materials of construction tested in accordance with the requirements set forth in UBC Standard 7-1 shall be rated for fire resistance in accordance with the results and conditions of such tests.

> **EXCEPTION:** The acceptance criteria of UBC Standard 7-1 for exterior-bearing walls shall not be required to be greater with respect to heat transmission and passage of flame or hot gases than would be required of a nonbearing wall in the same building with the same distance to the property line. The fire exposure time period, water pressure and duration of application for the hose stream test shall be based on the fire-resistive rating determined by this exception.

Fire-resistive assemblies tested under UBC Standard 7-1 shall not be considered to be restrained unless evidence satisfactory to the building official is furnished by the person responsible for the structural design showing that the construction qualifies for a restrained classification in accordance with UBC Standard 7-1. Restrained construction shall be identified on the plans.

703.3 Calculating Fire Resistance. The fire-resistive rating of a material or assembly may be established by calculations. The procedures used for such calculations shall be in accordance with UBC Standard 7-7.

703.4 Standards of Quality. In addition to all the other requirements of this code, fire-resistive materials shall meet the requirements for fire-resistive construction given in this chapter.

The standards listed below labeled a "UBC standard" are also listed in Chapter 35, Part II, and are part of this code. The standards listed below labeled an "Adopted Standard" are also listed in Chapter 35, Part III, and are part of this code. The other standards listed below are recognized standards. (See Sections 3503 and 3504.)

1. UBC Standard 7-1, Fire Tests of Building Construction and Materials

2. UBC Standard 7-2, Fire Tests of Door Assemblies

3. UBC Standard 7-3, Tinclad Fire Doors

4. UBC Standard 7-4, Fire Tests of Window Assemblies

5. UBC Standard 7-5, Fire Tests of Through-penetration Fire Stops

6. UBC Standard 7-6, Thickness, Density Determination and Cohesion/Adhesion for Spray-applied Fire-resistive Fireproofing

7. UBC Standard 7-7, Methods for Calculating Fire Resistance of Steel, Concrete, Wood, Concrete Masonry and Clay Masonry Construction

8. ASTM C 516, Vermiculite Loose-fill Insulation

9. ASTM C 549, Perlite Loose-fill Insulation

10. ANSI/NFPA 80, Standard for Fire Doors and Fire Windows

11. ASTM C 587 and C 588, Gypsum Base for Veneer Plaster and Gypsum Veneer

12. ASTM C 332, Lightweight Aggregates for Insulating Concrete

13. ASTM C 331, Lightweight Aggregates for Concrete Masonry Units

14. UL 555, Fire Dampers

15. UL 555C, Ceiling Dampers

16. UL 555S, Leakage Rated Dampers for Use in Smoke Control Systems

17. UL 33, Heat Response Links for Fire Protection Service

18. UL 353, Limit Controls

19. ASTM E 1399, Cyclic Movement and Measuring the Minimum and Maximum Joint Widths of Architectural Joint Systems

20. Adopted standard—Fire-Resistance Design Manual, Fourteenth Edition

21. Adopted standard—ASTM C 330, Lightweight Aggregates for Structural Concrete

22. Adopted standard—CPSC 16 CFR, Parts 1209 and 1404 Test Standard for Cellulose Insulation

SECTION 709 — WALLS AND PARTITIONS

709.1 General. Fire-resistive walls and partitions shall be assumed to have the fire-resistance ratings set forth in Table 7-B.

Where materials, systems or devices are incorporated into the assembly that have not been tested as part of the assembly, sufficient data shall be made available to the building official to show that the required fire-resistive rating is not reduced. Materials and methods of construction used to protect joints and penetrations in fire-resistive, fire-rated building assemblies shall not reduce the required fire-resistive rating.

709.2 Combustible Members. Combustible members framed into a wall shall be protected at their ends by not less than one half the required fire-resistive thickness of such wall.

709.3 Exterior Walls.

709.3.1 Extension through attics and concealed spaces. In fire-resistive exterior wall construction, the fire-resistive rating shall be maintained for such walls passing through attic areas or other areas containing concealed spaces.

709.3.2 Vertical fire spread at exterior walls.

709.3.2.1 General. The provisions of this section are intended to restrict the passage of smoke, flame and hot gases from one floor to another at exterior walls. See Section 710 for floor penetrations.

709.3.2.2 Interior. When fire-resistive floor or floor-ceiling assemblies are required, voids created at the intersection of the exterior wall assemblies and such floor assemblies shall be sealed with an approved material. Such material shall be securely installed and capable of preventing the passage of flame and hot gases sufficient to ignite cotton waste when subjected to UBC Standard 7-1 time-temperature fire conditions under a minimum positive pressure differential of 0.01 inch of water column (2.5 Pa) for the time period at least equal to the fire-resistance rating of the floor assembly.

709.3.2.3 Exterior. When openings in an exterior wall are above and within 5 feet (1524 mm) laterally of an opening in the story below, such openings shall be separated by an approved flame barrier extending 30 inches (762 mm) beyond the exterior wall in the plane of the floor or by approved vertical flame barriers not less than 3 feet (914 mm) high measured vertically above the top of the lower opening. Flame barriers shall have a fire resistance of not less than three-fourths hour.

> **EXCEPTIONS:** 1. Flame barriers are not required in buildings equipped with an approved automatic sprinkler system throughout.
>
> 2. This section shall not apply to buildings of three stories or less in height.
>
> 3. Flame barriers are not required on Group S, Division 4 Occupancies.

709.4 Parapets.

709.4.1 General. Parapets shall be provided on all exterior walls of buildings.

> **EXCEPTION:** A parapet need not be provided on an exterior wall when any of the following conditions exist:
>
> 1. The wall is not required to be of fire-resistive construction.
>
> 2. The wall, due to location on property line, may have unprotected openings.
>
> 3. The building has an area of not more than 1,000 square feet (93 m²) on any floor.
>
> 4. Walls that terminate at roofs of not less than two-hour fire-resistive construction or roofs constructed entirely of noncombustible materials.
>
> 5. One-hour fire-resistive exterior walls may terminate at the underside of the roof sheathing, deck or slab, provided:
>
> 5.1 Where the roof-ceiling framing elements are parallel to the walls, such framing and elements supporting such framing shall not be of less than one-hour fire-resistive construction for a width of 5 feet (1524 mm) measured from the interior side of the wall for Groups R and U Occupancies and 10 feet (3048 mm) for all other occupancies.
>
> 5.2 Where roof-ceiling framing elements are not parallel to the wall, the entire span of such framing and elements supporting such framing shall not be of less than one-hour fire-resistive construction.
>
> 5.3 Openings in the roof shall not be located within 5 feet (1524 mm) of the one-hour fire-resistive exterior wall for Groups R and U Occupancies and 10 feet (3048 mm) for all other occupancies.
>
> 5.4 The entire building shall be provided with not less than a Class B roofing assembly.

709.4.2 Construction. Parapets shall have the same degree of fire resistance required for the wall upon which they are erected, and on any side adjacent to a roof surface, shall have noncombustible faces for the uppermost 18 inches (457 mm), including counterflashing and coping materials. The height of the parapet shall not be less than 30 inches (762 mm) above the point where the roof surface and the wall intersect. Where the roof slopes toward a parapet at slopes greater than 2 units vertical in 12 units horizontal (16.7% slope), the parapet shall extend to the same height as any portion of the roof that is within the distance where protection of wall openings would be required, but in no case shall the height be less than 30 inches (762 mm).

709.5 Nonsymmetrical Wall Construction. Walls and partitions of nonsymmetrical construction shall be tested with both faces exposed to the furnace, and the assigned fire-resistive rating will be the shortest duration obtained from the two tests conducted in conformance with UBC Standard 7-1. When evidence is furnished to show that the wall was tested with the least fire-resistive side exposed to the furnace, subject to acceptance of the building official, the wall need not be subjected to tests from the opposite side.

709.6 Through Penetrations.

709.6.1 General. Through penetrations of the fire-resistive walls shall comply with Section 709.6.2 or 709.6.3.

> **EXCEPTION:** Where the penetrating items are steel, ferrous or copper pipes or steel conduits, the annular space shall be permitted to be protected as follows:
>
> 1. In concrete or masonry walls where the penetrating items are a maximum 6-inch (152 mm) nominal diameter and the opening is a maximum 144 square inches (92 903 mm²) concrete, grout or mortar shall be permitted when installed the full thickness of the wall or the thickness required to maintain the fire rating, or

> 2. The material used to fill the annular space shall prevent the passage of flame and hot gases sufficient to ignite cotton waste when subjected to UBC Standard 7-1 time-temperature fire conditions under a minimum positive pressure differential of 0.01 inch of water column (2.5 Pa) at the location of the penetration for the time period equivalent to the fire rating of the construction penetrated.

709.6.2 Fire-rated assembly. Penetrations shall be installed as tested in the approved UBC Standard 7-1 rated assembly.

709.6.3 Penetration firestop system. Penetrations shall be protected by an approved penetration firestop system installed as tested in accordance with UBC Standard 7-5 and shall have an F rating of not less than the required rating of the wall penetrated.

709.7 Membrane Penetrations. Membrane penetrations of the fire-resistive walls shall comply with Section 709.6.

> **EXCEPTIONS:** 1. Steel electrical boxes that do not exceed 16 square inches (10 323 mm²) in area, provided that the area of such openings does not exceed 100 square inches for any 100 square feet (694 mm²/m²) of wall area. Outlet boxes on opposite sides of the wall shall be separated by a horizontal distance of not less than 24 inches (610 mm). Membrane penetrations for electrical outlet boxes of any material are permitted, provided that such boxes are tested for use in fire-resistive assemblies and installed in accordance with the tested assembly.
>
> 2. The annular space created by the penetration of a fire sprinkler shall be permitted to be unprotected, provided such space is covered by a metal escutcheon plate.

Noncombustible penetrating items shall not be connected to combustible materials on both sides of the membrane unless it can be confirmed that the fire-resistive integrity of the wall is maintained in accordance with UBC Standard 7-1.

709.8 Joints. The protection of joints shall comply with the requirements of Section 706.

MATERIAL	ITEM NUMBER	CONSTRUCTION	MINIMUM FINISHED THICKNESS FACE-TO-FACE[2] (inches) × 25.4 for mm			
			4 Hr.	3 Hr.	2 Hr.	1 Hr.
1. Brick of clay or shale	1-1.1	Solid units (at least 75 percent solid).	8		6[3]	4
	1-2.1	Hollow brick units[4] (at least 71 percent solid).		8		
	1-2.2	Hollow brick units (at least 60 percent solid, cells filled with loose-fill insulation).	8			
	1-2.3	Hollow brick units at least 64 percent solid.	12			
	1-2.4	Hollow brick, not filled.	5.0	4.3	3.4	2.3
	1-2.5	Hollow brick unit wall, grout or filled with perlite vermiculite or expanded shale aggregate.	6.6	5.5	4.4	3.0
	1-3.1	Hollow (rowlock[5]).	12		8	
	1-4.1	Cavity wall consisting of two 3″ (76 mm) (actual) solid clay brick units separated by 2″ (51 mm) air space, joint reinforcement every 16″ (406 mm) on center vertically.		8		
	1-4.2	Cavity wall consisting of two 4″ (100 mm) nominal solid clay brick units separated by 2″ (51 mm) air space, 1/4″ (6.4 mm) metal ties for 3 square feet (0.28 m²) of wall area.	10			
	1-5.1	4″ (102 mm) nominal thick units at least 75 percent solid backed with a hat-shaped metal furring channel 3/4″ (19 mm) thick formed from 0.021″ (0.53 mm) sheet metal attached to the brick wall on 24″ (610 mm) centers with approved fasteners, and 1/2″ (12.7 mm) Type X gypsum wallboard[7] attached to the metal furring strips with 1″ (25 mm) long Type S screws spaced 8″ (203 mm) on center.			5[6]	
2. Hollow clay tile, nonload-bearing	2-1.1	Two cells in wall thickness, units at least 40 percent solid.				8
	2-1.2	Two cells in wall thickness, units at least 43 percent solid.				8
	2-1.3	Two cells in wall thickness, units at least 46 percent solid.				8
	2-1.4	Two cells in wall thickness, units at least 49 percent solid.			8	
	2-1.5	Three or four cells in wall thickness, units at least 40 percent solid.				8
	2-1.6	Three or four cells in wall thickness, units at least 43 percent solid.			8	
	2-1.7	Three or four cells in wall thickness, units at least 48 percent solid.			8	
	2-1.8	Three or four cells in wall thickness, units at least 53 percent solid.		8		
	2-1.9	Three cells in wall thickness, units at least 40 percent solid.			12	
	2-1.10	Three cells in wall thickness, units at least 45 percent solid.		12		
	2-1.11	Three cells in wall thickness, units at least 49 percent solid.		12		
	2-1.12	Two units and three or four cells in wall thickness, units at least 40 percent solid.		12		
	2-1.13	Two units and three or four cells in wall thickness, units at least 45 percent solid.	12			
	2-1.14	Two units and three or four cells in wall thickness, units at least 53 percent solid.	12			
	2-1.15	Two or three units and four or five cells in wall thickness, units at least 40 percent solid.	16			
3. Structural clay tile, load-bearing	3-1.1	One cell in wall thickness, units at least 40 percent solid.[8,9]				4
	3-1.2	One cell in wall thickness, units at least 30 percent solid.[8,9]			6	
	3-1.3	Two cells in wall thickness, units at least 45 percent solid.[10]				6
	3-1.4	One cell in wall thickness, units at least 40 percent solid.[9,10]				4
	3-1.5	One cell in wall thickness, units at least 30 percent solid.[9,10]			6	
4. Hollow structural clay tile, load-bearing	4-1.1	Two cells in wall thickness, units at least 40 percent solid.				8
	4-1.2	Two cells in wall thickness, units at least 49 percent solid.			8	
	4-1.3	Three or four cells in wall thickness, units at least 53 percent solid.		8		
	4-1.4	Two cells in wall thickness, units at least 46 percent solid.				8
	4-1.5	Three cells in wall thickness, units at least 40 percent solid.			12	
	4-1.6	Two units and three cells in wall thickness, units at least 40 percent solid.		12		
	4-1.7	Two units and three or four cells in wall thickness, units at least 45 percent solid.	12			
	4-1.8	Three cells in wall thickness, units at least 45 percent solid.			12	
	4-1.9	Three cells in wall thickness, units at least 49 percent solid.			12	
	4-1.10	Two units and four cells in wall thickness, units at least 43 percent solid.	16			
	4-1.11	Two or three units and four or five cells in wall thickness, units at least 40 percent solid.	16			

(Continued)

MATERIAL	ITEM NUMBER	CONSTRUCTION	4 Hr.	3 Hr.	2 Hr.	1 Hr.
5. Combination of clay brick and load-bearing hollow clay tile	5.1.1	4″ (102 mm) solid brick and 4″ (102 mm) tile (at least 40 percent solid).		8		
	5.1.2	4″ (102 mm) solid brick and 8″ (203 mm) tile (at least 40 percent solid).	12			
6. Concrete masonry units	6-1.1[11,12]	Expanded slag or pumice.	4.7	4.0	3.2	2.1
	6-1.2[11,12]	Expanded clay, shale or slate.	5.1	4.4	3.6	2.6
	6-1.3[11]	Limestone, cinders or air-cooled slag.	5.9	5.0	4.0	2.7
	6-1.4[11,12]	Calcareous or siliceous gravel.	6.2	5.3	4.2	2.8
7. Solid concrete[13,14]	7-1.1	Siliceous aggregate concrete.	7.0	6.2	5.0	3.5
		Carbonate aggregate concrete.	6.6	5.7	4.6	3.2
		Sand-lightweight concrete.	5.4	4.6	3.8	2.7
		Lightweight concrete.	5.1	4.4	3.6	2.5
8. Glazed or unglazed facing tile, nonload-bearing	8-1.1	One 2″ (51 mm) unit cored 15 percent maximum and one 4″ (102 mm) unit cored 25 percent maximum with $3/4″$ (19 mm) mortar-filled collar joint. Unit positions reversed in alternate courses.		$6^{3}/_{8}$		
	8-1.2	One 2″ (51 mm) unit cored 15 percent maximum and one 4″ (102 mm) unit cored 40 percent maximum with $3/8″$ (9.5 mm) mortar-filled collar joint. Plastered one side with $3/4″$ (19 mm) gypsum plaster. Two wythes tied together every fourth course with No. 22 gage corrugated metal ties.		$6^{3}/_{4}$		
	8-1.3	One unit with three cells in wall thickness, cored 29 percent maximum.			6	
	8-1.4	One 2″ (51 mm) unit cored 22 percent maximum and one 4″ (102 mm) unit cored 41 percent maximum with $1/4″$ (6 mm) mortar-filled collar joint. Two wythes tied together every third course with 0.030 inch (0.76 mm) (No. 22 galvanized sheet steel gage) corrugated metal ties.			6	
	8-1.5	One 4″ (102 mm) unit cored 25 percent maximum with $3/4″$ (19 mm) gypsum plaster on one side.			$4^{3}/_{4}$	
	8-1.6	One 4″ (102 mm) unit with two cells in wall thickness, cored 22 percent maximum.				4
	8-1.7	One 4″ (102 mm) unit cored 30 percent maximum with $3/4″$ (19 mm) vermiculite gypsum plaster on one side.			$4^{1}/_{2}$	
	8-1.8	One 4″ (102 mm) unit cored 39 percent maximum with $3/4″$ (19 mm) gypsum plaster on one side.				$4^{1}/_{2}$

MINIMUM FINISHED THICKNESS FACE-TO-FACE[2] (inches) × 25.4 for mm

[a]Generic fire-resistance ratings (those not designated as PROPRIETARY* in the listing) in the *Fire-Resistance Design Manual,* Fourteenth Edition, dated April 1994, as published by the Gypsum Association, may be accepted as if herein listed.

[1]Staples with equivalent holding power and penetration may be used as alternate fasteners to nails for attachment to wood framing.

[2]Thickness shown for brick and clay tile are nominal thicknesses unless plastered, in which case thicknesses are net. Thickness shown for concrete masonry and hollow clay or shale brick is equivalent thickness defined as the average thickness of solid material in the wall and is represented by the formula:

$$T_E = \frac{V}{L \times H}$$

WHERE:

H = height of block or brick using specified dimensions as defined in Chapter 21, in inches (mm).

L = length of block or brick using specified dimensions as defined in Chapter 21, in inches (mm).

T_E = equivalent thickness, in inches (mm).

V = net volume (gross volume less volume of voids), in cubic inches (mm^3).

When all cells are solid grouted or filled with silicone-treated perlite loose-fill insulation; vermiculite loose-fill insulation; or expanded clay, shale or slate lightweight aggregate, the equivalent thickness shall be the thickness of the block or brick using specified dimensions as defined in Chapter 21. Equivalent thickness may also include the thickness of applied plaster and lath or gypsum wallboard, where specified.

[3]Single-wythe brick.

[4]Hollow brick units 4-inch by 8-inch by 12-inch (102 mm by 203 mm by 305 mm) nominal with two interior cells having a $1^{1}/_{2}$-inch (38 mm) web thickness between cells and $1^{3}/_{4}$-inch-thick (44.5 mm) face shells.

[5]Rowlock design employs clay brick with all or part of bricks laid on edge with the bond broken vertically.

[6]Shall be used for nonbearing purposes only.

[7]For all of the construction with gypsum wallboard described in this table, gypsum base for veneer plaster of the same size, thickness and core type may be substituted for gypsum wallboard, provided attachment is identical to that specified for the wallboard, and the joints on the face layer are reinforced and the entire surface is covered with a minimum of $1/_{16}$-inch (1.6 mm) gypsum veneer plaster.

[8]Ratings are for hard-burned clay or shale tile.

[9]Cells filled with tile, stone, slag, cinders or sand mixed with mortar.

[10]Ratings are for medium-burned clay tile.

[11]The fire-resistive time period for concrete masonry units meeting the equivalent thicknesses required for a two-hour fire-resistive rating in Item 6, and having a thickness of not less than $7^{5}/_{8}$ inches (194 mm) is four hours when cores which are not grouted are filled with silicone-treated perlite loose-fill insulation; vermiculite loose-fill insulation; or expanded clay, shale or slate lightweight aggregate, sand or slag having a maximum particle size of $3/_{8}$ inch (9.5 mm).

[12]For determining the fire-resistance rating of concrete masonry units composed of a combination of aggregate types or where plaster is applied directly to the concrete masonry, see UBC Standard 7-7, Part III. Lightweight aggregates shall have a maximum combined density of 65 pounds per cubic foot (1049 kg/m^3).

[13]See also Footnote 2. The equivalent thickness may include the thickness of cement plaster or 1.5 times the thickness of gypsum plaster applied in accordance with the requirements of Chapter 25.

(Continued)

TABLE 7-B
UBC 1-80, 84

1997 UBC - FIRE RESISTANCE

55

[14]Concrete walls shall be reinforced with horizontal and vertical temperature reinforcement as required by Sections 1914.3.2 and 1914.3.3.

[15]Studs are welded truss wire studs with 0.18 inch (4.57 mm) (No. 7 B.W. gage) flange wire and 0.18 inch (4.57 mm) (No. 7 B.W. gage) truss wires.

[16]Nailable metal studs consist of two channel studs spot welded back to back with a crimped web forming a nailing groove.

[17]Wood structural panels may be installed between the fire protection and the wood studs on either the interior or exterior side of the wood-frame assemblies in this table, provided the length of the fasteners used to attach the fire protection are increased by an amount at least equal to the thickness of the wood structural panel.

[18]The design stress of studs shall be reduced to 78 percent of allowable F'_c with the maximum not greater than 78 percent of the calculated stress with studs having a slenderness ratio l_e/d of 33.

[19]For properties of cooler or wallboard nails, see approved nationally recognized standards.

Appendix Chapter 21
PRESCRIPTIVE MASONRY CONSTRUCTION IN HIGH-WIND AREAS

SECTION 2112 — GENERAL

2112.1 Purpose. The provisions of this chapter are intended to promote public safety and welfare by reducing the risk of wind-induced damages to masonry construction.

2112.2 Scope. The requirements of this chapter shall apply to masonry construction in buildings when all of the following conditions are met:

1. The building is located in an area with a basic wind speed from 80 through 110 miles per hour (mph) (129 km/h through 177 km/h).

2. The building is located in Seismic Zone 0, 1 or 2.

3. The building does not exceed two stories.

4. Floor and roof joists shall be wood or steel or of precast hollowcore concrete planks with a maximum span of 32 feet (9754 mm) between bearing walls. Masonry walls shall be provided for the support of steel joists or concrete planks.

5. The building is of regular shape.

2112.3 General. The requirements of Chapter 21 are applicable except as specifically modified by this chapter. Other methods may be used provided a satisfactory design is submitted showing compliance with the provisions of Chapter 16, Part II, and other applicable provisions of this code.

Wood floor, roof and interior walls shall be constructed as specified in Appendix Chapter 23 and as further regulated in this section.

In areas where the wind speed exceeds 110 mph (177 km/h), masonry buildings shall be designed in accordance with Chapter 16, Part II, and other applicable provisions of this code.

Buildings of unusual shape or size, or split-level construction, shall be designed in accordance with Chapter 16, Part II, and other applicable provisions of this code.

In addition to the other provisions of this chapter, foundations for buildings in areas subject to wave action or tidal surge shall be designed in accordance with approved national standards.

All metal connectors and fasteners used in exposed locations or in areas otherwise subject to corrosion shall be of corrosion-resistant or noncorrosive material. When the terms "corrosion resistant" or "noncorrosive" are used in this chapter, they shall mean having a corrosion resistance equal to or greater than a hot-dipped galvanized coating of 1.5 ounces of zinc per square foot (3.95 g/m^2) of surface area. When an element is required to be corrosion resistant or noncorrosive, all of its parts, such as screws, nails, wire, dowels, bolts, nuts, washers, shims, anchors, ties and attachments, shall be corrosion resistant.

2112.4 Materials.

2112.4.1 General. All masonry materials shall comply with Section 2102.2 as applicable for standards of quality.

2112.4.2 Hollow-unit masonry.

1. Exterior concrete block shall be a minimum of Grade N-II with a compressive strength of not less than 1,900 pounds per square inch (psi) (13 091 kPa) on the net area.

2. Interior concrete block shall be a minimum of Grade S-II with a compressive strength of not less than 700 psi (4823 kPa) on the gross area.

3. Exterior clay or shale hollow brick shall have a compressive strength of not less than 2,500 psi (17 225 kPa) on the net area. Such hollow brick shall be at least Grade MW except that where subject to severe freezing it shall be Grade SW.

4. Interior clay or shale hollow brick shall be Grade MW with a compressive strength of 2,000 psi (13 780 kPa) on the net area.

2112.4.3 Solid masonry.

1. Exterior clay or shale bricks shall have a compressive strength of not less than 2,500 psi (17 225 kPa) on the net area.

2. Exterior clay or shale bricks shall be Grade MW, except that where subject to severe freezing they shall be Grade SW.

3. Interior clay or shale bricks shall have a compressive strength of not less than 2,000 psi (13 780 kPa).

2112.4.4 Grout. Grout shall achieve a compressive strength of not less than 2,000 psi (13 780 kPa).

2112.4.5 Mortar. Mortar for exterior walls and for interior shear walls shall be Type M or Type S.

2112.5 Construction Requirements. Grouted cavity wall and block wall construction shall comply with Section 2104.

Unburned clay masonry and stone masonry shall not be used.

2112.6 Foundations. Footings shall have a thickness of not less than 8 inches (203 mm) and shall comply with Tables A-21-A-1 and A-21-A-2 for width. See Figure A-21-1 for other applicable details.

Footings shall extend 18 inches (457 mm) below the undisturbed ground surface or the frost depth, whichever is deeper.

Foundation stem walls shall be as wide as the wall they support. They shall be reinforced with reinforcing bar sizes and spacing to match the reinforcement of the walls they support.

Basement and other below-grade walls shall comply with Table A-21-B.

2112.7 Drainage. Basement walls and other walls or portions thereof retaining more than 3 feet (914 mm) of earth and enclosing interior spaces or floors below grade shall have a minimum 4-inch-diameter (102 mm) footing drain as illustrated in Table A-21-B and Figure A-21-3.

The finish elevations around the building shall be graded to provide a slope away from the building of not less than $^1/_4$ unit vertical in 12 units horizontal (2% slope).

2112.8 Wall Construction.

2112.8.1 Minimum thickness. Reinforced exterior bearing walls shall have a minimum 8-inch (203 mm) nominal thickness. Interior masonry nonbearing walls shall have a minimum 6-inch (152 mm) nominal thickness. Unreinforced grouted brick walls shall have a minimum 10-inch (254 mm) thickness. Unreinforced hollow-unit and solid masonry shall have a minimum 8-inch (203 mm) nominal thickness.

EXCEPTION: In buildings not more than two stories or 26 feet (7924.8 mm) in height, masonry walls may be of 8-inch (203 mm) nominal thickness. Solid masonry walls in one-story buildings may be of 6-inch (152 mm) nominal thickness when not over 9 feet (2743 mm) in height, provided that when gable construction is used an additional 6 feet (1829 mm) are permitted to the peak of the gable.

2112.8.2 Lateral support and height. All walls shall be laterally supported at the top and bottom. The maximum unsupported height of bearing walls or other masonry walls shall be 12 feet (3658 mm). Gable-end walls may be 15 feet (4572 mm) at their peak.

Wood-framed gable-end walls on buildings shall comply with Table A-21-I and Figure A-21-17 or A-21-18.

2112.8.3 Walls in Seismic Zone 2 and use of stack bond. In Seismic Zone 2, walls shall comply with Figure A-21-2 as a minimum. Walls with stack bond shall be designed.

2112.8.4 Lintels. The span of lintels over openings shall not exceed 12 feet (3658 mm), and lintels shall be reinforced. The reinforcement bars shall extend not less than 2 feet (610 mm) beyond the edge of opening and into lintel supports.

Lintel reinforcement shall be within fully grouted cells in accordance with Table A-21-E.

2112.8.5 Reinforcement. Walls shall be reinforced as shown in Tables A-21-C-1 through A-21-C-5 and Figure A-21-2.

2112.8.6 Anchorage of walls to floors and roofs. Anchors between walls and floors or roofs shall be embedded in grouted cells or cavities and shall conform to Section 2112.9.

2112.9 Floor and Roof Systems. The anchorage of wood roof systems which are supported by masonry walls shall comply with Appendix Sections 2337.5.1 and 2337.5.8, Table A-21-D and Figure A-21-7.

Wood roof and floor systems which are supported by ledgers at the inside face of masonry walls shall comply with Table A-21-D, Part I.

The ends of joist girders shall extend a distance of not less than 6 inches (152 mm) over masonry or concrete supports and be attached to a steel bearing plate. This plate is to be located not more than $1/2$ inch (12.7 mm) from the face of the wall and is to be not less than 9 inches (229 mm) wide perpendicular to the length of the joist girder. Ends of joist girders resting on steel bearing plates on masonry or structural concrete shall be attached thereto with a minimum of two $1/4$-inch (6.4 mm) fillet welds 2 inches (51 mm) long, or with two $3/4$-inch (19 mm) bolts.

Ends of joist girders resting on steel supports shall be connected thereto with a minimum of two $1/4$-inch (6.4 mm) fillet welds 2 inches (51 mm) long, or with two $3/4$-inch (19 mm) bolts. In steel frames, joist girders at column lines shall be field bolted to the columns to provide lateral stability during construction.

Steel joist roof and floor systems shall be anchored in accordance with Table A-21-H.

Wall ties spaced as shown in Table A-21-D, Part II, shall connect to framing or blocking at roofs and walls. Wall ties shall enter grouted cells or cavities and shall be $1^1/8$-inch (29 mm) minimum width by 0.036 inch (0.91 mm) (No. 20 galvanized sheet gage) sheet steel.

Roof and floor hollow-core precast plank systems shall be anchored in accordance with Table A-21-G.

Roof uplift anchorage shall enter a grouted bond beam reinforced with horizontal bars as shown in Tables A-21-C-1 through A-21-C-5 and Figure A-21-7.

2112.10 Lateral Force Resistance.

2112.10.1 Complete load path and uplift resistance. Strapping, approved framing anchors, and mechanical fasteners, bond beams, and vertical reinforcement shall be installed to provide a continuous tie from the roof to the foundation system. (See Figure A-21-8.) In addition, roof and floor systems, masonry shear walls, or masonry or wood cross walls shall provide lateral stability.

2112.10.2 Floor and roof diaphragms. Floor and roof diaphragms shall be connected to masonry walls as shown in Table A-21-F, Part II.

Gabled and sloped roof members not supported at the ridge shall be tied by ceiling joists or equivalent lateral ties located as close to where the roof member bears on the wall as is practically possible, at not more than 48 inches (1219 mm) on center. Collar ties shall not be used for these lateral ties. (See Figure A-21-17 and Table A-21-I.)

2112.10.3 Walls. Masonry walls shall be provided around all sides of floor and roof systems in accordance with Figure A-21-9 and Table A-21-F.

The cumulative length of exterior masonry walls along each side of the floor or roof systems shall be at least 20 percent of the parallel dimension. Required elements shall be without openings and shall not be less than 48 inches (1219 mm) in width.

Interior cross walls (nonbearing) at right angles to bearing walls shall be provided when the length of the building perpendicular to the span of the floor or roof framing exceeds twice the distance between shear walls or 32 feet (9754 mm), whichever is greater. Cross walls, when required, shall conform to Section 2112.10.4.

2112.10.4 Interior cross walls. When required by Table A-21-F, Part I, masonry walls shall be at least 6 feet (1829 mm) long and reinforced with 9 gage wire joint reinforcement spaced not more than 16 inches (406 mm) on center. Cross walls shall comply with Footnote 3 of Table A-21-F, Part I.

Interior wood stud walls may be used to resist the wind load from one-story masonry buildings in areas where the basic wind speed is 100 mph (161 km/h), Exposure C or less, and 110 mph (177 km/h), Exposure B. When wood stud walls are so used, they shall:

1. Be perpendicular to exterior masonry walls at 15 feet (4572 mm) or less on center.

2. Be at least 8 feet (2438 mm) long without openings and be sheathed on at least one side with $15/32$-inch (12 mm) wood structural panel nailed with 8d common or galvanized box nails at 6 inches (152 mm) on center edge and field nailing. All unsupported edges of wood structural panels shall be blocked.

3. Be connected to wood blocking or wood joists below with two 16d nails at 16 inches (406 mm) on center through their sill plates. They shall be connected to footings with $1/2$-inch-diameter (12.7 mm) bolts at 3 feet 6 inches (1067 mm) on center.

4. Connect to wood roof systems as outlined in Table A-21-F, Part II, as a cross wall. Wood structural panel roof sheathing shall have all unsupported edges blocked.

TABLE A-21-A-1—EXTERIOR FOUNDATION REQUIREMENTS FOR MASONRY BUILDINGS WITH 6- AND 8-INCH-THICK WALLS
(Wood or Steel Framing)
(Width of Footings in Inches)[1,2,3]
See Figure A-21-1 for typical details.

		ONE-STORY BUILDINGS			TWO-STORY BUILDINGS					
		Roof Live Load[4]			Roof Live Load[4] (psf)					
		$\times$ 0.0479 for kN/m^2			$\times$ 0.0479 for kN/m^2					
					20		30		40	
					Plus Floor Live Load[5] (psf)					
					$\times$ 0.0479 for kN/m^2					
WALL HEIGHT (feet)	SPAN TO BEARING WALLS (feet)	20 psf (inches)	30 psf (inches)	40 psf (inches)	50	100	50	100	50	100
$\times$ 304.8 for mm		Minimum Width of Footing (inches) $\times$ 25.4 for mm								
8	8	12			12	12	12	12	12	12
	16				12	14	12	14	12	14
	24				14	18	14	18	16	18
	32				16	20	18	20	18	20
10	8	12			12	12	12	12	12	12
	16				14	16	14	16	14	16
	24				16	20	16	18	16	20
	32				20	24	20	22	20	24
12	8	12	12	12	12	14	12	14	12	14
	16	12	12	12	16	18	16	16	14	16
	24	12	12	14	18	20	18	20	18	20
	32	12	14	16	20	22	22	22	22	24

[1]For buildings with under-floor space or basements, footing thickness is to be a minimum of 12 inches (305 mm). It shall be reinforced with No. 4 bars at 24 inches (610 mm) on center when its width is required to be 18 inches (457 mm) or larger and it supports more than the roof and one floor.

[2]Soil to be at least Class 4 as shown in Table 18-I-A.

[3]Footings are a minimum of 10 inches (254 mm) thick for a one-story building and 12 inches (305 mm) thick for a two-story building. Bottom of footing to be 18 inches (457 mm) below grade or the frost depth, whichever is deeper. Footing to be reinforced with No. 4 bars at 24 inches (610 mm) on center when supporting more than the roof and one floor.

[4]From Table 21-C or local snow load tables. For areas without snow loads use 20 pounds per square foot (0.96 kN/m^2).

[5]From Table 21-A. For intermediate floor loads go to next higher value.

TABLE A-21-A-2—INTERIOR FOUNDATION REQUIREMENTS FOR
MASONRY BUILDINGS WITH 6- AND 8-INCH-THICK WALLS
(Wood or Steel Framing)
(Width of Footings in Inches)[1,2,3,4]
See Figure A-21-1 for typical details.

WALL HEIGHT (feet)	SPAN TO BEARING WALLS (feet)	ONE-STORY BUILDINGS Roof Live Load[5] $\times$ 0.0479 for kN/m^2			TWO-STORY BUILDINGS Roof Live Load[5] (psf) $\times$ 0.0479 for kN/m^2 20 — Plus Floor Live Load[6] (psf) $\times$ 0.0479 for kN/m^2		30		40	
		20 psf (inches)	30 psf (inches)	40 psf (inches)	50	100	50	100	50	100
$\times$ 304.8 for mm		Minimum Width of Footing (inches) $\times$ 25.4 for mm								
8	8	12	12	12	12	14	12	14	12	14
	16	12	12	12	16	20	18	20	18	22
	24	12	12	14	20	26	22	28	22	28
	32	14	14	16	24	28	26	32	28	34
10	8	12	12	12	14	16	14	16	14	16
	16	12	12	12	20	24	20	22	20	22
	24	12	14	14	22	28	22	28	22	28
	32	14	14	16	26	34	26	32	28	34
12	8	12	12	12	14	16	16	18	16	18
	16	12	14	16	20	24	20	22	20	22
	24	14	14	16	24	28	22	28	24	28
	32	16	16	18	28	30	28	32	28	34

[1]For buildings with under-floor space or basements, footing thickness is to be a minimum of 12 inches (305 mm). It shall be reinforced with No. 4 bars at 24 inches (610 mm) on center when its width is required to be 18 inches (457 mm) or larger and it supports more than the roof and one floor.

[2]Soil to be at least Class 4 as shown in Table 18-I-A.

[3]Footings are 10 inches (254 mm) thick for up to 24 inches (610 mm) wide and 12 inches (305 mm) thick for up to 34 inches (864 mm) wide. Footings shall be reinforced with No. 4 bars at 24 inches (610 mm) on center when supporting more than the roof and one floor.

[4]These interior footings support roof-ceiling or floors or both for a distance on each side equal to the span length shown. A tributary width equal to the span length may be used.

[5]From Table 16-C or local snow load tables. For areas without snow loads use 20 pounds per square foot (0.96 kN/m^2).

[6]From Table 16-A. For intermediate floor loads go to next higher value.

TABLE A-21-B—VERTICAL REINFORCEMENT AND TOP RESTRAINT FOR VARIOUS HEIGHTS OF BASEMENT AND OTHER BELOW-GRADE WALLS

DESIGN ASSUMPTIONS
A. Materials:
1. **Concrete Masonry Units**—Grade hollow load-bearing units conforming to Section 2112.4.2 for strength of units should not be less than that required for applicable f'_m.
2. **Mortar**—Type M, 2,500 psi (17 240 kPa) strength.
3. **Corefill**—Fine or coarse grout (UBC Standard 21-19) with an ultimate strength (28 days) of at least 2,500 psi. (17 240 kPa)
4. **Reinforcement**—Deformed billet-steel bars.
5. 1,500 psf (71.8 kPa) soil bearing required.[1]
B. Allowable stresses in accordance with Section 2106 and Table 21-M.

Soil Equiv.-fluid wt. = 30 pcf[1] (4.71 kN/m³)			Vertical Reinforcement with Axial Compressive Load (P) Equal to or Less than 5,000 lb./lin. ft. (72.92 kN/m)		
	Floor Connection[2,3]		f'_m = 1,500 psi (10 335 kPa)		
Wall Depth below Grade h (feet)	Wood Floor		Spacing of Reinforcement (inches)[4]		
× 304.8 for mm	Bolt and Spacing	Angle Clip Spacing	× 25.4 for mm		
8-Inch Walls			No. 3	No. 4	No. 5
× 25.4 for mm					
4	$1/2''$ at 60″	48″ o.c.	24	40	56
5	$1/2''$ at 40″	32″ o.c.	16	24	40
6	$5/8''$ at 32″	20″ o.c.	—	16	24
10-Inch Walls			Spacing of Reinforcement (inches)		
			× 25.4 for mm		
× 25.4 for mm			No. 4	No. 5	No. 6	No. 7
6	$5/8''$ at 32″	20″ o.c.	40	56	64	72
7	$5/8''$ at 24″	16″ o.c.	24	40	48	56
9	$3/4''$ at 20″	2 at 24″ o.c.	16	24	32	40
12-Inch Walls			Spacing of Reinforcement (inches)		
			× 25.4 for mm		
× 25.4 for mm			No. 4	No. 5	No. 6	No. 7
7	$5/8''$ at 24″	16″ o.c.	40	56	80	80
8	$3/4''$ at 20″	2 at 24″ o.c.	32	48	56	64
9	$7/8''$ at 18″	2 at 18″ o.c.	24	40	48	48
10	1″ at 16″	2 at 16″ o.c.	16	32	40	40

[1]Soil type is at least Class 4 as shown in Table 18-I-A.
[2]There shall be no backfill placed until after the wall is anchored to the floor and seven days have passed after grouting.
[3]For Figure A-21-4 only.
[4]See Figure A-21-5 for placement of reinforcement.

TABLE A-21-C-1—VERTICAL REINFORCING STEEL REQUIREMENTS FOR 6-INCH-THICK (153 mm) MASONRY WALLS[1] IN AREAS WHERE BASIC WIND SPEEDS ARE 80 MILES PER HOUR (129 km/h) OR GREATER[2,3,4,5]
(Wood or Steel Roof and Floor Framing)

Criteria: Roof Live Load = 20 psf to 40 psf (0.96 kN/m² to 1.9 kN/m²);
Floor Live Load = 50 psf (2.4 kN/m²); enclosed building[6]

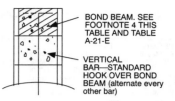

BOND BEAM. SEE FOOTNOTE 4 THIS TABLE AND TABLE A-21-E

VERTICAL BAR—STANDARD HOOK OVER BOND BEAM (alternate every other bar)

EXPO-SURE	STORIES	UNSUPPORTED HEIGHT (feet) × 304.8 for mm	80 MPH (× 1.61 for km/h) 8	16	24	32	90 MPH 8	16	24	32	100 MPH 8	16	24	32	110 MPH 8	16	24	32
			Span between Bearing Walls (feet) × 304.8 for mm / Size of Rebar and Spacing (inches) × 25.4 for mm															
B	One-story building	8	NR*								No. 4 80	No. 4 80	No. 4 80	No. 4 80	No. 4 64	No. 4 64	No. 4 72	No. 4 88
		10	No. 4 80	No. 4 88	No. 4 96	No. 4 96	No. 4 64	No. 4 64	No. 4 72	No. 4 80	No. 4 48	No. 4 48	No. 4 48	No. 4 56	No. 4 40	No. 4 40	No. 4 40	No. 4 48
		12	No. 4 48	No. 4 48	No. 4 56	No. 4 64	No. 4 40	No. 4 40	No. 4 48	No. 4 48	No. 5 48	No. 5 48	No. 5 56	No. 5 56	No. 5 40	No. 5 40	No. 5 40	No. 5 40
	Two-story building		Design required or use 8-inch or larger units for two-story condition.															
C	One-story building	8	No. 4 72	No. 4 72	No. 4 72	No. 4 96	No. 4 56	No. 4 56	No. 4 56	No. 4 56	No. 4 40	No. 4 40	No. 4 48	No. 4 48	No. 4 32	No. 4 32	No. 4 32	No. 4 40
		10	No. 4 40	No. 4 40	No. 4 40	No. 4 48	No. 4 32	No. 4 32	No. 4 32	No. 4 32	No. 5 40	No. 5 40	No. 5 40	No. 5 48	No. 5 32	No. 5 32	No. 5 32	No. 5 40
		12	No. 5 40	No. 5 48	No. 5 48	No. 5 48	No. 5 32	No. 5 32	No. 5 32	No. 5 40	Use 8-inch or larger units.							
	Two-story building		Design required or use 8-inch or larger units for two-story condition.															
D	One-story building	8	No. 4 56	No. 4 56	No. 4 64	No. 4 80	No. 4 48	No. 4 48	No. 4 48	No. 4 48	No. 4 32	No. 4 40	No. 4 40	No. 4 40	No. 4 32	No. 4 32	No. 4 32	No. 4 32
		10	No. 4 32	No. 4 32	No. 4 32	No. 4 40	No. 5 40	No. 5 40	No. 5 48	No. 5 48	No. 5 48	No. 5 32	No. 5 32	No. 5 40				
		12	No. 5 32	No. 5 40	No. 5 40	No. 5 40	Use 8-inch or larger units											
	Two-story building		Design required or use 8-inch or larger units for two-story condition.															

*NR — No vertical reinforcement required. However, see Table A-21-F for shear wall reinforcement.

[1]These values are for walls with running bond. For stack bond see Section 2112.8.3.

[2]The figure on top of the listed data is the bar size; the figure below it is the maximum spacing in inches (mm). Reinforcing bar strength shall be A 615 Grade 60. The vertical bars are centered in the middle of the wall.

[3]Roof load is assumed to be concentrically loaded on the wall. For roofs which hang on ledgers, a design is required.

[4]Minimum horizontal reinforcement shall be one No. 4 at the ledger and foundation. Also, see Table A-21-E for lintels and Table A-21-F for shear wall reinforcing where applicable.

[5]Hook vertical bars over bond beam bars as shown. Extend bars into footing using lap splices where necessary.

[6]Design required for open buildings of 6-inch-thick (153 mm) masonry.

To use this table, check criteria by the following method:

    6.1Choose proper roof live load from Table 16-C or snow load criteria for the locality in which the building is located.

    6.2Check if building is enclosed or partially enclosed by the procedure in Chapter 16, Part III.

    6.3Choose proper floor load from Table 16-A. [For loads less than 50 pounds per square foot (psf) (2.4 kN/m²), use 50 psf (2.4 kN/m²), and for loads between 50 psf (2.4 kN/m²) and 100 psf (4.8 kN/m²), use 100 psf (4.8 kN/m²).]

    6.4Find proper wind speed and exposure for the site—see Figure 16-1, Chapter 16, Sections 1619 and 1620.

    6.5Within the proper vertical column, choose appropriate span-to-bearing wall and appropriate height and story.

    6.6Read proper size and spacing of reinforcement for the thickness of the wall mentioned in the title of the table. (Equivalent area of steel, taking spacing into account, may be substituted.)

    6.7For buildings in Seismic Zone 2 (see Figure 16-2 in Chapter 16), use minimum reinforcement in Figure A-21-2 if it is more restrictive than the table values.

TABLE A-21-C-2—VERTICAL REINFORCING STEEL REQUIREMENTS FOR 8-INCH-THICK (203 mm) MASONRY WALLS[1] IN AREAS WHERE BASIC WIND SPEEDS ARE 80 MILES PER HOUR (129 km/h) OR GREATER[2,3,4,5]
(Wood or Steel Roof and Floor Framing)

BOND BEAM. SEE FOOTNOTE 4 THIS TABLE AND TABLE A-21-E

VERTICAL BAR—STANDARD HOOK OVER BOND BEAM (alternate every other bar)

Criteria: Roof Live Load = 20 psf to 40 psf (0.96 kN/m² to 1.9 kN/m²); Floor Live Load = 50 psf (2.4 kN/m²); enclosed building

EXPOSURE	STORIES	UNSUPPORTED HEIGHT (feet) × 304.8 for mm	80 MPH 8	16	24	32	90 MPH 8	16	24	32	100 MPH 8	16	24	32	110 MPH 8	16	24	32
B	One-story building or top story of two-story building	8	NR*				NR*				NR*				No. 3 56	No. 3 56	No. 3 64	No. 3 64
		10	NR*				No. 4 80	No. 4 80	No. 4 88	No. 4 88	No. 4 64	No. 4 64	No. 4 64	No. 4 72	No. 4 48	No. 4 48	No. 4 56	No. 4 56
		12	No. 4 64	No. 4 72	No. 4 72	No. 4 72	No. 4 56	No. 4 56	No. 4 56	No. 4 56	No. 4 40	No. 4 40	No. 5 64	No. 5 64	No. 5 56	No. 5 56	No. 5 56	No. 5 56
	First story of a two-story building	8	No. 3 96	No. 3 96	No. 3 96	No. 3 96	No. 3 96	No. 3 96	No. 3 96	No. 3 96	No. 3 96	No. 3 88	No. 3 80	No. 3 72	No. 3 72	No. 3 72	No. 3 64	No. 3 64
		10	No. 3 88	No. 3 80	No. 3 72	No. 3 64	No. 3 64	No. 3 64	No. 3 56	No. 3 56	No. 4 72	No. 4 72	No. 4 64	No. 4 64	No. 4 64	No. 4 56	No. 4 56	No. 4 56
		12	No. 4 80	No. 4 72	No. 4 64	No. 4 64	No. 4 64	No. 4 56	No. 4 56	No. 4 48	No. 4 48	No. 4 48	No. 5 64	No. 5 56	No. 5 56	No. 5 56	No. 5 48	No. 5 48
C	One-story building or top story of two-story building	8	NR*				No. 3 48	No. 3 48	No. 3 48	No. 3 56	No. 4 64	No. 4 64	No. 4 72	No. 4 72	No. 4 48	No. 4 56	No. 4 56	No. 4 56
		10	No. 4 56	No. 4 56	No. 4 64	No. 4 64	No. 4 48	No. 4 48	No. 4 48	No. 4 48	No. 5 56	No. 5 56	No. 5 56	No. 5 56	No. 5 48	No. 5 48	No. 5 48	No. 5 48
		12	No. 5 56	No. 5 64	No. 5 64	No. 5 64	No. 5 48	No. 5 48	No. 5 48	No. 5 48	No. 6 56	No. 6 56	No. 6 56	No. 6 56	No. 6 40	No. 6 40	No. 6 40	No. 6 48
	First story of a two-story building	8	No. 3 80	No. 3 80	No. 3 56	No. 3 72	No. 3 56	No. 3 56	No. 3 56	No. 3 56	No. 4 72	No. 4 72	No. 4 72	No. 4 64	No. 4 56	No. 4 56	No. 4 56	No. 4 56
		10	No. 4 72	No. 4 64	No. 4 64	No. 4 56	No. 4 56	No. 4 48	No. 4 48	No. 4 48	No. 5 64	No. 5 56	No. 5 56	No. 5 56	No. 5 48	No. 5 48	No. 5 48	No. 5 48
		12	No. 5 64	No. 5 64	No. 5 56	No. 5 56	No. 5 48	No. 5 48	No. 5 48	No. 5 48	No. 6 56	No. 6 56	No. 6 48	No. 6 48	No. 6 48	No. 6 40	No. 6 40	No. 6 40
D	One-story building or top story of two-story building	8	No. 3 48	No. 3 48	No. 3 56	No. 3 56	No. 4 64	No. 4 72	No. 4 72	No. 4 80	No. 4 56	No. 4 56	No. 4 56	No. 4 56	No. 4 40	No. 4 48	No. 4 48	No. 4 48
		10	No. 4 48	No. 4 48	No 4 48	No. 4 56	No. 5 56	No. 5 64	No. 5 64	No. 5 64	No. 5 48	No. 5 48	No. 5 48	No. 5 48	No. 5 40	No. 5 40	No. 5 40	No. 5 40
		12	No. 5 48	No. 5 48	No. 5 56	No. 5 56	No. 6 56	No. 6 56	No. 6 56	No. 6 56	No. 6 48	No. 6 48	No .6 48	No. 6 48	No. 6 40	No. 6 40	No. 6 40	No. 6 40
	First story of a two-story building	8	No. 3 64	No. 3 64	No. 3 64	No. 3 56	No. 4 80	No. 4 80	No. 4 72	No. 4 72	No. 4 64	No. 4 56	No. 4 56	No. 4 56	No. 4 48	No. 4 48	No. 4 48	No. 4 48
		10	No. 4 56	No. 4 56	No. 4 56	No. 4 48	No. 4 48	No. 5 64	No. 5 64	No. 5 56	No. 5 48	No. 5 48	No. 5 48	No. 5 48	No. 5 40	No. 5 40	No. 5 40	No. 5 32
		12	No. 5 56	No. 5 56	No. 5 56	No. 5 48	No. 6 56	No. 6 56	No. 6 56	No. 6 48	No. 6 48	No. 6 48	No. 6 40	No. 6 40	No. 6 40	No. 6 32	No. 6 32	No. 6 32

*NR — No vertical reinforcement required. However, see Table A-21-F for shear wall reinforcement.

[1]These values are for walls with running bond. For stack bond see Section 2112.8.3.
[2]The figure on top of the listed data is the bar size; the figure below it is the maximum spacing in inches (mm). Reinforcing bar strength shall be A 615 Grade 60.
[3]Roof load is assumed to be concentrically loaded on the wall. For roofs which hang on ledgers, a design is required.
[4]Minimum horizontal reinforcement shall be one No. 4 at the ledger and foundation. Also, see Table A-21-E for lintels and Table A-21-F for shear wall reinforcing where applicable.
[5]Hook vertical bars over bond beam as shown. Extend bars into footing using lap splices where necessary. Where second-story bar spacing does not match those on the first story, hook bars around floor bond beam also.

To use this table, check criteria by the following method:
[5.1]Choose proper roof live load from Table 16-C or snow load criteria for the locality in which the building is located.
[5.2]Check if building is enclosed or partially enclosed by the procedure in Chapter 16, Part III.
[5.3]Choose proper floor load from Table 16-A. [For loads less that 50 psf (2.4 kN/m²), use 50 psf (2.4 kN/m²), and for loads between 50 psf (2.4 kN/m²) and 100 psf (4.8 kN/m²), use 100 psf (4.8 kN/m²).]
[5.4]Find proper wind speed and exposure for the site—see Figure 16-1, Chapter 16, Sections 1619 and 1620.
[5.5]Within the proper vertical column, choose appropriate span-to-bearing wall and appropriate height and story.
[5.6]Read proper size and spacing of reinforcement for the thickness of the wall mentioned in the title of the table. (Equivalent area of steel, taking spacing into account, may be substituted.)
[5.7]For buildings in Seismic Zone 2 (see Figure 16-2 in Chapter 16), use minimum reinforcement in Figure A-21-2 if it is more restrictive than the table values.

TABLE A-21-C-3—VERTICAL REINFORCING STEEL REQUIREMENTS FOR 8-INCH-THICK (203 mm) MASONRY WALLS[1] IN AREAS WHERE BASIC WIND SPEEDS ARE 80 MILES PER HOUR (129 km/h) OR GREATER[2,3,4,5]
(Wood or Steel Roof and Floor Framing)

BOND BEAM. SEE FOOTNOTE 4 THIS TABLE AND TABLE A-21-E

VERTICAL BAR—STANDARD HOOK OVER BOND BEAM (alternate every other bar)

Criteria: Roof Live Load = 20 psf to 40 psf (0.96 kN/m² to 1.9 kN/m²); Floor Live Load = 100 psf (4.8 kN/m²); enclosed building

Span between Bearing Walls (feet): × 304.8 for mm · × 1.61 for km/h · Size of Rebar and Spacing (inches): × 25.4 for mm · Unsupported Height (feet): × 304.8 for mm

The figure on top of the listed data is the bar size; the figure below it is the maximum spacing in inches (mm).

Exposure	Stories	Unsup. Height (feet)	80 MPH 8	16	24	32	90 MPH 8	16	24	32	100 MPH 8	16	24	32	110 MPH 8	16	24	32
B	One-story building or top story of two-story building	8	NR*	NR*	NR*	NR*	NR*	NR*	NR*	NR*	NR*	NR*	NR*	NR*	No. 3 / 56	No. 3 / 56	No. 3 / 64	No. 3 / 64
		10	NR*	NR*	NR*	NR*	No. 4 / 64	No. 4 / 64	No. 4 / 64	No. 4 / 72	No. 4 / 64	No. 4 / 64	No. 4 / 64	No. 4 / 72	No. 4 / 48	No. 4 / 48	No. 4 / 56	No. 4 / 56
		12	No. 4 / 64	No. 4 / 72	No. 4 / 72	No. 4 / 72	No. 4 / 56	No. 4 / 56	No. 4 / 56	No. 4 / 56	No. 4 / 40	No. 4 / 40	No. 4 / 40	No. 4 / 40	No. 5 / 56	No. 5 / 56	No. 5 / 56	No. 5 / 56
	First story of a two-story building	8	No. 3 / 96	No. 3 / 96	No. 3 / 80	No. 3 / 64	No. 3 / 96	No. 3 / 88	No. 3 / 72	No. 3 / 56	No. 3 / 80	No. 3 / 64	No. 3 / 56	No. 3 / 48	No. 3 / 64	No. 3 / 56	No. 3 / 48	No. 4 / 64
		10	No. 3 / 72	No. 3 / 64	No. 3 / 56	No. 3 / 48	No. 3 / 56	No. 3 / 48	No. 4 / 64	No. 4 / 56	No. 4 / 72	No. 4 / 64	No. 4 / 56	No. 4 / 48	No. 4 / 56	No. 4 / 48	No. 4 / 48	No. 5 / 56
		12	No. 4 / 72	No. 4 / 64	No. 4 / 56	No. 4 / 48	No. 4 / 56	No. 4 / 48	No. 4 / 48	No. 4 / 40	No. 4 / 48	No. 4 / 40	No. 5 / 48	No. 5 / 48	No. 5 / 56	No. 5 / 48	No. 5 / 48	No. 5 / 40
C	One-story building or top story of two-story building	8	NR*	NR*	NR*	NR*	No. 3 / 48	No. 3 / 48	No. 3 / 48	No. 3 / 56	No. 4 / 64	No. 4 / 64	No. 4 / 72	No. 4 / 72	No. 4 / 48	No. 4 / 56	No. 4 / 56	No. 4 / 56
		10	No. 4 / 56	No. 4 / 56	No. 4 / 64	No. 4 / 64	No. 4 / 48	No. 4 / 48	No. 4 / 48	No. 4 / 48	No. 5 / 56	No. 5 / 56	No. 5 / 56	No. 5 / 56	No. 5 / 48	No. 5 / 48	No. 5 / 48	No. 5 / 48
		12	No. 4 / 40	No. 5 / 64	No. 5 / 64	No. 5 / 64	No. 5 / 48	No. 5 / 48	No. 5 / 48	No. 5 / 48	No. 5 / 40	No. 6 / 56	No. 6 / 56	No. 6 / 56	No. 6 / 40	No. 6 / 40	No. 6 / 40	No. 5 / 32
	First story of a two-story building	8	No. 3 / 72	No. 3 / 64	No. 3 / 56	No. 3 / 48	No. 3 / 56	No. 3 / 48	No. 4 / 64	No. 4 / 56	No. 4 / 72	No. 4 / 64	No. 4 / 56	No. 4 / 48	No. 4 / 56	No. 4 / 48	No. 4 / 48	No. 5 / 56
		10	No. 4 / 64	No. 4 / 56	No. 4 / 48	No. 4 / 48	No. 4 / 48	No. 4 / 48	No. 5 / 56	No. 5 / 56	No. 5 / 56	No. 5 / 56	No. 5 / 48	No. 5 / 48	No. 5 / 48	No. 5 / 48	No. 5 / 40	No. 5 / 40
		12	No. 4 / 40	No. 5 / 56	No. 5 / 48	No. 5 / 48	No. 5 / 48	No. 5 / 40	No. 5 / 40	No. 5 / 48	No. 6 / 40	No. 6 / 48	No. 6 / 48	No. 6 / 40	No. 6 / 40	No. 6 / 40	No. 6 / 40	No. 6 / 32
D	One-story building or top story of two-story building	8	No. 3 / 48	No. 3 / 48	No. 3 / 56	No. 3 / 56	No. 3 / 64	No. 4 / 72	No. 4 / 72	No. 4 / 80	No. 4 / 56	No. 4 / 56	No. 4 / 56	No. 4 / 56	No. 4 / 40	No. 4 / 48	No. 4 / 48	No. 4 / 48
		10	No. 4 / 48	No. 4 / 48	No 4 / 48	No. 4 / 56	No. 5 / 56	No. 5 / 64	No. 5 / 64	No. 5 / 64	No. 5 / 48	No. 5 / 48	No. 5 / 48	No. 5 / 48	No. 5 / 40	No. 5 / 40	No. 5 / 40	No. 5 / 40
		12	No. 5 / 48	No. 5 / 48	No. 5 / 56	No. 5 / 56	No. 5 / 40	No. 5 / 40	No. 6 / 56	No. 6 / 56	No. 6 / 48	No. 6 / 48	No. 6 / 48	No. 5 / 32	No. 6 / 40	No. 6 / 40	No. 6 / 40	No. 6 / 40
	First story of a two-story building	8	No. 3 / 56	No. 3 / 56	No. 3 / 48	No. 4 / 64	No. 3 / 48	No. 4 / 64	No. 4 / 56	No. 4 / 56	No. 4 / 56	No. 4 / 56	No. 4 / 48	No. 5 / 64	No. 4 / 48	No. 5 / 64	No. 5 / 56	No. 5 / 56
		10	No. 4 / 56	No. 4 / 48	No. 4 / 48	No. 5 / 56	No. 5 / 64	No. 5 / 56	No. 5 / 48	No. 5 / 48	No. 5 / 48	No. 5 / 48	No. 5 / 40	No. 5 / 40	No. 5 / 40	No. 6 / 40	No. 6 / 48	No. 6 / 48
		12	No. 5 / 48	No. 5 / 48	No. 5 / 40	No. 5 / 40	No. 5 / 40	No. 5 / 40	No. 6 / 48	No. 6 / 40	No. 6 / 48	No. 6 / 40	No. 6 / 40	No. 6 / 32	No. 6 / 32	No. 6 / 32	No. 6 / 32	No. 6 / 24

*NR — No vertical reinforcement required. However, see Table A-21-F for shear wall reinforcement.

[1]These values are for walls with running bond. For stack bond see Section 2112.8.3.

[2]The figure on top of the listed data is the bar size; the figure below it is the maximum spacing in inches (mm). Reinforcing bar strength shall be A 615 Grade 60.

[3]Roof load is assumed to be concentrically loaded on the wall. For roofs which hang on ledgers, a design is required.

[4]Minimum horizontal reinforcement shall be one No. 4 at the ledger and foundation. Also, see Table A-21-E for lintels and Table A-21-F for shear wall reinforcing where applicable.

[5]Hook vertical bars over bond beam as shown. Extend bars into footing using lap splices where necessary. Where second-story bar spacing does not match those on the first story, hook bars around floor bond beam also.

To use this table, check criteria by the following method:

5.1 Choose proper roof live load from Table 16-C or snow load criteria for the locality in which the building is located.

5.2 Check if building is enclosed or partially enclosed by the procedure in Chapter 16, Part III.

5.3 Choose proper floor load from Table 16-A. [For loads less than 50 psf (2.4 kN/m²), use 50 psf (2.4 kN/m²), and for loads between 50 psf (2.4 kN/m²) and 100 psf (4.8 kN/m²), use 100 psf (4.8 kN/m²).]

5.4 Find proper wind speed and exposure for the site—see Figure 16-1, Chapter 16, Sections 1619 and 1620.

5.5 Within the proper vertical column, choose appropriate span-to-bearing wall and appropriate height and story.

5.6 Read proper size and spacing of reinforcement for the thickness of the wall mentioned in the title of the table. (Equivalent area of steel, taking spacing into account, may be substituted.)

5.7 For buildings in Seismic Zone 2 (see Figure 16-2 in Chapter 16), use minimum reinforcement in Figure A-21-2 if it is more restrictive than the table values.

TABLE A-21-C-4—VERTICAL REINFORCING STEEL REQUIREMENTS FOR 8-INCH-THICK (203 mm) MASONRY WALLS[1] IN AREAS WHERE BASIC WIND SPEEDS ARE 80 MILES PER HOUR (129 km/h) OR GREATER[2,3,4,5]
(Wood or Steel Roof and Floor Framing)

BOND BEAM. SEE FOOTNOTE 4 THIS TABLE AND TABLE A-21-E

VERTICAL BAR—STANDARD HOOK OVER BOND BEAM (alternate every other bar)

Criteria: Roof Live Load = 20 psf to 40 psf (0.96 kN/m² to 1.9 kN/m²); Floor Live Load = 50 psf (2.4 kN/m²); partially enclosed building

			80 MPH				90 MPH				100 MPH				110 MPH			
			× 1.61 for km/h															
			Span between Bearing Walls (feet) — × 304.8 for mm															
EXPO-SURE	STORIES	UNSUP-PORTED HEIGHT (feet) × 304.8 for mm	8	16	24	32	8	16	24	32	8	16	24	32	8	16	24	32
			Size of Rebar and Spacing (inches) — × 25.4 for mm															
B	One-story building or top story of two-story building	8	No.4 96	No.4 96	No.3 80	No.3 88	No.3 56	No.3 56	No.3 64	No.3 64	No.3 40	No.3 48	No.3 48	No.3 48	No.4 64	No.4 64	No.4 64	No.4 72
		10	No.4 64	No.4 64	No.4 72	No.4 72	No.4 48	No.4 56	No.4 56	No.4 56	No.4 40	No.4 40	No.4 40	No.4 40	No.5 48	No.5 56	No.5 56	No.5 56
		12	No.4 40	No.4 48	No.4 48	No.4 48	No.5 56	No.5 56	No.5 56	No.5 56	No.5 40	No.5 40	No.5 40	No.5 40	No.6 48	No.6 48	No.6 48	No.6 48
	First story of a two-story building	8	No.3 96	No.3 96	No.3 88	No.3 80	No.3 72	No.3 72	No.3 64	No.3 64	No.3 48	No.3 48	No.3 48	No.3 48	No.4 64	No.4 64	No.4 64	No.4 64
		10	No.3 48	No.3 48	No.3 48	No.3 48	No.4 64	No.4 56	No.4 56	No.4 56	No.4 48	No.4 48	No.4 48	No.4 40	No.4 40	No.5 56	No.5 56	No.5 48
		12	No.4 48	No.4 48	No.4 48	No.4 40	No.4 40	No.5 56	No.5 56	No.5 48	No.5 48	No.5 40	No.5 40	No.5 40	No.6 48	No.6 48	No.6 48	No.6 48
C	One-story building or top story of two-story building	8	No.3 40	No.3 40	No.3 40	No.3 40	No.4 56	No.4 56	No.4 56	No.4 56	No.4 40	No.4 40	No.4 40	No.4 48	No.5 56	No.5 56	No.5 56	No.5 56
		10	No.4 40	No.4 40	No.4 40	No.4 40	No.5 48	No.5 48	No.5 48	No.5 48	No.5 32	No.5 32	No.5 40	No.5 40	No.6 40	No.6 40	No.6 40	No.6 40
		12	No.5 40	No.5 40	No.5 40	No.5 40	No.6 40	No.6 48	No.6 48	No.6 48	No.6 32	No.6 32	No.6 32	No.6 32	Use 10-inch or larger units			
	First story of a two-story building	8	No.3 48	No.3 48	No.3 48	No.3 48	No.4 56	No.4 56	No.4 56	No.4 56	No.4 48	No.4 48	No.4 40	No.4 40	No.5 56	No.5 56	No.5 56	No.5 48
		10	No.4 40	No.4 40	No.4 40	No.4 40	No.5 48	No.5 48	No.5 48	No.5 48	No.5 40	No.5 40	No.5 40	No.5 48	No.6 48	No.6 48	No.6 40	No.6 40
		12	No.5 40	No.5 40	No.5 40	No.6 48	No.6 48	No.6 48	No.6 40	No.6 40	No.6 32	No.6 32	No.6 32	No.6 32	Use 10-inch or larger units			
D	One-story building or top story of two-story building	8	No.4 56	No.4 56	No.4 56	No.4 64	No.4 40	No.4 48	No.4 48	No.4 48	No.5 56	No.5 56	No.5 56	No.5 56	No.5 48	No.5 48	No.5 48	No.5 48
		10	No.5 48	No.5 48	No.5 48	No.5 56	No.5 40	No.5 40	No.5 40	No.5 40	No.6 56	No.6 48	No.6 48	No.6 48	No.6 32	No.6 32	No.6 40	No.6 40
		12	No.6 48	No.6 48	No.6 48	No.6 48	No.6 40	No.6 40	No.6 40	No.6 40	No.6 24	No.6 32	No.6 32	No.6 32	Use 10-inch or larger units			
	First story of a two-story building	8	No.4 64	No.4 64	No.4 64	No.4 56	No.4 48	No.4 48	No.4 48	No.4 48	No.4 40	No.4 40	No.5 56	No.5 56	No.5 48	No.5 48	No.5 48	No.5 40
		10	No.5 56	No.5 56	No.5 48	No.5 48	No.5 40	No.5 40	No.5 40	No.5 40	No.6 48	No.6 48	No.6 48	No.6 40	No.6 40	No.6 40	No.6 32	No.6 32
		12	No.6 48	No.6 48	No.6 48	No.6 48	No.6 40	No.6 40	Use 10-inch or larger units										

*NR — No vertical reinforcement required. However, see Table A-21-F for shear wall reinforcement.

[1] These values are for walls with running bond. For stack bond see Section 2112.8.3.

[2] The figure on top of the listed data is the bar size; the figure below it is the maximum spacing in inches (mm). Reinforcing bar strength shall be A 615 Grade 60.

[3] Roof load is assumed to be concentrically loaded on the wall. For roofs which hang on ledgers, a design is required.

[4] Minimum horizontal reinforcement shall be one No. 4 at the ledger and foundation. Also, see Table A-21-E for lintels and Table A-21-F for shear wall reinforcing where applicable.

[5] Hook vertical bars over bond beam as shown. Extend bars into footing using lap splices where necessary.

To use this table, check criteria by the following method:

5.1 Choose proper roof live load from Table 16-C or snow load criteria for the locality in which the building is located.

5.2 Check if building is enclosed or partially enclosed by the procedure in Chapter 16, Part III.

5.3 Choose proper floor load from Table 16-A. [For loads less than 50 psf (2.4 kN/m²), use 50 psf (2.4 kN/m²), and for loads between 50 psf (2.4 kN/m²) and 100 psf (4.8 kN/m²), use 100 psf (4.8 kN/m²).]

5.4 Find proper wind speed and exposure for the site—see Figure 16-1, Chapter 16, Sections 1619 and 1620.

5.5 Within the proper vertical column, choose appropriate span-to-bearing wall and appropriate height and story.

5.6 Read proper size and spacing of reinforcement for the thickness of the wall mentioned in the title of the table. (Equivalent area of steel, taking spacing into account, may be substituted.)

5.7 For buildings in Seismic Zone 2 (see Figure 16-2 in Chapter 16), use minimum reinforcement in Figure A-21-2 if it is more restrictive than the table values.

TABLE A-21-C-5—VERTICAL REINFORCING STEEL REQUIREMENTS FOR 8-INCH-THICK (203 mm) MASONRY WALLS[1] IN AREAS WHERE BASIC WIND SPEEDS ARE 80 MILES PER HOUR (129 km/h) OR GREATER[2,3,4,5]
(Wood or Steel Roof and Floor Framing)

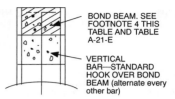

BOND BEAM. SEE FOOTNOTE 4 THIS TABLE AND TABLE A-21-E

VERTICAL BAR—STANDARD HOOK OVER BOND BEAM (alternate every other bar)

Criteria: Roof Live Load = 20 psf to 40 psf (0.96 kN/m^2 to 1.9 kN/m^2);
Floor Live Load = 100 psf (4.8 kN/m^2); partially enclosed building

EXPO-SURE	STORIES	UNSUP-PORTED HEIGHT (feet) × 304.8 for mm	80 MPH × 304.8 for mm — 8	16	24	32	90 MPH 8	16	24	32	100 MPH 8	16	24	32	110 MPH 8	16	24	32
			\multicolumn{16}{ } Size of Rebar and Spacing (inches) × 25.4 for mm															
B	One-story building or top story of two-story building	8	No. 3 / 72	No. 4 / 96	No. 3 / 80	No. 3 / 88	No. 3 / 56	No. 3 / 56	No. 3 / 64	No. 3 / 64	No. 4 / 80	No. 4 / 80	No. 4 / 80	No. 4 / 88	No. 4 / 64	No. 4 / 64	No. 4 / 64	No. 4 / 72
		10	No. 4 / 64	No. 4 / 64	No. 4 / 72	No. 4 / 72	No. 4 / 48	No. 4 / 56	No. 4 / 56	No. 4 / 56	No. 4 / 40	No. 4 / 40	No. 4 / 40	No. 4 / 40	No. 5 / 48	No. 5 / 56	No. 5 / 56	No. 5 / 56
		12	No. 4 / 40	No. 4 / 48	No. 4 / 48	No. 4 / 48	No. 5 / 56	No. 5 / 56	No. 5 / 56	No. 5 / 56	No. 5 / 40	No. 5 / 40	No. 5 / 40	No. 5 / 40	No. 6 / 48	No. 6 / 48	No. 6 / 48	No. 6 / 48
	First story of a two-story building	8	No. 3 / 88	No. 3 / 96	No. 3 / 56	No. 3 / 72	No. 3 / 64	No. 3 / 56	No. 3 / 64	No. 3 / 64	No. 4 / 80	No. 4 / 72	No. 4 / 64	No. 4 / 56	No. 4 / 64	No. 4 / 56	No. 4 / 48	No. 4 / 48
		10	No. 4 / 72	No. 4 / 64	No. 4 / 56	No. 4 / 48	No. 4 / 56	No. 4 / 48	No. 4 / 48	No. 4 / 40	No. 4 / 48	No. 4 / 40	No. 5 / 56	No. 5 / 48	No. 5 / 56	No. 5 / 48	No. 5 / 48	No. 5 / 40
		12	No. 4 / 48	No. 4 / 40	No. 4 / 40	No. 5 / 48	No. 5 / 56	No. 5 / 48	No. 5 / 48	No. 5 / 40	No. 5 / 40	No. 5 / 40	No. 6 / 48	No. 6 / 48	No. 6 / 48	No. 6 / 48	No. 6 / 40	No. 6 / 40
C	One-story building or top story of two-story building	8	No. 4 / 64	No. 4 / 64	No. 4 / 72	No. 4 / 72	No. 4 / 56	No. 4 / 56	No. 4 / 56	No. 4 / 56	No. 4 / 40	No. 4 / 40	No. 4 / 40	No. 4 / 48	No. 5 / 56	No. 5 / 56	No. 5 / 56	No. 5 / 56
		10	No. 5 / 56	No. 5 / 56	No. 5 / 40	No. 5 / 40	No. 5 / 48	No. 5 / 48	No. 5 / 48	No. 5 / 48	No. 6 / 32	No. 6 / 56	No. 6 / 56	No. 5 / 40	No. 6 / 40	No. 6 / 40	No. 6 / 40	No. 6 / 40
		12	No. 5 / 40	No. 5 / 40	No. 5 / 40	No. 5 / 40	No. 6 / 40	No. 6 / 48	No. 6 / 48	No. 6 / 48	No. 6 / 32	No. 6 / 32	No. 6 / 32	No. 6 / 32	\multicolumn{4}{ } Use 10-inch or larger units			
	First story of a two-story building	8	No. 4 / 72	No. 4 / 64	No. 4 / 56	No. 4 / 48	No. 4 / 56	No. 4 / 48	No. 4 / 48	No. 4 / 40	No. 4 / 40	No. 4 / 40	No. 4 / 40	No. 5 / 48	No. 5 / 56	No. 5 / 48	No. 5 / 48	No. 5 / 40
		10	No. 5 / 64	No. 5 / 56	No. 5 / 48	No. 5 / 48	No. 5 / 48	No. 5 / 48	No. 5 / 40	No. 5 / 40	No. 5 / 40	No. 6 / 48	No. 6 / 48	No. 6 / 40	No. 6 / 40	No. 6 / 40	No. 6 / 40	No. 6 / 32
		12	No. 5 / 40	No. 6 / 48	No. 6 / 48	No. 6 / 40	No. 6 / 40	No. 6 / 40	No. 6 / 40	No. 6 / 32	No. 6 / 32	No. 6 / 32	No. 6 / 32	\multicolumn{5}{ } Use 10-inch or larger units				
D	One-story building or top story of two-story building	8	No. 4 / 56	No. 4 / 56	No. 4 / 56	No. 4 / 64	No. 4 / 40	No. 5 / 72	No. 5 / 72	No. 5 / 72	No. 5 / 56	No. 5 / 56	No. 5 / 56	No. 5 / 56	No. 5 / 48	No. 5 / 48	No. 5 / 48	No. 5 / 48
		10	No. 5 / 48	No. 5 / 48	No. 5 / 48	No. 5 / 56	No. 5 / 40	No. 5 / 40	No. 5 / 40	No. 5 / 40	No. 6 / 56	No. 6 / 48	No. 6 / 48	No. 6 / 48	No. 6 / 32	No. 6 / 32	No. 6 / 40	No. 6 / 40
		12	No. 6 / 48	No. 6 / 48	No. 6 / 48	No. 6 / 48	No. 6 / 40	No. 6 / 40	No. 6 / 40	No. 6 / 40	No. 6 / 24	No. 6 / 32	No. 6 / 32	No. 6 / 32	\multicolumn{4}{ } Use 10-inch or larger units			
	First story of a two-story building	8	No. 4 / 64	No. 4 / 56	No. 4 / 48	No. 4 / 48	No. 4 / 48	No. 5 / 64	No. 5 / 56	No. 5 / 56	No. 5 / 56	No. 5 / 48	No. 5 / 48	No. 5 / 48	No. 5 / 48	No. 5 / 40	No. 5 / 40	No. 5 / 40
		10	No. 5 / 56	No. 5 / 48	No. 5 / 48	No. 5 / 40	No. 5 / 40	No. 5 / 40	No. 6 / 48	No. 6 / 48	No. 6 / 48	No. 6 / 40	No. 6 / 40	No. 6 / 40	No. 6 / 40	No. 6 / 32	No. 6 / 32	No. 6 / 32
		12	No. 6 / 48	No. 6 / 48	No. 6 / 40	No. 6 / 40	No. 6 / 40	No. 6 / 32	No. 6 / 32	No. 6 / 32	\multicolumn{8}{ } Use 10-inch or larger units							

*NR — No vertical reinforcement required. However, see Table A-21-F for shear wall reinforcement.

[1] These values are for walls with running bond. For stack bond see Section 2112.8.3.

[2] The figure on top of the listed data is the bar size; the figure below it is the maximum spacing in inches (mm). Reinforcing bar strength shall be A 615 Grade 60.

[3] Roof load is assumed to be concentrically loaded on the wall. For roofs which hang on ledgers, a design is required.

[4] Minimum horizontal reinforcement shall be one No. 4 at the ledger and foundation. Also, see Table A-21-E for lintels and Table A-21-F for shear wall reinforcing where applicable.

[5] Hook vertical bars over bond beam as shown. Extend bars into footing using lap splices where necessary.

To use this table, check criteria by the following method:

5.1 Choose proper roof live load from Table 16-C or snow load criteria for the locality in which the building is located.

5.2 Check if building is enclosed or partially enclosed by the procedure in Chapter 16, Part III.

5.3 Choose proper floor load from Table 16-A. [For loads less than 50 psf (2.4 kN/m^2), use 50 psf (2.4 kN/m^2), and for loads between 50 psf (2.4 kN/m^2) and 100 psf (4.8 kN/m^2), use 100 psf (4.8 kN/m^2).]

5.4 Find proper wind speed and exposure for the site—see Figure 16-1, Chapter 16, Sections 1619 and 1620.

5.5 Within the proper vertical column, choose appropriate span-to-bearing wall and appropriate height and story.

5.6 Read proper size and spacing of reinforcement for the thickness of the wall mentioned in the title of the table. (Equivalent area of steel, taking spacing into account, may be substituted.)

5.7 For buildings in Seismic Zone 2 (see Figure 16-2 in Chapter 16), use minimum reinforcement in Figure A-21-2 if it is more restrictive than the table values.

TABLE A-21-D—ANCHORAGE OF WOOD MEMBERS TO EXTERIOR WALLS FOR VERTICAL AND UPLIFT FORCES
[In areas where basic wind speeds are 80 miles per hour (129 km/h) or greater]
See Figure A-21-7 for details
Part I—Anchor bolt size and spacing [in inches (mm)][1,2,3] on wood ledgers carrying vertical loads from roofs and floors[4,5]
Douglas fir-larch, California redwood (close grain) and southern pine[6,7]

		2-INCH (51 mm) × LEDGER				3-INCH (76 mm) × LEDGER				4-INCH (102 mm) × LEDGER			
		Span between Bearing Walls (feet)											
		× 304.8 for mm											
TYPE OF LOADING	LIVE LOAD[8,9] psf	8	16	24	32	8	16	24	32	8	16	24	32
	× 0.0479 for kN/m²	× 25.4 for mm											
Roof	20	$^{1}/_{2}$ 32	$(2)^{1}/_{2}$ 16	$^{5}/_{8}$ 16	$^{7}/_{8}$ 16	$^{1}/_{2}$ 32	$^{1}/_{2}$ 16	$(2)^{1}/_{2}$ 32	$^{7}/_{8}$ 16	—	$^{5}/_{8}$ 32	$^{7}/_{8}$ 32	$(2)^{5}/_{8}$ 32
	30	$(2)^{1}/_{2}$ 32	$^{1}/_{2}$ 16	$^{3}/_{4}$ 16	$^{7}/_{8}$ 16	$^{1}/_{2}$ 16	$(2)^{7}/_{8}$ 32	$^{7}/_{8}$ 16	$^{7}/_{8}$ 16	—	$(2)^{1}/_{2}$ 32	$^{5}/_{8}$ 16	$^{3}/_{4}$ 16
	40	$^{1}/_{2}$ 16	$^{5}/_{8}$ 16	$^{3}/_{4}$ 8	—	$^{5}/_{8}$ 16	$(2)^{5}/_{8}$ 32	$^{7}/_{8}$ 16	1 16	$^{5}/_{8}$ 32	$^{5}/_{8}$ 16	$^{3}/_{4}$ 16	$^{7}/_{8}$ 16
Floor[10]	50	$^{1}/_{2}$ 16	1 12	—	—	$^{5}/_{8}$ 24	$^{3}/_{4}$ 32	$^{3}/_{4}$ 12	$1^{1}/_{4}$ 12	$^{5}/_{8}$ 24	$^{7}/_{8}$ 24	$^{7}/_{8}$ 16	$^{7}/_{8}$ 12
	100	1 16	$(2)^{3}/_{4}$ 12	—	—	$^{5}/_{8}$ 16	1 12	$(2)^{3}/_{4}$ 12	$(2)1$ 12	$^{7}/_{8}$ 16	$^{3}/_{4}$ 12	1 12	$(2)^{3}/_{4}$ 12

[1]Closer spacing may be used.

[2]Use two bolts, one above the other, at splices and locate them away from the splice end by $3^{1}/_{2}$ inches (89 mm) for $^{1}/_{2}$-inch (13 mm) diameter, $4^{1}/_{2}$ inches (114.3 mm) for $^{5}/_{8}$-inch (15.9 mm) diameter, $5^{1}/_{4}$ inches (133 mm) for $^{3}/_{4}$-inch (19 mm) diameter, $6^{1}/_{4}$ inches (158 mm) for $^{7}/_{8}$-inch (22.2 mm) diameter and 7 inches (178 mm) for 1-inch (25.4 mm) diameter.

[3]See Table A-21-F for lateral force requirements (when applicable).

[4]Tabulated values are based on short-term loading due to roof loads (25 percent) or snow loads (15 percent), whichever controls. No increase is allowed for floor loads.

[5]See details in Figure A-21-7 for location relative to other construction. Note that roofs are concentrically loaded.

[6]See Chapter 23, Division III, Part I, for other species. Adjust spacing in direct proportion to the perpendicular-to-grain values for the applicable ledger and bolt sizes shown using the procedure described in Chapter 23, Division III, Part I. No increase is allowed for special inspection.

[7]Values on top are bolt sizes and underneath are spacing. Multiple bolts are shown in parenthesis: example (2) = two.

[8]See Table 16-C or Appendix Chapter 16, Division I, for values.

[9]Joist spacing is limited to 30 inches (762 mm) on center maximum.

[10]Where two bolts are required they shall be staggered at half the spacing shown or be placed one above the other.

Part II—Uplift anchors[1] for wood roof members [number of common nails in a 0.036 inch (0.91 mm) (No. 20 galvanized sheet gage) by $1^{1}/_{8}$-inch (28.6 mm) tie strap embedded 5 inches (127 mm) into a masonry bond beam[2]]

		80 MPH				90 MPH				100 MPH				110 MPH			
		× 1.61 for km/h															
		Span between Bearing Walls (feet)[5]															
		× 304.8 for mm															
ENCLOSURE[3]	EXPOSURE[4]	8	16	24	32	8	16	24	32	8	16	24	32	8	16	24	32
Enclosed	B	NR	NR	NR	NR	NR	NR	NR	NR	NR	NR	NR	NR	NR	NR	2-8d	2-8d
	C	NR	NR	NR	NR	NR	2-8d	3-8d	4-8d	2-8d	4-8d	5-10d	5-10d	2-10d	4-10d	3-10d 24″	4-10d 24″
	D	NR	2-8d	3-8d	4-8d	2-8d	4-8d	4-10d	5-10d	3-8d	5-8d	5-10d	4-10d 24″	3-10d	5-10d	4-10d 24″	5-10d 24″
Open	B	NR	NR	NR	NR	NR	NR	2-8d	2-8d	NR	2-8d	4-8d	5-10d	2-8d	4-8d	5-8d	6-10d
	C	2-8d	4-8d	5-8d	5-10d	3-8d	5-8d	3-10d 24″	4-10d 24″	3-10d	5-10d	5-10d 24″	5-10d 16″	5-8d	4-10d 24″	5-10d 16″	6-10d 16″
	D	2-8d	5-8d	5-10d	5-10d 24″	4-8d	5-10d	4-10d 24″	5-10d 24″	5-8d	4-10d 24″	6-10d 24″	6-10d 16″	4-8d	5-10d 24″	6-10d 16″	6-10d 12″

NR — No requirements; use Table 23-II-B-1 minimum.

[1]Tie straps are at 48 inches (1219 mm) on center unless otherwise stated. See Figure A-21-7 for illustration of tie straps.

[2]Bond beam to be at least 48 inches (1219 mm) deep nominal and shall be reinforced as shown in Table A-21-E for lintels, or Tables A-21-C-1 through A-21-C-5 for walls in general where they are more restrictive.

[3]See Chapter 21, Part II, for definitions.

[4]See Section 1616 for definitions.

[5]For flat roofs connected to interior walls, the span shall be one half the larger distance on either side of the wall.

TABLE A-21-E—LINTEL REINFORCEMENT OVER EXTERIOR OPENINGS[1,2]—WOOD AND STEEL FRAMING[3]
[Lintels larger than 12 feet 0 inch (3658 mm) shall be designed.][4]
8-INCH (203 mm) MASONRY UNITS[5]

Part I—Roof Loads[5]

ANY WALL HEIGHT (feet)	SPAN TO BEARING WALLS (feet)[9]	SECOND STORY OF A TWO-STORY OR ONE-STORY BUILDINGS ROOF LIVE LOAD[6,7,8]					
		20-30 psf			40 psf		
		× 0.0479 for kN/m²					
		Width of Opening[9] (feet)					
		× 304.8 for mm					
		4	8	12	4	8	12
		Lintel depth (inches) number and size of rebar					
× 304.8 for mm		× 25.4 for mm					
Any (up to 12′)	8	8 1 No. 3	8 1 No. 3	16 1 No. 4 (B)	8 1 No. 3	8 1 No. 4	16 1 No. 4 (B)
	16	8 1 No. 3	8 1 No. 3	16 1 No. 4 (B)	8 1 No. 3	8 2 No. 4 (A)	16 2 No. 5 (B)
	24	8 1 No. 3	8 1 No. 4	16 1 No. 4 (B)	8 1 No. 3	16 1 No. 4 (B)	24 2 No. 5 (B)
	32	8 1 No. 3	16 1 No. 4 (B)	16 1 No. 5 (B)	8 1 No. 3	16 1 No. 5 (B)	24 2 No. 5 (C)

Part II—Floor and Roof Loads[5]

WALL HEIGHT	SPAN TO BEARING[9,11] WALLS (feet)	FIRST STORY OF TWO-STORY BUILDINGS FLOOR LIVE LOAD[10]					
		50 psf			100 psf		
		× 0.0479 for kN/m²					
		Width of Opening[9] (feet)					
		× 304.8 for mm					
		4	8	12	4	8	12
		Lintel depth (inches) number and size of rebar					
× 304.8 for mm		× 25.4 for mm					
Any (up to 12′)	8	8 1 No. 3	8 1 No. 4	16 1 No. 4 (B)	8 1 No. 3	8 2 No. 4 (A)	16 2 No. 5 (B)
	16	8 1 No. 3	8 2 No. 4 (A)	16 2 No. 5 (B)	8 1 No. 3	16 1 No. 4 (B)	24 2 No. 4 (C)
	24	8 1 No. 3	16 1 No. 4 (B)	24 2 No. 5 (B)	8 1 No. 3 (A)	16 1 No. 5 (B)	24 3 No. 5 (C)
	32	8 1 No. 3	16 1 No. 5 (B)	24 2 No. 5 (C)	8 1 No. 4 (B)	24 2 No. 5 (C)	Design Required

[1]The values shown are number and size of A 615, 60 grade steel reinforcement bars: Example—2 No. 4 is two $^1/_2$-inch-diameter (13 mm) deformed reinforcing bars. See also Figure A-21-8 for continuous load path.

[2]Stirrup spacing requirements: A = No. 3 at 8 inches (203 mm) on center, B = No. 3 at 4 inches (102 mm) on center, C = No. 4 at 8 inches (203 mm) on center. None are required unless specifically mentioned in the table.

[3]Design required for lintels supporting precast planks.

[4]Lintels are 8-inch (203 mm) nominal depth where supporting roof loads only and 16-inch (406 mm) nominal depth where supporting floor and roofs unless otherwise stated. All lintels are solidly grouted.

[5]Wall weight is included.

[6]The stirrup size and spacing, where required, as indicated in parenthesis below the reinforcing bar requirements.

[7]All exposure categories are included for wind uplift on the lintel. See Footnote 4 of Tables A-21-C-1 through A-21-C-5 as a minimum bond beam. Table A-21-F may also control.

[8]Two No. 5 vertical bars minimum are required on each side of the lintel for 100 and 110 miles per hour (161 and 177 km/h), Exposure D. Bar to extend 25 inches (635 mm) beyond opening or hook over top bars.

[9]For spans between the figures shown, go to next higher span width.

[10]From Table 21-A. For other floor loads go to next higher value. Where required floor load exceeds 100 pounds per square foot (4.8 kN/m²), a design is required.

[11]When interior walls support floors from each side, these values may be used if the spans on each side are less than 16 feet 0 inch (4877 mm) each. Enter the table with the total of both span widths.

TABLE A-21-F—MASONRY SHEAR WALL[1,2,3] AND DIAPHRAGM REQUIREMENTS IN HIGH-WIND AREAS[4]

Part I—Minimum wall length and horizontal bar reinforcement required for exterior shear walls and cross walls[5] (all wall heights). [Design criteria: 20 psf to 40 psf (0.96 kN/m^2 to 1.9 kN/m^2) roof load; 50 psf or 100 psf (2.4 kN/m^2 or 4.8 kN/m^2) floor load; open or enclosed buildings.]

		8-INCH (203 mm) WALLS[6]		
Wind Speed	Exposure	Distance between Shear Resisting Walls[7] "L" or "b" (feet)	One-story Building or Second Story of a Two-story Building	First Story of a Two-story Building
× 1.61 for km/h		× 304.8 for mm	inch × 25.4 for mm foot × 304.8 mm	
80 mph	B	32	NSR	9'-4"
		48	NSR	5'-4" DBL (D)
		64	10'-0"	7'-6" DBL (C)
	C	32	NSR	5'-4" DBL (C)
		48	11'-0"	8'-8" DBL (C)
		64	13'-4"	15'-0" (D)
	D	32	8'-8"	7'-0" (C)
		48	9'-4" (C)	10'-8" (D)
		64	10'-0" (D)	13'-8" (D)
90 mph	B	32	NSR	7'-8" DBL (C)
		48	NSR	8'-0" (D)
		64	12'-8"	12'-0" (D)
	C	32	NSR	14'-8"
		48	13'-8"	10'-0" (D)
		64	10'-8" (C)	15'-6" DBL (B)
	D	32	7'-8" (C)	11'-8" (D)
		48	12'-0" (C)	12'-8" DBL (B)
		64	11'-8" (D)	18'-4" DBL (C)
100 mph	B	32	NSR	5'-4" DBL (C)
		48	10'-0"	10'-0" (D)
		64	15'-4"	64'-8" DBL (C)

(Continued)

Wind Speed × 1.61 for km/h	Exposure	Distance between Shear Resisting Walls[7] "L" or "b" (feet) × 304.8 for mm	One-story Building or Second Story of a Two-story Building	First Story of a Two-story Building
			inch × 25.4 for mm foot × 304.8 mm	
100 mph (cont.)	C	32	5'-4" (D)	11'-8" (D)
		48	12'-8" (C)	12'-8" DBL (C)
		64	12'-4" (D)	19'-8" DBL (C)
	D	32	5'-4" DBL (B)	9'-4" DBL (C)
		48	9'-4" (D)	14'-8" DBL (C)
		64	17'-4" (D)	21'-0" DBL (C)
110 mph	B	32	NSR	6'-0" DBL (C)
		48	12'-0"	10'-0" DBL (C)
		64	12'-8" (C)	14'-0" (D)
	C	32	5'-4" DBL (B)	9'-8" (D)
		48	12'-0" (C)	15'-4" (D)
		64	16'-8" (C)	18'-8" DBL (C)
	D	32	8'-8" (C)	11'-4" (D)
		48	12'-4" (C)	18'-0" (D)
		64	18'-8" (C)	20'-8" DBL (C)

*NSR—No special horizontal reinforcement required for shear resistance if 5 feet 4 inches (1626 mm) long minimum.

[1]Cumulative shear wall length is to be at least as long as is shown in this table. However, see Figure A-21-9. The top figure is the minimum length. When required, the figure below it in parenthesis is the spacing of steel reinforcing wire installed as shown in Figure A-21-10, below. (A) = two 0.148 inch (3.76 mm) (No. 9 B.W. gage) at 16 inches (406 mm) on center, (B) = two $^3/_{16}$ inch (4.76 mm) at 16 inches (406 mm) on center, (C) = two 0.148 inches (3.76 mm) (No. 9 B.W. gage) at 8 inches (203 mm) on center, (D) = two $^3/_{16}$ inch (4.76 mm) at 8 inches (203 mm) on center. The symbol DBL means double these amounts. Equivalent areas of reinforcing bars spaced not over 4 feet 0 inch (1219 mm) on center may be used.

[2]All bearing and shear walls are to be in-plane with vertical reinforcement, when required, extending from one floor to the other as dictated in Tables A-21-C-1 through A-21-C-5.

[3]Minimum bond beam shall be 100 miles per hour (mph) (161 km/h), Exposure B; 90 mph (145 km/h), Exposure B, and 80 mph (129 km/h), Exposures B and C, one No. 4; 100 mph (161 km/h), Exposure C; 80 and 90 mph (129 and 145 km/h); Exposures C and D, two No. 4; all others two No. 5.

[4]Table is adjusted to include provisions for Seismic Zones 0, 1 and 2.

[5]Cross walls are to be at least twice as long as shown in the table for shear walls. The tributary width (L/2) shall be the distance used in the third column above to find minimum reinforcement and length.

[6]For walls which width is equal to or less than half its height, add an extra No. 5 vertical bar at each end.

[7]Use 32-foot (9753 mm) requirements for distances less than 32 feet (9754 mm). Also use it for bearing walls used as shear walls.

(Continued)

Part II—Wood floor and roof diaphragms and connections[8,9]
[All wall heights 8 feet to 12 feet (2438 mm to 3657 mm).]

Wind Speed	Exposure	Distance between Shear Walls[10] "L" or "b" (feet)	Minimum Wood Structural Panel/Particleboard Size[9] and Nailing[11,12]		
			Thickness (inches)	Common Nail Size (penny)	Nail Spacing (inches)
× 1.61 for km/h		× 304.8 for mm	× 25.4 for mm		× 25.4 for mm
80 mph	B	16	$5/16$	6	6 o.c.
		32	$3/8$	6	6 o.c.
		48	$3/8$	8	6 o.c.
		64	$3/8$	8	6 o.c.
	C	16	$3/8$	8	6 o.c.
		32	$1/2$ or $15/32$	8	6 o.c.
		48	$1/2$ or $15/32$	10	6 o.c.
		64	$5/8$ or $19/32$	10	6 o.c.
	D	16	$1/2$ or $15/32$	8	6 o.c.
		32	$5/8$ or $19/32$	10	6 o.c.
		48	$1/2$ or $15/32$ blocked	8	4/6 o.c.
		64	$1/2$ or $15/32$ blocked	8	4/6 o.c.
90 mph	B	16	$5/16$	6	6 o.c.
		32	$3/8$	8	6 o.c.
		48	$3/8$	8	6 o.c.
		64	$3/8$	8	6 o.c.
	C	16	$1/2$ or $15/32$	10	6 o.c.
		32	$3/8$ blocked	8	4/6 o.c.
		48	$3/8$ blocked	8	4/6 o.c.
		64	$5/8$ or $19/32$ blocked	10	6 o.c.
	D	16	$5/8$ or $19/32$	10	6 o.c.
		32	$1/2$ or $15/32$ blocked	10	4/6 o.c.
		48	$1/2$ or $15/32$ blocked	10	4/6 o.c.
		64	Design required or provide extra cross walls		
100 mph	B	16	$3/8$	8	6 o.c.
		32	$1/2$ or $15/32$	8	6 o.c.
		48	$1/2$ or $15/32$	8	6 o.c.
		64	$5/8$ or $19/32$	10	6 o.c.

(Continued)

Wind Speed × 1.61 for km/h	Exposure	Distance between Shear Walls[10] "L" or "b" (feet) × 304.8 for mm	Minimum Wood Structural Panel/Particleboard Size[9] and Nailing[11,12]		
			Thickness (inches) × 25.4 for mm	Common Nail Size (penny)	Nail Spacing (inches) × 25.4 for mm
100 mph (cont.)	C	16	$^3/_8$ blocked	8	4/6 o.c.
		32	$^5/_8$ or $^{19}/_{32}$ blocked	10	4/6 o.c.
		48	$^5/_8$ or $^{19}/_{32}$ blocked	10	4/6 o.c.
		64	Design required or provide extra cross walls		
	D	16	$^1/_2$ or $^{15}/_{32}$ blocked	10	4/6 o.c.
		32	$^5/_8$ or $^{19}/_{32}$ blocked	10	4/6 o.c.
		48	Design required or provide extra cross walls		
		64	Design required or provide extra cross walls		
110 mph	B	16	$^1/_2$ or $^{15}/_{32}$	8	6 o.c.
		32	$^1/_2$ or $^{15}/_{32}$	10	6 o.c.
		48	$^5/_8$ or $^{19}/_{32}$	10	6 o.c.
		64	$^1/_2$ or $^{15}/_{32}$ blocked	8	4/6 o.c.
	C	16	$^1/_2$ or $^{15}/_{32}$ blocked	8	4/6 o.c.
		32	$^5/_8$ or $^{19}/_{32}$ blocked	10	4/6 o.c.
		48	Design required or provide extra cross walls		
		64	Design required or provide extra cross walls		
	D	16	$^5/_8$ or $^{19}/_{32}$ blocked	10	4/6 o.c.
		32	Design required or provide extra cross walls		
		48	Design required or provide extra cross walls		
		64	Design required or provide extra cross walls		

[8]These requirements represent the maximum values for a diaphragm which is within a maximum 32-foot-by-64-foot (9.75 m by 19.5 m) module surrounded by shear walls, cross walls or bearing walls. (See Figure A-21-9.)

[9]See Tables 23-II-E-1 and 23-II-E-2 for minimum sizes depending on span between joists.

[10]See Figure A-21-9 for "L" and "b."

[11]The wood structural panel/particleboard (all grades) thickness is given first. The nailing size and boundary/supported edge spacing is shown next. Blocking of unsupported edges is stated where required. Twelve-inch (305 mm) spacing required in the field of the roof/floor. Boundary nailing is required over interior walls [see Figure A-21-12 (b)].

[12]Use Case 1 for unblocked diaphragms and any case for blocked diaphragms.

TABLE A-21-G—MINIMUM WALL CONNECTION REQUIREMENTS IN HIGH-WIND AREAS
Precast Hollow-core Plank Floors and Roofs

Spacing of No. 4 bent reinforcing bar in block or brick walls connected to precast concrete planks[1,2]

WIND SPEED AND EXPOSURE	EXTERIOR WALLS	INTERIOR WALLS
	× 25.4 for mm	
90 mph (145 km/h) Exposure C and less 100 mph (161 km/h) Exposure B	32″ o.c.	16″ o.c.
90 mph (145 km/h) Exposure D 100 mph (161 km/h) Exposure C 110 mph (177 km/h) Exposure B	24″ o.c.	12″ o.c.
100 mph (161 km/h) Exposure D 110 mph (177 km/h) Exposures C and D	16″ o.c.	12″ o.c.

[1]This table assumes maximum wall height of 12 feet (3.7 m) and a width-to-length ratio of diaphragm between shear walls of 3:1 or less.
[2]The precast planks shall be designed as shall the walls and footings supporting them.

TABLE A-21-H—MINIMUM HOLD-DOWN REQUIREMENTS IN HIGH-WIND AREAS
Steel Floors and Roofs

WIND SPEED AND EXPOSURE	MAXIMUM SPACING OF ROOF JOISTS WITH CONNECTION SHOWN[1,2,3]
	× 25.4 for mm
100 mph (161 km/h) Exposure B 90 mph (145 km/h) Exposures B and C 80 mph (129 km/h) Exposures B, C and D	48″
110 mph (177 km/h) Exposure B 100 mph (161 km/h) Exposure C	30″
110 mph (177 km/h) Exposures C and D 100 mph (161 km/h) Exposure D	Design required

[1]Maximum span is 32 feet (9.75 m) to bearing walls.
[2]Joists and decking to be designed.
[3]Bottom chord of joists to be braced for reversal of stresses caused by wind uplift.

TABLE A-21-I—DIAGONAL BRACING REQUIREMENTS
FOR GABLE-END WALL[1,2] ROOF PITCH 3:12 to 5:12

EXPOSURE	BASIC WIND SPEED (mph)							
	× 1.61 for km/h							
	80		90		100		110	
	3:12 (25%)	4:12 (33%) and 5:12 (42%)	3:12 (25%)	4:12 (33%) and 5:12 (42%)	3:12 (25%)	4:12 (33%) and 5:12 (42%)	3:12 (25%)	4:12 (33%) and 5:12 (42%)
	× 25.4 for mm							
B	I at 48″ o.c.	III at 48″ o.c.	I at 48″ o.c.	III at 48″ o.c.	I at 24″ o.c.	III at 24″ o.c.	I at 24″ o.c.	III at 24″ o.c.
C	I at 24″ o.c.	III at 48″ o.c.	I at 24″ o.c.	III at 24″ o.c.	II at 24″ o.c.	IV at 24″ o.c.	II at 24″ o.c.	IV at 24″ o.c.
D	I at 24″ o.c.	III at 48″ o.c.	II at 24″ o.c.	IV at 24″ o.c.	II at 24″ o.c.	IV at 24″ o.c.	Two-II at 24″ o.c.	Two-III at 24″ o.c.

[1] I = 2-inch-by-4-inch brace, one clip angle (51 mm × 102 mm).

   II = 2-inch-by-4-inch brace, two clip angles (one each side) (51 mm × 102 mm).

  III = 3-inch-by-4-inch brace, one clip angle (76 mm × 102 mm).

  IV = 3-inch-by-4-inch brace, two clip angles (one each side) (76 mm × 102 mm).

The spacing requirements of the brace are shown below the symbol.

[2]See Figures A-21-17 and A-21-18 for details and size of clip angles.

NOTE: Horizontal and vertical reinforcement to be determined by Tables A-21-C-1 through A-21-C-5 and A-21-F.

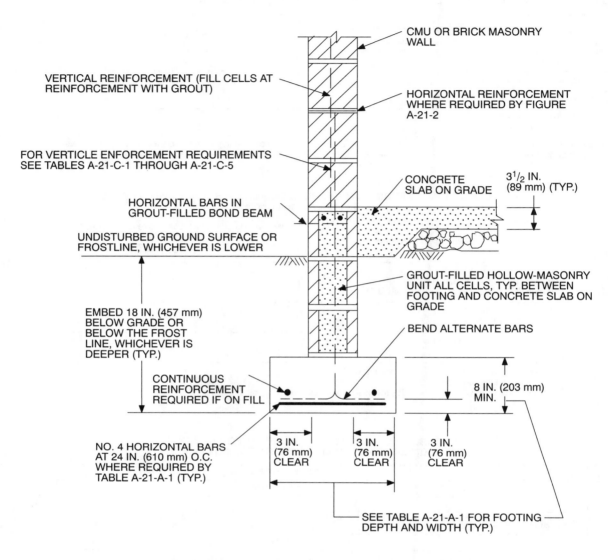

VERTICAL REINFORCEMENT (FILL CELLS AT REINFORCEMENT WITH GROUT)

FOR VERTICLE ENFORCEMENT REQUIREMENTS SEE TABLES A-21-C-1 THROUGH A-21-C-5

HORIZONTAL BARS IN GROUT-FILLED BOND BEAM

UNDISTURBED GROUND SURFACE OR FROSTLINE, WHICHEVER IS LOWER

EMBED 18 IN. (457 mm) BELOW GRADE OR BELOW THE FROST LINE, WHICHEVER IS DEEPER (TYP.)

CONTINUOUS REINFORCEMENT REQUIRED IF ON FILL

NO. 4 HORIZONTAL BARS AT 24 IN. (610 mm) O.C. WHERE REQUIRED BY TABLE A-21-A-1 (TYP.)

CMU OR BRICK MASONRY WALL

HORIZONTAL REINFORCEMENT WHERE REQUIRED BY FIGURE A-21-2

CONCRETE SLAB ON GRADE

3¹/₂ IN. (89 mm) (TYP.)

GROUT-FILLED HOLLOW-MASONRY UNIT ALL CELLS, TYP. BETWEEN FOOTING AND CONCRETE SLAB ON GRADE

BEND ALTERNATE BARS

8 IN. (203 mm) MIN.

3 IN. (76 mm) CLEAR

3 IN. (76 mm) CLEAR

3 IN. (76 mm) CLEAR

SEE TABLE A-21-A-1 FOR FOOTING DEPTH AND WIDTH (TYP.)

HOLLOW-MASONRY UNIT EXTERIOR FOUNDATION WALL

FIGURE A-21-1—VARIOUS DETAILS OF FOOTINGS
(See Tables A-21-A-1 and A-21-A-2 for widths.)

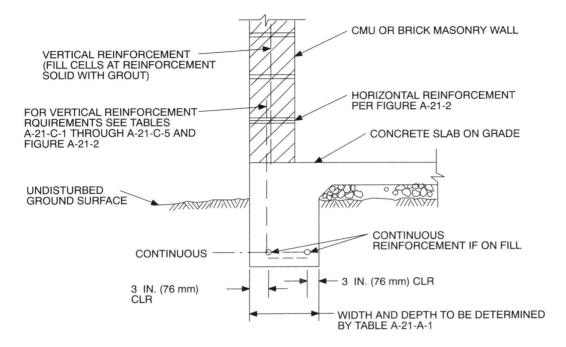

VERTICAL REINFORCEMENT (FILL CELLS AT REINFORCEMENT SOLID WITH GROUT)

CMU OR BRICK MASONRY WALL

FOR VERTICAL REINFORCEMENT RQUIREMENTS SEE TABLES A-21-C-1 THROUGH A-21-C-5 AND FIGURE A-21-2

HORIZONTAL REINFORCEMENT PER FIGURE A-21-2

CONCRETE SLAB ON GRADE

UNDISTURBED GROUND SURFACE

CONTINUOUS REINFORCEMENT IF ON FILL

CONTINUOUS

3 IN. (76 mm) CLR

3 IN. (76 mm) CLR

WIDTH AND DEPTH TO BE DETERMINED BY TABLE A-21-A-1

GRADE BEAM OR CONTINUOUS CONCRETE SLAB—TURN DOWN

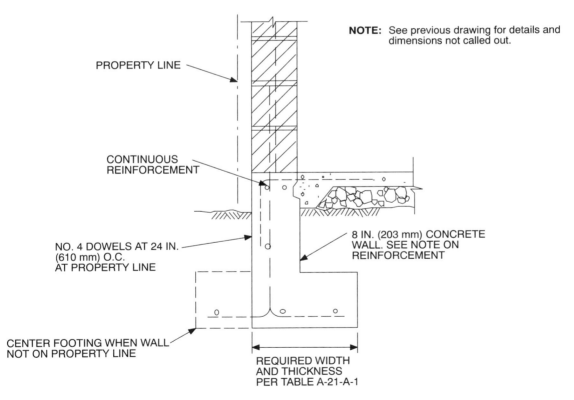

NOTE: See previous drawing for details and dimensions not called out.

PROPERTY LINE

CONTINUOUS REINFORCEMENT

NO. 4 DOWELS AT 24 IN. (610 mm) O.C. AT PROPERTY LINE

8 IN. (203 mm) CONCRETE WALL. SEE NOTE ON REINFORCEMENT

CENTER FOOTING WHEN WALL NOT ON PROPERTY LINE

REQUIRED WIDTH AND THICKNESS PER TABLE A-21-A-1

HOLLOW-MASONRY UNIT CONCRETE EXTERIOR FOUNDATION WALL

FIGURE A-21-1—VARIOUS DETAILS OF FOOTINGS—(Continued)
(See Tables A-21-A-1 and A-21-A-2 for widths.)

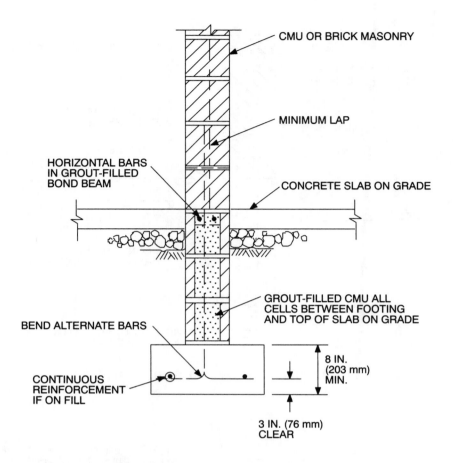

HOLLOW–MASONRY UNIT INTERIOR FOUNDATION WALL

NOTE: See previous drawing for details and dimensions not called out.

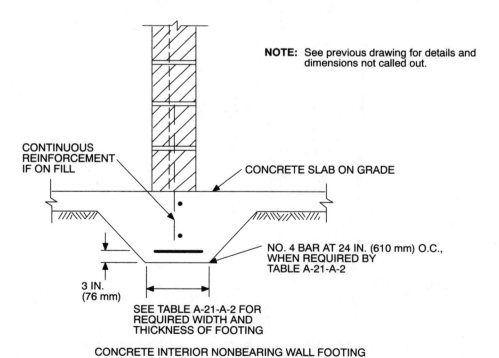

CONCRETE INTERIOR NONBEARING WALL FOOTING

FIGURE A-21-1—VARIOUS DETAILS OF FOOTINGS—(Continued)
(See Tables A-21-A-1 and A-21-A-2 for widths.)

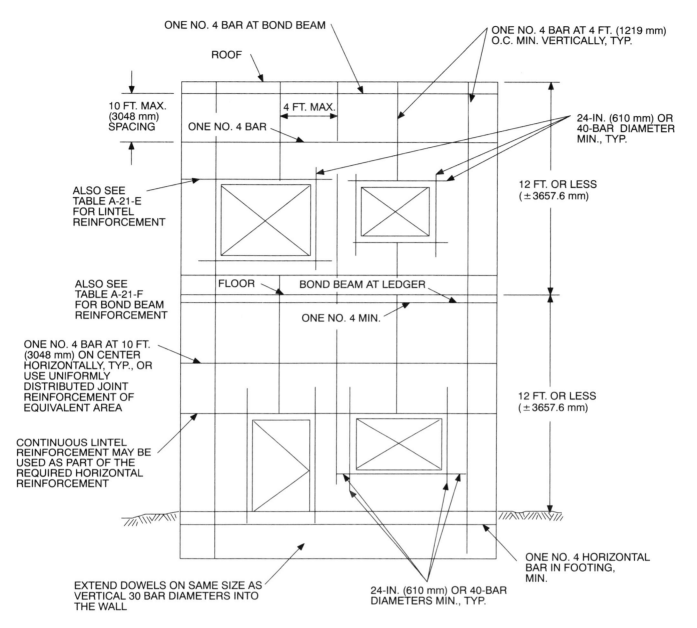

ONE NO. 4 BAR AT BOND BEAM

ROOF

ONE NO. 4 BAR AT 4 FT. (1219 mm) O.C. MIN. VERTICALLY, TYP.

10 FT. MAX. (3048 mm) SPACING

4 FT. MAX.

ONE NO. 4 BAR

24-IN. (610 mm) OR 40-BAR DIAMETER MIN., TYP.

ALSO SEE TABLE A-21-E FOR LINTEL REINFORCEMENT

12 FT. OR LESS (±3657.6 mm)

ALSO SEE TABLE A-21-F FOR BOND BEAM REINFORCEMENT

FLOOR   BOND BEAM AT LEDGER

ONE NO. 4 MIN.

ONE NO. 4 BAR AT 10 FT. (3048 mm) ON CENTER HORIZONTALLY, TYP., OR USE UNIFORMLY DISTRIBUTED JOINT REINFORCEMENT OF EQUIVALENT AREA

12 FT. OR LESS (±3657.6 mm)

CONTINUOUS LINTEL REINFORCEMENT MAY BE USED AS PART OF THE REQUIRED HORIZONTAL REINFORCEMENT

ONE NO. 4 HORIZONTAL BAR IN FOOTING, MIN.

EXTEND DOWELS ON SAME SIZE AS VERTICAL 30 BAR DIAMETERS INTO THE WALL

24-IN. (610 mm) OR 40-BAR DIAMETERS MIN., TYP.

FIGURE A-21-2—MINIMUM MASONRY WALL REQUIREMENTS IN SEISMIC ZONE 2

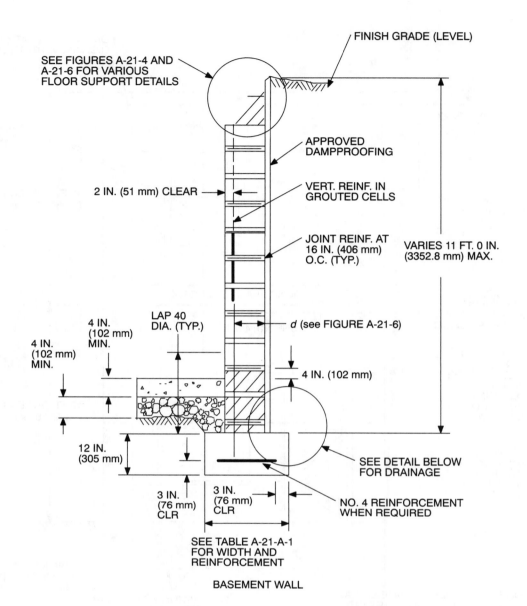

SEE FIGURES A-21-4 AND A-21-6 FOR VARIOUS FLOOR SUPPORT DETAILS

FINISH GRADE (LEVEL)

APPROVED DAMPPROOFING

2 IN. (51 mm) CLEAR

VERT. REINF. IN GROUTED CELLS

JOINT REINF. AT 16 IN. (406 mm) O.C. (TYP.)

VARIES 11 FT. 0 IN. (3352.8 mm) MAX.

4 IN. (102 mm) MIN.

4 IN. (102 mm) MIN.

LAP 40 DIA. (TYP.)

d (see FIGURE A-21-6)

4 IN. (102 mm)

12 IN. (305 mm)

3 IN. (76 mm) CLR

3 IN. (76 mm) CLR

SEE DETAIL BELOW FOR DRAINAGE

NO. 4 REINFORCEMENT WHEN REQUIRED

SEE TABLE A-21-A-1 FOR WIDTH AND REINFORCEMENT

BASEMENT WALL

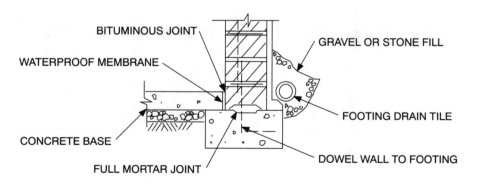

BITUMINOUS JOINT

GRAVEL OR STONE FILL

WATERPROOF MEMBRANE

FOOTING DRAIN TILE

CONCRETE BASE

FULL MORTAR JOINT

DOWEL WALL TO FOOTING

FIGURE A-21-3—BELOW-GRADE WALL AND DRAINAGE DETAILS

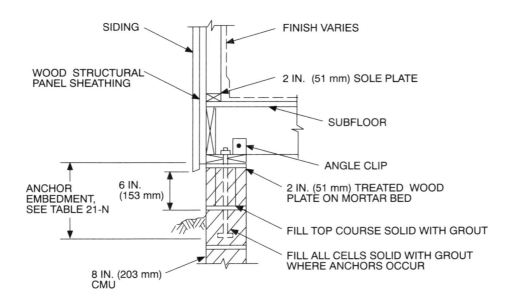

SIDING

FINISH VARIES

WOOD STRUCTURAL PANEL SHEATHING

2 IN. (51 mm) SOLE PLATE

SUBFLOOR

ANGLE CLIP

2 IN. (51 mm) TREATED WOOD PLATE ON MORTAR BED

ANCHOR EMBEDMENT, SEE TABLE 21-N

6 IN. (153 mm)

FILL TOP COURSE SOLID WITH GROUT

FILL ALL CELLS SOLID WITH GROUT WHERE ANCHORS OCCUR

8 IN. (203 mm) CMU

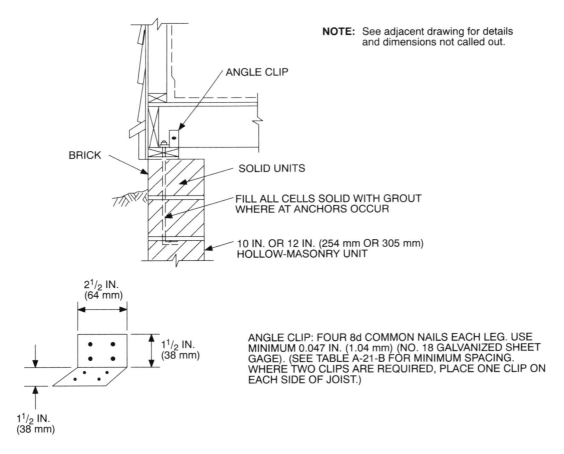

NOTE: See adjacent drawing for details and dimensions not called out.

ANGLE CLIP

BRICK

SOLID UNITS

FILL ALL CELLS SOLID WITH GROUT WHERE AT ANCHORS OCCUR

10 IN. OR 12 IN. (254 mm OR 305 mm) HOLLOW-MASONRY UNIT

$2^1/_2$ IN. (64 mm)

$1^1/_2$ IN. (38 mm)

$1^1/_2$ IN. (38 mm)

ANGLE CLIP: FOUR 8d COMMON NAILS EACH LEG. USE MINIMUM 0.047 IN. (1.04 mm) (NO. 18 GALVANIZED SHEET GAGE). (SEE TABLE A-21-B FOR MINIMUM SPACING. WHERE TWO CLIPS ARE REQUIRED, PLACE ONE CLIP ON EACH SIDE OF JOIST.)

FIGURE A-21-4—HOLLOW-MASONRY UNIT FOUNDATION WALL—WOOD FLOOR

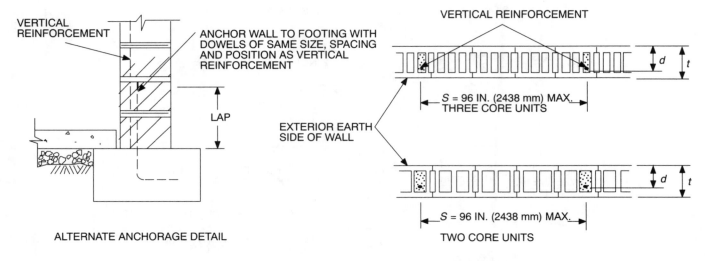

8 IN. (203 mm) WALLS: $t = 7^5/_8$ IN. (194 mm) $d = 5$ IN. (127 mm)
10 IN. (254 mm) WALLS: $t = 9^5/_8$ IN. (245 mm) $d = 7$ IN. (153 mm)
12 IN. (305 mm) WALLS: $t = 11^5/_8$ IN. (295 mm) $d = 8^3/_4$ IN. (225 mm)

FIGURE A-21-5—PLACEMENT OF REINFORCEMENT

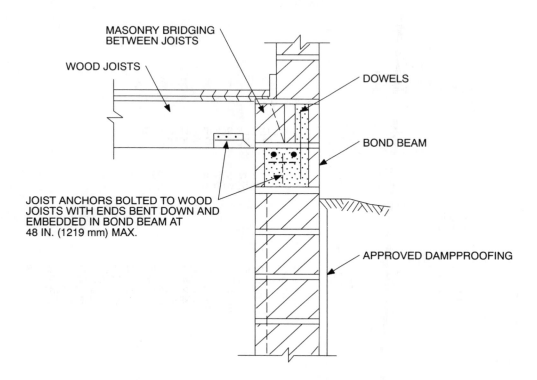

(A) HOLLOW—MASONRY UNIT WALL—WOOD FLOOR

FIGURE A-21-6—VARIOUS CONNECTIONS OF FLOORS TO BASEMENT WALLS

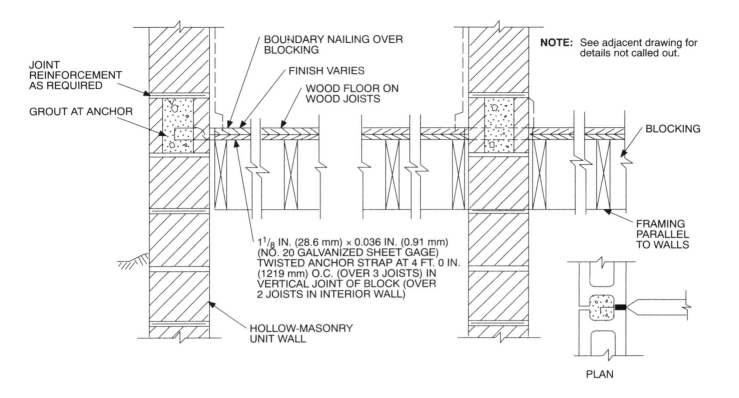

JOINT
REINFORCEMENT
AS REQUIRED

GROUT AT ANCHOR

BOUNDARY NAILING OVER
BLOCKING

FINISH VARIES

WOOD FLOOR ON
WOOD JOISTS

NOTE: See adjacent drawing for
details not called out.

BLOCKING

FRAMING
PARALLEL
TO WALLS

$1^{1}/_{8}$ IN. (28.6 mm) × 0.036 IN. (0.91 mm)
(NO. 20 GALVANIZED SHEET GAGE)
TWISTED ANCHOR STRAP AT 4 FT. 0 IN.
(1219 mm) O.C. (OVER 3 JOISTS) IN
VERTICAL JOINT OF BLOCK (OVER
2 JOISTS IN INTERIOR WALL)

HOLLOW-MASONRY
UNIT WALL

PLAN

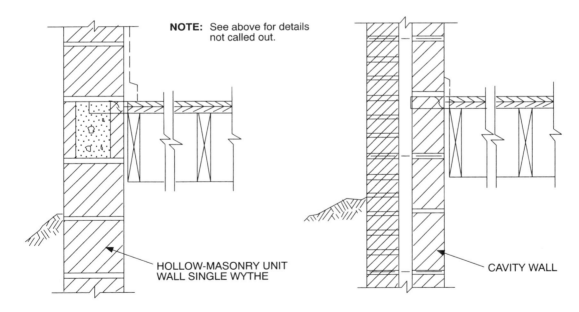

NOTE: See above for details
not called out.

HOLLOW-MASONRY UNIT
WALL SINGLE WYTHE

CAVITY WALL

(B) WOOD FLOOR, JOISTS PARALLEL TO WALL

FIGURE A-21-6—VARIOUS CONNECTIONS OF FLOORS TO BASEMENT WALLS—(Continued)

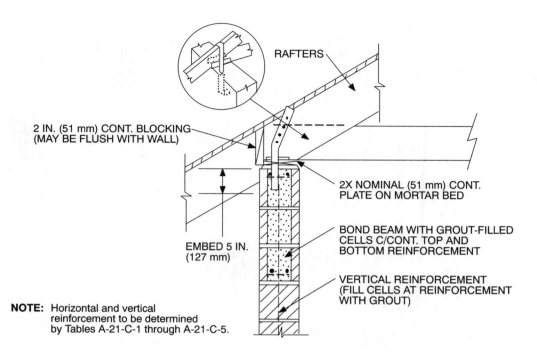

2 IN. (51 mm) CONT. BLOCKING
(MAY BE FLUSH WITH WALL)

RAFTERS

2X NOMINAL (51 mm) CONT.
PLATE ON MORTAR BED

EMBED 5 IN.
(127 mm)

BOND BEAM WITH GROUT-FILLED
CELLS C/CONT. TOP AND
BOTTOM REINFORCEMENT

VERTICAL REINFORCEMENT
(FILL CELLS AT REINFORCEMENT
WITH GROUT)

NOTE: Horizontal and vertical
reinforcement to be determined
by Tables A-21-C-1 through A-21-C-5.

ROOF WITH OVERHANG

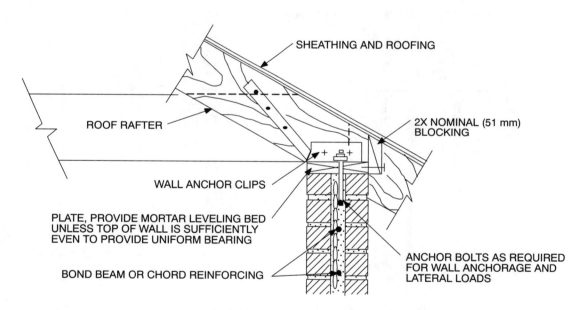

SHEATHING AND ROOFING

2X NOMINAL (51 mm)
BLOCKING

ROOF RAFTER

WALL ANCHOR CLIPS

PLATE, PROVIDE MORTAR LEVELING BED
UNLESS TOP OF WALL IS SUFFICIENTLY
EVEN TO PROVIDE UNIFORM BEARING

ANCHOR BOLTS AS REQUIRED
FOR WALL ANCHORAGE AND
LATERAL LOADS

BOND BEAM OR CHORD REINFORCING

**FIGURE A-21-7—VARIOUS DETAILS ASSOCIATED
WITH TABLE A-21-D (Uplift Resistance)**

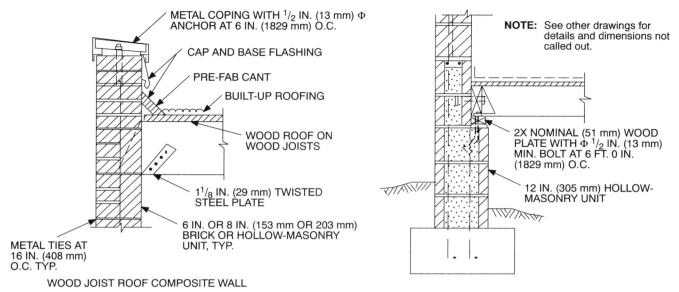

METAL COPING WITH $^1/_2$ IN. (13 mm) Φ ANCHOR AT 6 IN. (1829 mm) O.C.

CAP AND BASE FLASHING

PRE-FAB CANT

BUILT-UP ROOFING

WOOD ROOF ON WOOD JOISTS

$1^1/_8$ IN. (29 mm) TWISTED STEEL PLATE

6 IN. OR 8 IN. (153 mm OR 203 mm) BRICK OR HOLLOW-MASONRY UNIT, TYP.

METAL TIES AT 16 IN. (408 mm) O.C. TYP.

WOOD JOIST ROOF COMPOSITE WALL

NOTE: See other drawings for details and dimensions not called out.

2X NOMINAL (51 mm) WOOD PLATE WITH Φ $^1/_2$ IN. (13 mm) MIN. BOLT AT 6 FT. 0 IN. (1829 mm) O.C.

12 IN. (305 mm) HOLLOW-MASONRY UNIT

HOLLOW-MASONRY UNIT FOUNDATION WALL—JOIST PERPENDICULAR

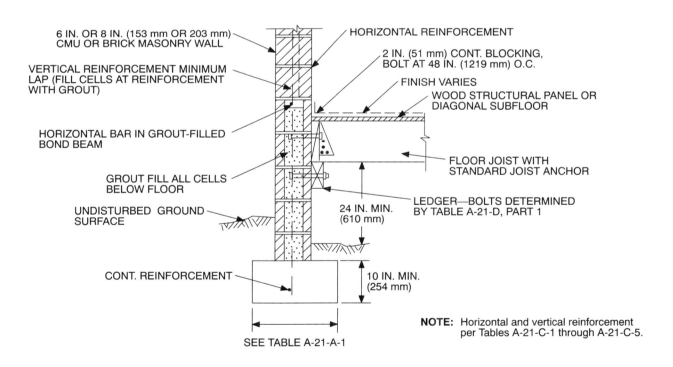

6 IN. OR 8 IN. (153 mm OR 203 mm) CMU OR BRICK MASONRY WALL

VERTICAL REINFORCEMENT MINIMUM LAP (FILL CELLS AT REINFORCEMENT WITH GROUT)

HORIZONTAL BAR IN GROUT-FILLED BOND BEAM

GROUT FILL ALL CELLS BELOW FLOOR

UNDISTURBED GROUND SURFACE

CONT. REINFORCEMENT

HORIZONTAL REINFORCEMENT

2 IN. (51 mm) CONT. BLOCKING, BOLT AT 48 IN. (1219 mm) O.C.

FINISH VARIES

WOOD STRUCTURAL PANEL OR DIAGONAL SUBFLOOR

FLOOR JOIST WITH STANDARD JOIST ANCHOR

LEDGER—BOLTS DETERMINED BY TABLE A-21-D, PART 1

24 IN. MIN. (610 mm)

10 IN. MIN. (254 mm)

SEE TABLE A-21-A-1

NOTE: Horizontal and vertical reinforcement per Tables A-21-C-1 through A-21-C-5.

HOLLOW-MASONRY UNIT FOUNDATION WALL—JOIST PERPENDICULAR—(Continued)

FIGURE A-21-7—VARIOUS DETAILS ASSOCIATED WITH TABLE A-21-D (Uplift Resistance)—(Continued)

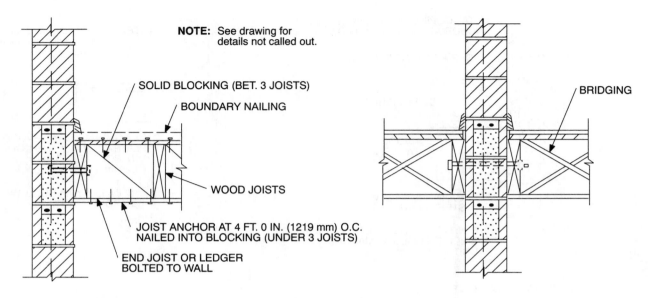

NOTE: See drawing for details not called out.

SOLID BLOCKING (BET. 3 JOISTS)

BOUNDARY NAILING

WOOD JOISTS

JOIST ANCHOR AT 4 FT. 0 IN. (1219 mm) O.C. NAILED INTO BLOCKING (UNDER 3 JOISTS)

END JOIST OR LEDGER BOLTED TO WALL

EXTERIOR WALL—JOIST PARALLEL

BRIDGING

INTERIOR WALL—JOIST PARALLEL

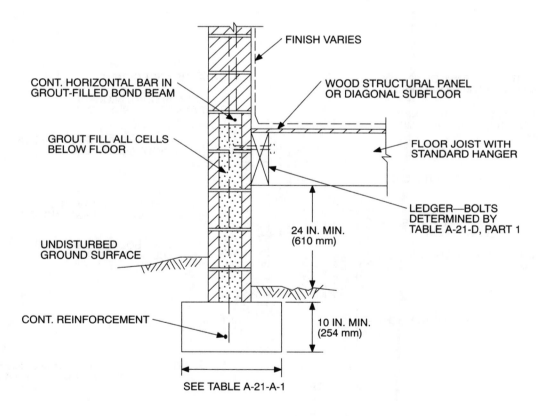

FINISH VARIES

CONT. HORIZONTAL BAR IN GROUT-FILLED BOND BEAM

WOOD STRUCTURAL PANEL OR DIAGONAL SUBFLOOR

GROUT FILL ALL CELLS BELOW FLOOR

FLOOR JOIST WITH STANDARD HANGER

LEDGER—BOLTS DETERMINED BY TABLE A-21-D, PART 1

UNDISTURBED GROUND SURFACE

24 IN. MIN. (610 mm)

CONT. REINFORCEMENT

10 IN. MIN. (254 mm)

SEE TABLE A-21-A-1

HOLLOW-MASONRY UNIT FOUNDATION WALL—JOIST PERPENDICULAR

FIGURE A-21-7—VARIOUS DETAILS ASSOCIATED WITH TABLE A-21-D (Uplift Resistance)—(Continued)

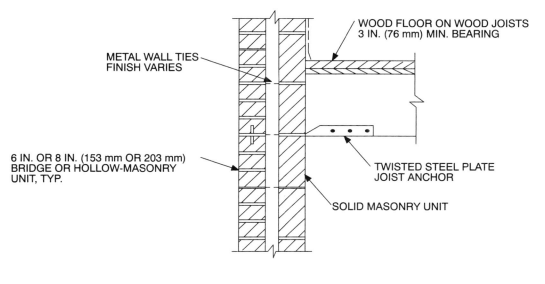

METAL WALL TIES
FINISH VARIES

WOOD FLOOR ON WOOD JOISTS
3 IN. (76 mm) MIN. BEARING

6 IN. OR 8 IN. (153 mm OR 203 mm)
BRIDGE OR HOLLOW-MASONRY
UNIT, TYP.

TWISTED STEEL PLATE
JOIST ANCHOR

SOLID MASONRY UNIT

WOOD FLOOR

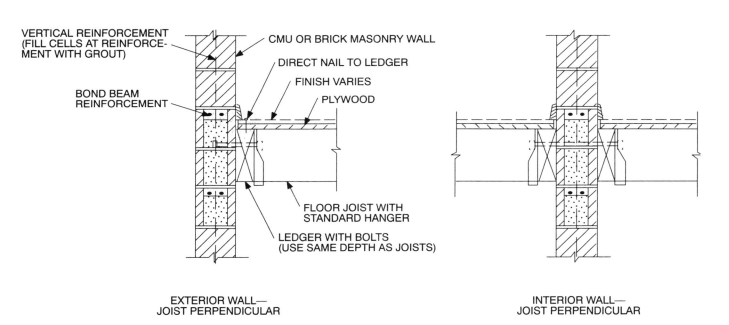

VERTICAL REINFORCEMENT
(FILL CELLS AT REINFORCE-
MENT WITH GROUT)

CMU OR BRICK MASONRY WALL

DIRECT NAIL TO LEDGER

FINISH VARIES

PLYWOOD

BOND BEAM
REINFORCEMENT

FLOOR JOIST WITH
STANDARD HANGER

LEDGER WITH BOLTS
(USE SAME DEPTH AS JOISTS)

EXTERIOR WALL—
JOIST PERPENDICULAR

INTERIOR WALL—
JOIST PERPENDICULAR

FIGURE A-21-7—VARIOUS DETAILS ASSOCIATED WITH TABLE A-21-D (Uplift Resistance)—(Continued)

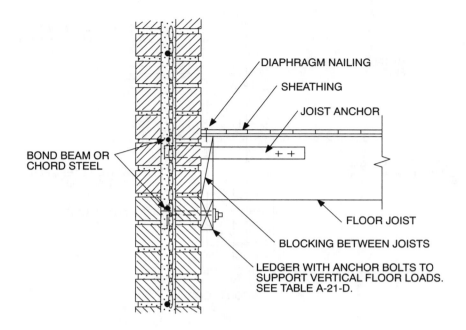

BOND BEAM OR
CHORD STEEL

DIAPHRAGM NAILING

SHEATHING

JOIST ANCHOR

+ +

FLOOR JOIST

BLOCKING BETWEEN JOISTS

LEDGER WITH ANCHOR BOLTS TO
SUPPORT VERTICAL FLOOR LOADS.
SEE TABLE A-21-D.

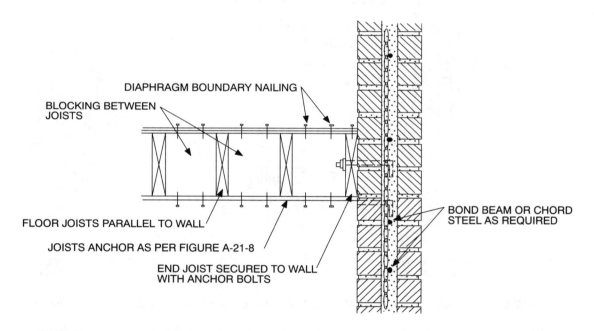

DIAPHRAGM BOUNDARY NAILING

BLOCKING BETWEEN
JOISTS

FLOOR JOISTS PARALLEL TO WALL

JOISTS ANCHOR AS PER FIGURE A-21-8

END JOIST SECURED TO WALL
WITH ANCHOR BOLTS

BOND BEAM OR CHORD
STEEL AS REQUIRED

FIGURE A-21-7—VARIOUS DETAILS ASSOCIATED WITH TABLE A-21-D (Uplift Resistance)—(Continued)

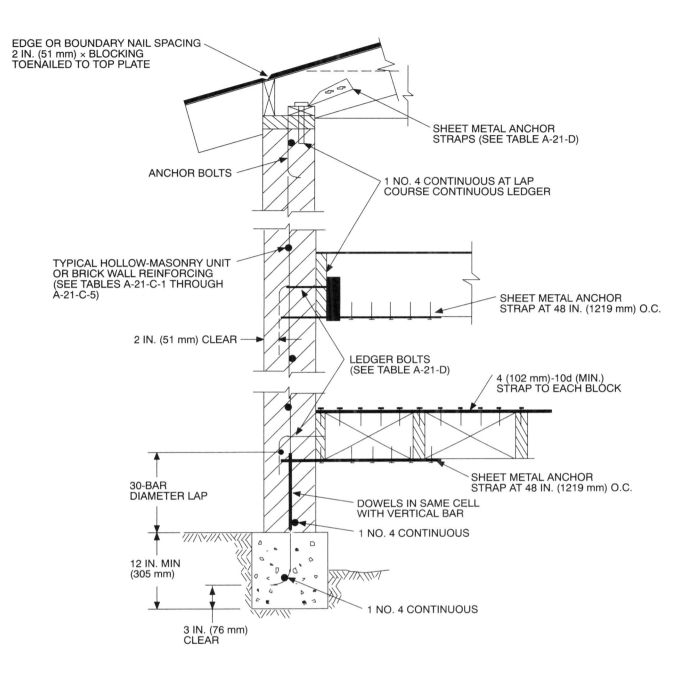

EDGE OR BOUNDARY NAIL SPACING
2 IN. (51 mm) × BLOCKING
TOENAILED TO TOP PLATE

SHEET METAL ANCHOR
STRAPS (SEE TABLE A-21-D)

ANCHOR BOLTS

1 NO. 4 CONTINUOUS AT LAP
COURSE CONTINUOUS LEDGER

TYPICAL HOLLOW-MASONRY UNIT
OR BRICK WALL REINFORCING
(SEE TABLES A-21-C-1 THROUGH
A-21-C-5)

SHEET METAL ANCHOR
STRAP AT 48 IN. (1219 mm) O.C.

2 IN. (51 mm) CLEAR

LEDGER BOLTS
(SEE TABLE A-21-D)

4 (102 mm)-10d (MIN.)
STRAP TO EACH BLOCK

30-BAR
DIAMETER LAP

SHEET METAL ANCHOR
STRAP AT 48 IN. (1219 mm) O.C.

DOWELS IN SAME CELL
WITH VERTICAL BAR

1 NO. 4 CONTINUOUS

12 IN. MIN
(305 mm)

1 NO. 4 CONTINUOUS

3 IN. (76 mm)
CLEAR

FIGURE A-21-8—CONTINUOUS LOAD PATH

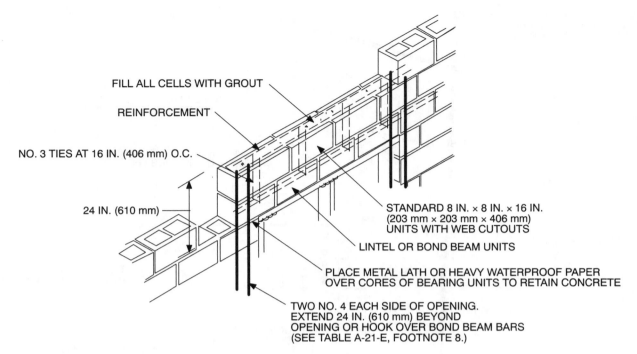

FILL ALL CELLS WITH GROUT

REINFORCEMENT

NO. 3 TIES AT 16 IN. (406 mm) O.C.

24 IN. (610 mm)

STANDARD 8 IN. × 8 IN. × 16 IN.
(203 mm × 203 mm × 406 mm)
UNITS WITH WEB CUTOUTS

LINTEL OR BOND BEAM UNITS

PLACE METAL LATH OR HEAVY WATERPROOF PAPER
OVER CORES OF BEARING UNITS TO RETAIN CONCRETE

TWO NO. 4 EACH SIDE OF OPENING.
EXTEND 24 IN. (610 mm) BEYOND
OPENING OR HOOK OVER BOND BEAM BARS
(SEE TABLE A-21-E, FOOTNOTE 8.)

REINFORCING DETAILS

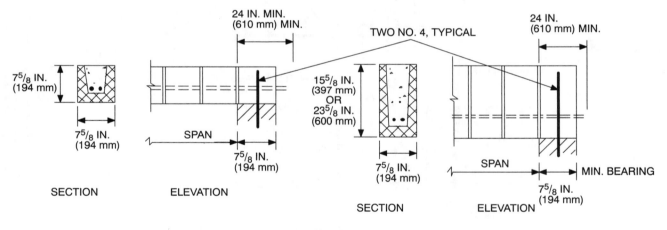

7⅝ IN.
(194 mm)

7⅝ IN.
(194 mm)

24 IN. MIN.
(610 mm) MIN.

TWO NO. 4, TYPICAL

24 IN.
(610 mm) MIN.

15⅝ IN.
(397 mm)
OR
23⅝ IN.
(600 mm)

SPAN

7⅝ IN.
(194 mm)

7⅝ IN.
(194 mm)

SPAN

MIN. BEARING

7⅝ IN.
(194 mm)

SECTION   ELEVATION

SECTION   ELEVATION

WITHOUT STIRRUPS

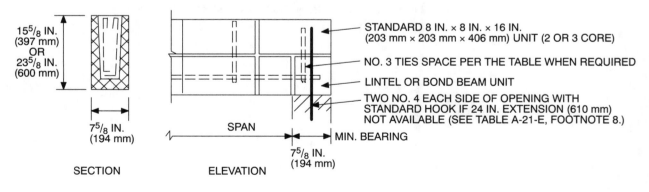

15⅝ IN.
(397 mm)
OR
23⅝ IN.
(600 mm)

STANDARD 8 IN. × 8 IN. × 16 IN.
(203 mm × 203 mm × 406 mm) UNIT (2 OR 3 CORE)

NO. 3 TIES SPACE PER THE TABLE WHEN REQUIRED

LINTEL OR BOND BEAM UNIT

TWO NO. 4 EACH SIDE OF OPENING WITH
STANDARD HOOK IF 24 IN. EXTENSION (610 mm)
NOT AVAILABLE (SEE TABLE A-21-E, FOOTNOTE 8.)

7⅝ IN.
(194 mm)

SPAN   MIN. BEARING

7⅝ IN.
(194 mm)

SECTION   ELEVATION

WITH STIRRUPS

FIGURE A-21-8—CONTINUOUS LOAD PATH—(Continued)

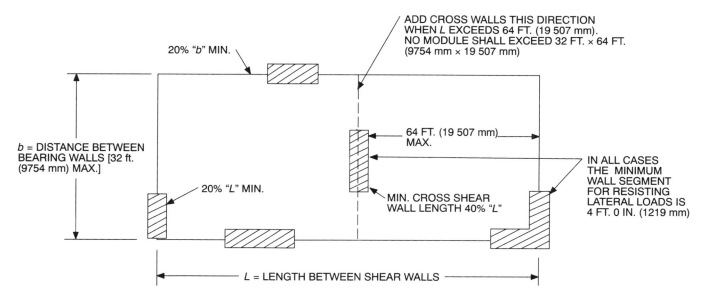

FIGURE A-21-9—SPACING AND LENGTHS OF SHEAR WALLS

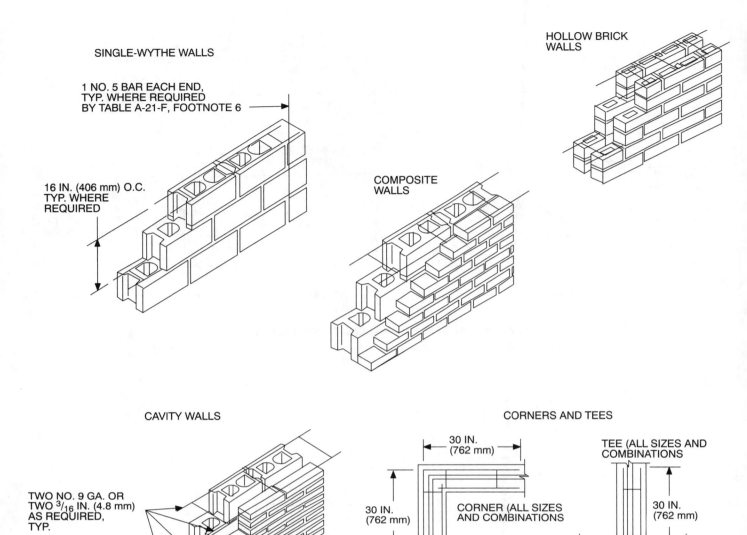

FIGURE A-21-10—SPACING OF STEEL REINFORCING WIRE

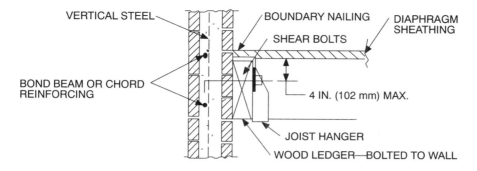

(a) FLOOR JOISTS PERPENDICULAR TO WALL JOIST HANGER SUPPORTS

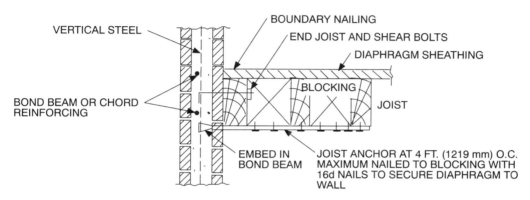

(b) FLOOR JOISTS PARALLEL TO WALL

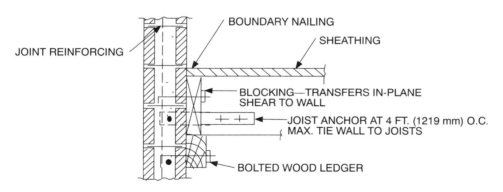

(c) WOOD LEDGER FLOOR JOIST SUPPORT

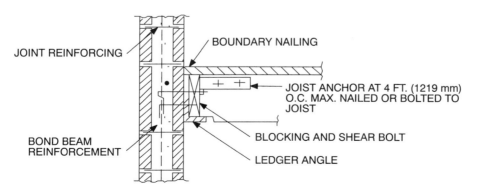

(d) STEEL LEDGER FLOOR JOIST SUPPORT

FIGURE A-21-11—FLOOR-TO-WALL CONNECTION DETAILS

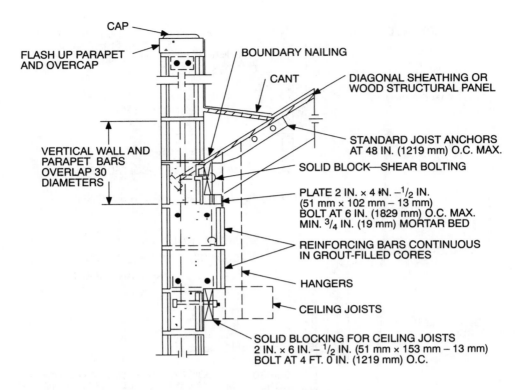

CAP

FLASH UP PARAPET
AND OVERCAP

BOUNDARY NAILING

CANT

DIAGONAL SHEATHING OR
WOOD STRUCTURAL PANEL

VERTICAL WALL AND
PARAPET BARS
OVERLAP 30
DIAMETERS

STANDARD JOIST ANCHORS
AT 48 IN. (1219 mm) O.C. MAX.

SOLID BLOCK—SHEAR BOLTING

PLATE 2 IN. × 4 IN. $-^1/_2$ IN.
(51 mm × 102 mm – 13 mm)
BOLT AT 6 IN. (1829 mm) O.C. MAX.
MIN. $^3/_4$ IN. (19 mm) MORTAR BED

REINFORCING BARS CONTINUOUS
IN GROUT-FILLED CORES

HANGERS

CEILING JOISTS

SOLID BLOCKING FOR CEILING JOISTS
2 IN. × 6 IN. $-^1/_2$ IN. (51 mm × 153 mm – 13 mm)
BOLT AT 4 FT. 0 IN. (1219 mm) O.C.

(a) EXTERIOR WALL SUPPORT

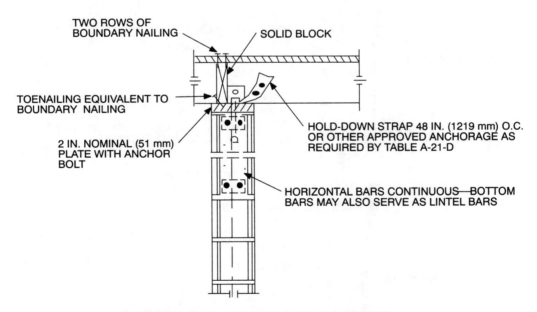

TWO ROWS OF
BOUNDARY NAILING

SOLID BLOCK

TOENAILING EQUIVALENT TO
BOUNDARY NAILING

2 IN. NOMINAL (51 mm)
PLATE WITH ANCHOR
BOLT

HOLD-DOWN STRAP 48 IN. (1219 mm) O.C.
OR OTHER APPROVED ANCHORAGE AS
REQUIRED BY TABLE A-21-D

HORIZONTAL BARS CONTINUOUS—BOTTOM
BARS MAY ALSO SERVE AS LINTEL BARS

(b) INTERIOR WALL SUPPORT BOND–BEAM SUPPORTS

FIGURE A-21-12—ROOF-TO-WALL CONNECTION DETAILS

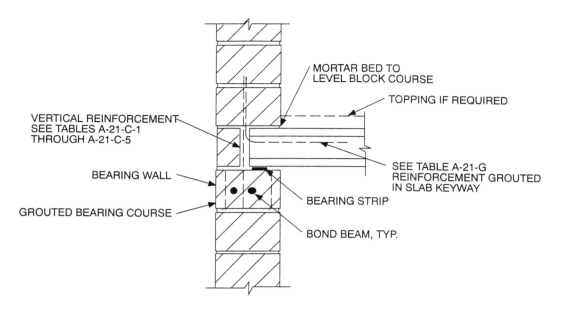

(a) SLAB PERPENDICULAR TO WALL

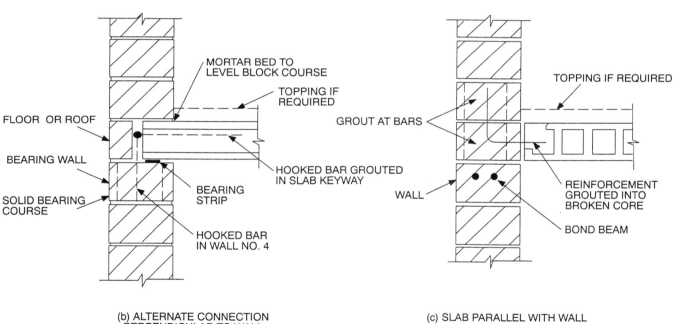

(b) ALTERNATE CONNECTION PERPENDICULAR TO WALL

(c) SLAB PARALLEL WITH WALL

FIGURE A-21-13—VARIOUS TYPES OF WALL CONNECTIONS

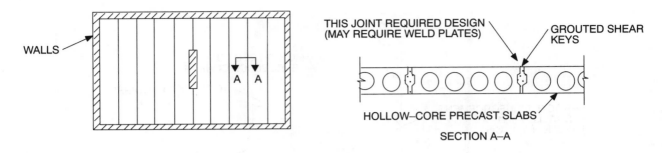

(d) PLAN VIEW OF FLOOR OR ROOF AND CROSS SECTION THROUGH PLANKS

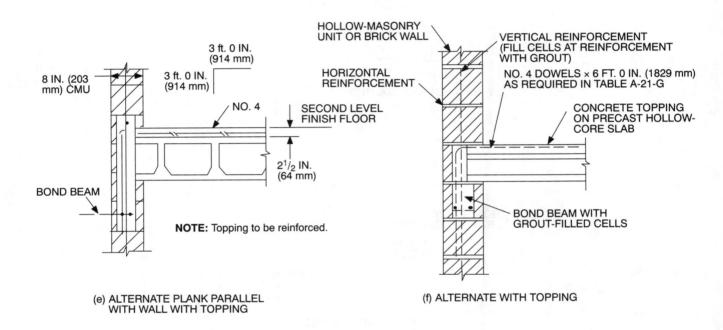

(e) ALTERNATE PLANK PARALLEL WITH WALL WITH TOPPING

(f) ALTERNATE WITH TOPPING

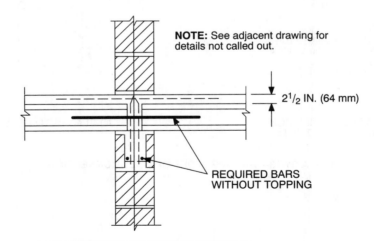

(g) INTERIOR WALL MINIMUM CONNECTION

FIGURE A-21-13—VARIOUS TYPES OF WALL CONNECTIONS—(Continued)

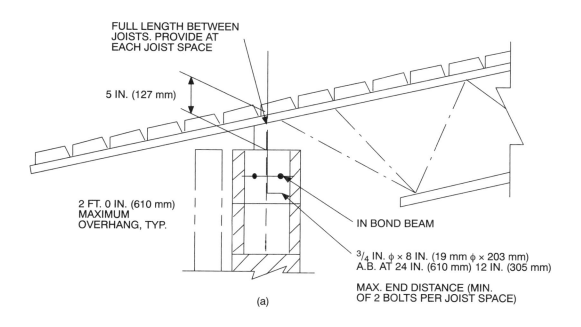

BENT ℝ 0.114 IN. (2.90 mm) (NO. 11 GALVANIZED SHEET GAGE) × 4 IN. (102 mm) / 4 IN. (102 mm)

FULL LENGTH BETWEEN JOISTS. PROVIDE AT EACH JOIST SPACE

5 IN. (127 mm)

2 FT. 0 IN. (610 mm) MAXIMUM OVERHANG, TYP.

IN BOND BEAM

$3/4$ IN. ϕ × 8 IN. (19 mm ϕ × 203 mm) A.B. AT 24 IN. (610 mm) 12 IN. (305 mm)

MAX. END DISTANCE (MIN. OF 2 BOLTS PER JOIST SPACE)

(a)

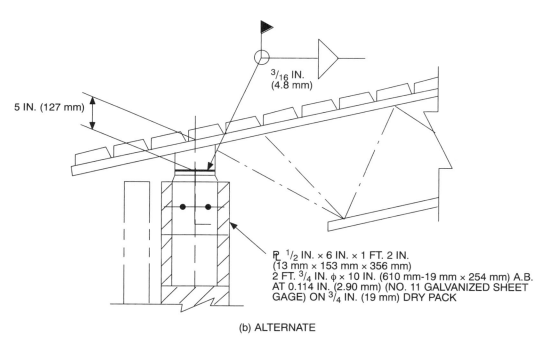

5 IN. (127 mm)

$3/16$ IN. (4.8 mm)

ℝ $1/2$ IN. × 6 IN. × 1 FT. 2 IN. (13 mm × 153 mm × 356 mm) 2 FT. $3/4$ IN. ϕ × 10 IN. (610 mm-19 mm × 254 mm) A.B. AT 0.114 IN. (2.90 mm) (NO. 11 GALVANIZED SHEET GAGE) ON $3/4$ IN. (19 mm) DRY PACK

(b) ALTERNATE

FIGURE A-21-14—EXTERIOR WALL DETAILS

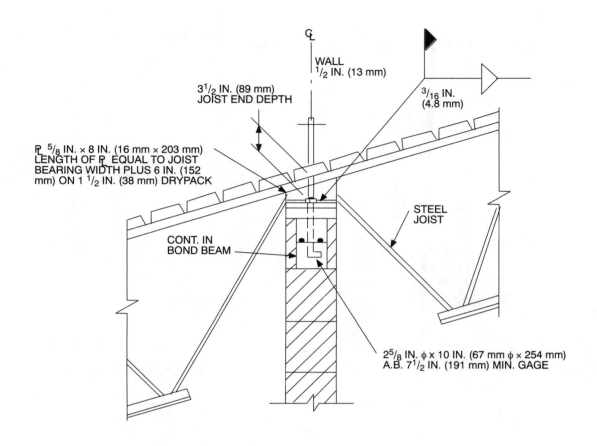

3 1/2 IN. (89 mm)
JOIST END DEPTH

℄

WALL
1/2 IN. (13 mm)

3/16 IN.
(4.8 mm)

P 5/8 IN. × 8 IN. (16 mm × 203 mm)
LENGTH OF P EQUAL TO JOIST
BEARING WIDTH PLUS 6 IN. (152
mm) ON 1 1/2 IN. (38 mm) DRYPACK

STEEL
JOIST

CONT. IN
BOND BEAM

2 5/8 IN. φ × 10 IN. (67 mm φ × 254 mm)
A.B. 7 1/2 IN. (191 mm) MIN. GAGE

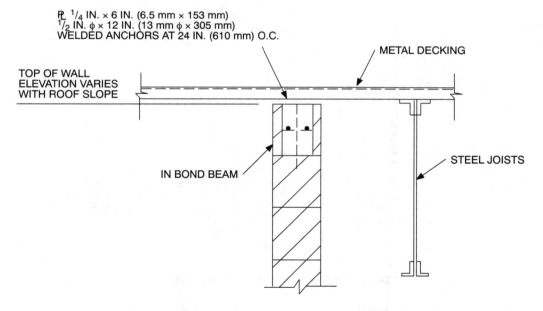

P 1/4 IN. × 6 IN. (6.5 mm × 153 mm)
1/2 IN. φ × 12 IN. (13 mm φ × 305 mm)
WELDED ANCHORS AT 24 IN. (610 mm) O.C.

METAL DECKING

TOP OF WALL
ELEVATION VARIES
WITH ROOF SLOPE

IN BOND BEAM

STEEL JOISTS

FIGURE A-21-15—INTERIOR WALL DETAILS

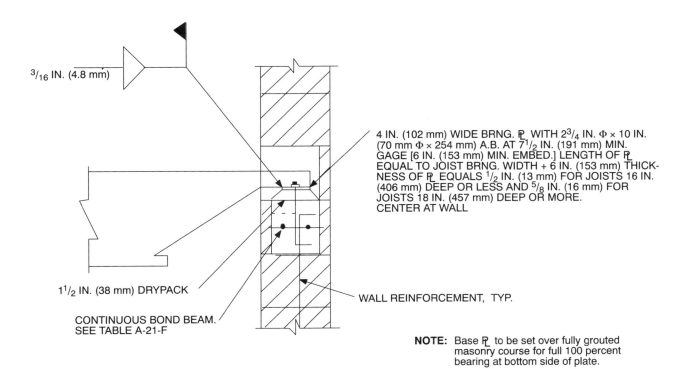

$^3/_{16}$ IN. (4.8 mm)

4 IN. (102 mm) WIDE BRNG. ℔ WITH 2$^3/_4$ IN. Φ × 10 IN. (70 mm Φ × 254 mm) A.B. AT 7$^1/_2$ IN. (191 mm) MIN. GAGE [6 IN. (153 mm) MIN. EMBED.] LENGTH OF ℔ EQUAL TO JOIST BRNG. WIDTH + 6 IN. (153 mm) THICKNESS OF ℔ EQUALS $^1/_2$ IN. (13 mm) FOR JOISTS 16 IN. (406 mm) DEEP OR LESS AND $^5/_8$ IN. (16 mm) FOR JOISTS 18 IN. (457 mm) DEEP OR MORE. CENTER AT WALL

1$^1/_2$ IN. (38 mm) DRYPACK

CONTINUOUS BOND BEAM. SEE TABLE A-21-F

WALL REINFORCEMENT, TYP.

NOTE: Base ℔ to be set over fully grouted masonry course for full 100 percent bearing at bottom side of plate.

FIGURE A-21-16—FLOOR DETAILS
(Design Required for Joists and Wall)

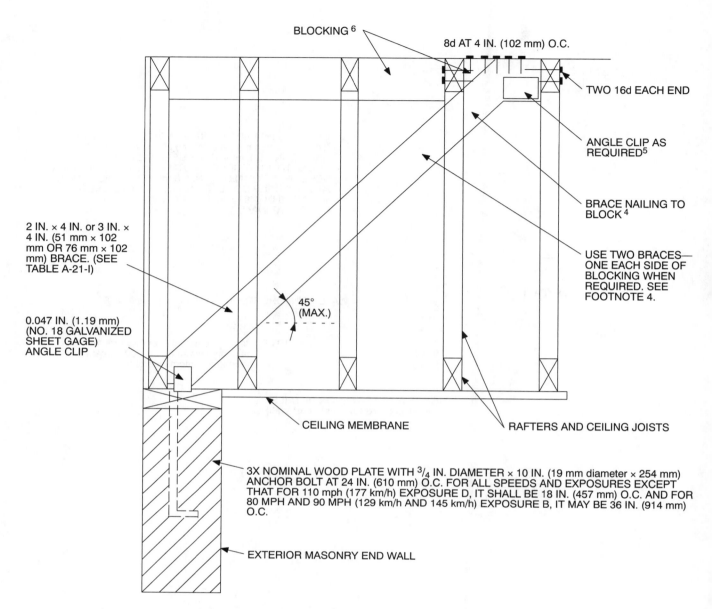

BLOCKING [6]

8d AT 4 IN. (102 mm) O.C.

TWO 16d EACH END

ANGLE CLIP AS REQUIRED[5]

BRACE NAILING TO BLOCK [4]

USE TWO BRACES— ONE EACH SIDE OF BLOCKING WHEN REQUIRED. SEE FOOTNOTE 4.

2 IN. × 4 IN. or 3 IN. × 4 IN. (51 mm × 102 mm OR 76 mm × 102 mm) BRACE. (SEE TABLE A-21-I)

0.047 IN. (1.19 mm) (NO. 18 GALVANIZED SHEET GAGE) ANGLE CLIP

45° (MAX.)

CEILING MEMBRANE

RAFTERS AND CEILING JOISTS

3X NOMINAL WOOD PLATE WITH $3/4$ IN. DIAMETER × 10 IN. (19 mm diameter × 254 mm) ANCHOR BOLT AT 24 IN. (610 mm) O.C. FOR ALL SPEEDS AND EXPOSURES EXCEPT THAT FOR 110 mph (177 km/h) EXPOSURE D, IT SHALL BE 18 IN. (457 mm) O.C. AND FOR 80 MPH AND 90 MPH (129 km/h AND 145 km/h) EXPOSURE B, IT MAY BE 36 IN. (914 mm) O.C.

EXTERIOR MASONRY END WALL

[1] For roof slopes up to 5 units vertical in 12 units horizontal (42%); see Table A-21-I.
[2] See Detail 2, Table A-21-B, for size of angle clip.
[3] Angle clip one side or both sides as required by Table A-21-I.
[4] Use six 16d nails to fasten brace to block, except use two braces and six 16d nails each for 110 miles per hour (mph) (177 km/h), Exposure D. Place on brace on each side block.
[5] Add angle clip each end of block for 90 mph (145 km/h), Exposure D, and 100 and 110 mph (161 and 177 km/h) for Exposures C and D.
[6] Use 2 in. × 6 in. (51 mm × 153 mm) block with 2 in. (51 mm) × brace, 2 in. × 8 in. (51 mm × 203 mm) block with 3 in. (76 mm) × brace.

FIGURE A-21-17—DIAGONAL BRACING OF GABLE-END WALL[1]

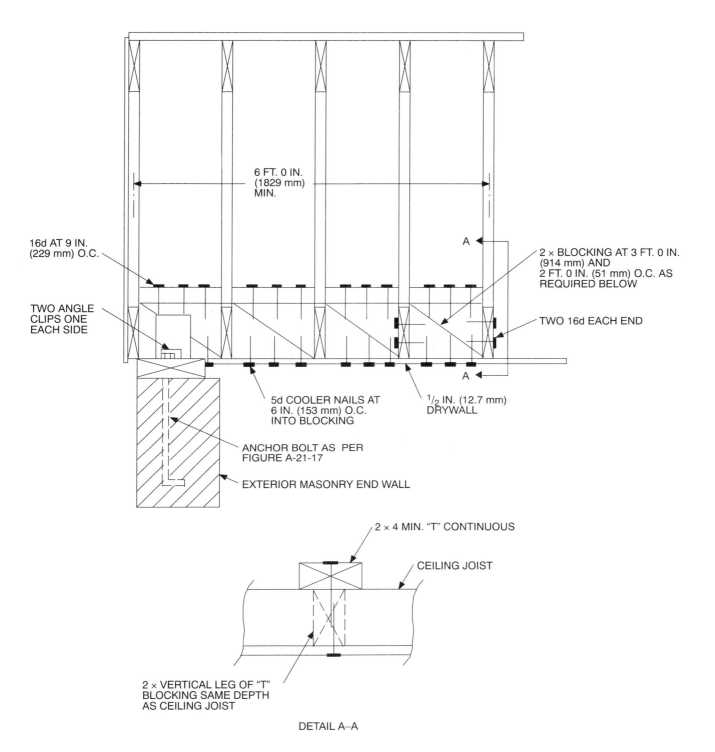

6 FT. 0 IN. (1829 mm) MIN.

16d AT 9 IN. (229 mm) O.C.

TWO ANGLE CLIPS ONE EACH SIDE

2 × BLOCKING AT 3 FT. 0 IN. (914 mm) AND 2 FT. 0 IN. (51 mm) O.C. AS REQUIRED BELOW

TWO 16d EACH END

A

A

5d COOLER NAILS AT 6 IN. (153 mm) O.C. INTO BLOCKING

$^1/_2$ IN. (12.7 mm) DRYWALL

ANCHOR BOLT AS PER FIGURE A-21-17

EXTERIOR MASONRY END WALL

2 × 4 MIN. "T" CONTINUOUS

CEILING JOIST

2 × VERTICAL LEG OF "T" BLOCKING SAME DEPTH AS CEILING JOIST

DETAIL A–A

NOTE: This detail may be used for flat roofs also, except use full height blocking connected to roof sheathing in lieu of "T."
2 × 4 "T" at 36 in. (914 mm) on center—90 miles per hour (mph) (145 km/h) Exposure C and less, and 100 mph and 110 mph (161 km/h and 177 km/h), Exposure B.
2 × 4 "T" at 24 in. (610 mm) on center—required for 90 mph (145 mm) exposure.
See Figure A-21-4 for details of clip angle and connections.

FIGURE A-21-18—ALTERNATE HORIZONTAL BRACING OF GABLE-END WALL

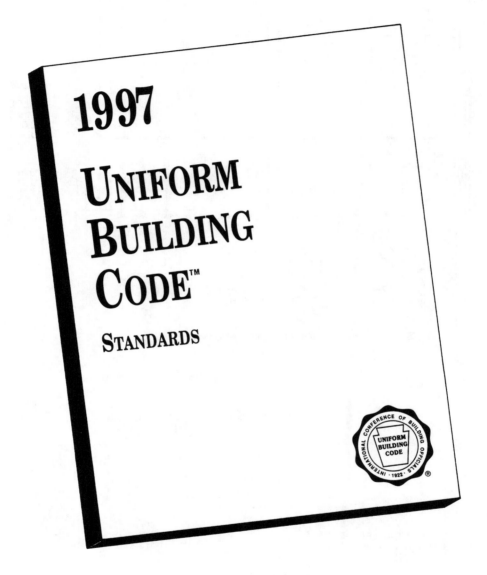

International Conference of Building Officials
5360 WORKMAN MILL ROAD
WHITTIER, CALIFORNIA 90601-2298
(800) 284-4406 • (562) 699-0541

PRINTED IN THE U.S.A.

Preface

The *Uniform Building Code*™ is dedicated to the development of better building construction and greater safety to the public by uniformity in building laws. The code is founded on broad-based principles that make possible the use of new materials and new construction systems.

The *Uniform Building Code* was first enacted by the International Conference of Building Officials at the Sixth Annual Business Meeting held in Phoenix, Arizona, October 18-21, 1927. Revised editions of this code have been published since that time at approximate three-year intervals. New editions incorporate changes approved since the last edition.

The *Uniform Building Code* is designed to be compatible with related publications to provide a complete set of documents for regulatory use. See the publications list following this preface for a listing of the complete family of Uniform Codes and related publications.

Code Changes. The ICBO code development process has been suspended by the Board of Directors and, because of this action, changes to the *Uniform Building Code* will not be processed. For more information, write to the International Conference of Building Officials, 5360 Workman Mill Road, Whittier, California 90601-2298. An analysis of changes between editions is published in the *Analysis of Revisions to the Uniform Codes*.

Marginal Markings. Solid vertical lines in the margins within the body of the code indicate a change from the requirements of the 1994 edition except where an entire chapter was revised, a new chapter was added or a change was minor. Where an entire chapter was revised or a new chapter was added, a notation appears at the beginning of that chapter. The letter **F** repeating in line vertically in the margin indicates that the provision is maintained under the code change procedures of the International Fire Code Institute. Deletion indicators (◆) are provided in the margin where a paragraph or item listing has been deleted if the deletion resulted in a change of requirements.

Three-Volume Set. Provisions of the *Uniform Building Code* have been divided into a three-volume set. Volume 1 accommodates administrative, fire- and life-safety, and field inspection provisions. Chapters 1 through 15 and Chapters 24 through 35 are printed in Volume 1 in their entirety. Any appendix chapters associated with these chapters are printed in their entirety at the end of Volume 1. Excerpts of certain chapters from Volume 2 are reprinted in Volume 1 to provide greater usability.

Volume 2 accommodates structural engineering design provisions, and specifically contains Chapters 16 through 23 printed in their entirety. Included in this volume are design standards that have been added to their respective chapters as divisions of the chapters. Any appendix chapters associated with these chapters are printed in their entirety at the end of Volume 2. Excerpts of certain chapters from Volume 1 are reprinted in Volume 2 to provide greater usability.

Volume 3 contains material, testing and installation standards.

Metrication. The *Uniform Building Code* was metricated in the 1994 edition. The metric conversions are provided in parenthesis following the English units. Where industry has made metric conversions available, the conversions conform to current industry standards.

Formulas are also provided with metric equivalents. Metric equivalent formulas immediately follow the English formula and are denoted by "For **SI**:" preceding the metric equivalent. Some formulas do not use dimensions and, thus, are not provided with a metric equivalent. Multiplying conversion factors have been provided for formulas where metric forms were unavailable. Tables are provided with multiplying conversion factors in subheadings for each tabulated unit of measurement.

METHODS FOR CALCULATING FIRE RESISTANCE OF STEEL, CONCRETE, WOOD, CONCRETE MASONRY AND CLAY MASONRY CONSTRUCTION

Standard of the International Conference of Building Officials

**Part I—Method for Calculating the
Fire Resistance of Steel Construction**

See Sections 703.3 and 703.4, *Uniform Building Code*

SECTION 7.703 — STRUCTURAL STEEL COLUMN PROTECTION

7.703.1 Procedures. These procedures establish a basis for determining the fire resistance of column assemblies as a function of the thickness of fire-resistive material, the weight (W) or cross-sectional area (A) of steel columns and the heated perimeter (D or P) of steel columns. As used in these sections, W is the average weight of a structural steel column in pounds per linear foot (kg/m) and A is the cross-sectional area of a structural steel column in square inches (mm^2). The heated perimeter (D) is the inside perimeter of the fire-resistive material in inches (mm) as illustrated in Figure 7-7-S-1.

Application of these procedures shall be limited to column assemblies in which the fire-resistive material is not designed to carry any of the load acting on the column. In the absence of substantiating fire-endurance test results, ducts, conduit, piping and similar mechanical, electrical and plumbing installations shall not be embedded in any required fire-resistive materials of assemblies designed in accordance with this standard. Table 7-7-S-A1 contains weight-to-heated-perimeter ratios (W/D) for both contour and box fire-protection profiles for the wide-flange shapes most often used as columns. For different fire-resistive design profiles or column cross sections, the weight-to-heated-perimeter ratios (W/D) and cross-sectional-area-to-heated-perimeter ratios (A/P) shall be determined in accordance with the definitions given in this section.

7.703.6 Concrete Masonry. The fire-resistance rating of structural steel columns protected with concrete masonry, as illustrated in Figure 7-7-S-7, may be determined in accordance with the following expression:

$$R = 0.401(A_s/p_s)^{0.7} + 0.285(T_{ea}{}^{1.6}/K^{0.2}) \times$$
$$[1.0 + 42.7\{(A_s/DT_{ea})/(0.25p + T_{ea})\}^{0.8}]$$

For **SI**:

$$R = 0.042(A_s/p_s)^{0.7} + 0.0018(T_{ea}{}^{1.6}/K^{0.2}) \times$$
$$[1.0 + 384\{(A_s/DT_{ea})/(0.25p + T_{ea})\}^{0.8}]$$

WHERE:

A_s = cross-sectional area of the steel column, square inches (mm^2).

D = density of the concrete masonry protection, pounds per cubic foot (kg/m^3).

K = ambient thermal conductivity of concrete masonry. See Table 7-7-S-J, Btu/hr.ft.°F (W/m·k).

p = inner perimeter of concrete masonry protection, inches (mm).

p_s = heated perimeter of steel column, inches (mm).

R = fire-resistance rating of the column assembly, hours.

T_{ea} = equivalent thickness of concrete masonry protection assembly, inches (mm).

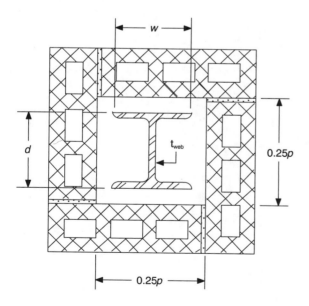

STANDARD UNITS

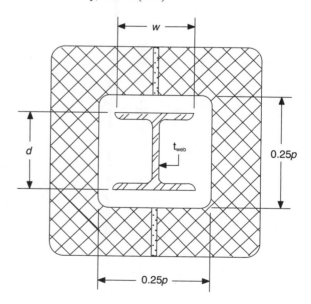

SPECIAL COLUMN COVER UNITS

FIGURE 7-7-S-7—CONCRETE MASONRY PROTECTED STRUCTURAL STEEL COLUMNS

Part III—Methods for Calculating the Fire-resistance Rating of Concrete Masonry

SECTION 7.711 — SCOPE

This part of this standard contains procedures by which the fire-resistance rating of concrete masonry assemblies can be established by calculations. It is applicable to concrete masonry walls, concrete masonry columns, concrete masonry lintels and steel columns protected with concrete masonry.

SECTION 7.712 — MATERIAL REQUIREMENTS

Materials used in accordance with this standard shall comply with the following:

7.712.1 Concrete Masonry Units.

UBC Standard 21-2, Standard Specification for Calcium Silicate Face Brick (Sand-lime Brick)

UBC Standard 21-3, Standard Specification for Concrete Building Brick

UBC Standard 21-4, Standard Specification for Hollow and Solid Load-bearing Concrete Masonry Units

UBC Standard 21-5 Standard Specification for Nonload-bearing Concrete Masonry Units

7.712.2 Mortar.

UBC Standard 21-15, Standard Specifications for Mortar for Unit Masonry and Reinforced Masonry Other Than Gypsum

7.712.3 Grout.

UBC Standard 21-19, Standard Specification for Grout for Masonry

7.712.4 Material for Filling Cells of Units.

Sand or slag having a maximum particle size of $3/8$ inch (9.5 mm).

ASTM C 33-86, Standard Specification for Aggregate

ASTM C 144-89, Standard Specification for Aggregate for Masonry Mortar

ASTM C 330-85 and C 332-83, Standard Specifications for Lightweight Aggregates for Structural and Insulating Concrete

ASTM C 549-81, Perlite Loose-fill Insulation (Type II)

ASTM C 516-80, Vermiculite Loose-fill Insulation (Type I and Type II)

7.712.5 Material for Surface Coverings.

ASTM C 28-76a, Gypsum Plasters

ASTM C 36-76a, Gypsum Wallboard

SECTION 7.713 — CONCRETE MASONRY WALLS

7.713.1 General. The fire-resistance rating of concrete masonry walls shall be determined in accordance with this section. The wall shall have the minimum equivalent thickness for the desired fire-resistive rating as specified in Table 7-7-M-A. The equivalent thickness of the wall may be increased by adding finishes in accordance with Section 7.713.3 and may be modified by combining more than one type of aggregate in the manufacture of the concrete masonry units in accordance with Section 7.713.4.

7.713.2 Determining Equivalent Thickness. Equivalent thickness of concrete masonry walls shall be determined in accordance

with Formula (13-1) for units composed of a single aggregate and by Formula (13-2) for units composed of combined aggregates. When a plaster or gypsum wallboard finish material is applied over an entire face of a concrete masonry wall, the equivalent thickness of the wall assembly shall be determined in accordance with Formula (13-1). Equivalent thickness of units filled with grout or 100 percent solid units shall be the specified thickness.

$$T_E = \frac{V}{(L \times H)} + T_F \qquad (13\text{-}1)$$

WHERE:

H = height of block or brick using specified dimensions as defined in Chapter 21, inches (mm).

L = length of block or brick using specified dimensions as defined in Chapter 21, inches (mm).

T_E = equivalent thickness of wall, inches (mm).

T_F = equivalent thickness of finishes in Table 7-7-M-B.

V = net volume of unit, cubic inch (mm^3) (See ASTM C 140).

7.713.3 Finishes. When a plaster or gypsum wallboard finish is applied over an entire face of the concrete masonry wall, the equivalent thickness of finish shall be determined in accordance with Table 7-7-M-B. The calculated equivalent thickness of the finish can then be added to the calculated equivalent thickness of the concrete masonry wall to determine the total equivalent thickness in accordance with Formula (13-1).

7.713.4 Minimum Required Equivalent Thickness for a Combination of Aggregates. The fire-resistance rating of concrete masonry units composed of a combination of aggregate types shall be based on equivalent thickness values determined as follows:

Determine equivalent thickness values for each tabular column of the desired fire-resistance rating in Table 7-7-M-A by interpolating between equivalent thickness values for aggregate types in proportion to the percentage by volume of each aggregate used in accordance with Formula (13-2).

$$T_R = T_1 \times V_1 + T_2 \times V_2 \dots T_n \times V_n \qquad (13\text{-}2)$$

WHERE:

$T_1, T_2 \dots$

T_n = equivalent thickness for each aggregate Type 1, 2, ... n, respectively, used as indicated in Table 7-7-M-A for the desired fire-resistance rating.

T_R = minimum required equivalent thickness corresponding to the desired fire-resistance rating as listed in Table 7-7-M-A for concrete masonry units manufactured with a particular combination of aggregate types.

$V_1, V_2, \dots$

V_n = percentage by volume of each aggregate Type 1, 2, ... n, respectively, which is used in the manufacture of the concrete masonry unit.

7.713.5 Fire-resistance Increase. When the calculated fire-resistance rating of the concrete masonry wall without fill materials or finishes is not less than two hours, the fire-resistance rating may be increased to four hours provided the cells are completely filled with any of the materials specified in the Building Code and the minimum specified thickness of the concrete masonry units is $7^5/8$ inches (193.7 mm) as determined in accordance with Chapter 21 of the Building Code.

7.713.6 Framing into Wall. Combustible members framed into a wall shall be protected at their ends by not less than one half the required equivalent thickness of such wall.

7.713.7 Multiwythe. The fire-resistance rating of multiwythe walls, such as illustrated in Figure 7-7-M-1, shall be based on the fire-resistance rating of each wythe and the continuous air space between each wythe in accordance with Formula (13-3).

$$R = (R_1^{0.59} + R_2^{0.59} + R_n^{0.59} + A_1, + A_2 + \ldots A_n)^{1.7} \quad (13\text{-}3)$$

WHERE:

$A_1, A_2, \ldots$

$\quad A_n$ = 0.30, factor for each continuous air space (1,2, ... n, respectively) having a depth of $1/2$ inch or more between wythes.

$R_1, R_2, \ldots$

$\quad R_n$ = fire-resistance rating of wythe 1, 2, ... n (hours), respectively.

SECTION 7.714 — CONTROL JOINTS

7.714.1 Design. Control joints installed in fire-resistance-rated concrete masonry walls may be designed in accordance with this section to maintain the fire-resistance rating of the wall in which they are installed.

7.714.2 Materials. The control joints shall be sealed with approved caulk, grout or gaskets in accordance with the details provided in Figure 7-7-M-2.

SECTION 7.715 — STEEL COLUMNS PROTECTED BY CONCRETE MASONRY

7.715.1 The fire-resistance rating of steel columns illustrated in Figure 7-7-S-7, protected by concrete masonry shall be determined in accordance with Part I of this standard.

SECTION 7.716 — CONCRETE MASONRY COLUMNS

7.716.1 Concrete masonry columns shall be designed and reinforced in accordance with the requirements of this code. The fire-resistance rating of concrete masonry columns shall be determined based on the least dimension of the column faces in accordance with the requirements of Table 7-7-M-C.

SECTION 7.717 — CONCRETE MASONRY LINTELS

7.717.1 The fire-resistance rating of concrete lintels shall be determined based on the nominal thickness of the lintel and the minimum thickness of concrete or concrete masonry or any combination thereof, covering the reinforcing steel as determined in accordance with Table 7-7-M-D.

TABLE 7-7-M-A—FIRE-RESISTANCE RATING OF CONCRETE MASONRY WALLS

AGGREGATE TYPE	MINIMUM REQUIRED EQUIVALENT THICKNESS, T_R (inches)			
	× 25.4 for mm			
	4 Hours	3 Hours	2 Hours	1 Hour
Calcareous or siliceous gravel	6.2	5.3	4.2	2.8
Limestone, cinders or slag	5.9	5.0	4.0	2.7
Expanded clay, shale or slate	5.1	4.4	3.6	2.6
Expanded slag or pumice	4.7	4.0	3.2	2.1

NOTE: The minimum required equivalent thickness of concrete masonry units made with a combination of aggregates shall be determined by linear interpolation of the values shown for each aggregate type in accordance with Formula (13-2) and Section 7.713.4.

TABLE 7-7-M-B—EQUIVALENT THICKNESS FOR EACH INCH OF FINISH THICKNESS (inches)

FINISH	AGGREGATE TYPE			
	Siliceous or Calcareous Gravel	Limestone Cinders or Slag	Expanded Shale, Clay or Slate	Expanded Slag or Pumice
	× 25.4 for mm			
Portland cement-sand plaster	1.00	0.75	0.75	0.50
Gypsum-sand plaster or gypsum wallboard	1.25	1.00	1.00	1.00
Gypsum-vermiculite or perlite plaster	1.75	1.50	1.25	1.25

TABLE 7-7-M-C—MINIMUM SIZES OF CONCRETE MASONRY COLUMNS

MINIMUM COLUMN DIMENSIONS, INCHES, FOR FIRE-RESISTANCE RATING OF			
1 Hour	2 Hours	3 Hours	4 Hours
× 25.4 for mm			
8	10	12	14

TABLE 7-7-M-D—MINIMUM COVER ON MAIN REINFORCING BARS FOR REINFORCED CONCRETE MASONRY LINTELS

LINTEL THICKNESS (inches) (Nominal)	COVER THICKNESS (inches) FOR FIRE-RESISTANCE RATING OF			
	× 25.4 for mm			
× 25.4 for mm	1 Hour	2 Hours	3 Hours	4 Hours
6	1	$1^{1}/_{4}$	—	—
8	1	1	$1^{3}/_{4}$	3
10 or more	1	1	1	$1^{3}/_{4}$

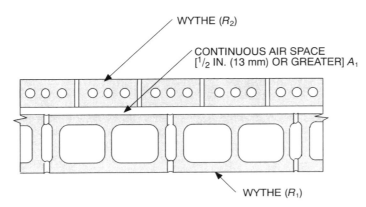

A_1 = FIRE-RESISTANCE RATING FACTOR OF AIR SPACE
R_1 = FIRE-RESISTANCE RATING OF WYTHE 1
R_2 = FIRE-RESISTANCE RATING OF WYTHE 2

FIGURE 7-7-M-1—MULTIWYTHE WALL

TABLE 7-7-M-D - FIGURE 7-7-M-1
UBC 3-80

1997 UBC - STANDARD 7-7

105

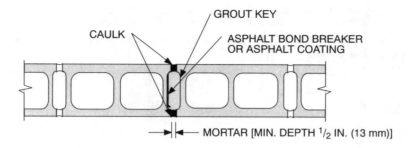

CAULK

GROUT KEY

ASPHALT BOND BREAKER
OR ASPHALT COATING

MORTAR [MIN. DEPTH $1/2$ IN. (13 mm)]

FOR RATINGS UP TO AND INCLUDING 4 HOURS

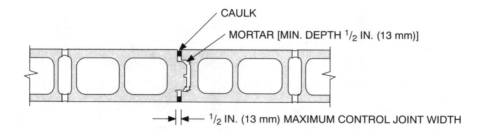

CAULK

MORTAR [MIN. DEPTH $1/2$ IN. (13 mm)]

$1/2$ IN. (13 mm) MAXIMUM CONTROL JOINT WIDTH

FOR RATINGS UP TO AND INCLUDING 4 HOURS

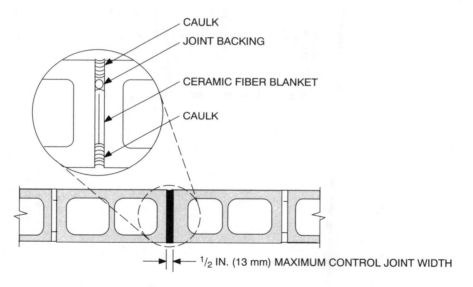

CAULK

JOINT BACKING

CERAMIC FIBER BLANKET

CAULK

$1/2$ IN. (13 mm) MAXIMUM CONTROL JOINT WIDTH

FOR RATINGS UP TO AND INCLUDING 4 HOURS

**FIGURE 7-7-M-2—TYPES OF CONTROL JOINTS FOR
FIRE-RESISTANCE-RATED CONCRETE MASONRY WALLS**

(Continued)

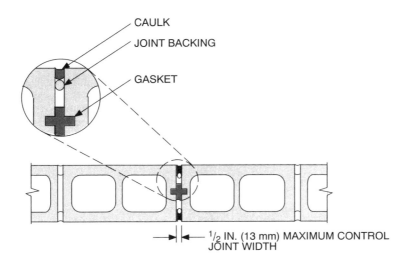

CAULK
JOINT BACKING
GASKET

$^1/_2$ IN. (13 mm) MAXIMUM CONTROL JOINT WIDTH

FOR RATINGS UP TO AND INCLUDING 2 HOURS

**FIGURE 7-7-M-2—TYPES OF CONTROL JOINTS FOR
FIRE-RESISTANCE-RATED CONCRETE MASONRY WALLS—(Continued)**

Part IV—Methods of Calculating the Fire-resistance Rating of Clay Masonry

SECTION 7.718 — SCOPE

This standard provides methods for calculating the fire-resistance-rating periods of clay and shale masonry. This standard is applicable to clay and tile masonry walls.

SECTION 7.719 — GENERAL

Clay masonry construction shall comply with the applicable requirements of this code.

SECTION 7.720 — CLAY MASONRY WALLS

7.720.1 General. The rated fire-resistive period of clay masonry walls shall be determined in accordance with this section. The fire-resistance periods of clay masonry units shall be determined from Tables 7-7-B-A, 7-7-B-B and 7-7-B-C. When sanded gypsum plaster is applied over the entire face of the clay masonry wall, the rated fire-resistive period shall be determined in accordance with Section 7.720.2. When continuous air spaces separate multiwythe walls, the rated fire-resistive period shall be determined in accordance with Section 7.720.4. The rated fire-resistive period of multiwythe walls shall be determined in accordance with Section 7.720.4. Hollow clay masonry walls shall have a minimum equivalent thickness for the desired fire-resistive rating as specified in Section 7.720.5.

7.720.2 Plaster Finishes. The fire-resistive rating period of sanded gypsum plastered clay masonry walls shall be based in accordance with Formula (20-1).

$$R = (R_n^{0.59} + Pl)^{1.7} \qquad (20\text{-}1)$$

WHERE:

Pl = thickness coefficient of sanded gypsum plaster.

R = fire-resistive rating of the assembly, hours.

R_n = fire-resistive period of wythe, hours.

Coefficients for thickness of sanded gypsum plaster shall be selected from Table 7-7-B-D, based on the actual thickness of plaster applied to the clay masonry wall and whether one or two sides of the wall are plastered.

7.720.3 Continuous Air Spaces. The fire-resistive rating period of multiwythe clay masonry walls separated by a continuous air space between each wythe shall be based in accordance with Formula (20-2).

$$R = (R_1^{0.59} + R_2^{0.59} + \ldots R_n^{0.59} + A_s)^{1.7} \qquad (20\text{-}2)$$

WHERE:

A_s = 0.30 factor for each continuous air space having a depth of $^1/_2$ inch to $3^1/_2$ inches (12.7 mm to 88.9 mm) between wythes.

R = fire-resistive rating of the assembly, hours.

$R_1, R_2,$

R_n = fire-resistive period of each individual wythe, hours.

7.720.4 Multiwythe Walls. The fire-resistive rating period of multiwythe walls consisting of two or more dissimilar wythes shall be based on the fire-resistive periods of each wythe and shall be based in accordance with Formula (20-3).

$$R = (R_1^{0.59} + R_2^{0.59} + R_n^{0.59})^{1.7} \qquad (20\text{-}3)$$

WHERE:

R = fire-resistive rating of the assembly, hours.

$R_1, R_2, \ldots$

R_n = fire-resistive period of each individual wythe, hours.

For walls which consist of two or more wythes of dissimilar materials (concrete or concrete masonry units) in combination with clay masonry units, the fire-resistive period of the dissimilar materials shall be based in accordance with Table 7-7-C-B for concrete, Table 7-7-M-A for concrete masonry units or Table 7-7-B-A, 7-7-B-B or 7-7-B-C for clay masonry units.

FIGURE 7-7-M-2 - SEC. 7.720.4
UBC 3-82

1997 UBC - STANDARD 7-7

107

7.720.5 Hollow Clay Masonry Walls. The rated fire-resistive period of hollow clay masonry units shall be based on the equivalent thickness in accordance with Formula (20-4).

$$T_E = \frac{V_n}{L \times H} \qquad (20\text{-}4)$$

WHERE:

H = height of brick using the specified dimensions as defined in Chapter 21 of the Building Code, inches (mm).

L = length of brick using the specified dimensions as defined in Chapter 21 of the Building Code, inches (mm).

T_E = equivalent thickness of wall, inches (mm).

V_n = net volume of unit, cubic inches (mm^3).

The fire-resistive rating for hollow clay brick shall be determined from Table 7-7-B-C based on the equivalent thickness. The fire-resistive rating determined from Table 7-7-B-C may be used in the calculated fire-resistance procedure of Sections 7.720.1, 7.720.2, 7.720.3 and 7.720.4.

**TABLE 7-7-B-A—FIRE-RESISTIVE PERIODS FOR
NONLOAD-BEARING AND LOAD-BEARING CLAY MASONRY WALLS[1]**

WALL OR PARTITION ASSEMBLY, MINIMUM NOMINAL THICKNESS × 25.4 for mm × 0.093 for m^2	FIRE-RESISTIVE PERIOD (hours)
CLAY OR SHALE, SOLID 4-inch brick 6-inch brick 8-inch brick	 1.25 2.55 4.00
CLAY OR SHALE, HOLLOW 8-inch brick, 71% solid 12-inch brick, 64% solid 8-inch brick, 60% solid, cells filled with loose fill insulation	 3.00 4.00 4.00
CLAY OR SHALE, ROLOK 8-inch Hollow Rolok 12-inch Hollow Rolok	 2.50 4.00
CAVITY WALLS, CLAY OR SHALE 8-inch wall; two 3-inch (actual) brick wythes separated by 2-inch air space; masonry joint reinforcement spaced 16 inches on center vertically 10-inch wall; two nominal 4-inch wythes separated by 2-inch air space; $^1/_4$-inch metal ties for each 3 square feet of wall area	 3.00 4.00
CLAY OR SHALE BRICK, METAL FURRING CHANNELS 5-inch wall, 4-inch nominal brick (75% solid) backed with a hat-shaped metal furring channel $^3/_4$ inch thick formed from 0.021 inch sheet metal attached to brick wall on 24 inch centers with approved fasteners; and $^1/_2$-inch Type X gypsum board attached to the metal furring strips with 1-inch-long Type S screws spaced 8 inches on center	 2.00
HOLLOW CLAY TILE, BRICK FACING 8-inch wall; 4-inch units (40% solid)[2] plus 4-inch solid brick 12-inch wall; 8-inch units (40% solid)[2] plus 4-inch solid brick	 3.50 4.00

[1]Units shall comply with the requirements of UBC Standard 21-1 or ASTM C 126.
[2]Units shall comply with the requirements of ASTM C 34.

WALL OR PARTITION ASSEMBLY, MINIMUM NOMINAL THICKNESS	FIRE-RESISTIVE PERIOD (hours)
× 25.4 for mm	
HOLLOW CLAY TILE	
8-inch unit; 2 cells in wall thickness, 40% solid	1.25
8-inch unit; 2 cells in wall thickness, 43% solid	1.50
8-inch unit; 2 cells in wall thickness, 46% solid	1.75
8-inch unit; 2 cells in wall thickness, 49% solid	2.00
8-inch unit; 3 or 4 cells in wall thickness, 40% solid	1.75
8-inch unit; 3 or 4 cells in wall thickness, 43% solid	2.00
8-inch unit; 3 or 4 cells in wall thickness, 48% solid	2.50
8-inch unit; 3 or 4 cells in wall thickness, 53% solid	3.00
12-inch unit; 3 cells in wall thickness, 40% solid	2.50
12-inch unit; 3 cells in wall thickness, 45% solid	3.00
12-inch unit; 3 cells in wall thickness, 49% solid	3.50
12-inch wall; 2 units with 3 or 4 cells in wall thickness, 40% solid	3.50
12-inch wall; 2 units with 3 or 4 cells in wall thickness, 45% solid	4.00
12-inch wall; 2 units with 3 or 4 cells in wall thickness, 53% solid	4.00
16-inch wall; 2 or 3 units with 4 or 5 cells in wall thickness, 40% solid	4.00
CLAY TILE	
4-inch unit; 1 cell in wall thickness, 40% solid[2,3]	1.25
6-inch unit; 1 cell in wall thickness, 30% solid[2,3]	2.00
6-inch unit; 2 cell in wall thickness, 45% solid[4]	1.00
4-inch unit; 1 cell in wall thickness, 40% solid[3,4]	1.25
6-inch unit; 1 cell in wall thickness, 40% solid[3,4]	2.00
HOLLOW STRUCTURAL CLAY TILE	
8-inch unit; 2 cells in wall thickness, 40% solid	1.25
8-inch unit; 2 cells in wall thickness, 49% solid	2.00
8-inch unit; 3 or 4 cells in wall thickness, 53% solid	3.00
8-inch unit; 2 cells in wall thickness, 46% solid	1.75
12-inch unit; 3 cells in wall thickness, 40% solid	2.50
12-inch wall; 2 units, with 3 cells in wall thickness, 40% solid	3.50
12-inch wall; 2 units with 3 or 4 cells in wall thickness, 45% solid	4.00
12-inch unit, 3 cells in wall thickness, 45% solid	3.00
12-inch unit, 3 cells in wall thickness, 49% solid	3.50
16-inch wall, 2 units with 4 cells in wall thickness, 43% solid	4.00
16-inch wall; 2 or 3 units with 4 or 5 cells in wall thickness, 40% solid	4.00

[1]Units shall comply with the requirements of ASTM C 34, C 56, C 212 or C 530.

[2]Ratings are for dense hard-burned clay or shale tile.

[3]Cells filled with tile, stone, slag, cinders or sand mixed with mortar.

[4]Ratings are for medium-burned clay tile.

TABLE 7-7-B-C — MINIMUM EQUIVALENT THICKNESS[1] (inches) OF LOAD-BEARING
OR NONLOAD-BEARING HOLLOW CLAY MASONRY WALLS[2,3,4]

TYPE OF MATERIAL	FIRE-RESISTIVE PERIOD (hours)			
	1	2	3	4
	× 25.4 for mm			
Brick of clay or shale, unfilled	2.3	3.4	4.3	5.0
Brick of clay or shale, grouted or filled with perlite, vermiculite or expanded shale aggregate	3.0	4.4	5.5	6.6

[1]Equivalent thickness as determined for UBC Standard 7-7, Section 7.720.5.

[2]Values between those shown can be determined by direct interpolation.

[3]Where combustible members are framed in the wall, the thickness of solid material between the end of each member and the opposite face of the wall, or between members set in from opposite sides, shall not be less than 93 percent of the thickness shown in the table.

[4]Units shall comply with the requirements of UBC Standard 21-1, Section 21.107.

TABLE 7-7-B-D — COEFFICIENTS FOR PLASTER (PI)[1]

THICKNESS OF PLASTER (inch)	ONE-SIDE	TWO-SIDE
× 25.4 for mm		
1/2	0.30	0.60
5/8	0.37	0.75
3/4	0.45	0.90

[1]Values listed are for 1:3 sanded gypsum plaster.

TABLE 7-7-B-B - TABLE 7-7-B-D
UBC 3-84

1997 UBC - STANDARD 7-7

109

BUILDING BRICK, FACING BRICK AND HOLLOW BRICK
(MADE FROM CLAY OR SHALE)

Based on Standard Specifications C 62-94a, C 216-92c, and C 652-94 of the American Society for Testing and Materials. Extracted, with permission, from the *Annual Book of ASTM Standards,* copyright American Society for Testing and Materials, 100 Barr Harbor Drive, West Conshohocken, PA 19428

See Section 2102.2, Item 4, *Uniform Building Code*

SECTION 21.101 — SCOPE

21.101.1 General. This standard covers brick made from clay or shale and subjected to heat treatment at elevated temperatures (firing), and intended for use in brick masonry. In addition, this standard covers dimension and distortion tolerances for facing brick and hollow brick to be used in masonry construction.

21.101.2 Definition.

BRICK is a solid clay masonry unit whose net cross-sectional area in any plane parallel to the surface containing the cores or cells is at least 75 percent of the gross cross-sectional area measured in the same plane.

21.101.3 Grades. Three grades of brick are covered.

Grade SW. Brick intended for use where a high and uniform resistance to damage caused by cyclic freezing is desired and the exposure is such that the brick may be frozen when saturated with water.

Grade MW. Brick intended for use where moderate resistance to cyclic freezing damage is permissible or where brick may be damp but not saturated with water when freezing occurs.

Grade NW. Brick with little resistance to cyclic freezing damage but which may be acceptable for applications protected from water absorption and freezing.

21.101.4 Grade Requirements for Face Exposure. The selection of the grade of brick for face exposure of vertical or horizontal surfaces shall conform to Table 21-1-A and Figure 21-1-1.

SECTION 21.102 — PHYSICAL PROPERTIES

21.102.1 Durability. The brick shall conform to the physical requirements for the grade specified, as prescribed in Table 21-1-B.

21.102.2 Substitution of Grades. Grades SW and MW may be used in lieu of Grade NW, and Grade SW in lieu of Grade MW.

21.102.3 Waiver of Saturation Coefficient. The saturation coefficient shall be waived provided the average cold-water absorption of a random sample of five bricks does not exceed 8 percent, no more than one brick of the sample exceeds 8 percent and its cold-water absorption must be less than 10 percent.

21.102.4 Freezing and Thawing. The requirements specified in this standard for water absorption (five-hour boiling) and saturation coefficient shall be waived, provided a sample of five bricks, meeting all other requirements, complies with the following requirements when subjected to 50 cycles of the freezing-and-thawing test:

Grade SW   No breakage and not greater than 0.5 percent loss in dry weight of any individual brick.

Brick is not required to conform to the provisions of this section, and these do not apply unless the sample fails to conform to the requirements for absorption and saturation coefficient prescribed in Table 21-1-B or the absorption requirements in Section 21.102.3.

A particular lot or shipment shall be given the same grading as a previously tested lot, without repeating the freezing-and-thawing test, provided the brick is made by the same manufacturer from similar raw materials and by the same method of forming; and provided also that a sample of five bricks selected from the particular lot has an average and individual minimum strength not less than a previously graded sample, and has average and individual maximum water absorption and saturation coefficient not greater than those of the previously tested sample graded according to the freezing-and-thawing test.

21.102.5 Waiver of Durability Requirements. If brick is intended for use exposed to weather where the weathering index is less than 50 (see Figure 21-1-1), unless otherwise specified, the requirements given in Section 21.102.1 for water absorption (five-hour boiling) and for saturation coefficient shall be waived and a minimum average strength of 2,500 pounds per square inch (17 200 kPa) shall apply.

SECTION 21.103 — SIZE, CORING AND FROGGING

21.103.1 Tolerances on Dimensions. The maximum permissible variation in dimensions of individual units shall not exceed those given in Table 21-1-C.

21.103.2 Coring. The net cross-sectional area of cored brick in any plane parallel to the surface containing the cores or cells shall be at least 75 percent of the gross cross-sectional area measured in the same plane. No part of any hole shall be less than $^3/_4$ inch (19.1 mm) from any edge of the brick.

21.103.3 Frogging. One bearing face of each brick may have a recess or panel frog and deep frogs. The recess or panel frog shall not exceed $^3/_8$ inch (9.5 mm) in depth and no part of the recess or panel frog shall be less than $^3/_4$ inch (19.1 mm) from any edge of the brick. In brick containing deep frogs, frogs deeper than $^3/_8$ inch (9.5 mm), any cross section through the deep frogs parallel to the surface containing the deep frogs shall conform to the requirements of Section 21.103.2.

SECTION 21.104 — VISUAL INSPECTION

21.104.1 General. The brick shall be free of defects, deficiencies and surface treatments, including coatings, that would interfere with the proper setting of the brick or significantly impair the strength or performance of the construction.

Minor indentations or surface cracks incidental to the usual method of manufacture, or the chipping resulting from the customary methods of handling in shipment and delivery should not be deemed grounds for rejection.

SECTION 21.105 — SAMPLING AND TESTING

21.105.1 Sampling and Testing. Brick shall be sampled and tested in accordance with ASTM C 67.

SECTION 21.106 — FACING BRICK

21.106.1 General. Facing brick shall be of Grade SW or MW and shall comply with the degree of mechanical perfection and size variations specified in this section. Grade SW may be used in lieu of Grade MW.

21.106.2 Types. Three types of facing brick are covered:

Type FBS. Brick for general use in exposed exterior and interior masonry walls and partitions where greater variation in sizes are permitted than are specified for Type FBX.

Type FBX. Brick for general use in exposed exterior and interior masonry walls and partitions where a high degree of mechanical perfection and minimum permissible variation in size are required.

Type FBA. Brick manufactured and selected to produce characteristic architectural effects resulting from nonuniformity in size and texture of individual units.

When the type is not specified, the requirements for Type FBS shall govern.

21.106.3 Tolerances on Dimensions. The brick shall not depart from the specified size to be used by more than the individual tolerance for the type specified set forth in Table 21-1-D. Tolerances on dimensions for Type FBA shall be as specified by the purchaser, but not more restrictive than Type FBS.

21.106.4 Warpage. Tolerances for distortion or warpage of face or edges of indivi-dual brick from a plane surface and from a straight line, respectively, shall not exceed the maximum for the type specified as set forth in Table 21-1-E. Tolerances on distortion for Type FBA shall be as specified by the purchaser.

21.106.5 Coring. Brick may be cored. The net cross-sectional area of cored brick in any plane parallel to the surface containing the cores or cells shall be at least 75 percent of the gross cross-sectional area measured in the same plane. No part of any hole shall be less than $^3/_4$ inch (19.1 mm) from any edge of the brick.

21.106.6 Frogging. One bearing face of each brick may have a recess or panel frog and deep frogs. The recess or panel frog shall not exceed $^3/_8$ inch (9.5 mm) in depth and no part of the recess or panel frog shall be less than $^3/_4$ inch (19.1 mm) from any edge of the brick. In brick containing deep frogs, frogs deeper than $^3/_8$ inch (9.5 mm), any cross section through the deep frogs parallel to the surface containing the deep frogs shall conform to the requirements of Section 21.106.5.

21.106.7 Visual Inspection. In addition to the requirements of Section 21.104, brick used in exposed wall construction shall have faces which are free of cracks or other imperfections detracting from the appearance of the designated sample when viewed from a distance of 15 feet (4600 mm) for Type FBX and a distance of 20 feet (6100 mm) for Types FBS and FBA.

SECTION 21.107 — HOLLOW BRICK

21.107.1 General. Hollow brick shall be of Grade SW or MW and comply with the physical requirements in Table 21-1-B and other requirements of this section. Grade SW may be used in lieu of Grade MW.

21.107.2 Definitions.

HOLLOW BRICK is a clay masonry unit whose net cross-sectional area (solid area) in any plane parallel to the surface, containing the cores, cells or deep frogs is less than 75 percent of its gross cross-sectional area measured in the same plane.

CORES are void spaces having a gross cross-sectional area equal to or less than $1^1/_2$ square inches (968 mm^2).

CELLS are void spaces having a gross cross-sectional area greater than $1^1/_2$ square inches (968 mm^2).

21.107.3 Types. Four types of hollow brick are covered:

Type HBS. Hollow brick for general use in exposed exterior and interior masonry walls and partitions greater variation in size are permitted than is specified for Type HBX.

Type HBX. Hollow brick for general use in exposed exterior and interior masonry walls and partitions where a high degree of mechanical perfection and minimum permissible variation in size are required.

Type HBA. Hollow brick manufactured and selected to produce characteristic architectural effects resulting from nonuniformity in size and texture of the individual units.

Type HBB. Hollow brick for general use in masonry walls and partitions where a particular color, texture, finish, uniformity, or limits on cracks, warpage, or other imperfections detracting from the appearance are not a consideration.

When the type is not specified, the requirements for Type HBS shall govern.

21.107.4 Class. Two classes of hollow brick are covered:

Class H40V. Hollow brick intended for use where void areas or hollow spaces greater than 25 percent, but not greater than 40 percent, of the gross cross-sectional area of the unit measured in any plane parallel to the surface containing the cores, cells or deep frogs are desired. The void spaces, web thicknesses and shell thicknesses shall comply with the requirements of Sections 21.107.5, 21.107.6 and 21.107.7.

Class H60V. Hollow brick intended for use where larger void areas are desired. The sum of these void areas shall be greater than 40 percent, but not greater than 60 percent, of the gross cross-sectional area of the unit measured in any plane parallel to the surface containing the cores, cells or deep frogs. The void spaces, web thicknesses and shell thicknesses shall comply with the requirements of Sections 21.107.5, 21.107.6 and 21.107.7 and to the minimum requirements of Table 21-1-F.

When the class is not specified, the requirements for Class H40V shall govern.

21.107.5 Hollow Spaces. Core holes shall not be less than $^5/_8$ inch (15.9 mm) from any edge of the brick, except for cored-shell hollow brick. Cored-shell hollow brick shall have a minimum shell thickness of $1^1/_2$ inches (38 mm). Cores greater than 1 square inch (645 mm^2) in cored shells shall not be less than $^1/_2$ inch (13 mm) from any edge. Cores not greater than 1 inch square (645 mm^2) in shells cored not more than 35 percent shall not be less than $^3/_8$ inch (9.5 mm) from any edge.

Cells shall not be less than $^3/_4$ inch (19.1 mm) from any edge of the brick except for double-shell hollow brick.

Double-shell hollow brick with inner and outer shells not less than $^1/_2$ inch (13 mm) thick may not have cells greater than $^5/_8$ inch (15.9 mm) in width or 5 inches (127 mm) in length between the inner and outer shell.

21.107.6 Webs. The thickness for webs between cells shall not be less than $^1/_2$ inch (13 mm), $^3/_8$ inch (9.5 mm) between cells and cores or $^1/_4$ inch (6 mm) between cores. The distance of voids from unexposed edges, which are recessed not less than $^1/_2$ inch (13 mm), shall not be less than $^1/_2$ inch (13 mm).

21.107.7 Frogging. One bearing face of each brick may have a recess or panel frog and deep frogs. The recess or panel frog shall not exceed $^3/_8$ inch (9.5 mm) in depth and no part of the recess or panel frog shall be less than $^5/_8$ inch (15.9 mm) from any edge of the brick. In brick containing deep frogs, frogs deeper than $^3/_8$ inch (9.5 mm), any cross section through the deep frogs parallel to the bearing surface shall conform to other requirements of Sections 21.107.2 and 21.107.4 for void area and Section 21.107.5 for hollow spaces.

21.107.8 Tolerances on Dimensions. The hollow brick shall not depart from the specified size by more than the individual tolerance for specified size by more than individual tolerances for the type specified as set forth in Table 21-1-G. Tolerances and dimensions for Type HBA shall be as specified by the purchaser.

21.107.9 Warpage. Tolerances for distortion or warpage of face or edges of individual hollow brick from a plane surface and from a straight line, respectively, shall not exceed the maximum for the type specified in Table 21-1-H. Tolerances on distortion for Type HBA shall be as specified by the purchaser.

21.107.10 Visual Inspection. In addition to the requirements of Section 21.104, brick used in exposed wall construction shall have faces which are free of cracks or other imperfections detracting from the appearance of a sample wall when viewed from a distance of 15 feet (4600 mm) for Type HBX and a distance of 20 feet (6100 mm) for Types HBS and HBA.

TABLE 21-1-A—GRADE REQUIREMENTS FOR FACE EXPOSURE

	WEATHERING INDEX		
EXPOSURE	Less than 50	50 to 500	500 and greater
In vertical surfaces: In contact with earth Not in contact with earth	MW MW	SW SW	SW SW
In other than vertical surfaces: In contact with earth Not in contact with earth	SW MW	SW SW	SW SW

TABLE 21-1-B—PHYSICAL REQUIREMENTS FOR TYPES OF UNIT MASONRY[5]

TYPE OF MASONRY	GRADE	MINIMUM FACE SHELL THICKNESS (inches)	MINIMUM[1] COMPRESSIVE STRENGTH PSI AVERAGE GROSS AREA × 6.89 for kPa Average of Five Tests	Individual	MAXIMUM WATER ABSORPTION By Five-hour Boiling (percent) Average of Five Tests	Individual	MAXIMUM SATURATION COEFFICIENT[2] Average of Five Test	Individual	WATER ABSORPTION Maximum Pounds per Cubic Foot × 16 for kg/m³	MOISTURE CONTENT Maximum Percentage of Total Absorption	MINIMUM MODULUS OF RUPTURE Average of Five Tests	Individual
24-1. Building brick made from clay or shale[3]	SW MW NW		(brick flatwise) 3,000 2,500 1,500	2,500 2,200 1,250	17 22 no limit	20 25	.78 .88	.80 .90			(brick flatwise) psi Average Gross Area × 6.89 for kPa	
Hollow Brick[3]	SW MW	See Table 21-1-F	(net area)[4] 3,000 2,500	2,500 2,000	17 22	20 25	.78 .88	.80 .90				
24-2. Sand-lime building brick	SW MW		4,500 2,500	3,500 2,000							600 450	400 300
24-14. Unburned clay masonry units			Based on Net Area (psi)[4] × 6.89 for kPa 300	250					Based on % of Dry Wt. 2.5%	4.0%	50	35

[1]Gross area of a unit shall be determined by multiplying the horizontal face dimension of the unit as placed in the wall by its thickness.

[2]The saturation coefficient is the ratio of absorption by 24-hour submersion in cold water to that after five-hour submersion in boiling water.

[3]If the average cold-water absorption of a random sample of five bricks does not exceed 8.0 percent, when no more than one brick unit of the sample exceeds 8.0 percent and its cold-water absorption must be less than 10.0 percent, the saturation coefficient shall be waived.

[4]Based on net area of a unit which shall be taken as the area of solid material in shells and webs actually carrying stresses in a direction parallel to the direction of loading.

[5]For the compressive strength requirements, test the unit with the compressive force perpendicular to the bed surface of the unit, with the unit in the stretcher position.

TABLE 21-1-C—TOLERANCES ON DIMENSIONS

SPECIFIED DIMENSION (inches)	MAXIMUM PERMISSIBLE VARIATION FROM SPECIFIED DIMENSION, PLUS OR MINUS (inch)
× 25.4 for mm	
Up to 3, incl.	3/32
Over 3 to 4, incl.	1/8
Over 4 to 6, incl.	3/16
Over 6 to 8, incl.	1/4
Over 8 to 12, incl.	5/16
Over 12 to 16, incl.	3/8

TABLE 21-1-D—TOLERANCES ON DIMENSIONS

SPECIFIED DIMENSION (inches)	MAXIMUM PERMISSIBLE VARIATION FROM SPECIFIED DIMENSION, PLUS OR MINUS (inch)	
	Type FBX	Type FBS
	× 25.4 for mm	
3 and under	1/16	3/32
Over 3 to 4, incl.	3/32	1/8
Over 4 to 6, incl.	1/8	3/16
Over 6 to 8, incl.	5/32	1/4
Over 8 to 12, incl.	7/32	5/16
Over 12 to 16, incl.	9/32	3/8

TABLE 21-1-E—TOLERANCES ON DISTORTION

MAXIMUM FACE DIMENSION (inches)	MAXIMUM PERMISSIBLE DISTORTION (inch)	
	Type FBX	Type FBS
	× 25.4 for mm	
8 and under	1/16	3/32
Over 8 to 12, incl.	3/32	1/8
Over 12 to 16, incl.	1/8	5/32

TABLE 21-1-F—HOLLOW BRICK (Class H60V) MINIMUM THICKNESS OF FACE SHELLS AND WEBS

NOMINAL WIDTH OF UNIT (inches)	FACE SHELL THICKNESS (inches)		END SHELLS OR WEBS (inches)	WEB THICKNESS PER FOOT, TOTAL (inches per foot)[1]
	Solid	Cored or Double Shell		
	× 25.4 for mm			× 83 for mm per m
3 and 4	3/4	—	3/4	1 5/8
6	1	1 1/2	1	2 1/4
8	1 1/4	1 1/2	1	2 1/4
10	1 3/8	1 5/8	1 1/8	2 1/2
12	1 1/2	2	1 1/8	2 1/2

[1]The sum of the measured thickness of all webs in the unit, multiplied by 12 (305 when using metric), and divided by the length of the unit. In the case of open-ended units where the open-end portion is solid grouted, the length of that open-ended portion shall be deducted from the overall length of the unit.

TABLE 21-1-C - TABLE 21-1-F
UBC 3-340

1997 UBC - STANDARD 21-1

113

TABLE 21-1-G—TOLERANCES ON DIMENSIONS

SPECIFIED DIMENSION (inches)	MAXIMUM PERMISSIBLE VARIATION FROM SPECIFIED DIMENSION, PLUS OR MINUS (inch)	
	Type HBX	Types HBS and HBB
	× 25.4 for mm	
3 and under	$1/16$	$3/32$
Over 3 to 4, incl.	$3/32$	$1/8$
Over 4 to 6, incl.	$1/8$	$3/16$
Over 6 to 8, incl.	$5/32$	$1/4$
Over 8 to 12, incl.	$7/32$	$5/16$
Over 12 to 16, incl.	$9/32$	$3/8$

TABLE 21-1-H—TOLERANCES ON DISTORTION

MAXIMUM FACE DIMENSION (inches)	MAXIMUM PERMISSIBLE DISTORTION (inch)	
	Type HBX	Types HBS and HBB
	× 25.4 for mm	
8 and under	$1/16$	$3/32$
Over 8 to 12, incl.	$3/32$	$1/8$
Over 12 to 16, incl.	$1/8$	$5/32$

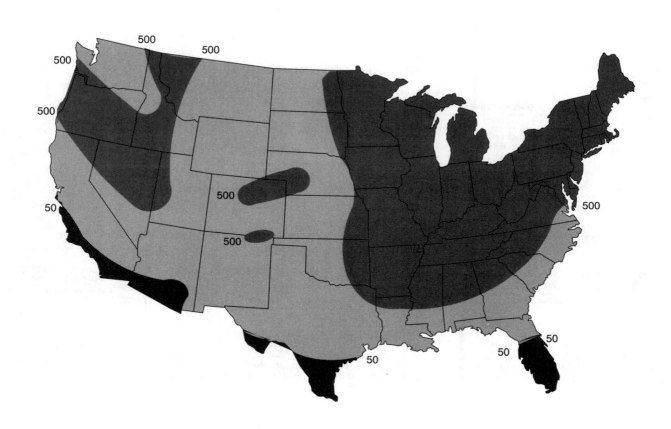

WEATHERING REGIONS

▓ NEGLIGIBLE WEATHERING

▒ MODERATE WEATHERING

■ SEVERE WEATHERING

FIGURE 21-1-1—WEATHERING INDEXES IN THE UNITED STATES

UNIFORM BUILDING CODE STANDARD 21-2
CALCIUM SILICATE FACE BRICK
(SAND-LIME BRICK)

**Based on Standard Specification C 73-95 of the American Society for Testing and Materials.
Extracted, with permission, from the *Annual Book of ASTM Standards,* copyright American Society for
Testing and Materials, 100 Barr Harbor Drive, West Conshohocken, PA 19428**

See Section 2102.2, Item 6, *Uniform Building Code*

SECTION 21.201 — SCOPE

21.201.1 Grades. This standard covers brick made from sand and lime and intended for use in brick masonry. Two grades of brick are covered:

21.201.1.1 Grade SW. Brick intended for use where exposed to temperatures below freezing in the presence of moisture.

21.201.1.2 Grade MW. Brick intended for use where exposed to temperature below freezing but unlikely to be saturated with water.

21.201.2 Definition. The term "brick" used in this standard shall mean brick or a solid sand-lime masonry unit.

SECTION 21.202 — PHYSICAL PROPERTIES

21.202.1 Durability. The brick shall conform to the physical requirements for the grade specified as prescribed in Table 21-2-A.

21.202.2 Substitution of Grades. Unless otherwise specified, brick of Grade SW shall be accepted in lieu of Grade MW.

SECTION 21.203 — SIZE

The size of the brick shall be as specified by the purchaser, and the average size of brick furnished shall approximate the size specified in the invitation for bids.

No overall dimension (width, height and length) shall differ by more than $1/8$ inch (3.2 mm) from the specified standard dimension. Standard dimensions of units are the manufacturer's designated dimensions.

SECTION 21.204 — VISUAL INSPECTION

Brick shall pass a visual inspection for soundness, compact structure, reasonably uniform shape, and freedom from the following: cracks, warpage, large pebbles, balls of clay, or particles of lime that would affect the serviceability or strength of the brick.

SECTION 21.205 — METHODS OF SAMPLING AND TESTING

The purchaser or the purchaser's authorized representative shall be accorded proper facilities to inspect and sample the units at the place of manufacture from the lots ready for delivery. At least 10 days should be allowed for completion of the tests.

Sample and test units in accordance with ASTM C 140.

TABLE 21-2-A—PHYSICAL REQUIREMENTS FOR SAND-LIME BUILDING BRICK

TYPE OF MASONRY	GRADE	MINIMUM COMPRESSIVE STRENGTH PSI AVERAGE GROSS AREA		MINIMUM MODULUS OF RUPTURE		WATER ABSORPTION MAX. lb./ft.^{3} (kg/m^3)
		Average of Five Tests	Individual	Average of Five Tests	Individual	
				(Brick Flatwise) psi Average Gross Area		
		$\times$ 6.89 for kPa				
Sand-lime	SW	4500	3500	600	400	10 (160)
Building brick	MW	2500	2000	450	300	13 (208)

Gross area of a unit shall be determined by multiplying the horizontal face dimension of the unit as placed in the wall by its thickness.

UNIFORM BUILDING CODE STANDARD 21-3
CONCRETE BUILDING BRICK

Based on Standard Specification C 55-95 of the American Society for Testing and Materials.
Extracted, with permission, from the *Annual Book of ASTM Standards,* copyright American Society for
Testing and Materials, 100 Barr Harbor Drive, West Conshohocken, PA 19428

See Section 2102.2, Item 5, *Uniform Building Code*

SECTION 21.301 — SCOPE

This standard covers concrete building brick and similar solid units made from portland cement, water and suitable mineral aggregates with or without the inclusion of other materials.

SECTION 21.302 — CLASSIFICATION

21.302.1 Types. Two types of concrete brick in each of two grades are covered, as follows:

21.302.1.1 Type I, moisture-controlled units. Concrete brick designated as Type I (Grades N-I and S-I) shall conform to all requirements of this standard, including the requirements of Table 21-3-A.

21.302.1.2 Type II, nonmoisture-controlled units. Concrete brick designated as Type II (Grades N-II and S-II) shall conform to all requirements of this standard except the requirements of Table 21-3-A.

21.302.2 Grades. Concrete brick manufactured in accordance with this standard shall conform to two grades as follows:

21.302.2.1 Grade N. For use as architectural veneer and facing units in exterior walls and for use where high strength and resistance to moisture penetration and severe frost action are desired.

21.302.2.2 Grade S. For general use where moderate strength and resistance to frost action and moisture penetration are required.

SECTION 21.303 — MATERIALS

21.303.1 Cementitious Materials. Materials shall conform to the following applicable standards:

1. Portland Cement—ASTM C 150 modified as follows:

   Limitation on insoluble residue—1.5 percent.
   Limitation on air content of mortar,
      Volume percent—22 percent maximum.
   Limitation on loss on ignition—7 percent maximum.
   Limestone with a minimum 85 percent calcium carbonate ($CaCO_3$) content may be added to the cement, provided the requirements of ASTM C 150 as modified above are met.

2. Blended Cements—ASTM C 595.

3. Hydrated Lime, Type S—UBC Standard 21-13.

21.303.2 Other Constituents. Air-entraining agents, coloring pigments, integral water repellents, finely ground silica, etc., shall be previously established as suitable for use in concrete or shall be shown by test or experience not to be detrimental to the durability of the concrete.

SECTION 21.304 — PHYSICAL REQUIREMENTS

At the time of delivery to the work site, the concrete brick shall conform to the physical requirements prescribed in Table 21-3-B.

At the time of delivery to the purchaser, the total linear drying shrinkage of Type II units shall not exceed 0.065 percent when tested in accordance with ASTM C 426.

The moisture content of Type I concrete brick at the time of delivery shall conform to the requirements prescribed in Table 21-3-A.

SECTION 21.305 — DIMENSIONS AND PERMISSIBLE VARIATIONS

Overall dimensions (width, height, or length) shall not differ by more than $1/8$ inch (3.2 mm) from the specified standard dimensions.

> **NOTE:** Standard dimensions of concrete brick are the manufacturer's designated dimensions. Nominal dimensions of modular-size concrete brick are equal to the standard dimensions plus $3/8$ inch (9.5 mm), the thickness of one standard mortar joint. Nominal dimensions of nonmodular size concrete brick usually exceed the standard dimensions by $1/8$ inch to $1/4$ inch (3.2 mm to 6.4 mm).

Variations in thickness of architectural units such as split-faced or slumped units will usually vary from the specified tolerances.

SECTION 21.306 — VISUAL INSPECTION

21.306.1 General. All concrete brick shall be sound and free of cracks or other defects that would interfere with the proper placing of the unit or impair the strength or permanence of the construction. Minor cracks incidental to the usual method of manufacture, or minor chipping resulting from customary methods of handling in shipment and delivery, shall not be deemed grounds for rejection.

21.306.2 Brick in Exposed Walls. Where concrete brick is to be used in exposed wall construction, the face or faces that are to be exposed shall be free of chips, cracks or other imperfections when viewed from 20 feet (6100 mm), except that if not more than 5 percent of a shipment contains slight cracks or small chips not larger than $1/2$ inch (13 mm), this shall not be deemed grounds for rejection.

SECTION 21.307 — METHODS OF SAMPLING AND TESTING

The purchaser or authorized representative shall be accorded proper facilities to inspect and sample the concrete brick at the place of manufacture from the lots ready for delivery. At least 10 days shall be allowed for completion of the test.

Sample and test concrete brick in accordance with ASTM C 140 and C 426, when applicable.

Total linear drying shrinkage shall be based on tests of concrete brick made with the same materials, concrete mix design, manufacturing process and curing method, conducted in accordance with ASTM C 426 not more than 24 months prior to delivery.

SECTION 21.308 — REJECTION

If the shipment fails to conform to the specific requirements, the manufacturer may sort it, and new specimens shall be selected by the purchaser from the retained lot and tested at the expense of the manufacturer. If the second set of specimens fails to conform to the test requirements, the entire lot shall be rejected.

TABLE 21-3-A—MOISTURE CONTENT REQUIREMENTS FOR TYPE I CONCRETE BRICK

LINEAR SHRINKAGE, PERCENT	MOISTURE CONTENT, MAX. PERCENT OF TOTAL ABSORPTION (Average of 3 Concrete Brick)		
	Humidity[1] Conditions at Jobsite or Point of Use		
	Humid	Intermediate	Arid
0.03 or less	45	40	35
From 0.03 to 0.045	40	35	30
0.045 to 0.065, max.	35	30	25

[1]Arid—Average annual relative humidity less than 50 percent.
Intermediate—Average annual relative humidity 50 to 75 percent.
Humid—Average annual relative humidity above 75 percent.

TABLE 21-3-B—STRENGTH AND ABSORPTION REQUIREMENTS

	COMPRESSIVE STRENGTH, MIN., psi (Concrete Brick Tested Flatwise)		WATER ABSORPTION, MAX., (Avg. of 3 Brick) WITH OVEN-DRY WEIGHT OF CONCRETE Lb./Ft.³		
	× 6.89 for kPa		× 16 for kg/m³		
	Average Gross Area		Weight Classification		
Grade	Avg. of 3 Concrete Brick	Individual Concrete Brick	Lightweight Less Than 105	Medium Weight Less Than 125 to 105	Normal Weight 125 or More
N-I	3,500	3,000	15	13	10
N-II	3,500	3,000	15	13	10
S-I	2,500	2,000	18	15	13
S-II	2,500	2,000	18	15	13

UNIFORM BUILDING CODE STANDARD 21-4
HOLLOW AND SOLID LOAD-BEARING
CONCRETE MASONRY UNITS

Based on Standard Specification C 90-95 of the American Society for Testing and Materials.
Extracted, with permission, from the *Annual Book of ASTM Standards,* copyright American Society for
Testing and Materials, 100 Barr Harbor Drive, West Conshohocken, PA 19428

SECTION 21.401 — SCOPE

This standard covers solid (units with 75 percent or more net area) and hollow load-bearing concrete masonry units made from portland cement, water and mineral aggregates with or without the inclusion of other materials.

SECTION 21.402 — CLASSIFICATION

21.402.1 Types. Two types of concrete masonry units in each of two grades are covered as follows:

21.402.1.1 Type I, moisture-controlled units. Units designated as Type I shall conform to all requirements of this standard including the moisture content requirements of Table 21-4-A.

21.402.1.2 Type II, nonmoisture-controlled units. Units designated as Type II shall conform to all requirements of this standard except the moisture content requirements of Table 21-4-A.

21.402.2 Grades. Concrete masonry units manufactured in accordance with this standard shall conform to two grades as follows:

21.402.2.1 Grade N. Units having a weight classification of 85 pcf (1360 kg/m^3) or greater, for general use such as in exterior walls below and above grade that may or may not be exposed to moisture penetration or the weather and for interior walls and backup.

21.402.2.2 Grade S. Units having a weight classification of less than 85 pcf (1360 kg/m^3), for uses limited to above-grade installation in exterior walls with weather-protective coatings and in walls not exposed to the weather.

SECTION 21.403 — MATERIALS

21.403.1 Cementitious Materials. Materials shall conform to the following applicable standards:

1. Portland Cement—ASTM C 150 modified as follows:

   Limitation on insoluble residue—1.5 percent maximum.
   Limitation on air content of mortar,
     Volume percent—22 percent maximum.
   Limitation on loss on ignition—7 percent maximum.
   Limestone with a minimum 85 percent calcium carbonate (C_aCO_3) content may be added to the cement, provided the requirements of ASTM C 150 as modified above are met.

2. Blended Cements—ASTM C 595.

3. Hydrated Lime, Type S—UBC Standard 21-13.

21.403.2 Other Constituents and Aggregates. Air-entraining agents, coloring pigments, integral water repellents, finely ground silica, aggregates, and other constituents, shall be previously established as suitable for use in concrete or shall be shown by test or experience to not be detrimental to the durability of the concrete.

SECTION 21.404 — PHYSICAL REQUIREMENTS

At the time of delivery to the work site, the units shall conform to the physical requirements prescribed in Table 21-4-B. The moisture content of Type I concrete masonry units at time of delivery shall conform to the requirements prescribed in Table 21-4-A.

At the time of delivery to the purchaser, the linear shrinkage of Type II units shall not exceed 0.065 percent.

SECTION 21.405 — MINIMUM FACE-SHELL AND WEB THICKNESSES

Face-shell (FST) and web (WT) thicknesses shall conform to the requirements listed in Table 21-4-C.

SECTION 21.406 — PERMISSIBLE VARIATIONS IN DIMENSIONS

21.406.1 Precision Units. For precision units, no overall dimension (width, height and length) shall differ by more than $^1/_8$ inch (3.2 mm) from the specified standard dimensions.

21.406.2 Particular Feature Units. For particular feature units, dimensions shall be in accordance with the following:

1. For molded face units, no overall dimension (width, height and length) shall differ by more than $^1/_8$ inch (3.2 mm) from the specified standard dimension. Dimensions of molded features (ribs, scores, hex-shapes, patterns, etc.) shall be within $^1/_{16}$ inch (1.6 mm) of the specified standard dimensions and shall be within $^1/_{16}$ inch (1.6 mm) of the specified placement of the unit.

2. For split-faced units, all non-split overall dimensions (width, height and length) shall differ by no more than $^1/_8$ inch (3.2 mm) from the specified standard dimensions. On faces that are split, overall dimensions will vary. Local suppliers should be consulted to determine dimensional tolerances achievable.

3. For slumped units, no overall height dimension shall differ by more than $^1/_8$ inch (3.2 mm) from the specified standard dimension. On faces that are slumped, overall dimensions will vary. Local suppliers should be consulted to determine dimension tolerances achievable.

> **NOTE:** Standard dimensions of units are the manufacturer's designated dimensions. Nominal dimensions of modular size units, except slumped units, are equal to the standard dimensions plus $^3/_8$ inch (9.5 mm), the thickness of one standard mortar joint. Slumped units are equal to the standard dimensions plus $^1/_2$ inch (13 mm), the thickness of one standard mortar joint. Nominal dimensions of nonmodular size units usually exceed the standard dimensions by $^1/_8$ inch to $^1/_4$ inch (3.2 mm to 6.4 mm).

SECTION 21.407 — VISUAL INSPECTION

All units shall be sound and free of cracks or other defects that would interfere with the proper placing of the unit or impair the strength or perm-anence of the construction. Units may have minor cracks incidental to the usual method of manufacture, or minor chipping resulting from customary methods of handling in shipment and delivery.

Units that are intended to serve as a base for plaster or stucco shall have a sufficiently rough surface to afford a good bond.

Where units are to be used in exposed wall construction, the face or faces that are to be exposed shall be free of chips, cracks or other imperfections when viewed from 20 feet (6100 mm), except that not more than 5 percent of a shipment may have slight cracks or small chips not larger than 1 inch (25.4 mm).

SECTION 21.408 — METHODS OF SAMPLING AND TESTING

The purchaser or authorized representative shall be accorded proper facilities to inspect and sample the units at the place of manufacture from the lots ready for delivery.

Sample and test units in accordance with ASTM C 140.

Total linear drying shrinkage shall be based on tests of concrete masonry units made with the same materials, concrete mix design, manufacturing process and curing method, conducted in accordance with ASTM C 426 and not more than 24 months prior to delivery.

SECTION 21.409 — REJECTION

If the samples tested from a shipment fail to conform to the specified requirements, the manufacturer may sort it, and new specimens shall be selected by the purchaser from the retained lot and tested at the expense of the manufacturer. If the second set of specimens fails to conform to the specified requirements, the entire lot shall be rejected.

TABLE 21-4-A—MOISTURE CONTENT REQUIREMENTS FOR TYPE I UNITS

	MOISTURE CONTENT, MAX. PERCENT OF TOTAL ABSORPTION (Average of 3 Units)		
	Humidity Conditions at Jobsite or Point of Use		
LINEAR SHRINKAGE, PERCENT	Humid[1]	Intermediate[2]	Arid[3]
0.03 or less	45	40	35
From 0.03 to 0.045	40	35	30
0.045 to 0.065, max.	35	30	25

[1]Average annual relative humidity above 75 percent.
[2]Average annual relative humidity 50 to 75 percent.
[3]Average annual relative humidity less than 50 percent.

TABLE 21-4-B—STRENGTH AND ABSORPTION REQUIREMENTS

COMPRESSIVE STRENGTH, MIN, psi (MPa)		WATER ABSORPTION, MAX, lb./ft. (kg/m) (Average of 3 Units)		
Average Net Area		Weight Classification—Oven-dry Weight of Concrete, lb./ft. (kg/m)		
Average of 3 Units	Individual Unit	Lightweight, Less than 105 (1680)	Medium Weight, 105 to less than 125 (1680-2000)	Normal Weight, 125 (2000) or more
1900 (13.1)	1700 (11.7)	18 (288)	15 (240)	13 (208)

TABLE 21-4-C—MINIMUM THICKNESS OF FACE-SHELLS AND WEBS

		WEB THICKNESS (WT)	
NOMINAL WIDTH (W) OF UNIT (inches)	FACE-SHELL THICKNESS (FST) MIN., (inches)[1, 4]	Webs[1] Min., (inches)	Equivalent Web Thickness, Min., In./Lin. Ft.[2]
	× 25.4 for mm		× 83 for mm/lin. m
3 and 4	$3/4$	$3/4$	$1^5/_8$
6	1	1	$2^1/_4$
8	$1^1/_4$	1	$2^1/_4$
10	$1^3/_8$	$1^1/_8$	$2^1/_2$
	$1^1/_4{}^3$		
12	$1^1/_2$	$1^1/_8$	$2^1/_2$
	$1^1/_4{}^3$		

[1]Average of measurements on three units taken at the thinnest point.
[2]Sum of the measured thickness of all webs in the unit, multiplied by 12 (305 when using metric), and divided by the length of the unit. In the case of open-ended units where the open-ended portion is solid grouted, the length of that open-ended portion shall be deducted from the overall length of the unit.
[3]This face-shell thickness (FST) is applicable where allowable design load is reduced in proportion to the reduction in thicknesses shown, except that allowable design load on solid-grouted units shall not be reduced.
[4]For split-faced units, a maximum of 10 percent of a shipment may have face-shell thicknesses less than those shown, but in no case less than $3/4$ inch (19 mm).

UNIFORM BUILDING CODE STANDARD 21-5
NONLOAD-BEARING CONCRETE MASONRY UNITS

Based on Standard Specification C 129-95 (1980) of the American Society for Testing and Materials.
Extracted, with permission, from the *Annual Book of ASTM Standards,* copyright American Society for
Testing and Materials, 100 Barr Harbor Drive, West Conshohocken, PA 19428

See Section 2102.2, Item 5, *Uniform Building Code*

SECTION 21.501 — SCOPE

This standard covers hollow and solid nonload-bearing concrete masonry units made from portland cement, water, and mineral aggregates with or without the inclusion of other materials. Such units are intended for use in nonload-bearing partitions but under certain conditions may be suitable for use in nonload-bearing exterior walls above grade, where effectively protected from the weather.

SECTION 21.502 — CLASSIFICATION

21.502.1 Weight Classifications. Nonload-bearing concrete masonry units manufactured in accordance with this standard shall conform to one of three weight classifications and two types as follows:

WEIGHT CLASSIFICATION	OVEN-DRY WEIGHT OF CONCRETE lb./cu.ft.
Lightweight	105 (1680 kg/m^3) max.
Medium weight	105 - 125 (1680 - 2000 kg/m^3)
Normal weight	125 (2000 kg/m^3) min.

21.502.2 Types. Nonload-bearing concrete masonry units shall be of two types as follows:

21.502.2.1 Type I, moisture-controlled units. Type I units shall conform to all requirements of this standard, including the requirements of Table 21-5-A.

21.502.2.2 Type II, nonmoisture-controlled units. Type II units shall conform to all requirements of this standard, except the requirements listed in Table 21-5-A.

SECTION 21.503 — MATERIALS

21.503.1 Cementitious Materials. Cementitious materials shall conform to the following applicable standards:

1. Portland Cement—ASTM C 150 modified as follows:

   Limitation on insoluble residue—1.5 percent.
   Limitation on air content of mortar,
      Volume percent—22 percent maximum.
   Limitation on loss on ignition—7 percent maximum.
   Limestone with a minimum 85 percent calcium carbonate (CaCO$_3$) content may be added to the cement, provided the requirements of ASTM C 150 as modified above are met.

2. Blended Cements—ASTM C 595.

3. Hydrated Lime, Type S—UBC Standard 21-13.

21.503.2 Other Constituents. Air-entraining agents, coloring pigments, integral water repellents, finely ground silica, etc., shall be previously established as suitable for use in concrete or shall be shown by test or experience not to be detrimental to the durability of the concrete.

SECTION 21.504 — PHYSICAL REQUIREMENTS

At the time of delivery to the work site, the units shall conform to the strength requirements prescribed in Table 21-5-B.

The moisture content of Type I concrete masonry units at the time of delivery shall conform to the requirements prescribed in Table 21-5-A.

At the time of delivery to the purchaser, the total linear drying of Type II units shall not exceed 0.065 percent.

SECTION 21.505 — DIMENSIONS AND PERMISSIBLE VARIATIONS

Minimum face-shell thickness shall not be less than $^1/_2$ inch (13 mm).

No overall dimension (width, height or length) shall differ by more than $^1/_8$ inch (3.2 mm) from the specified standard dimensions.

> **NOTE:** Standard dimensions of units are the manufacturer's designated dimensions. Nominal dimensions of modular-size units are equal to the standard dimensions plus $^3/_8$ inch (9.5 mm), the thickness of one standard mortar joint. Nominal dimensions of nonmodular size units usually exceed the standard dimensions by $^1/_8$ inch to $^1/_4$ inch (3.2 mm to 6.4 mm).

Variations in thickness of architectural units such as split-faced or slumped units will usually exceed the specified tolerances.

SECTION 21.506 — VISUAL INSPECTION

21.506.1 General. All units shall be sound and free of cracks or other defects that would interfere with the proper placing of the units or impair the strength or permanence of the construction. Units may have minor cracks incidental to the usual method of manufacture, or minor chipping resulting from customary methods of handling in shipment and delivery.

21.506.2 Exposed Units. Where units are to be used in exposed wall construction, the face or faces that are to be exposed shall be free of chips, cracks or other imperfections when viewed from 20 feet (6100 mm), except that not more than 5 percent of a shipment may have slight cracks or small chips not larger than 1 inch (25 mm).

21.506.3 Identification. Nonloading concrete masonry units shall be clearly marked in a manner to preclude their use as load-bearing units.

SECTION 21.507 — METHODS OF SAMPLING AND TESTING

The purchaser or authorized representative shall be accorded proper facilities to inspect and sample the units at the place of manufacture from the lots ready for delivery. At least 10 days shall be allowed for the completion of the tests.

Sample and test units in accordance with ASTM C 140 and ASTM C 426 when applicable.

Total linear drying shrinkage shall be based on tests of concrete masonry units made with the same materials, concrete mix design, manufacturing process and curing method, conducted in accordance with ASTM C 426 and not more than 24 months prior to delivery.

SECTION 21.508 — REJECTION

If the shipment fails to conform to the specified requirements, the manufacturer may sort it, and new specimens shall be selected by the purchaser from the retained lot and tested at the expense of the manufacturer. If the second set of specimens fails to conform to the specified requirements, the entire lot shall be rejected.

TABLE 21-5-A—MOISTURE CONTENT REQUIREMENTS FOR TYPE I UNITS

LINEAR SHRINKAGE, PERCENT	MOISTURE CONTENT, MAX. PERCENT OF TOTAL ABSORPTION (Average of 3 Units)		
	Humidity[1] Conditions at Jobsite or Point of Use		
	Humid	Intermediate	Arid
0.03 or less	45	40	35
From 0.03 to 0.045	40	35	30
0.045 to 0.065, max.	35	30	25

[1]Arid—Average annual relative humidity less than 50 percent.

Intermediate—Average annual relative humidity 50 to 75 percent.

Humid—Average annual relative humidity above 75 percent.

TABLE 21-5-B—STRENGTH REQUIREMENTS

	COMPRESSIVE STRENGTH (Average Net Area) Min., psi
	× 6.89 for kPa
Average of 3 units	600
Individual units	500

UNIFORM BUILDING CODE STANDARD 21-6
IN-PLACE MASONRY SHEAR TESTS

Test Standard of the International Conference of Building Officials

See Appendix Chapter 1, Sections A106.3.3 and A107.2,
Uniform Code for Building Conservation

SECTION 21.601 — SCOPE

This standard applies when the *Uniform Code for Building Conservation* requires in-place testing of the quality of masonry mortar.

SECTION 21.602 — PREPARATION OF SAMPLE

The bed joints of the outer wythe of the masonry shall be tested in shear by laterally displacing a single brick relative to the adjacent bricks in the same wythe. The head joint opposite the loaded end of the test brick shall be carefully excavated and cleared. The brick adjacent to the loaded end of the test brick shall be carefully removed by sawing or drilling and excavating to provide space for a hydraulic ram and steel loading blocks.

SECTION 21.603 — APPLICATION OF LOAD AND DETERMINATION OF RESULTS

Steel blocks, the size of the end of the brick, shall be used on each end of the ram to distribute the load to the brick. The blocks shall not contact the mortar joints. The load shall be applied horizontally, in the plane of the wythe, until either a crack can be seen or slip occurs. The strength of the mortar shall be calculated by dividing the load at the first cracking or movement of the test brick by the nominal gross area of the sum of the two bed joints.

TESTS OF ANCHORS IN UNREINFORCED MASONRY WALLS

Test Standard of the International Conference of Building Officials

See Appendix Chapter 1, Section A107.3 and A107.4,
Uniform Code for Building Conservation

SECTION 21.701 — SCOPE

Shear and tension anchors in existing masonry construction shall be tested in accordance with this standard when required by the *Uniform Code for Building Conservation.*

SECTION 21.702 — DIRECT TENSION TESTING OF EXISTING ANCHORS AND NEW BOLTS

The test apparatus shall be supported by the masonry wall. The distance between the anchor and the test apparatus support shall not be less than one half the wall thickness for existing anchors and 75 percent of the embedment for new embedded bolts. Existing wall anchors shall be given a preload of 300 pounds (1335 N) prior to establishing a datum for recording elongation. The tension test load reported shall be recorded at $^1/_8$ inch (3.2 mm) relative movement of the existing anchor and the adjacent masonry surface. New embedded tension bolts shall be subject to a direct tension load of not less than 2.5 times the design load but not less than 1,500 pounds (6672 N) for five minutes (10 percent deviation).

SECTION 21.703 — TORQUE TESTING OF NEW BOLTS

Bolts embedded in unreinforced masonry walls shall be tested using a torque-calibrated wrench to the following minimum torques:

$^1/_2$-inch-diameter (13 mm) bolts—40 foot pounds (54.2 N·m)

$^5/_8$-inch-diameter (16 mm) bolts—50 foot pounds (67.8 N·m)

$^3/_4$-inch-diameter (19 mm) bolts—60 foot pounds (81.3 N·m)

SECTION 21.704 — PREQUALIFICATION TEST FOR BOLTS AND OTHER TYPES OF ANCHORS

This section is applicable when it is desired to use tension or shear values for anchors greater than those permitted by Table A-1-E of the *Uniform Code for Building Conservation.* The direct-tension test procedure set forth in Section 21.702 for existing anchors may be used to determine the allowable tension values for new embedded or through bolts, except that no preload is required. Bolts shall be installed in the same manner and using the same materials as will be used in the actual construction. A minimum of five tests for each bolt size and type shall be performed for each class of masonry in which they are proposed to be used. The allowable tension values for such anchors shall be the lesser of the average ultimate load divided by a factor of safety of 5.0 or the average load of which $^1/_8$ inch (3.2 mm) elongation occurs for each size and type of bolt and class of masonry.

Shear bolts may be similarly prequalified. The test procedure shall comply with ASTM E 488-90 or another approved procedure.

The allowable values determined in this manner may exceed those set forth in Table A-1-E of the *Uniform Code for Building Conservation.*

SECTION 21.705 — REPORTS

Results of all tests shall be reported. The report shall include the test results as related to anchor size and type, orientation of loading, details of the anchor installation and embedment, wall thickness, and joist orientation.

POINTING OF UNREINFORCED MASONRY WALLS

Construction Specification of the International Conference of Building Officials

See Appendix Chapter 1, Section A106.3.3.2,
Uniform Code for Building Conservation

SECTION 21.801 — SCOPE

Pointing of deteriorated mortar joints when required by the *Uniform Code for Building Conservation* shall be in accordance with this standard.

SECTION 21.802 — JOINT PREPARATION

The old or deteriorated mortar joint shall be cut out, by means of a toothing chisel or nonimpact power tool, to a uniform depth of $^3/_4$ inch (19 mm) until sound mortar is reached. Care shall be taken not to damage the brick edges. After cutting is complete, all loose material shall be removed with a brush, air or water stream.

SECTION 21.803 — MORTAR PREPARATION

The mortar mix shall be Type N or Type S proportioned as required by the construction specifications. The pointing mortar shall be prehydrated by first thoroughly mixing all ingredients dry and then mixing again, adding only enough water to produce a damp unworkable mix which will retain its form when pressed into a ball. The mortar shall be kept in a damp condition for one and one-half hours; then sufficient water shall be added to bring it to a consistency that is somewhat drier than conventional masonry mortar.

SECTION 21.804 — PACKING

The joint into which the mortar is to be packed shall be damp but without freestanding water. The mortar shall be tightly packed into the joint in layers not exceeding $^1/_4$ inch (6.4 mm) in depth until it is filled; then it shall be tooled to a smooth surface to match the original profile.

UNBURNED CLAY MASONRY UNITS AND STANDARD METHODS OF SAMPLING AND TESTING UNBURNED CLAY MASONRY UNITS

Test Standard of the International Conference of Building Officials

See Section 2102.2, Item 6, *Uniform Building Code*

Part I—Unburned Clay Masonry

SECTION 21.901 — SCOPE

This standard covers unburned clay masonry units made from a suitable mixture of soil, clay and stabilizing agent, and intended for use in brick masonry.

SECTION 21.902 — COMPOSITION OF UNITS

21.902.1 Soil. The soil used shall contain not less than 25 percent and not more than 45 percent of material passing a No. 200 mesh (75 μm) sieve. The soil shall contain sufficient clay to bind the particles together, but shall contain not more than 0.2 percent of water-soluble salts.

21.902.2 Stabilizer. The stabilizing agent shall be emulsified asphalt. The stabilizing agent shall be uniformly mixed with the soil in amounts sufficient to provide the required resistance to absorption.

SECTION 21.903 — PHYSICAL REQUIREMENTS

The units shall conform to the physical requirements prescribed in Table 21-1-B of UBC Standard 21-1.

SECTION 21.904 — SHRINKAGE CRACKS

No units shall contain more than three shrinkage cracks, and no shrinkage crack shall exceed 3 inches (76 mm) in length or $1/8$ inch (3.2 mm) in width.

Part II—Sampling and Testing of Unburned Clay Masonry Units

SECTION 21.905 — SCOPE

These methods cover procedures for the sampling and testing of unburned clay masonry units for compressive strength, modulus of rupture, absorption and moisture content.

Sampling

SECTION 21.906 — TEST SPECIMENS

For each of the tests prescribed in this standard, five sample units shall be selected at random from each lot of 5,000 units or fraction thereof.

SECTION 21.907 — IDENTIFICATION

Each specimen shall be marked so that it may be identified at any time. Markings shall not cover more than 5 percent of the superficial area of the specimen.

Compressive Strength

SECTION 21.908 — PROCEDURE

Five full-size specimens shall be tested for compressive strength according to the following procedure:

1. Dry the specimens at a temperature of 85°F ± 15°F (29°C ± 9°C) in an atmosphere having a relative humidity of not more than 50 percent. Weigh the specimens at one-day intervals until constant weight is attained.

2. Test the specimens in the position in which the unburned clay masonry unit is designed to be used, and bed on and cap with a felt pad not less than $1/8$ inch (3.2 mm) nor more than $1/4$ inch (6.4 mm) in thickness.

3. The specimens may be suitably capped with calcined gypsum mortar or the bearing surfaces of the tile may be planed or rubbed smooth and true. When calcined gypsum is used for capping, conduct the test after the capping has set and the specimen has been dried to constant weight in accordance with Item 1 of this section.

4. The loading head shall completely cover the bearing area of the specimen and the applied load shall be transmitted through a spherical bearing block of proper design. The speed of the moving head of the testing machine shall not be more than 0.05 inch (1.27 mm) per minute.

5. Calculate the average compressive strength of the specimens tested and report this as the compressive strength of the block.

Modulus of Rupture

SECTION 21.909 — PROCEDURE

Five full-size specimens shall be tested for modulus of rupture according to the following procedure:

1. Cured specimen shall be positioned on cylindrical supports 2 inches (51 mm) in diameter, located 2 inches (51 mm) from each end, and extending across the full width of the specimen.

2. A cylinder 2 inches (51 mm) in diameter shall be positioned on the specimen midway between and parallel to the cylindrical supports.

3. Load shall be applied to the cylinder at the rate of 500 pounds (2224 N) per minute until failure occurs.

4. Calculate modulus of rupture from the formula $S = \dfrac{3WL}{2Bd^2}$

WHERE:

B = width of specimen.

d = thickness of specimen.

L = distance between supports.

S = modulus of rupture, psi (kPa).

W = load at failure.

SECTION 21.910 — PROCEDURE

A 4-inch (102 mm) cube cut from a sample unit shall be tested for absorption according to the following procedure:

1. Dry specimen to a constant weight in a ventilated oven at 212°F to 239°F (100°C to 115°C).

2. Place specimen on a constantly water-saturated porous surface for seven days. Weigh specimen.

3. Calculate absorption as a percentage of the initial dry weight.

SECTION 21.911 — PROCEDURE

Five representative specimens shall be tested for moisture content according to the following procedure:

1. Obtain weight of each specimen immediately upon receiving.

2. Dry all specimens to constant weight in a ventilated oven at 212°F to 239°F (100°C to 115°C) and obtain dry weight.

3. Calculate moisture content as a percentage of the initial dry weight.

UNIFORM BUILDING CODE STANDARD 21-10
JOINT REINFORCEMENT FOR MASONRY

Specification Standard of the International Conference of Building Officials

See Sections 2102.2; 2104 and 2106.1.12.4, Item 2, *Uniform Building Code*

Part I—Joint Reinforcement for Masonry

SECTION 21.1001 — SCOPE

This standard covers joint reinforcement fabricated from cold-drawn steel wire for reinforcing masonry.

SECTION 21.1002 — DESCRIPTION

Joint reinforcement consists of deformed longitudinal wires welded to cross wires (Figure 21-10-1) in sizes suitable for placing in mortar joints between masonry courses.

SECTION 21.1003 — CONFIGURATION AND SIZE OF LONGITUDINAL AND CROSS WIRES

21.1003.1 General. The distance between longitudinal wires and the configuration of cross wires connecting the longitudinal wires shall conform to the design and the requirements of Figure 21-10-1.

21.1003.2 Longitudinal Wires. The diameter of longitudinal wires shall not be less than 0.148 inch (3.76 mm) or more than one half the mortar joint thickness.

21.1003.3 Cross Wires. The diameter of cross wires shall not be less than (No. 9 gage) 0.148-inch (3.76 mm) diameter nor more than the diameter of the longitudinal wires. Cross wires shall not project beyond the outside longitudinal wires by more than $^1/_8$ inch (3.2 mm).

21.1003.4 Width. The width of joint reinforcement shall be the out-to-out distance between outside longitudinal wires. Variation in the width shall not exceed $^1/_8$ inch (3.2 mm).

21.1003.5 Length. The length of pieces of joint reinforcement shall not vary more than $^1/_2$ inch (13 mm) or 1.0 percent of the specified length, whichever is less.

SECTION 21.1004 — MATERIAL REQUIREMENTS

21.1004.1 Tensile Properties. Wire of the finished product shall meet the following requirements:

Tensile strength, minimum	75,000 psi (517 MPa)
Yield strength, minimum	60,000 psi (414 MPa)
Reduction of area, minimum	30 percent

For wire testing over 100,000 psi (689 MPa), the reduction of area shall not be less than 25 percent.

21.1004.2 Bend Properties. Wire shall not break or crack along the outside diameter of the bend when tested in accordance with Section 21.1008.

21.1004.3 Weld Shear Properties. The least weld shear strength in pounds shall not be less than 25,000 (11.3 Mg) multiplied by the specified area of the smaller wire in square inches.

SECTION 21.1005 — FABRICATION

Wire shall be fabricated and finished in a workmanlike manner, shall be free from injurious imperfections and shall conform to this standard.

The wires shall be assembled by automatic machines or by other suitable mechanical means which will assure accurate spacing and alignment of all members of the finished product.

Longitudinal and cross wires shall be securely connected at every intersection by a process of electric-resistance welding.

Longitudinal wires shall be deformed. One set of four deformations shall occur around the perimeter of the wire at a maximum spacing of 0.7 times the diameter of the wire but not less than eight sets per inch (25.4 mm) of length. The overall length of each deformation within the set shall be such that the summation of gaps between the ends of the deformations shall not exceed 25 percent of the perimeter of the wire. The height or depth of the deformations shall be 0.012 inch (0.305 mm) for $^3/_{16}$ inch (4.76 mm) diameter or larger wire, 0.011 (0.28 mm) for 0.162-inch (4.11 mm) diameter wire and 0.009 inch (0.23 mm) for 0.148-inch (3.76 mm) diameter wire.

SECTION 21.1006 — TENSION TESTS

Tension tests shall be made on individual wires cut from the finished product across the welds.

Tension tests across a weld shall have the welded joint located approximately at the center of the wire being tested.

Tensile strength shall be the average of four test values determined by dividing the maximum test load by the specified cross-sectional area of the wire.

Reduction of area shall be determined by measuring the ruptured section of a specimen which has been tested.

SECTION 21.1007 — WELD SHEAR STRENGTH TESTS

Test specimens shall be obtained from the finished product by cutting a section of wire which includes one weld.

Weld shear strength tests shall be conducted using a fixture of such design as to prevent rotation of the cross wire. The cross wire shall be placed in the anvil of the testing device which is secured in the tensile machine and the load then applied to the longitudinal wire.

Weld shear strength shall be the average test load in pounds of four tests.

SECTION 21.1008 — BEND TESTS

Test specimens shall be obtained from the finished product by cutting a section of wire without welds.

The test specimens shall be bent cold through 180 degrees around a pin, the diameter of which is equal to the diameter of the specimen.

The specimen shall not break nor shall there be visual cracks on the outside diameter of the bend.

SECTION 21.1009 — FREQUENCY OF TESTS

One set of tension tests, weld strength shear tests and bend tests shall be performed for each 2,000,000 lineal feet (610 000 m) of joint reinforcement, but not less than monthly.

SECTION 21.1010 — CORROSION PROTECTION

When corrosion protection of joint reinforcement is provided, it shall be in accordance with one of the following:

21.1010.1 Brite Basic. No coating.

21.1010.2 Mill Galvanized. Zinc coated, by the hot-dipped method, with no minimum thickness of zinc coating. The coating may be applied before fabrication.

21.1010.3 Class I Mill Galvanized. Zinc coated, by the hot-dipped method, with a minimum of 0.40 ounce of zinc per square foot (0.12 kg/m^2) of surface area. The coating may be applied before fabrication.

21.1010.4 Class III Mill Galvanized. Zinc coated, by the hot-dipped method, with a minimum of 0.80 ounce of zinc per square foot (0.24 kg/m^2) of surface area. The coating may be applied before fabrication.

21.1010.5 Hot-dipped Galvanized. Zinc coated, by the hot-dipped method, with a minimum of 1.50 ounces of zinc per square foot (0.45 kg/m^2) of surface area. The coating shall be applied after fabrication.

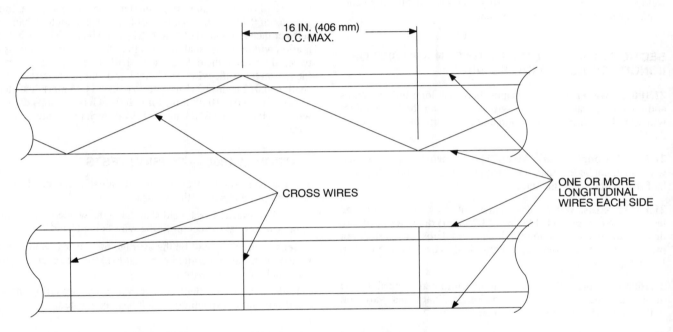

FIGURE 21-10-1—JOINT REINFORCEMENT

Part II—Cold-drawn Steel Wire for Concrete Reinforcement

Based on Standard Specification A 82-90a of the American Society for Testing and Materials. Extracted, with permission, from the _Annual Book of ASTM Standards,_ copyright American Society for Testing and Materials, 100 Barr Harbor Drive, West Conshohocken, PA 19428

See Sections 2101.3; 2104 and 2106.1.12.4, Item 2, _Uniform Building Code_

SECTION 21.1011 — SCOPE

This standard covers cold-drawn steel wire to be used as such or in fabricated form, for the reinforcement as follows:

SIZE NUMBER	NOMINAL DIAMETER (inch) (× 25.4 for mm)	NOMINAL AREA (square inch) (× 645 for mm²)
W 31	0.628	0.310
W 30	0.618	0.300
W 28	0.597	0.280
W 26	0.575	0.260
W 24	0.553	0.240
W 22	0.529	0.220
W 20	0.505	0.200
W 18	0.479	0.180
W 16	0.451	0.160
W 14	0.422	0.140
W 12	0.391	0.120
W 10	0.357	0.100
W 8	0.319	0.080
W 6	0.276	0.060
W 5.5	0.265	0.055
W 5	0.252	0.050
W 4.5	0.239	0.045
W 4	0.226	0.040
W 3.5	0.211	0.035
W 2.9	0.192	0.029
W 2.5	0.178	0.025
W 2	0.160	0.020
W 1.4	0.134	0.014
W 1.2	0.124	0.012
W 0.5	0.080	0.005

SECTION 21.1012 — PROCESS

The steel shall be made by one or more of the following processes: open hearth, electric furnace or basic oxygen.

The wire shall be cold drawn from rods that have been hot rolled from billets.

Unless otherwise specified, the wire shall be "as cold drawn," except wire smaller than size number W 1.2 for welded fabric, which shall be galvanized at finish size.

SECTION 21.1013 — TENSILE PROPERTIES

The material, except as specified in this section, shall conform to the following tensile property requirements based on nominal area of wire:

Tensile strength, minimum, psi	80,000 (552 MPa)
Yield strength, minimum, psi	70,000 (483 MPa)
Reduction of area, minimum, percent	30

For material testing over 100,000 pounds per square inch (689 MPa) tensile strength, the reduction of area shall not be less than 25 percent.

For material to be used in the fabrication of welded fabric, the following tensile and yield strength properties based on nominal area of wire shall apply:

	SIZE W. 1.2 AND LARGER	SMALLER THAN SIZE W 1.2
Tensile strength, minimum, psi	75,000 (517 MPa)	70,000 (483 MPa)
Yield strength, minimum, psi	65,000 (448 MPa)	56,000 (386 MPa)

The yield strength shall be determined at an extension of 0.005 inch per inch (0.005 mm per mm) of gage length.

The material shall not exhibit a definite yield point as evidenced by a distinct drop of the beam or halt in the gage of the testing machine prior to reaching ultimate tensile load.

SECTION 21.1014 — BENDING PROPERTIES

The bend test specimen shall stand being bent cold through 180 degrees without cracking on the outside of the bent portion, as follows:

SIZE NUMBER OF WIRE	BEND TEST
W 7 and smaller	Bend around a pin, the diameter of which is equal to the diameter of the specimen.
Larger than W 7	Bend around a pin, the diameter of which is equal to twice the diameter of the specimen.

SECTION 21.1015 — TEST SPECIMENS

Tension and bend test specimens shall be of the full section of the wire and shall be obtained from ends of wire coils.

SECTION 21.1016 — NUMBER OF TESTS

One tension test and one bend test shall be made from each 10 tons (89 kN) or less of each size of wire or fraction thereof in a lot, or a total of seven samples, whichever is less. A lot shall consist of all the coils of a single size offered for delivery at the same time.

If any test specimen shows imperfections or develops flaws, it may be discarded and another specimen substituted.

SECTION 21.1017 — PERMISSIBLE VARIATIONS IN WIRE DIAMETER

The permissible variation in the diameter of the wire shall conform to the following:

SIZE NUMBER	PERMISSIBLE VARIATION PLUS AND MINUS (inch) (× 25.4 for mm)
Smaller than W 5	0.003
W 5 to W 12, inclusive	0.004
Over W 12 to W 20, inclusive	0.006
Over W 20	0.008

The difference between the maximum and minimum diameter, as measured on any given cross section of the wire, shall be more than the tolerances shown above for the given wire size.

SECTION 21.1018 — FINISH

The wire shall be free from injurious imperfections and shall have a workmanlike finish with smooth surface.

Galvanized wire shall be completely covered in a workmanlike manner with a zinc coating.

CEMENT, MASONRY

Based on Standard Specification C 91-93 of the American Society for Testing and Materials.
Extracted, with permission, from the *Annual Book of ASTM Standards,* copyright American Society for
Testing and Materials, 100 Barr Harbor Drive, West Conshohocken, PA 19428

See Section 2102.2, Item 2 and Table 21-A, *Uniform Building Code*

SECTION 21.1101 — SCOPE

This standard covers three types of masonry cement for use in masonry mortars.

SECTION 21.1102 — CLASSIFICATIONS

21.1102.1 General. Masonry cement complying with this standard shall be classified as one of the types set forth in this section.

21.1102.2 Type N. Type N cement is for use as the cementitious material in the prep-aration of UBC Standard 21-15 Type N and Type O mortars. It is for use in combination with portland or blended hydraulic cements in the preparation of Type S or Type M mortars.

21.1102.3 Type S. Type S cement is for use as the cementitious material in the preparation of UBC Standard 21-15 Type S mortar.

21.1102.4 Type M. Type M cement is for use as the cementitious material in the preparation of UBC Standard 21-15 Type M mortar.

SECTION 21.1103 — PHYSICAL REQUIREMENTS

Masonry cement shall conform to the requirements set forth in Table 21-11-A for its classifications.

SECTION 21.1104 — PACKAGE LABELING

Masonry cement packages shall carry a statement indicating that the product conforms to requirements of this standard and shall include the brand, name of manufacturer, type of masonry cement and net weight of the package in pounds.

SECTION 21.1105 — CERTIFICATION

Certification shall be submitted upon request of the building official and shall certify compliance with the requirements of this standard.

SECTION 21.1106 — SAMPLING AND TESTING

Every 90 days, each masonry cement producer shall retain an approved agency to obtain a random sample from a local point of supply in the market area served by the producer.

The agency shall test the masonry cement for compliance with the physical requirements of Table 21-11-A.

Upon request of the building official, the producer shall furnish (at no cost) test results to the building official, architect, structural engineer, general contractor and masonry contractor.

SECTION 21.1107 — TEMPERATURE AND HUMIDITY

The temperature of the air in the vicinity of the mixing slab and dry materials, molds, base plates and mixing bowl shall be maintained between 68°F and 81.5°F (20°C and 27.5°C). The temperature of the mixing water, moist cabinet or moist room, and water in the storage tank shall not vary from 73.4°F (23°C) by more than 3°F (1.7°C).

The relative humidity of the laboratory air shall not be less than 50 percent. The moist cabinet or moist room atmosphere shall have a relative humidity of not less than 90 percent.

The moist cabinet or moist room shall conform to applicable standards.

SECTION 21.1108 — FINENESS

The fineness of the cement shall be determined from the residue on the No. 325 (45 μm) sieve.

SECTION 21.1109 — NORMAL CONSISTENCY

Determine normal consistency by the Vicat apparatus.

SECTION 21.1110 — AUTOCLAVE EXPANSION

The autoclave expansion shall be determined. After molding, store the bars in the moist cabinet or room for 48 hours ± 30 minutes before removal from the molds for measurement and test in the autoclave. Calculate the difference in the lengths of the test specimen before and after autoclaving to the nearest 0.01 percent of the effective gage length and report as the autoclave expansion of the masonry cement.

SECTION 21.1111 — TIME OF SETTING

The time of setting shall be determined by the Gillmore needle method.

SECTION 21.1112 — DENSITY

The density of the masonry cement shall be determined by using kerosene as the liquid. Use the density so determined in the calculation of the air content of the mortars.

SECTION 21.1113 — APPARATUS FOR MORTAR TESTS

The apparatus for mortar tests shall be in accordance with applicable standards.

SECTION 21.1114 — BLENDED SAND

The sand shall be a blend of equal parts by weight of graded standard sand and Standard 20-30 sand.

SECTION 21.1115 — PREPARATION OF MORTAR

21.1115.1 Proportions for Mortar. Mortar for air entrainment, compressive strength and water-retention tests shall be proportioned to contain the weight of cement, in grams, equal to six times the printed bag weight in pounds (13.228 times the printed bag weight in kilograms) and 1,440 grams of sand. The sand shall consist of 720 grams of graded Ottawa sand and 720 grams of Standard 20-30 sand. The quantity of water, measured in milliliters, shall be such as to produce a flow of 110 ± 5 as determined by the flow table.

21.1115.2 Mixing of Mortars. The mortar shall be mixed in accordance with the applicable standards.

21.1115.3 Determination of Flow. The flow shall be determined in accordance with applicable standards.

SECTION 21.1116 — AIR ENTRAINMENT

21.1116.1 Procedure. If the mortar has the correct flow, use a separate portion of the mortar for the determination of entrained air. Determine the mass of 400 ml of the mortar.

21.1116.2 Calculation. Calculate the air content of the mortar and report it to the nearest 0.1 percent as follows:

$$D = (W_1 + W_2 + V_w) [(W_1/S_1) + (W_2/S_2) + V_w]$$

$$A = 100 - (w_m/4D)$$

WHERE:

A = volume percent of entrained air.

D = density of air-free mortar, g/ml.

S_1 = density of cement, g/ml.

S_2 = density of standard sand, 2.65 g/ml.

V_w = milliliters-grams of water used.

W_m = mass of 400 ml.

W_1 = mass of cement, g.

W_2 = mass of sand, g.

SECTION 21.1117 — COMPRESSIVE STRENGTH

21.1117.1 Test Specimens.

21.1117.1.1 Molding. Immediately after determining the flow and the mass of 400 ml of mortar, return all the mortar to the mixing bowl and remix for 15 seconds at the medium speed. Then mold test specimens in accordance with applicable standards, except that elapsed time for mixing mortar, determining flow, determining air entrainment and starting the molding of cubes shall be within eight minutes.

21.1117.1.2 Storage. Store all test specimens immediately after molding in the molds on plane plates in a moist cabinet or moist room for 48 to 52 hours, in such a manner that the upper surfaces shall be exposed to the moist air. Then remove the cubes from the molds and place in the moist cabinet or moist room for five days in such a manner as to allow free circulation of air around at least five faces of the specimens. At the age of seven days, immerse the cubes for the 28-day tests in saturated lime water in storage tanks of noncorrodible materials.

21.1117.2 Procedure. Test the cube specimens immediately after their removal from the moist cabinet or moist room for seven-day specimens, and immediately after their removal from storage water for all other specimens. If more than one specimen at a time is removed from the moist cabinet or moist room for seven-day tests, cover these cubes with a damp cloth until time of testing. If more than one specimen at a time is removed from the storage water for testing, place these cubes in a pan of water at a temperature

of 73.4°F ± 3°F (23°C ± 1.7°C), and of sufficient depth to completely immerse each cube until time of testing.

The remainder of the testing procedure shall conform to applicable standards.

SECTION 21.1118 — WATER RETENTION

21.1118.1 Apparatus. The water-retention test shall conform to applicable standards.

21.1118.2 Procedure. Adjust the mercury relief column to maintain a vacuum of 51 ± 3 mm as indicated by the manometer. Seat the perforated dish on the greased gasket or greased rim of the funnel. Place a wetted filter paper in the bottom of the dish. Turn the stopcock to apply the vacuum to the funnel and check the apparatus for leaks and to determine that the required vacuum is obtained. Then turn the stopcock to shut off the vacuum from the funnel.

Mix the mortar to a flow of 110 ± 5 percent in accordance with applicable standards. Immediately after making the flow test, return the mortar on the flow table to the mixing bowl and remix the entire batch for 15 seconds at medium speed. Immediately after remixing the mortar, fill the perforated dish with the mortar to slightly above the rim. Tamp the mortar 15 times with the tamper. Apply 10 of the tamping strokes at approximately uniform spacing adjacent to the rim of the dish and with the long axis of the tamping face held at right angles to the radius of the dish. Apply the remaining five tamping strokes at random points distributed over the central area of the dish. The tamping pressure shall be just sufficient to ensure filling of the dish. On completion of the tamping, the top of the mortar will extend slightly above the rim of the dish. Smooth off the mortar by drawing the flat side of the straightedge (with the leading edge slightly raised) across the top of the dish. Then cut off the mortar to a plane surface flush with the rim of the dish by drawing the straightedge with a sawing motion across the top of the dish in two cutting strokes, starting each cut from near the center of the dish. If the mortar is pulled away from the side of the dish during the process of drawing the straightedge across the dish, gently press the mortar back into contact with the side of the dish using the tamper.

Turn the stopcock to apply the vacuum to the funnel. The time elapsed from the start of mixing the cement and water to the time of applying the vacuum shall not exceed eight minutes. After suction for 60 seconds, quickly turn the stopcock to expose the funnel to atmospheric pressure. Immediately slide the perforated dish off from the funnel, touch it momentarily on a damp cloth to remove droplets of water, and set the dish on the table. Then, using the bowl scraper, plow and mix the mortar in the dish for 15 seconds. Upon completion of mixing, place the mortar in the flow mold and determine the flow. The entire operation shall be carried out without interruption and as quickly as possible, and shall be completed within an elapsed time of 11 minutes after the start of mixing the cement and water for the first flow determination. Both flow determinations shall be made in accordance with applicable standards.

21.1118.3 Calculation. Calculate the water-retention value for the mortar as follows:

$$\text{Water-retention value} = (A/B) \times 100$$

WHERE:

A = flow after suction.

B = flow immediately after mixing.

TABLE 21-11-A—PHYSICAL REQUIREMENTS

MASONRY CEMENT TYPE	N	S	M
Fineness, residue on a No. 325 (45 µm) sieve, maximum percent	24	24	24
Soundness: Autoclave expansion, maximum, percent	1.0	1.0	1.0

Time of setting, Gilmore method:
    Initial set, minimum, hour . . . .
    Final set, maximum, hour . . . .
	2
	24
		$1^1/_2$
		24
			$1^1/_2$
			24

Compressive strength
    (average of 3 cubes):
    Initial compressive strength of mortar cubes, composed of 1 part
    cement and 3 parts blended sand (half Graded Ottawa sand, and half
    Standard 20-30 Ottawa sand) by volume, prepared and tested in
    accordance with this specification shall be equal to or higher than
    the values specified for the ages indicated below:
        7 days, psi
        28 days, psi . . . .
	500 (3445 kPa)
	900 (6201 kPa)
		1,300 (8957 kPa)
		2,100 (14 469 kPa)
			1,800 (12 402 kPa)
			2,900 (19 981 kPa)

Air content of mortar:
    Minimum percent by volume . . . .
    Maximum percent by volume . . . .
	8
	21
		8
		19
			8
			19

Water retention, flow after suction, minimum, percent of original flow	70	70	70

QUICKLIME FOR STRUCTURAL PURPOSES

**Based on Standard Specification C 5-79 (Reapproved 1992) of the American Society for Testing and Materials.
Extracted, with permission, from the *Annual Book of ASTM Standards,* copyright American Society for
Testing and Materials, 100 Barr Harbor Drive, West Conshohocken, PA 19428**

See Section 2102.2, Item 3, *Uniform Building Code*

SECTION 21.1201 — SCOPE

This standard covers all classes of quicklime, such as crushed lime, granular lime, ground lime, lump lime, pebble lime and pulverized lime, used for structural purposes.

SECTION 21.1202 — GENERAL REQUIREMENTS

Quicklime shall be slaked and aged in accordance with the printed directions of the manufacturer. The resulting lime putty shall be stored until cool.

SECTION 21.1203 — CHEMICAL COMPOSITION

The quicklime shall conform to the following requirements as to chemical composition, calculated to the nonvolatile basis:

	CALCIUM LIME	MAGNESIUM LIME
Calcium oxide, minimum, percent	75	—
Magnesium oxide, minimum, percent	—	20
Calcium and magnesium oxides, minimum, percent	95	95
Silica, alumina, and oxide of iron, maximum, percent	5	5
Carbon dioxide, maximum, percent:		
If sample is taken at the place of manufacture	3	3
If sample is taken at any other place	10	10

SECTION 21.1204 — RESIDUE

The quicklime shall not contain more than 15 percent by weight of residue.

SECTION 21.1205 — QUALITY CONTROL

Every 90 days, each lime producer shall retain an approved agency to obtain a random sample from a local point of supply in the market area served by the producer.

The agency shall test the lime for compliance with the physical requirements of Section 21.1204.

Upon request of the building official, the producer shall furnish (at no cost) test results to the building official, architect, structural engineer, general contractor and masonry contractor.

UNIFORM BUILDING CODE STANDARD 21-13
HYDRATED LIME FOR MASONRY PURPOSES

Based on Standard Specification C 207-91 (Reapproved 1992) of the American Society for Testing and Materials. Extracted, with permission, from the *Annual Book of ASTM Standards,* copyright American Society for Testing and Materials, 100 Barr Harbor Drive, West Conshohocken, PA 19428

See Section 2102.2, Item 3, *Uniform Building Code*

SECTION 21.1301 — SCOPE

This standard covers four types of hydrated lime. Types N and S are suitable for use in mortar, in the scratch and brown coats of cement plaster, for stucco, and for addition to portland-cement concrete. Types NA and SA are air-entrained hydrated limes that are suitable for use in any of the above uses where the inherent properties of lime and air entrainment are desired. The four types of lime sold under this specification shall be designated as follows:

Type N—Normal hydrated lime for masonry purposes.

Type S—Special hydrated lime for masonry purposes.

Type NA—Normal air-entraining hydrated lime for masonry purposes.

Type SA—Special air-entraining hydrated lime for masonry purposes.

> **NOTE:** Type S, special hydrated lime, and Type SA, special air-entraining hydrated lime, are differentiated from Type N, normal hydrated lime, and Type NA, normal air-entraining hydrated lime, principally by their ability to develop high, early plasticity and higher water retentivity and by a limitation on their unhydrated oxide content.

SECTION 21.1302 — DEFINITION

HYDRATED LIME. The hydrated lime covered by Type N or S in this standard shall contain no additives for the purpose of entraining air. The air content of cement-lime mortars made with Type N or S shall not exceed 7 percent. Types NA and SA shall contain an air-entraining additive as specified by Section 21.1305. The air content of cement-lime mortars made with Type NA or SA shall have a minimum of 7 percent and a maximum of 14 percent.

SECTION 21.1303 — ADDITIONS

Types NA and SA hydrated lime covered by this standard shall contain additives for the purpose of entraining air.

SECTION 21.1304 — MANUFACTURER'S STATEMENT

Where required, the nature, amount and identity of the air-entraining agent used and of any processing addition that may have been used shall be provided, as well as test data showing compliance of such air-entraining addition.

SECTION 21.1305 — CHEMICAL REQUIREMENTS COMPOSITION

Hydrated lime for masonry purposes shall conform to the requirements as to chemical composition set forth in Table 21-13-A.

SECTION 21.1306 — RESIDUE, POPPING AND PITTING

The four types of hydrated lime for masonry purposes shall conform to one of the following requirements:

1. The residue retained on a No. 30 (600 μm) sieve shall not be more than 0.5 percent, or

2. If the residue retained on a No. 30 (600 μm) sieve is over 0.5 percent, the lime shall show no pops and pits when tested.

SECTION 21.1307 — PLASTICITY

The putty made from Type S, special hydrate, or Type SA, special air-entraining hydrate, shall have a plasticity figure of not less than 200 within 30 minutes after mixing with water, when tested.

SECTION 21.1308 — WATER RETENTION

Hydrated lime mortar made with Type N, normal hydrated lime, or Type NA, normal air-entraining hydrated lime, after suction for 60 seconds, shall have a water-retention value of not less than 75 percent when tested in a standard mortar made from the dry hydrate or from putty made from the hydrate which has been soaked for a period of 16 to 24 hours.

Hydrated lime mortar made with Type S, special hydrated lime, or Type SA, special air-entraining hydrated lime, after suction for 60 seconds, shall have a water-retention value of not less than 85 percent when tested in a standard mortar made from the dry hydrate.

SECTION 21.1309 — SPECIAL MARKING

When Type NA or SA air-entraining hydrated lime is delivered in packages, the type under this standard and the words "air-entraining" shall be plainly indicated thereon or, in case of bulk shipments, so indicated on shipping notices.

SECTION 21.1310 — QUALITY CONTROL

Every 90 days, each lime producer shall retain an approved agency to obtain a random sample from a local point of supply in the market area served by the producer.

The agency shall test the lime for compliance with the physical requirements of Sections 21.1306, 21.1307 and 21.1308.

Upon request of the building official, the producer shall furnish (at no cost) test results to the building official, architect, structural engineer, general contractor and masonry contractor.

TABLE 21-13-A—CHEMICAL REQUIREMENTS

	HYDRATE TYPES			
	N	NA	S	SA
Calcium and magnesium oxides (nonvolatile basis), min. percent	95	95	95	95
Carbon dioxide (as-received basis), max. percent				
If sample is taken at place of manufacture	5	5	5	5
If sample is taken at any other place	7	7	7	7
Unhydrated oxides (as-received basis), max. percent	—	—	8	8

TABLE 21-13-A
UBC 3-370

1997 UBC - STANDARD 21-13

135

SECTION 21.1401 — SCOPE

This standard covers mortar cement for use in masonry mortars.

SECTION 21.1402 — CLASSIFICATIONS

There are three types of mortar cement:

1. **Type N.** For use as the cementitious material in the preparation of UBC Standard 21-15 Type N and Type O mortars. For use in combination with portland or blended hydraulic cements in the preparation of Type S or Type M mortars.

2. **Type S.** For use as the cementitious material in the preparation of UBC Standard 21-15 Type S mortar.

3. **Type M.** For use as the cementitious material in the preparation of UBC Standard 21-15 Type M mortar.

SECTION 21.1403 — PHYSICAL REQUIREMENTS

Mortar cement shall conform to the requirements set forth in Table 21-14-A for its classifications.

SECTION 21.1404 — CONSTITUENT MATERIALS

Upon request of the building official, the constituent materials shall be provided to the building official and engineer of record.

SECTION 21.1405 — RESTRICTED MATERIALS

Materials used in mortar cement shall conform to the requirements set forth in Table 21-14-B.

SECTION 21.1406 — DELETERIOUS MATERIAL

Materials listed in Table 21-14-C shall not be used in mortar cement.

SECTION 21.1407 — PACKAGE LABELING

Mortar cement packages shall carry a statement indicating that the product conforms to requirements of this standard and shall include the brand, name of manufacturer, type of mortar cement and net weight of the package in pounds.

SECTION 21.1408 — CERTIFICATION

Certification shall be submitted upon request of the building official and shall certify compliance with the requirements of this standard.

SECTION 21.1409 — SAMPLING AND TESTING

Every 90 days, each mortar cement producer shall retain an approved agency to obtain a random sample from a local point of supply in the market area served by the producer.

The agency shall test the mortar cement for compliance with the physical requirements of Table 21-14-A.

Upon request of the building official, the producer shall furnish (at no cost) test results to the building official, architect, structural engineer, general contractor and masonry contractor.

SECTION 21.1410 — TEMPERATURE AND HUMIDITY

The temperature of the air in the vicinity of the mixing slab and dry materials, molds, base plates and mixing bowl shall be maintained between 68°F and 81.5°F (20°C and 27.5°C). The temperature of the mixing water, moist cabinet or moist room, and water in the storage tank shall not vary from 73.4°F (23°C) by more than 3°F (1.7°C).

The relative humidity of the laboratory air shall not be less than 50 percent. The moist cabinet or moist room atmosphere shall have a relative humidity of not less than 90 percent.

The moist cabinet or moist room shall conform to applicable standards.

SECTION 21.1411 — FINENESS

Determine the residue on the No. 325 (45 μm) sieve.

SECTION 21.1412 — NORMAL CONSISTENCY

Determine normal consistency by the Vicat apparatus.

SECTION 21.1413 — AUTOCLAVE EXPANSION

Determine autoclave expansion. After molding, store bars in the moist cabinet or room for 48 hours, plus or minus 30 minutes, before removal from the molds for measurement and test in the autoclave. Calculate the difference in length of the test specimen before and after autoclaving to the nearest 0.01 percent of the effective gauge length and report as the autoclave expansion of the mortar cement.

SECTION 21.1414 — TIME OF SETTING

Determine the time of setting by the Gillmore needle method.

SECTION 21.1415 — DENSITY

Determine the density of the mortar cement using kerosene as the liquid. Use the density so determined in the calculation of the air content of the mortars.

SECTION 21.1416 — APPARATUS FOR MORTAR TESTS

Apparatus shall be in accordance with applicable standards.

SECTION 21.1417 — BLENDED SAND

The sand shall be a blend of equal parts by weight of graded Ottawa sand and Standard 20-30 Ottawa sand.

SECTION 21.1418 — PREPARATION OF MORTAR

21.1418.1 Proportions for Mortar. Mortar for air entrainment, compressive strength and water-retention tests shall be proportioned to contain the weight of cement, in grams, equal to six times the printed bag weight in pounds (13.228 times the printed bag weight in kilograms) and 1,440 grams of sand. The sand shall consist of 720 grams of graded Ottawa sand and 720 grams of Standard 20-30 sand. The quantity of water, measured in milliliters, shall be such as to produce a flow of 110 ± 5 as determined by the flow table.

21.1418.2 Mixing of Mortars. Mix the mortar in accordance with applicable standards.

21.1418.3 Determination of Flow. Determine the flow in accordance with applicable standards.

SECTION 21.1419 — AIR ENTRAINMENT

21.1419.1 Procedure. If the mortar has the correct flow, use a separate portion of the mortar for the determination of entrained air. Determine the weight of 400 cm^3 of mortar.

21.1419.2 Calculation. Calculate the air content of the mortar and report it to the nearest 0.1 percent as follows:

$$D = (W_1 + W_2 + V_w)/[(W_1/S_1) + (W_2/S_2) + V_w]$$
$$A = 100 - (W_m/4D)$$

WHERE:

A = volume percent of entrained air.

D = density of air-free mortar, g/cm^3.

S_1 = density of cement, g/cm^3.

S_2 = density of standard sand, 2.65 g/cm^3.

V_w = milliliters-grams of water used.

W_m = mass of 400 ml of mortar, g.

W_1 = weight of cement, g.

W_2 = weight of sand, g.

SECTION 21.1420 — COMPRESSIVE STRENGTH OF TEST SPECIMENS

21.1420.1 Molding. Immediately after determining the flow and the weight of 400 cm^3 or mortar, return all the mortar to the mixing bowl and remix for 15 seconds at the medium speed. Then mold test specimens in accordance with applicable standards, except that the elapsed time for mixing mortar, determining flow, determining air entrainment and starting the molding of cubes shall be within eight minutes.

21.1420.2 Storage. Store all test specimens immediately after molding in the molds on plane plates in a moist cabinet maintained at a relative humidity of 90 percent or more for 48 to 52 hours in such a manner that the upper surfaces shall be exposed to the moist air. Then remove the cubes from the molds and place in the moist cabinet for five days in such a manner as to allow free circulation of air around at least five faces of the specimens. At the age of seven days, immerse the cubes for the 28-day tests in saturated lime water in storage tanks of noncorrodible materials.

SECTION 21.1421 — PROCEDURE

Test the cube specimens immediately after their removal from the moist cabinet for seven-day specimens, and immediately after their removal from storage water for all other specimens. If more than one specimen at a time is removed from the moist closet for seven-day tests, cover these cubes with a damp cloth until time of testing. If more than one specimen at a time is removed from the storage water for testing, place these cubes in a pan of water at a temperature of 73.4°F ± 3°F (23°C ± 1.7°C), and of sufficient depth to completely immerse each cube until time of testing.

The remainder of the testing procedure shall conform to applicable standards.

SECTION 21.1422 — WATER RETENTION

21.1422.1 Water-retention Apparatus. For the water-retention test, and apparatus essentially the same as that shown in Figure 21-14-1 shall be used. This apparatus consists of a water aspirator or other source of vacuum controlled by a mercury-relief column and connected by way of a three-way stopcock to a funnel upon which rests a perforated dish. The perforated dish shall be made of metal not attacked by masonry mortar. The metal in the base of the dish shall have a thickness of 1.7 to 1.9 mm and shall conform to the requirements given in Figure 21-14-1. The bore of the stopcock shall have a 4 mm plus or minus 0.5 mm diameter, and the connecting glass tubing shall have a minimum inside diameter of 4 mm. A mercury manometer, connected as shown in Figure 21-14-1, indicates the vacuum. The contact surface of the funnel and perforated dish shall be plane and shall be lapped to ensure intimate contact. An airtight seal shall be maintained between the funnel and the dish during a test. This shall be accomplished by either of the following procedures: (1) a synthetic (grease-resistant) rubber gasket may be permanently sealed to the top of the funnel, using petrolatum or light grease to ensure a seal between the funnel and dish, or (2) the top of the funnel may be lightly coated with petrolatum or light grease to ensure a seal between the funnel and dish. Care should be taken to ensure that none of the holes in the perforated dish are clogged from the grease. Hardened, very smooth, not rapid filter paper shall be used. It shall be of such diameter that it will lie flat and completely cover the bottom of the dish.

A steel straightedge not less than 8 inches (203 mm) long and not less than $^1/_{16}$ inch (1.6 mm) nor more than $^1/_8$-inch (3.2 mm) thickness shall be used.

Other apparatus required for the water-retention tests shall conform to the applicable requirements of Section 21.1416.

21.1422.2 Procedure. Adjust the mercury-relief column to maintain a vacuum of 50.8 mm as measured on the manometer. Seat the perforated dish on the greased gasket of the funnel. Place a wetted filter paper in the bottom of the dish. Turn the stopcock to apply the vacuum to the funnel and check the apparatus for leaks and to determine that the required vacuum is obtained. Then turn the stopcock to shut off the vacuum from the funnel.

Mix the mortar to a flow of 110 plus or minus 5 percent in accordance with applicable standards. Immediately after making the flow test, return the mortar on the flow table to the mixing bowl and remix the entire batch for 15 seconds at medium speed. Immediately after remixing the mortar, fill the perforated dish with the mortar to slightly above the rim. Tamp the mortar 15 times with the tamper. Apply 10 of the tamping strokes at approximately uniform spacing adjacent to the rim of the dish and with the long axis of the tamping face held at right angles to the radius of the dish. Apply the remaining five tamping strokes at random points distributed over the central area of the dish. The tamping pressure shall be just sufficient to ensure filling of the dish. On completion of the tamping, the top of the mortar should extend slightly above the rim of the dish. Smooth off the mortar by drawing the flat side of the straightedge (with the leading edge slightly raised) across the top of the dish. Then cut off the mortar to a plane surface flush with the rim of the dish by drawing the straightedge with a sawing motion across the top of the dish in two cutting strokes, starting each cut from near the center of the dish. If the mortar is pulled away from the side of the dish during the process of drawing the straightedge across the dish, gently press the mortar back into contact with the side of the dish using the tamper.

Turn the stopcock to apply the vacuum to the funnel. The time elapsed from the start of mixing the cement and water to the time of applying the vacuum shall not exceed eight minutes. After suction for 60 seconds, quickly turn the stopcock to expose the funnel to atmospheric pressure. Immediately slide the perforated dish off from the funnel, touch it momentarily on a damp cloth to remove droplets of water, and set the dish on the table. Then, using the bowl scraper, in accordance with applicable standards, plow and mix the mortar in the dish for 15 seconds. Upon completion of mixing, place the mortar in the flow mold and determine the flow.

The entire operation shall be carried out without interruption and as quickly as possible, and shall be completed within an elapsed time of 11 minutes after the start of mixing the cement and water for the first flow determination. Both flow determinations shall be made in accordance with applicable standards.

21.1422.3 Calculation. Calculate the water-retention value for the mortar as follows:

$$\text{Water-retention value} = (a/b) \times 100$$

WHERE:

a = flow after suction.

b = flow immediately after mixing.

TABLE 21-14-A—PHYSICAL REQUIREMENTS

MORTAR CEMENT TYPE	N	S	M
Fineness, residue on a No. 325 (45 μm) sieve Maximum percent	24	24	24
Autoclave expansion Maximum, percent	1.0	1.0	1.0

Time of setting, Gillmore method:
    Initial set, minimum, hour
    Final set, maximum, hour
	2
	24
		$1^1/_2$
		24
			$1^1/_2$
			24

Compressive strength[1]
    7 days, minimum psi
    28 days, minimum psi
	500 (3445 kPa)
	900 (6201 kPa)
		1300 (8957 kPa)
		2100 (14 469 kPa)
			1800 (12 402 kPa)
			2900 (19 981 kPa)

Flexural bond strength[2] 28 days, minimum psi	71 (489 kPa)	104 (717 kPa)	116 (799 kPa)

Air content of mortar
    Minimum percent by volume
    Maximum percent by volume
	8
	16
		8
		14
			8
			14

Water retention Minimum, percent	70	70	70

[1]Compressive strength shall be based on the average of three mortar cubes composed of one part mortar cement and three parts blended sand (one half graded Ottawa sand, and one half Standard 20-30 Ottawa sand) by volume and tested in accordance with this standard.

[2]Flexural bond strength shall be determined in accordance with UBC Standard 21-20.

TABLE 21-14-B—RESTRICTED MATERIALS

MATERIAL	MAXIMUM LIMIT (percentage)
Chloride salts	0.06
Carboxylic acids	0.25
Sugars	1.00
Glycols	1.00
Lignin and derivatives	0.50
Stearates	0.50
Fly ash	No limit
Clay (except fireclay)	5.00

TABLE 21-14-C—DELETERIOUS MATERIALS NOT PERMITTED IN MORTAR CEMENT

Epoxy resins and derivatives Phenols Asbestos fiber Fireclays

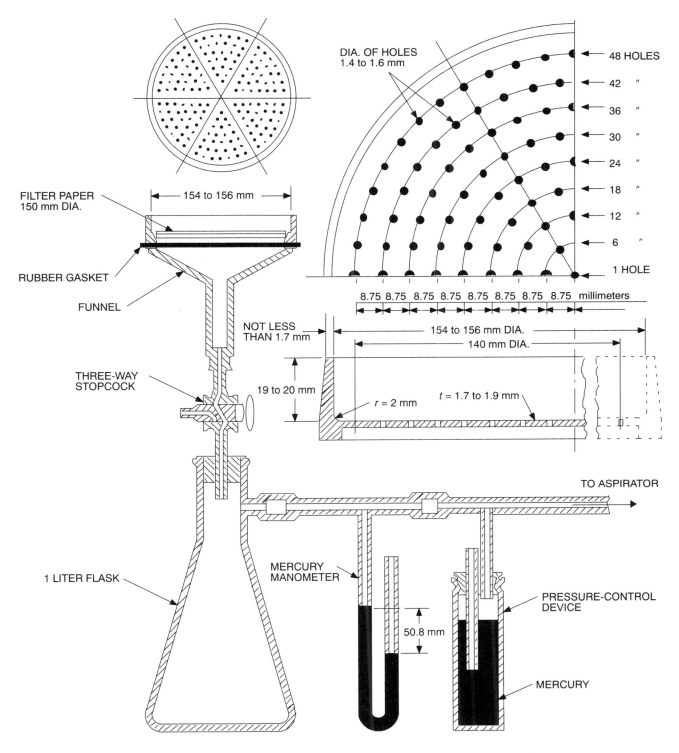

FILTER PAPER
150 mm DIA.

RUBBER GASKET

FUNNEL

154 to 156 mm

DIA. OF HOLES
1.4 to 1.6 mm

48 HOLES
42 "
36 "
30 "
24 "
18 "
12 "
6 "
1 HOLE

8.75 8.75 8.75 8.75 8.75 8.75 8.75 8.75 millimeters

NOT LESS
THAN 1.7 mm

154 to 156 mm DIA.
140 mm DIA.

19 to 20 mm

$r = 2$ mm $t = 1.7$ to 1.9 mm

**THREE-WAY
STOPCOCK**

TO ASPIRATOR

**MERCURY
MANOMETER**

**PRESSURE-CONTROL
DEVICE**

50.8 mm

MERCURY

1 LITER FLASK

FIGURE 21-14-1—APPARATUS ASSEMBLY FOR THE WATER-RETENTION TEST

FIGURE 21-14-1
UBC 3-374

1997 UBC - STANDARD 21-14

139

UNIFORM BUILDING CODE STANDARD 21-15
MORTAR FOR UNIT MASONRY AND REINFORCED
MASONRY OTHER THAN GYPSUM

Based on Standard Specification C 270-95 of the American Society for Testing and Materials.
Extracted, with permission, from the *Annual Book of ASTM Standards,* copyright American Society for
Testing and Materials, 100 Barr Harbor Drive, West Conshohocken, PA 19428

See Section 2102.2, Item 8, *Uniform Building Code*

SECTION 21.1501 — SCOPE

These specifications cover the required properties of mortars determined by laboratory tests for use in the construction of reinforced brick masonry structures and unit masonry structures. Two alternative specifications are covered as follows:

21.1501.1 Property specifications. Property specifications are those in which the acceptability of the mortar is based on the properties of the ingredients (materials) and the properties (water retention and compressive strength) of samples of the mortar mixed and tested in the laboratory.

21.1501.2 Proportion specifications. Proportion specifications are those in which the acceptability of the mortar is based on the properties of the ingredients (materials) and a definite composition of the mortar consisting of fixed proportions of these ingredients.

Unless data are presented to show that the mortar meets the requirements of the physical property specifications, the proportion specifications shall govern. For field tests of grout and mortars see UBC Standard 21-16.

Property Specifications

SECTION 21.1502 — MATERIALS

21.1502.1 General. Materials used as ingredients in the mortar shall conform to the requirements specified in the pertinent UBC Standards.

21.1502.2 Cementitious Materials. Cementitious materials shall conform to the following specifications:

1. **Portland cement.** Type I, IA, II, IIA, III or IIIA of ASTM C 150.

2. **Blended hydraulic cement.** Type IS, IS-A, S, S-A, IP, IP-A, I(PM) or I(PM)-A of ASTM C 1157.

3. **Plastic cement.** Plastic cement conforming to the requirements of UBC Standard 25-1 and UBC Standard 21-11, when used in lieu of masonry cement.

4. **Mortar cement.** UBC Standard 21-14.

5. **Masonry cements.** UBC Standard 21-11.

6. **Quicklime.** UBC Standard 21-12.

7. **Hydrated lime.** UBC Standard 21-13.

21.1502.3 Water. Water shall be clean and free of deleterious amounts of acids, alkalies or organic materials.

21.1502.4 Admixtures or Mortar Colors. Admixtures or mortar colors shall not be added to the mortar at the time of mixing unless provided for in the contract specifications and, after the material is so added, the mortar shall conform to the requirements of the property specifications.

Only pure mineral mortar colors shall be used.

21.1502.5 Antifreeze Compounds. No antifreeze liquid, salts or other substances shall be used in the mortar to lower the freezing point.

21.1502.6 Storage of Materials. Cementitious materials and aggregates shall be stored in such a manner as to prevent deterioration or intrusion of foreign material. Any material that has become unsuitable for good construction shall not be used.

SECTION 21.1503 — MIXING MORTAR

Mortar blended on the jobsite shall be mixed for a minimum period of three minutes, with the amount of water required to produce the desired workability, in a drum-type batch mixer. Factory-dry blended mortar shall be mixed with water in a mechanical mixer until workable but not to exceed 10 minutes.

SECTION 21.1504 — MORTAR

21.1504.1 Mortar for Unit Masonry. Mortar conforming to the proportion specifications shall consist of a mixture of cementitious material and aggregate conforming to the requirements of Section 21.1502, and the measurement and mixing requirements of Section 21.1503, and shall be proportioned within the limits given in Table 21-15-B for each mortar type specified.

21.1504.2 Mortar for Reinforced Masonry. In mortar used for reinforced masonry the following special requirements shall be met: Sufficient water has been added to bring the mixture to a plastic state. The volume of aggregate in mortar shall be at least two and one-fourth times but not more than three times the volume of cementitious materials.

21.1504.3 Aggregate Ratio. The volume of damp, loose aggregate in mortar used in brick masonry shall be not less than two and one-fourth times or more than three times the total separate volumes of cementitious materials used.

21.1504.4 Water Retention. Mortar shall conform to the water retention requirements of Table 21-15-A.

21.1504.5 Air Content. Mortar shall conform to the air content requirements of Table 21-15-A.

SECTION 21.1505 — COMPRESSIVE STRENGTH

The average compressive strength of three 2-inch (51 mm) cubes of mortar (before thinning) shall not be less than the strength given in Table 21-15-A for the mortar type specified.

Proportion Specifications

SECTION 21.1506 — MATERIALS

21.1506.1 General. Materials used as ingredients in the mortar shall conform to the requirements of Section 21.1502 and to the requirements of this section.

21.1506.2 Portland Cement. Portland cement shall conform to the requirements of ASTM C 150.

21.1506.3 Blended Hydraulic Cements. Blended hydraulic cements of Type IS, IS-A, IP, IP-A, I(PM) or I(PM)-A shall conform to the requirements of ASTM C 595, when used in lieu of masonry cement.

21.1506.4 Plastic Cement. Plastic cement conforming to the requirements of UBC Standard 25-1 and UBC Standard 21-11.

21.1506.5 Mortar Cement. Mortar cement shall conform to the requirements of UBC Standard 21-14.

21.1506.6 Masonry Cement. Masonry cement shall conform to the requirements of UBC Standard 21-11.

21.1506.7 Hydrated Lime. Hydrated lime shall conform to either of the two following requirements:

1. The total free (unhydrated) calcium oxide (CaO) and magnesium oxide (MgO) shall not be more than 8 percent by weight (calculated on the as-received basis for hydrates).

2. When the hydrated lime is mixed with portland cement in the proportion set forth in Table 21-15-B, the mixture shall give an autoclave expansion of not more than 0.50 percent.

Hydrated lime intended for use when mixed dry with other mortar ingredients shall have a plasticity figure of not less than 200 when tested 15 minutes after adding water.

21.1506.8 Lime Putty. Lime putty made from either quicklime or hydrated lime shall be soaked for a period sufficient to produce a plasticity figure of not less than 200 and shall conform to either the requirements for limitation on total free oxides of calcium and magnesium or the autoclave test specified for hydrated lime in Section 21.1506.5.

SECTION 21.1507 — MORTAR

Mortar shall consist of a mixture of cementitious materials and aggregate conforming to the requirements specified in Section 21.1504, mixed in one of the proportions shown in Table 21-15-B, to which sufficient water has been added to reduce the mixture to a plastic state.

TABLE 21-15-A—PROPERTY SPECIFICATIONS FOR MORTAR[1]

MORTAR	TYPE	AVERAGE COMPRESSIVE STRENGTH OF 2-INCH (51 mm) CUBES AT 28 DAYS (Min., psi) × 6.89 for kPa	WATER RETENTION (Min., percent)	AIR CONTENT (Max., percent)[2]	AGGREGATE MEASURED IN A DAMP, LOOSE CONDITION
Cement-lime or mortar cement	M	2,500	75	12	Not less than 2 1/4 and not more than 3 1/2 times the sum of the separate volumes of cementitious materials
	S	1,800	75	12	
	N	750	75	14[3]	
	O	350	75	14[3]	
Masonry cement	M	2,500	75	18	
	S	1,800	75	18	
	N	750	75	18	
	O	350	75	18	

[1]Laboratory-prepared mortar only.

[2]Determined in accordance with applicable standards.

[3]When structural reinforcement is incorporated in cement-lime mortar or mortar-cement mortar, the maximum air content shall be 12 percent.

TABLE 21-15-B—MORTAR PROPORTIONS FOR UNIT MASONRY

MORTAR	TYPE	Portland Cement or Blended Cement[1]	Masonry Cement[2] M	Masonry Cement[2] S	Masonry Cement[2] N	Mortar Cement[3] M	Mortar Cement[3] S	Mortar Cement[3] N	Hydrated Lime or Lime Putty[1]	AGGREGATE MEASURED IN A DAMP, LOOSE CONDITION
Cement-lime	M	1	—	—	—	—	—	—	1/4	Not less than 2 1/4 and not more than 3 times the sum of the separate volumes of cementitious materials
	S	1	—	—	—	—	—	—	over 1/4 to 1/2	
	N	1	—	—	—	—	—	—	over 1/2 to 1 1/4	
	O	1	—	—	—	—	—	—	over 1 1/4 to 2 1/2	
Mortar cement	M	1	—	—	—	—	—	1	—	
	M	—	—	—	—	1	—	—	—	
	S	1/2	—	—	—	—	—	1	—	
	S	—	—	—	—	—	1	—	—	
	N	—	—	—	—	—	—	1	—	
Masonry cement	M	1	—	—	1	—	—	—	—	
	M	—	1	—	—	—	—	—	—	
	S	1/2	—	—	1	—	—	—	—	
	S	—	—	1	—	—	—	—	—	
	N	—	—	—	1	—	—	—	—	
	O	—	—	—	1	—	—	—	—	

[1]When plastic cement is used in lieu of portland cement, hydrated lime or putty may be added, but not in excess of one tenth of the volume of cement.

[2]Masonry cement conforming to the requirements of UBC Standard 21-11.

[3]Mortar cement conforming to the requirements of UBC Standard 21-14.

UNIFORM BUILDING CODE STANDARD 21-16
FIELD TESTS SPECIMENS FOR MORTAR
Test Standard of the International Conference of Building Officials
See Section 2102.2, Item 8, *Uniform Building Code*

SECTION 21.1601 — FIELD COMPRESSIVE TEST SPECIMEN FOR MORTAR

Spread mortar on the masonry units $1/2$ inch to $5/8$ inch (13 mm to 16 mm) thick, and allow to stand for one minute, then remove mortar and place in a 2-inch by 4-inch (51 mm by 102 mm) cylinder in two layers, compressing the mortar into the cylinder using a flat-end stick or fingers. Lightly tap mold on opposite sides, level off and immediately cover molds and keep them damp until taken to the laboratory. After 48 hours' set, have the laboratory remove molds and place them in the fog room until tested in damp condition.

SECTION 21.1602 — REQUIREMENTS

Each such mortar test specimen shall exhibit a minimum ultimate compressive strength of 1,500 pounds per square inch (10 304 kPa).

TEST METHOD FOR COMPRESSIVE STRENGTH OF MASONRY PRISMS

Based on Standard Test Method E 447-92 of the American Society for Testing and Materials. Extracted, with permission, from the *Annual Book of ASTM Standards,* copyright American Society for Testing and Materials, 100 Barr Harbor Drive, West Conshohocken, PA 19428

See Sections 2102.2, Item 6.4; 2105.3.2; and 2105.3.3, *Uniform Building Code*

SECTION 21.1701 — SCOPE

This standard covers procedures for masonry prism construction, testing and procedures for determining the compressive strength of masonry.

SECTION 21.1702 — CONSTRUCTION OF PRISMS

Prisms shall be constructed on a flat, level base. Masonry units used in the prism shall be representative of the units used in the corresponding construction. Each prism shall be built in an opened moisture-tight bag which is large enough to enclose and seal the completed prism. The orientation of units, where top and bottom cross sections vary due to taper of the cells, or where the architectural surface of either side of the unit varies, shall be the same orientation as used in the corresponding construction. Prisms shall be a single wythe in thickness and laid up in stack bond (see Figure 21-17-1).

The length of masonry prisms may be reduced by saw cutting; however, prisms composed of regular shaped hollow units shall have at least one complete cell with one full-width cross web on either end. Prisms composed of irregular-shaped units shall be cut to obtain as symmetrical a cross section as possible. The minimum length of saw-cut prisms shall be 4 inches (102 mm).

Masonry prisms shall be laid in a full mortar bed (mortar bed both webs and face shells). Mortar shall be representative of that used in the corresponding construction. Mortar joint thickness, the tooling of joints and the method of positioning and aligning units shall be representative of the corresponding construction.

Prisms shall be a minimum of two units in height, but the total height shall not be less than 1.3 times the least actual thickness or more than 5.0 times the least actual thickness. Immediately following the construction of the prism, the moisture-tight bag shall be drawn around the prism and sealed.

Where the corresponding construction is to be solid grouted, prisms shall be solid grouted. Grout shall be representative of that used in the corresponding construction. Grout shall be placed not less than one day nor more than two days following the construction of the prism. Grout consolidation shall be representative of that used in the construction. Additional grout shall be placed in the prism after reconsolidation and settlement due to water loss, but prior to the grout setting. Excess grout shall be screeded off level with the top of the prism. Where open-end units are used, additional masonry units shall be used as forms to confine the grout during placement. Masonry unit forms shall be sufficiently braced to prevent displacement during grouting. Immediately following the grouting operation, the moisture-tight bag shall be drawn around the prism and resealed.

Where the corresponding construction is to be partially grouted, two sets of prisms shall be constructed; one set shall be grouted solid and the other set shall not be grouted.

Where the corresponding construction is of multiwythe composite masonry, masonry prisms representative of each wythe shall be built and tested separately.

Prisms shall be left undisturbed for at least two days after construction.

SECTION 21.1703 — TRANSPORTING MASONRY PRISMS

Prior to transporting each prism, strap or clamp the prism together to prevent damage during handling and transportation. Secure prism to prevent jarring, bouncing or falling over during transporting.

SECTION 21.1704 — CURING

Prisms shall remain sealed in the moisture-tight bag until two days prior to testing; the moisture-tight bag shall then be removed and curing continued in laboratory air maintained at a temperature of 75°F ± 15°F (24°C ± 8°C). Prisms shall be tested at 28 days after constructing the prism or at test age designated.

SECTION 21.1705 — PREPARATION FOR TESTING

21.1705.1 Capping the Prism. Cap top and bottom of the prism prior to testing with sulfur-filled capping or with high-strength gypsum plaster capping (such as "Hydrostone" or "Hyprocal White"). Sulfur-filled capping material shall be 40 to 60 percent by weight sulfur, the remainder being ground fireclay or other suitable inert material passing a No. 100 (150 μm) sieve, with or without a plasticizer. Spread the capping material over a level surface which is plane within 0.003 inch (0.076 mm) in 16 inches (406 mm). Bring the surface to be capped into contact with the capping paste; firmly press down the specimen, holding it so that its axis is at right angles to the capping surfaces. The average thickness of the cap shall not exceed $^1/_8$ inch (3.2 mm). Allow caps to age at least two hours before testing.

21.1705.2 Measurement of the Prism. Measure the length and thickness of the prism to the nearest 0.01 inch (0.25 mm) by averaging three measurements taken at the center and quarter points of the height of the specimen. Measure the height of the prism, including caps, to the nearest 0.1 inch (2.54 mm).

SECTION 21.1706 — TEST PROCEDURE

21.1706.1 Test Apparatus. The test machine shall have an accuracy of plus or minus 1.0 percent over the load range. The upper bearing shall be spherically seated, hardened metal block firmly attached at the center of the upper head of the machine. The center of the sphere shall lie at the center of the surface held in its spherical seat, but shall be free to turn in any direction, and its perimeter shall have at least $^1/_4$-inch (6.4 mm) clearance from the head to allow for specimens whose bearing surfaces are not exactly parallel. The diameter of the bearing surface shall be at least 5 inches (127 mm). A hardened metal bearing block may be used beneath the specimen to minimize wear of the lower platen of the machine. The bearing block surfaces intended for contact with the specimen shall have a hardness not less than 60 HRC (620 HB). These surfaces shall not depart from plane surfaces by more than

0.001 inch (0.0254 mm) in any 6-inch (153 mm) dimension. When the bearing area of the spherical bearing block is not sufficient to cover the area of the specimen, a steel plate with surfaces machined to true planes within plus or minus 0.001 inch (0.0254 mm) in any 6-inch (153 mm) dimension, and with a thickness equal to at least the distance from the edge of the spherical bearings to the most distant corner, shall be placed between the spherical bearing block and the capped specimen.

21.1706.2 Installing the Prism in the Test Machine. Wipe clean the bearing faces of the upper and lower platens or bearing blocks and of the test specimen and place the test specimen on the lower platen or bearing block. Align both centroidal axes of the specimen with the center of thrust of the test machine. As the spherically seated block is brought to bear on the specimen, rotate its movable portion gently by hand so that uniform seating is obtained.

21.1706.3 Loading. Apply the load, up one half of the expected minimum load, at any convenient rate, after which adjust the controls of the machine so that the remaining load is applied at a uniform rate in not less than one or more than two minutes.

21.1706.4 Observations. Describe the mode of failure as fully as possible or illustrate crack patterns, spalling, etc., on a sketch, or both. Note whether failure occurred on one side or one end of the prism prior to failure of the opposing side or end of the prism.

SECTION 21.1707 — CALCULATIONS

Calculations of test results shall be as follows:

21.1707.1 Net cross-sectional area. Determine the net cross-sectional area [square inches (mm^2)] of solid grouted prisms by multiplying the average measured width dimension [inches (mm)] by the average measured length dimension [inches (mm)]. The net cross-sectional area of ungrouted prisms shall be taken as the net cross-sectional area of masonry units determined from a representative sample of units.

21.1707.2 Masonry prism strength. Determine the compressive strength of each prism [psi (kPa)] by dividing the maximum compressive load sustained [pounds (N)] by the net cross-sectional area of the prism [square inches (mm^2 × 1,000,000)].

21.1707.3 Compressive strength of masonry. The compressive strength of masonry [psi (kPa)] for each set of prisms shall be the lesser of the average strength of the prisms in the set, or 1.25 times the least prism strength multiplied by the prism height-to-thickness correction factor from Table 21-17-A. Where a set of grouted and nongrouted prisms are tested, the compressive strength of masonry shall be determined for the grouted set and for the nongrouted set separately. Where a set of prisms is tested for each wythe of a multiwythe wall, the compressive strength of masonry shall be determined for each wythe separately.

SECTION 21.1708 — MASONRY PRISM TEST REPORT

The test report shall include the following:

1. Name of testing laboratory and name of professional engineer responsible for the tests.

2. Designation of each prism tested and description of prism, including width, height and length dimensions, mortar type, grout and masonry unit used in the construction.

3. Age of prism at time of test.

4. Maximum compressive load sustained by each prism, net cross-sectional area of each prism and net area compressive strength of each prism.

5. Test observations for each prism in accordance with Section 21.1706.

6. Compressive strength of masonry for each set of prisms.

TABLE 21-17-A—PRISM HEIGHT-TO-THICKNESS CORRECTION FACTORS

Prisms h/t_p[1]	1.30	1.50	2.00	2.50	3.00	4.00	5.00
Correction factor	0.75	0.86	1.00	1.04	1.07	1.15	1.22

[1]h/t_p—ratio of prism height to least actual lateral dimension of prism.

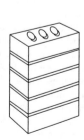
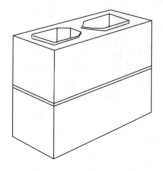

FIGURE 21-17-1—CONSTRUCTION OF PRISMS

METHOD OF SAMPLING AND TESTING GROUT

Based on Standard Method C 1019-89a (93) of the American Society for Testing and Materials.
Extracted, with permission, from the *Annual Book of ASTM Standards,* copyright American Society for Testing and Materials, 100 Barr Harbor Drive, West Conshohocken, PA 19428

See Section 2102.2, Item 9; and Table 21-B, *Uniform Building Code*

SECTION 21.1801 — SCOPE

This method covers procedures for both field and laboratory sampling and compression testing of grout used in masonry construction.

SECTION 21.1802 — APPARATUS

21.1802.1 Maximum-Minimum Thermometer.

21.1802.2 Straightedge. A steel straightedge not less than 6 inches (152.4 mm) long and not less than $^1/_{16}$ inch (1.6 mm) in thickness.

21.1802.3 Tamping Rod. A nonabsorbent smooth rod, either round or square in cross section nominally $^5/_8$ inch (15.9 mm) in dimension with ends rounded to hemispherical tips of the same diameter. The rod shall be a minimum length of 12 inches (304.8 mm).

21.1802.4 Wooden Blocks. Wooden squares with side dimensions equal to one-half the desired grout specimen height, within a tolerance of 5 percent, and of sufficient quantity or thickness to yield the desired grout specimen height, as shown in Figures 21-18-1 and 21-18-2.

Wooden blocks shall be soaked in limewater for 24 hours, sealed with varnish or wax, or covered with an impermeable material prior to use.

SECTION 21.1803 — SAMPLING

21.1803.1 Size of Sample. Grout samples to be used for slump and compressive strength tests shall be a minimum of $^1/_2$ ft.^{3} (0.014 m^3).

21.1803.2 Field Sample. Take grout samples as the grout is being placed into the wall. Field samples may be taken at any time except for the first and last 10 percent of the batch volume.

SECTION 21.1804 — TEST SPECIMEN AND SAMPLE

21.1804.1 Each grout specimen shall be a square prism, nominally 3 inches (76.2 mm) or larger on the sides and twice as high as its width. Dimensional tolerances shall be within 5 percent of the nominal width selected.

21.1804.2 Three specimens constitute one sample.

SECTION 21.1805 — PROCEDURE

21.1805.1 Select a level location where the molds can remain undisturbed for 48 hours.

21.1805.2 Mold Construction.

21.1805.2.1 The mold space should simulate the grout location in the wall. If the grout is placed between two different types of masonry units, both types should be used to construct the mold.

21.1805.2.2 Form a square prism space, nominally 3 inches (76.2 mm) or larger on each side and twice as high as its width, by stacking masonry units of the same type and moisture condition as those being used in the construction. Place wooden blocks, cut to proper size and of the proper thickness or quantity, at the bottom of the space to achieve the necessary height of specimen. Tolerance on space and specimen dimensions shall be within 5 percent of the specimen width. See Figures 21-18-1 and 21-18-2.

21.1805.2.3 Line the masonry surfaces that will be in contact with the grout specimen with a permeable material, such as paper towel, to prevent bond to the masonry units.

21.1805.3 Measure and record the slump of the grout.

21.1805.4 Fill the mold with grout in two layers. Rod each layer 15 times with the tamping rod penetrating $^1/_2$ inch (12.7 mm) into the lower layer. Distribute the strokes uniformly over the cross section of the mold.

21.1805.5 Level the top surface of the specimen with a straightedge and cover immediately with a damp absorbent material such as cloth or paper towel. Keep the top surface of the sample damp by wetting the absorbent material and do not disturb the specimen for 48 hours.

21.1805.6 Protect the sample from freezing and variations in temperature. Store an indicating maximum-minimum thermometer with the sample and record the maximum and minimum temperatures experienced prior to the time the specimens are placed in the moist room.

21.1805.7 Remove the masonry units after 48 hours. Transport field specimens to the laboratory, keeping the specimens damp and in a protective container.

21.1805.8 Store in a moist room conforming to nationally recognized standards.

21.1805.9 Cap the specimens in accordance with the applicable requirements of UBC Standard 21-17.

21.1805.10 Measure and record the width of each face at midheight. Measure and record the height of each face at midwidth. Measure and record the amount out of plumb at midwidth of each face.

21.1805.11 Test the specimens in a damp condition in accordance with applicable requirements of UBC Standard 21-17.

SECTION 21.1806 — CALCULATIONS

The report shall include the following:

1. Mix design.

2. Slump of the grout.

3. Type and number of units used to form mold for specimens.

4. Description of the specimens—dimensions, amount out of plumb—in percent.

5. Curing history, including maximum and minimum temperatures and age of specimen, when transported to laboratory and when tested.

6. Maximum load and compressive strength of the sample.

7. Description of failure.

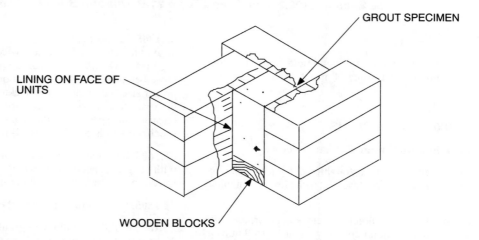

FIGURE 21-18-1—GROUT MOLD [UNITS 6 INCHES (152 mm) OR LESS IN HEIGHT, 2¹/₂-INCH-HIGH (63.5 mm) BRICK SHOWN]

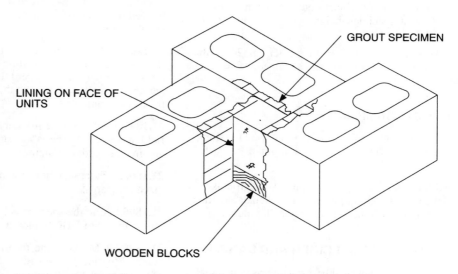

FIGURE 21-18-2—GROUT MOLD [UNITS GREATER THAN 6 INCHES (152 mm) IN HEIGHT, 8-INCH-HIGH (203 mm) CONCRETE MASONRY UNIT SHOWN]

UNIFORM BUILDING CODE STANDARD 21-19
GROUT FOR MASONRY

Based on Standard Specification C 476-91 of the American Society for Testing and Materials.
Extracted, with permission, from the *Annual Book of ASTM Standards,* copyright American Society for Testing and Materials, 100 Barr Harbor Drive, West Conshohocken, PA 19428

See Section 2102.2, Item 9, *Uniform Building Code*

SECTION 21.1901 — SCOPE

This standard covers grout for use in the construction of reinforced and nonreinforced masonry structures.

SECTION 21.1902 — MATERIALS

Materials used as ingredients in grout shall conform to the following:

21.1902.1 Cementitious Materials. Cementitious materials shall conform to one of the following standards:

   A. Portland Cement—Types I, II and III of ASTM C 150.
   B. Blended Cement—Type IS, IS(MS) or IP of ASTM C 595.
   C. Quicklime—UBC Standard 21-12.
   D. Hydrated lime—Type S of UBC Standard 21-13.

21.1902.2 Water. Water shall be clean and potable.

21.1902.3 Admixtures. Additives and admixtures to grout shall not be used unless approved by the building official.

21.1902.4 Antifreeze Compounds. No antifreeze liquids, chloride salts or other substances shall be used in grout.

21.1902.5 Storage of Materials. Cementitious materials and aggregates shall be stored in such a manner as to prevent deterioration or intrusion of foreign material or moisture. Any material that has become unsuitable for good construction shall not be used.

SECTION 21.1903 — MEASUREMENT OF MATERIALS

The method of measuring materials for the grout used in construction shall be such that the specified proportions of the grout materials can be controlled and accurately maintained.

SECTION 21.1904 — GROUT

Grout shall consist of cementitious material and aggregate that have been mixed thoroughly for a minimum of five minutes in a mechanical mixer with sufficient water to bring the mixture to the desired consistency. The grout proportions and any additives shall be based on laboratory or field experience considering the grout ingredients and the masonry units to be used, or the grout shall be proportioned within the limits given in Table 21-B of this code, or the grout shall have a minimum compressive strength when tested in accordance with UBC Standard 21-18 equal to its specified strength, but not less than 2,000 psi (13 800 kPa).

> **EXCEPTION:** Dry mixes for grout which are blended in the factory and mixed at the jobsite shall be mixed in mechanical mixers until workable, but not to exceed 10 minutes.

UNIFORM BUILDING CODE STANDARD 21-20
STANDARD TEST METHOD FOR FLEXURAL
BOND STRENGTH OF MORTAR CEMENT

Test Standard of the International Conference of Building Officials

See Section 2102.2, Item 8, *Uniform Building Code*, and
UBC Standard 21-14, Table 21-14-A

SECTION 21.2001 — SCOPE

This method covers the laboratory evaluation of the flexural bond strength of a standardized mortar and a standardized masonry unit.

SECTION 21.2002 — APPARATUS

The test apparatus consists of a metal frame designed to support a prism as shown in Figures 21-20-1 and 21-20-2. The prism support system shall be adjustable to support prisms ranging in height from two to seven masonry units. The upper clamping bracket that is clamped to the top masonry unit of the prism shall not come into contact with the lower clamping bracket during the test. An alignment jig, mortar template, and drop hammer as shown in Figures 21-20-3, 21-20-4 and 21-20-5 are used in the fabrication of prism specimens for testing.

SECTION 21.2003 — MATERIALS

21.2003.1 Masonry units used shall be standard masonry units selected for the purpose of determining the flexural bond strength properties of mortar cement mortars. The standard unit shall be in accordance with the following requirements:

1. Dimensions of units shall be $3^5/_8$ inches (92 mm) wide by $2^1/_4$ inches (57 mm) high by $7^5/_8$ inches (194 mm) long within a tolerance of plus or minus $1/_8$ inch (3.2 mm) and shall be 100 percent solid.

2. The unit material shall be concrete masonry manufactured with the following material proportions by volume:

   One part portland cement to eight parts aggregate

3. Aggregate used in the manufacture of the unit shall be as follows:

Bulk Specific Gravity Gradation	2.6 to 2.7 Percent Retained by Weight
$3/_8$-inch (9.5 mm) sieve	0
No. 4 (4.75 mm) sieve	0 to 5
No. 8 (2.36 mm) sieve	20 to 30
No. 16 (1.18 mm) sieve	20 to 30
No. 30 (600 µm) sieve	15 to 25
No. 50 (300 µm) sieve	5 to 15
No. 100 (150 µm) sieve	5 to 10
Pan	5 to 10

4. Density of the unit shall be 125 to 135 pounds per cubic foot (2000 to 2160 kg/m^3).

5. Unit shall be cured in a 100 percent relative humidity environment at 140°F ± 10°F (60°C ± 5.6°C) at atmospheric pressure for 10 to 20 hours. Additional curing, under covered atmospheric conditions, shall continue for at least 28 days. Unit shall be loose stacked in the cube (separated by a $1/_4$-inch (6.4 mm) gap) to allow air to circulate during drying.

6. At the time of fabricating the prisms, units shall have a moisture content in the range of 25 percent to 35 percent.

7. Upon delivery units shall be stored in the laboratory at normal temperature and humidity. Units shall not be wetted or surface treated prior to or during prism fabrication.

21.2003.2 Mortar. Mortar shall be prepared in accordance with the following:

1. Mortar proportions shall be in accordance with Table 21-20-A. The aggregate shall consist of a blend of one-half graded Ottawa sand and one-half Standard 20-30 Ottawa sand.

2. Mortar materials shall be mixed in a drum-type batch mixer for five minutes.

3. Determine mortar flow in accordance with applicable standards and adjust water until a flow of 125 ± 5 is achieved.

4. Determine mortar density, air content and initial cone penetration immediately after mixing the mortar in accordance with applicable standards. Mortar shall not be used when cone penetration is less than 80 percent of the initial cone penetration value.

SECTION 21.2004 — TEST SPECIMENS

21.2004.1 Number. Test specimens shall consist of one set of six prisms constructed with the mortar cement mortar. Each prism shall be six units in height.

21.2004.2 Prism Construction. (1) Each prism shall be built in an opened moisture-tight bag which is large enough to enclose and seal the completed prism. Set the first unit on a $1/_2$-inch (13 mm) plywood pallet in an alignment jig as shown in Figure 21-20-3. (2) Place the mortar template shown in Figure 21-20-4 on the unit such that the mortar bed depth prior to compaction is $1/_2$ inch (13 mm). Place mortar in template and strike off excess mortar with straight edge. (3) Remove template and immediately place the next unit on the mortar bed in contact with the three alignment bolts for that course using a bulls-eye level to assure uniform initial contact of the unit surface and bed mortar. Carefully position drop hammer apparatus shown in Figure 21-20-5 on top of unit and drop its 4-pound (1.81 kg) weight, round end down, once from a height of 1.5 inches (38 mm). (4) Repeat (2) and (3) until the prisms are complete. (5) Joints shall be cut flush after the prism is completely built. Joints shall not be tooled. (6) One hour, ± 15 minutes after completion of construction, place two masonry units of the type used to construct the prism upon the top course. (7) Identify all prisms using a water-resistant marker. (8) Draw and seal the moisture-tight bag around the prism. (9) All prisms should be cured for 28 days. Two days prior to testing remove the moisture-tight bag and continue curing in the laboratory air, maintained at a temperature of 75°F ± 15°F (23.9°C ± 8.3°C), with a relative humidity between 30 to 70 percent.

SECTION 21.2005 — TEST PROCEDURE

Place the prism vertically in the support frame as shown in Figure 21-20-1 and clamp firmly into a locked position using the lower clamping bracket. Orient the prism so that the face of the joint intended to be subjected to flexural tension is on the same side of the specimen as the clamping screws. The prism shall be positioned at the required elevation that results in a single unit projecting above the lower clamping bracket. A soft bearing material (for example, polystyrene) at least $1/_2$-inch (13 mm) thick shall be placed between the bottom of the prism and the adjustable prism base support.

Attach the upper clamping bracket to the top unit as shown in Figure 21-20-1. Tighten each clamping bolt using a torque not greater than 20 inch-pounds (2.26 N·m).

Apply the load at a uniform rate so that the total load is applied in not less than one minute or more than three minutes. Measure load to an accuracy of ± 2 percent with maximum error of five pounds (22.2 N).

SECTION 21.2006 — CALCULATIONS

Calculate the modulus of rupture of each mortar joint as follows:

$$f_r = \frac{6(PL + P_1 L_1)}{bd^2} - \frac{(P + P_1)}{bd}$$

For **SI:**
$$f_r = \frac{6(PL + P_1 L_1)}{1000\, bd^2} - \frac{(P + P_1)}{1000\, bd}$$

WHERE:

b = average width of cross section of failure surface, inches (mm).

d = average thickness of cross section of failure surface, inches (mm).

f_r = modulus of rupture, psi (kPa).

L = distance from center of prism to loading point, inches (mm).

L_1 = distance from center of prism to centroid of loading arm, inches (mm).

P = maximum applied load, pounds (N).

P_1 = weight of loading arm, pounds (N).

The flexural bond strength of mortar shall be determined as the average modulus of rupture of 30 joints minus 1.28 times the standard deviation of the sample which yields a value that a mortar joint's modulus of rupture will equal or exceed nine out of 10 times.

SECTION 21.2007 — REPORT

The report shall include the manufacturer of the mortar cement being evaluated, the source of manufacture, type of mortar cement, date of testing, laboratory name and laboratory personnel.

Report mortar density, air content, flow and cone penetration test data. Report the following data for the mortar cement mortar being evaluated:

PRISM NO.	PRISM WEIGHT (lbs.) (kg)	JOINT NO.	TEST LOAD (lbs.) (N)	MOMENT (in.-lbs.) (N·m)	MODULUS OF RUPTURE			
					f_r psi (kPa)	Mean psi (kPa)	Std. Dev. psi[1] (kPa)	COV %
1	—	1	—	—	—	—	—	—
		2	—	—	—			
		3	—	—	—			
		4	—	—	—			
		5	—	—	—			

[1]Also, report the standard deviation for all six prisms (30 joints).

Report the flexural bond strength (determined in accordance with Section 21.2006) of the mortar cement mortar.

TABLE 21-20-A—MORTAR PROPORTIONS BY VOLUME FOR EVALUATING FLEXURAL BOND

MORTAR	MORTAR CEMENT TYPE	PROPORTIONS	
		Mortar Cement	Aggregate
Type N	N	1	3
Type S	S	1	3
Type M	M	1	3

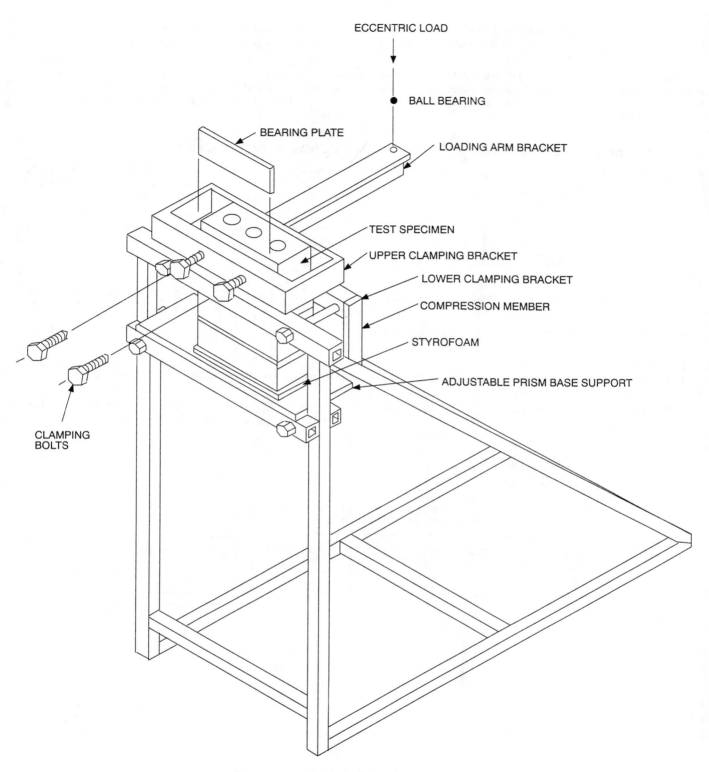

ECCENTRIC LOAD

BALL BEARING

BEARING PLATE

LOADING ARM BRACKET

TEST SPECIMEN

UPPER CLAMPING BRACKET

LOWER CLAMPING BRACKET

COMPRESSION MEMBER

STYROFOAM

ADJUSTABLE PRISM BASE SUPPORT

CLAMPING BOLTS

FIGURE 21-20-1—BOND WRENCH TEST APPARATUS

EXCERPTS FROM

International Conference of Building Officials
5360 WORKMAN MILL ROAD
WHITTIER, CALIFORNIA 90601-2298
(800) 284-4406 • (562) 699-0541

PRINTED IN THE U.S.A.

INTRODUCTION

The California Building Code (C.B.C.) is another name for the body of regulations known as the California Code of Regulations (C.C.R.), Title 24, Part 2—a portion of the "California Building Standards Code," as defined in the "California Building Standards Law" commencing with Section 18901 of the Health and Safety Code.

Title 24 is one of twenty-six titles of the C.C.R. Each numbered title is assigned to a specific state agency and contains all the regulations promulgated by that agency except for the California Building Standards Commission.

Title 24 is assigned to the California Building Standards Commission, which by law, is responsible for coordinating all building standards. Under law, all building standards must be centralized in Title 24 or they are unenforceable.

The 1995 edition of the C.B.C. incorporates by reference the 1994 edition of the Uniform Building Code, with necessary California amendments.

To differentiate between model code language and incorporated California amendments, all California amendments will appear in italics. Symbols in the margin and text indicate the status of code changes as follows:

C
A
C *This symbol indicates that a State of California amendment has been made to the Uniform*
A *Building Code.*

|| *This symbol indicates that a change has been made to a California amendment.*

✳ *This symbol in the margin indicates that model code language has been omitted as a California amendment.*

✳✳✳ *This symbol in the text indicates that model code language has been omitted as a California amendment.*

L
L
L *This symbol is primarily for the benefit of local agencies, such as cities' and counties' build-*
L *ing departments, who by law must enforce selected sections of the C.B.C. Such sections are*
L *signaled by an "L" next to the "CA" in the outside margin.*

Users are urged to express their comments and criticisms of this new publication's format, including suggestions for future changes or improvements in format, by writing directly to:

California Building Standards Commission
428 J Street, Suite 450
Sacramento, California 95814

Preface

The *Uniform Building Code* is dedicated to the development of better building construction and greater safety to the public by uniformity in building laws. The code is founded on broad-based principles that make possible the use of new materials and new construction systems.

The *Uniform Building Code* was first enacted by the International Conference of Building Officials at the Sixth Annual Business Meeting held in Phoenix, Arizona, October 18-21, 1927. Revised editions of this code have been published since that time at approximate three-year intervals. New editions incorporate changes approved since the last edition.

The *Uniform Building Code* is designed to be compatible with related publications to provide a complete set of documents for regulatory use. See the publications list following this preface for a listing of the complete family of Uniform Codes and related publications.

Code Changes. Anyone may propose amendments to this code. For more information, write to the International Conference of Building Officials, 5360 Workman Mill Road, Whittier, California 90601-2298. Changes to the code are processed each year and published as supplements in a form allowing ready adoption by local communities. These changes are carefully reviewed in public hearings by experts in the field of building construction and fire and life safety. An analysis of changes between editions is published in the *Analysis of Revisions to the Uniform Codes*.

Marginal Markings. Solid vertical lines in the margins within the body of the code indicate a change from the requirements of the 1991 edition except where an entire chapter was revised, a new chapter was added or a change was minor. Where an entire chapter was revised or a new chapter was added, a notation appears at the beginning of that chapter. The letter **F** repeating in line vertically in the margin indicates that the provision is maintained under the code change procedures of the International Fire Code Institute. Deletion indicators (◗) are provided in the margin where a paragraph or item listing has been deleted if the deletion resulted in a change of requirements.

Common Code Format. The provisions of the 1994 edition of the *Uniform Building Code* have been reformatted into the common code format established by the Council of American Building Officials. The new format establishes a common format of chapter designations for the three model building codes published in the United States. Apart from those changes approved by the conference membership, this reformatting has not changed the technical content of the code.

The chart on the page following this preface indicates how the new chapters are grouped, lists the new chapter designations and indicates the general location of the provisions from the 1991 edition. Cross-reference tables are available to assist in locating provisions of the 1991 edition in the 1994 edition.

Three-Volume Set. Provisions of the *Uniform Building Code* and the *U.B.C. Standards* have been divided into a three-volume set. Volume 1 accommodates administrative, fire- and life-safety and field inspection provisions. Chapters 1 through 15 and Chapters 24 through 35 are printed in Volume 1 in their entirety. Any appendix chapters associated with these chapters are printed in their entirety at the end of Volume 1. Excerpts of certain chapters from Volume 2 are reprinted in Volume 1 to provide greater usability.

Volume 2 accommodates structural engineering design provisions, and specifically contains Chapters 16 through 23 printed in their entirety. Included in this volume are design standards previously published in the *U.B.C. Standards*. Design standards have been added to their respective chapters as divisions of the chapters. Any appendix chapters associated with these chapters are printed in their entirety at the end of Volume 2. Excerpts of certain chapters from Volume 1 are reprinted in Volume 2 to provide greater usability.

Volume 3 contains material, testing and installation standards.

Metrication. The *Uniform Building Code* has been metricated for the 1994 edition. The metric conversions are provided in parenthesis following the English units. Where industry has made metric conversions available, the conversions conform to current industry standards.

Formulas are also provided with metric equivalents. Metric equivalent formulas immediately follow the English formula and are denoted by "For **SI:**" preceding the metric equivalent. Some formulas do not use dimensions and, thus, are not provided with a metric equivalent. Multiplying conversion factors have been provided for formulas where metric forms were unavailable. Tables are provided with multiplying conversion factors in subheadings for each tabulated unit of measurement. Metricated tables of the Uniform Codes are available from the Conference.

Chapter 21A *[DSA/SS, OSHPD 1]*
MASONRY
NOTE: This chapter has been reformatted in its entirety.

SECTION 2101A — GENERAL

2101A.1 Scope. The materials, design, construction and quality control of masonry shall be in accordance with this chapter.

2101A.2 Design Methods. Masonry shall comply with the provisions of one of the following design methods in this chapter as well as the requirements of Sections 2101A through 2105A.

2101A.2.1 Working stress design. Masonry designed by the working stress design method shall comply with the provisions of Sections 2106A and 2107A.

2101A.2.2 Strength design. Masonry designed by the strength design method shall comply with the provisions of Sections 2106A and 2108A.

2101A.2.3 *Not adopted by the State of California.*

2101A.2.4 Glass masonry. Glass masonry shall comply with the provisions of Section 2110A.

2101A.3 Definitions. For the purpose of this chapter, certain terms are defined as follows:

AREAS:

Bedded Area is the area of the surface of a masonry unit which is in contact with mortar in the plane of the joint.

Effective Area of Reinforcement is the cross-sectional area of reinforcement multiplied by the cosine of the angle between the reinforcement and the direction for which effective area is to be determined.

Gross Area is the total cross-sectional area of *any plane encompassed by the outer periphery of any* specified section.

Net Area is the gross cross-sectional area *at any plane* minus the area of ungrouted cores, notches, cells and unbedded areas, *etc*. Net area is the actual surface area of a cross section of masonry.

Transformed Area is the equivalent area of one material to a second based on the ratio of moduli of elasticity of the first material to the second.

BOND:

Adhesion Bond is the adhesion between masonry units and mortar or grout.

Reinforcing Bond is the adhesion between steel reinforcement and mortar or grout.

BOND BEAM is a horizontal grouted element within masonry in which reinforcement is embedded.

CELL is a void space having *minimum dimensions of 2 inches (51 mm) by 3 inches (76 mm). The minimum dimension for high-lift grouting shall be 3 inches (76 mm).*

CLEANOUT is an opening *at* the bottom of *cells or walls to be grouted which is* of sufficient size and spacing to allow the removal of *all* debris *from the supporting surface and of obstructions from the wall which might prevent proper grouting.*

COLLAR JOINT is the vertical, longitudinal, mortar or grouted joint between wythes.

COLUMN, REINFORCED, is a vertical structural member in which both the steel and masonry resist compression.

COLUMN, UNREINFORCED, is a vertical structural member whose horizontal dimension measured at right angles to the thickness does not exceed three times the thickness.

DIMENSIONS:

Actual Dimensions are the measured dimensions of a designated item. The actual dimension shall not vary from the specified dimension by more than the amount allowed in the appropriate standard of quality in Section 2102A.

Nominal Dimensions of masonry units are equal to its specified dimensions plus the thickness of the joint with which the unit is laid.

Specified Dimensions are the dimensions specified for the manufacture or construction of masonry, masonry units, joints or any other component of a structure.

GROUT LIFT is an increment of grout height within the total pour; a pour may consist of one or more grout lifts.

GROUT POUR is the total height of masonry wall to be grouted prior to the erection of additional masonry. A grout pour will consist of one or more grout lifts.

GROUTED MASONRY:

Grouted Hollow-unit Masonry is that form of grouted masonry construction in which *all* * * * cells of hollow units are continuously filled with grout. C A

Grouted Multiwythe Masonry is that form of grouted masonry construction in which the space between the wythes is solidly or periodically filled with grout.

JOINTS:

Bed Joint is the mortar joint that is horizontal at the time the masonry units are placed.

Collar Joint is the vertical *space separating a wythe of masonry from another wythe or from another continuous material and filled with mortar or grout.* C A C A C A C

Head Joint is the mortar joint *between units in the same wythe, usually vertical.*

MASONRY UNIT is brick, tile, stone, glass block or concrete block conforming to the requirements specified in Section *2102A.2, Items 4, 5 and 6.* C A

Hollow-masonry Unit is a masonry unit whose net cross-sectional area in every plane parallel to the bearing surface is less than 75 percent of the gross cross-sectional area in the same plane.

Solid-masonry Unit is a masonry unit whose net cross-sectional area in every plane parallel to the bearing surface is 75 percent or more of the gross cross-sectional area in the same plane.

PRISM is an assemblage of masonry units, mortar *and grout* used as a test specimen for determining properties of the masonry. C A

REINFORCED MASONRY is that form of masonry construction in which reinforcement acting in conjunction with the masonry is used to resist forces.

SHELL is the outer portion of a hollow masonry unit as placed in masonry.

WALLS:

Bonded Wall is a masonry wall in which two or more *of its* wythes *of masonry are adequately* bonded *together* to act as a structural unit. C A C A C

Hollow-unit Masonry Wall is that type of construction made with hollow masonry units in which the units are laid and set in mortar, reinforced and grouted solid except as provided in Section 2112A. C A C A C A C A C

Cavity Wall is a wall containing continuous air space with a minimum width of 2 inches (51 mm) and a maximum width of $4^1/_2$ inches (114 mm) between wythes which are tied with metal ties.

WALL TIE is a mechanical metal fastener which connects wythes of masonry to each other or to other materials.

WEB is an interior solid portion of a hollow-masonry unit as placed in masonry.

WYTHE is the portion of a wall which is one masonry unit in thickness. A collar joint is not considered a wythe.

2101A.4 Notations.

A_b = *gross* cross-sectional area of anchor bolt, square inches (mm^2).

A_e = effective area of masonry.

A_g = gross area of wall.

A_{jh} = total area of special horizontal reinforcement through a wall frame joint, square inches (mm^2).

A_{mv} = net area of masonry section bounded by wall thickness and length of section in the direction of shear force considered, square inches (mm^2).

A_p = area of tension (pullout) cone of an embedded anchor bolt projected onto the surface of masonry, square inches (mm^2).

A_s = effective cross-sectional area of reinforcement in a column or flexural member.

A_{se} = effective area of steel, square inches (mm^2).

A_{sh} = total cross-sectional area of rectangular tie reinforcement for the confined core.

A_v = area of steel required for shear reinforcement perpendicular to the longitudinal reinforcement.

A'_s = effective cross-sectional area of compression reinforcement in a flexural member.

a = depth of equivalent rectangular stress block for strength design.

B_{sn} = nominal shear strength of an anchor bolt, pounds (N).

B_t = allowable tension force on anchor bolts, pounds (N).

B_{tn} = nominal tensile strength of an anchor bolt, pounds (N).

B_v = allowable shear force on anchor bolts, pounds (N).

b = effective width of rectangular member or width of flange for T and I sections.

b_{su} = factored shear force supported by an anchor bolt, pounds (N).

b_t = computed tension force on anchor bolts, pounds (N).

b_{tu} = factored tensile force supported by an anchor bolt, pounds (N).

b_v = computed shear force on anchor bolts, pounds (N).

b' = width of web in T and I member.

C_d = masonry shear strength coefficient as obtained from Table 21A-K.

c = distance from the neutral axis to extreme fiber.

D = dead loads, or related internal moments and forces.

d = distance from the compression face of a flexural member to the centroid of longitudinal tensile reinforcement.

d_b = diameter of the reinforcing bar.

d_{bb} = diameter of largest beam longitudinal reinforcing bar passing through, or anchored in, the joint, inches (mm).

d_{bp} = diameter of largest pier longitudinal reinforcing bar passing through the joint, inches (mm).

E = load effects of earthquake, or related internal moments and forces.

E_m = modulus of elasticity of masonry.

E_s = modulus of elasticity of steel, 29,000,000 psi (200 GPa).

e = eccentricity of P_{uf}.

e_{mu} = maximum useable compressive strain of masonry.

F = loads due to weight and pressure of fluids or related moments and forces.

F_a = allowable average axial compressive stress for centroidally applied axial load only.

F_b = allowable flexural compressive stress if members were carrying bending load only.

F_{br} = allowable bearing stress.

F_s = allowable stress in reinforcement.

F_{sc} = allowable compressive stress in column reinforcement.

F_t = allowable flexural tensile stress in masonry.

F_v = allowable shear stress in masonry.

f_a = computed axial compressive stress due to design axial load.

f_b = computed flexural stress in the extreme fiber due to design bending loads only.

f_{md} = computed compressive stress in masonry due to dead load only.

f_r = modulus of rupture.

f_s = computed stress in reinforcement due to design loads.

f_v = computed shear stress due to design load.

f_y = tensile yield stress of reinforcement.

f'_g = specified compressive strength of grout at the age of 28 days.

f'_m = specified compressive strength of masonry at the age of 28 days.

G = shear modulus of masonry.

H = loads due to weight and pressure of soil, water in soil or related internal moments and forces.

h = height of wall between points of support.

h_b = beam depth, inches (mm).

h_c = cross-sectional dimension of grouted core measured center to center of confining reinforcement.

h_p = pier depth in the plane of the wall frame, inches (mm).

h' = effective height of a wall or column.

I = moment of inertia about the neutral axis of the cross-sectional area.

I_e = effective moment of inertia.

I_g, I_{cr} = gross, cracked moment of inertia of the wall cross section.

j = ratio or distance between centroid of flexural compressive forces and centroid of tensile forces to depth, d.

K = the reinforcement cover or the clear spacing, whichever is less.

k = the ratio of depth of the compressive stress in a flexural member to the depth, d.

L = live loads, or related internal moments and forces.

L_w = length of wall.

l = length of a wall or segment.

l_b = embedment depth of anchor bolts, inches (mm).

l_{be} = anchor bolt edge distance, the least length measured from the edge of masonry to the surface of the anchor bolt, inches (mm).

l_d = required development length of reinforcement.

M = design moment.

M_a = maximum moment in the member at the stage deflection is computed.

M_c = moment capacity of compression steel in a flexural member about the centroid of the tensile force.

M_{cr} = nominal cracking moment strength.

M_m = the moment of the compressive force in the masonry about the centroid of the tensile force in the reinforcement.

M_n = nominal moment strength.

M_s = the moment of the tensile force in the reinforcement about the centroid of the compressive force in the masonry.

M_{ser} = service moment at the midheight of the panel, including $P\Delta$ effects.

M_u = factored moment.

n = modular ratio

= E_s / E_m.

P = design axial load.

P_a = allowable centroidal axial load for reinforced masonry columns.

P_b = nominal balanced design axial strength.

P_f = load from tributary floor or roof area.

P_n = nominal axial strength, pounds (N).

P_o = nominal axial load strength without bending, pounds (N).

P_u = factored axial load, pounds (N).

P_{uf} = factored load from tributary floor or roof loads.

P_{uw} = factored weight of the wall tributary to the section under consideration.

P_w = weight of the wall tributary to section under consideration.

r = radius of gyration (based on specified unit dimensions or Tables 21A-H-1, 21A-H-2 and 21A-H-3), inches (mm).

r_b = ratio of the area of bars cut off to the total area of bars at the section.

S = section modulus.

s = spacing of stirrups or of bent bars in a direction parallel to that of the main reinforcement.

T = effects of temperature, creep, shrinkage and differential settlement.

t = effective thickness of a wythe, wall or column.

U = required strength to resist factored loads, or related internal moments and forces.

u = bond stress per unit of surface area of bar.

V = total design shear force.

V_{jh} = total horizontal joint shear, calculated in accordance with Section 2108A.2.6.2.9, Item 1.

V_m = nominal shear strength provided by masonry, pounds (N).

V_n = nominal shear strength, pounds (N).

V_s = nominal shear strength provided by shear reinforcement, pounds (N).

W = wind load, or related internal moments in forces.

w_u = factored distributed lateral load.

Δ_s = horizontal deflection at midheight under factored load.

ρ = ratio of the area of flexural tensile reinforcement, A_s, to the area bd.

ρ_b = *reinforcement ratio-producing balanced strain conditions.*

ρ_n = ratio of distributed shear reinforcement on a plane perpendicular to plane of A_{mv}.

Σ_o = sum of the perimeters of all the longitudinal reinforcement.

ϕ = strength-reduction factor.

SECTION 2102A — MATERIAL STANDARDS

2102A.1 Quality. Materials used in masonry shall conform to the requirements stated herein. If no requirements are specified in this section for a material, quality shall be based on generally accepted good practice, subject to the approval of the *enforcement agency.*

Reclaimed or previously used masonry units shall meet the applicable requirements as for new masonry units of the same material for their intended use.

2102A.2 Standards of Quality. The standards listed below labeled a "U.B.C. standard" are also listed in Chapter 35, Part II, and are part of this code. The other standards listed below are *American Society for Testing and Materials (ASTM) or American Concrete Institute (ACI)* standards *and are part of this code (*see Sections 3502 and 3503).

1. **Aggregates.**

    1.1 ASTM C 144, Aggregates for Masonry Mortar

    1.2 ASTM C 404, Aggregates for Grout

2. **Cement.**

    2.1 *Not adopted by the State of California.*

    2.2 U.B.C. Standard 19-1, Portland Cement and Blended Hydraulic Cements, *Cement for mortar and grout shall be Type I, II or III portland cement or Type I-A, II-A or III-A air-entraining portland cement as set forth in U.B.C. Standard 26-1, Part I. Masonry cement or plastic cement shall not be used in mortar or grout unless specifically approved by the enforcement agency.*

    2.3 U.B.C. Standard 21-14, Mortar Cement, *except the maximum limit for flash indicated in Table 24-19-B shall be 15 percent.* ◀

3. **Lime.**

    3.1 U.B.C. Standard 21-12, Quick Lime for Structural Purposes

    3.2 U.B.C. Standard 21-13, Hydrated Lime for Masonry Purposes. When Types N and NA hydrated lime are used in masonry mortar, they shall comply with the provisions of U.B.C. Standard 21-15, Section 21.1506.7, excluding the plasticity requirement.

4. **Masonry units of clay or shale.**

    4.1 *Not adopted by the State of California.*

    4.2 *Not adopted by the State of California.*

    4.3 U.B.C. Standard 21-1, Section 21.101, Building Brick (solid units), *Grade NW units shall not be used. Where severe frost action occurs in the presence of moisture, brick shall be at least Grade SW.*

    4.4 ASTM C 126, Ceramic Glazed Structural Clay Facing Tile, Facing Brick and Solid Masonry Units. Load-bearing glazed brick shall conform to the weathering and structural requirements of U.B.C. Standard 21-1, Section 21.106, Facing Brick

    4.5 *Not adopted by the State of California.*

    4.6 U.B.C. Standard 21-1, Section 21.106, Facing Brick (solid units)

    4.7 U.B.C. Standard 21-1, Section 21.107, Hollow Brick

4.8 ASTM C 67, Sampling and Testing Brick

4.9 *Not adopted by the State of California.*

4.10 *Not adopted by the State of California.*

5. Masonry units of concrete.

5.1 U.B.C. Standard 21-3, Concrete Building Brick. *Grade N only.*

5.2 U.B.C. Standard 21-4, Hollow and Solid Load-bearing Concrete Masonry Units, *Grade N-1 only. Open-end concrete masonry units shall be grouted solid and need not comply with the web thickness requirements of Table 21-4-C, U.B.C. Standard 21-4.*

5.3 U.B.C. Standard 21-5, Nonload-bearing Concrete Masonry Units

5.4 ASTM C 140, Sampling and Testing Concrete Masonry Units

5.5 ASTM C 426, Standard Test Method for Drying Shrinkage of Concrete Block

6. Masonry units of other materials.

6.1 *Not adopted by the State of California.*

6.2 *Not adopted by the State of California.*

6.3 ACI-704, Cast Stone. *Every concrete unit more than 18 inches (457 mm) in any dimension shall conform to the requirements of Chapter 19A for concrete.*

6.4 U.B.C. Standard 21-17, Test Method for Compressive Strength of Masonry Prisms

6.5 *Glass Block*

7. Connectors.

7.1 Wall ties and anchors made from steel wire shall conform to U.B.C. Standard 21-10, Part II, and other steel wall ties and anchors shall conform to A 36 in accordance with U.B.C. Standard 22-1. Wall ties and anchors made from copper, brass or other nonferrous metal shall have a minimum tensile yield strength of 30,000 psi (207 MPa).

7.2 All such items not fully embedded in mortar or grout shall either be corrosion resistant or shall be coated after fabrication with copper, zinc or a metal having at least equivalent corrosion-resistant properties.

8. Mortar.

8.1 U.B.C. Standard 21-15, Mortar for Unit Masonry

8.2 U.B.C. Standard 21-16, Field Test Specimens for Mortar

8.3 U.B.C. Standard 21-20, Standard Test Method for Flexural Bond Strength of Mortar Cement

9. Grout.

9.1 U.B.C. Standard 21-9, Grout for Masonry *and Section 2103A.4.*

9.2 U.B.C. Standard 21-18, Method of Sampling and Testing Grout

10. Reinforcement.

10.1 U.B.C. Standard 21-10, Part I, Joint Reinforcement for Masonry

10.2 ASTM A 615, A 616, A 617, A 706, A 767, and A 775, Deformed and Plain Billet-steel Bars, Rail-steel Deformed and Plain Bars, Axle-steel Deformed and Plain Bars, and Deformed Low-alloy Bars for Concrete Reinforcement

10.3 U.B.C. Standard 21-10, Part II, Cold-drawn Steel Wire for Concrete Reinforcement

▌SECTION 2103A — MORTAR AND GROUT

2103A.1 General. Mortar and grout shall comply with the provisions of this section. Special mortars, grouts or bonding systems may be used, subject to satisfactory evidence of their capabilities when approved by the *enforcement agency.*

2103A.2 Materials. Materials used as ingredients in mortar and grout shall conform to the applicable requirements in Section 2102A. Cementitious materials for grout shall be one or both of the following: lime and portland cement. Cementitious materials for mortar shall be one or more of the following: lime, masonry cement, portland cement and mortar cement. Cementitious materials or additives shall not contain epoxy resins and derivatives, phenols, asbestos fibers or fireclays.

Water used in mortar or grout shall be clean and free of deleterious amounts of acid, alkalies or organic material or other harmful substances.

2103A.3 Mortar.

2103A.3.1 General. Mortar shall ✳ ✳ ✳ *conform to the proportions shown in Table 21A-A for Type S mortar.*

Lime shall be the last material added to the mixer. Materials for mortar and grout shall be measured in suitable calibrated devices. Shovel measurements will not be accepted. Aggregates for mortar shall conform to the provisions set forth in ASTM C 144, Aggregates for Masonry Mortar.

2103A.3.2 Selecting proportions. Mortar *shall attain a minimum compressive strength of 1,500 psi (10.34 MPa) at 28 days for masonry with $f'_m = 1,500$ psi (10.34 MPa) for field test specimens prepared according to U.B.C. Standard 21-16.*

> NOTE: *See exception to Section 2105A.3.1 for specified values of f'_m in excess of 1,500 psi (10.34 MPa).*

2103A.4 Grout.

2103A.4.1 General. Grout shall consist of a mixture of cementitious materials and aggregate to which water has been added such that the mixture will flow without segregation of the constituents. Grout shall have a minimum compressive strength f'_g of 2,000 psi (13.79 MPa).

2103A.4.2 Selecting proportions. Water content shall be adjusted to provide proper workability and to enable proper placement under existing field conditions, without segregation. Grout shall be specified by one of the following methods:

1. Proportions of ingredients and any additives shall be based on laboratory or field experience with the grout ingredients and the masonry units to be used. The grout shall be specified by the proportion of its constituents in terms of parts by volume, or

2. Minimum compressive strength which will produce the required prism strength, or

3. Proportions by grout type shall be used as given in Table 21A-B.

2103A.4.3 Aggregate. *Aggregate for grout shall conform to the requirements set forth in ASTM C 404, Aggregates for Grout. Coarse grout shall be used in grout spaces 2 inches (51 mm) or more in width and in all filled-cell masonry construction.*

> NOTE: *See exception to Section 2105A.3.1 for specified values in excess of 1,500 psi (10.34 MPa).*

2103A.5 Additives and Admixtures.

2103A.5.1 General. Additives and admixtures to mortar or grout shall not be used unless approved by the *enforcement agency.*

2103A.5.2 Antifreeze compounds. Antifreeze liquids, chloride salts or other such substances shall not be used in mortar or grout.

2103A.5.3 Air entrainment. Air-entraining substances shall not be used in mortar or grout unless tests are conducted to determine compliance with the requirements of this code.

2103A.5.4 Colors. Only pure mineral oxide, carbon black or synthetic colors may be used. Carbon black shall be limited to a maximum of 3 percent of the weight of the cement.

SECTION 2104A — CONSTRUCTION

2104A.1 General. Masonry shall be constructed according to the provisions of this section.

2104A.2 Materials: Handling, Storage and Preparation. All materials shall comply with applicable requirements of Section 2102A. Storage, handling and preparation at the site shall conform also to the following:

1. Masonry materials shall be stored so that at the time of use the materials are clean and structurally suitable for the intended use.

2. All metal reinforcement shall be free from loose rust and other coatings that would inhibit reinforcing bond.

3. At the time of laying, burned clay units and sand lime units shall have an initial rate of absorption not exceeding 0.035 ounce per square inch (1.6 L/m^2) during a period of one minute. In the absorption test, the surface of the unit shall be held $^1/_8$ inch (3 mm) below the surface of the water.

4. Concrete masonry units shall not be wetted unless otherwise approved.

5. Materials shall be stored in a manner such that deterioration or intrusion of foreign materials is prevented and that the material will be capable of meeting applicable requirements at the time of mixing.

6. The method of measuring materials for mortar and grout shall be such that proportions of the materials can be controlled.

7. Mortar or grout mixed at the jobsite shall be mixed for a period of time not less than three minutes or more than 10 minutes in a mechanical mixer with the amount of water required to provide the desired workability. Hand mixing of small amounts of mortar is permitted. Mortar may be retempered. Mortar or grout which has hardened or stiffened due to hydration of the cement shall not be used. In no case shall mortar be used two and one-half hours, nor grout used one and one-half hours, after the initial mixing water has been added to the dry ingredients at the jobsite.

> **EXCEPTION:** Dry mixes for mortar and grout which are blended in the factory and mixed at the jobsite shall be mixed in mechanical mixers until workable, but not to exceed 10 minutes.

2104A.3 Cold-weather Construction.

2104A.3.1 General. All materials shall be delivered in a usable condition and stored to prevent wetting by capillary action, rain and snow.

The tops of all walls not enclosed or sheltered shall be covered with a strong weather-resistive material at the end of each day or shutdown.

Partially completed walls shall be covered at all times when work is not in progress. Covers shall be draped over the wall and extend a minimum of 2 feet (610 mm) down both sides and shall be securely held in place, except when additional protection is required in Section 2104A.3.4.

2104A.3.2 Preparation. If ice or snow has inadvertently formed on a masonry bed, it shall be thawed by application of heat carefully applied until top surface of the masonry is dry to the touch.

A section of masonry deemed frozen and damaged shall be removed before continuing construction of that section.

2104A.3.3 Construction. Masonry units shall be dry at time of placement. Wet or frozen masonry units shall not be laid.

Special requirements for various temperature ranges are as follows:

1. Air temperature 40°F. to 32°F. (4.5°C. to 0°C.): Sand or mixing water shall be heated to produce mortar temperatures between 40°F. and 120°F. (4.5°C. and 49°C.).

2. Air temperature 32°F. to 25°F. (0°C. to –4°C.): Sand and mixing water shall be heated to produce mortar temperatures between 40°F. and 120°F. (4.5°C. and 49°C.). Maintain temperatures of mortar on boards above freezing.

3. Air temperature 25°F. to 20°F. (–4°C. to –7°C.): Sand and mixing water shall be heated to produce mortar temperatures between 40°F. and 120°F. (4.5°C. and 49°C.). Maintain mortar tempera-

tures on boards above freezing. Salamanders or other sources of heat shall be used on both sides of walls under construction. Windbreaks shall be employed when wind is in excess of 15 miles per hour (24 km/h).

4. Air temperature 20°F. (–7°C.) and below: Sand and mixing water shall be heated to provide mortar temperatures between 40°F. and 120°F. (4.5°C. and 49°C.). Enclosure and auxiliary heat shall be provided to maintain air temperature above 32°F. (0°C.). Temperature of units when laid shall not be less than 20°F. (–7°C.).

2104A.3.4 Protection. *No masonry shall be laid when the temperature of the outside air is below 40°F. (4.5°C.) unless approved methods are used during the construction to prevent damage to the masonry. Such methods shall include protection of the masonry for a period of at least 48 hours where Type I or II portland cement is used in the mortar and grout and for a period of 24 hours where Type III portland cement is used.* C A C A C A C A C ✳

2104A.3.5 Placing grout and protection of grouted masonry. When air temperatures fall below 40°F. (4.5°C.), grout mixing water and aggregate shall be heated to produce grout temperatures between 40°F. and 120°F. (4.5°C. and 49°C.).

Masonry to be grouted shall be maintained above freezing during grout placement and for at least 24 hours after placement.

When atmospheric temperatures fall below 20°F. (–7°C.), enclosures shall be provided around the masonry during grout placement and for at least 24 hours after placement. *No grout shall be placed when the atmospheric temperature falls below 20°F. (–7°C.).* C A C

2104A.4 Placing Masonry Units.

2104A.4.1 Mortar. The mortar shall be sufficiently plastic and units shall be placed with sufficient pressure to extrude mortar from the joint and produce a tight joint. Deep furrowing which produces voids shall not be used.

The initial bed joint thickness shall not be less than $^1/_4$ inch (6 mm) or more than 1 inch (25 mm); subsequent bed joints shall not be less than $^1/_4$ inch (6 mm) or more than $^5/_8$ inch (16 mm) in thickness.

2104A.4.2 Surfaces. Surfaces to be in contact with mortar or grout shall be clean and free of deleterious materials.

2104A.4.3 Solid masonry units. Solid masonry units shall have full head and bed joints.

2104A.4.4 Hollow-masonry units. All head and bed joints shall be filled solidly with mortar for a distance in from the face of the unit not less than the thickness of the shell.

Head joints of open-end units with beveled ends that are to be fully grouted need not be mortared. The beveled ends shall form a grout key which permits grout within $^5/_8$ inch (16 mm) of the face of the unit. The units shall be tightly butted to prevent leakage of grout.

2104A.4.5 Corbeling. *Corbels may be built only into solid-masonry walls 12 inches (305 mm) or more in thickness. The projection for each course in such corbel shall not exceed 1 inch (25 mm), and the maximum projection shall not exceed one third the total thickness of the wall when used to support structural members, and not more than 6 inches (152 mm) when used to support a chimney built into the wall. The top course of all corbels shall be a header course.* C A C A C A C A C

2104A.5 Reinforcement Placing. Reinforcement details shall conform to the requirements of this chapter. Metal reinforcement shall be located in accordance with the plans and specifications. Reinforcement shall be secured against displacement prior to grouting by wire positioners or other suitable devices at intervals not exceeding 200 bar diameters.

Tolerances for the placement of reinforcement in walls and flexural elements shall be plus or minus $^1/_2$ inch (13 mm) for *d* equal to 8 inches (203 mm) or less, ± 1 inch (± 25 mm) for *d* equal to

24 inches (610 mm) or less but greater than 8 inches (203 mm), and ± 1$^1/_4$ inches (32 mm) for d greater than 24 inches (610 mm).

Tolerance for longitudinal location of reinforcement shall be ± 2 inches (51 mm).

2104A.6 Grouted Masonry.

2104A.6.1 General conditions. Grouted masonry shall be constructed in such a manner that all elements of the masonry act together as a structural element.

Prior to grouting, the grout space shall be clean so that all spaces to be filled with grout do not contain mortar projections greater than $^1/_4$ inch *(6.4 mm)*, mortar droppings *and* other foreign material. Grout shall be placed so that all spaces to be grouted *do not contain voids.*

Grout materials and water content shall be controlled to provide adequate fluidity for placement without segregation of the constituents, and shall be mixed thoroughly. *Reinforcement shall be clean, properly positioned and solidly embedded in the grout.*

The grouting of any section of wall shall be completed in one day with no interruptions greater than one hour.

Between grout pours, a horizontal construction joint shall be formed by stopping all wythes at the same elevation and with the grout stopping a minimum of 1$^1/_2$ inches (38 mm) below a mortar joint, except at the top of the wall. Where bond beams occur, the grout pour shall be stopped a minimum of $^1/_2$ inch (13 mm) below the top of the masonry.

2104A.6.1.1 Reinforced grouted masonry.

2104A.6.1.1.1 General. Reinforced grouted masonry is that form of construction made with clay or shale brick or made with solid concrete building brick in which interior joints of masonry are filled by pouring grout around reinforcing therein as the work progresses.

At the time of laying, all masonry units shall be free of dust and dirt.

> *NOTES: 1. For rate of absorption, see Section 2104A.2. All units in a masonry assembly shall have a compatible absorption rate.*
>
> *2. For mortar, see Section 2103A.*
>
> *3. See Section 2105A.3 for assumed masonry strength.*

2104A.6.1.1.2 Low-lift grouted construction. Requirements for construction shall be as follows:

1. All units in the two outer tiers shall be laid with full-shoved head joint and bed mortar joints. Masonry headers shall not project into the grout space.

2. The minimum grout space for low-lift grout masonry shall be 2$^1/_2$ inches (64 mm). Floaters shall be used where the grout space exceeds 5 inches (127 mm) in width. The thickness of grout between masonry units and floaters shall be a minimum of 1 inch (25 mm). Floaters shall be worked into fresh puddled grout using a vibrating motion until half of the floater is embedded in the grout. All reinforcing and wire ties shall be embedded in the grout. The thickness of the grout between masonry units and reinforcing shall be a minimum of one bar diameter.

3. One tier of a grouted reinforced masonry wall may be carried up 12 inches (305 mm) before grouting, but the other tier shall be laid up and grouted in lifts not to exceed one masonry unit in height. All grout shall be puddled with a mechanical vibrator or wood stick immediately after placing so as to completely fill all voids and to consolidate the grout. All vertical and horizontal steel shall be held firmly in place by a frame or suitable devices.

4. If the work is stopped for one hour or more, the horizontal construction joints shall be formed by stopping all tiers at the same elevation, and with the grout $^1/_2$ inch (13 mm) below the top.

5. Toothing of masonry walls is prohibited. Racking is to be held to a minimum.

2104A.6.1.1.3 High-lift grouted construction. Where high-lift grouting is used, the method shall be subject to the approval of the enforcement agency. Requirements for construction shall be as follows:

1. All units in the two tiers shall be laid with full head and bed mortar joints.

2. The two tiers shall be bonded together with wall ties. Ties shall not be less than No. 9 wire in the form of rectangles 4 inches (102 mm) wide and 2 inches (51 mm) in length less than the overall wall thickness. Kinks, water drips, or deformations shall not be permitted in the ties. One tier of the wall shall be built up not more than 16 inches (406 mm) ahead of the other tier. Ties shall be laid not to exceed 24 inches (610 mm) on center horizontally and 16 inches (406 mm) on center vertically for running bond, and not more than 24 inches (610 mm) on center horizontally and 12 inches (305 mm) on center vertically for stack bond.

3. Cleanouts shall be provided for each pour by leaving out every other unit in the bottom tier of the section being poured or by cleanout openings in the foundation. The foundation or other horizontal construction joints shall be cleaned of all loose material and mortar droppings before each pour. The cleanouts shall be sealed before grouting, after inspection.

4. The grout space in high-lift grouted masonry shall be a minimum of $3^1/_2$ inches (89 mm). All reinforcing and wire ties shall be embedded in the grout. The thickness of the grout between masonry units and reinforcing shall be a minimum of one bar diameter.

5. Vertical grout barriers or dams shall be built of solid masonry across the grout space the entire height of the wall to control the flow of the grout horizontally. Grout barriers shall not be more than 30 feet (9144 mm) apart.

6. An approved admixture of a type that reduces early water loss and produces an expansive action shall be used in high-lift grout.

7. Grouting shall be done in a continuous pour in lifts not exceeding 4 feet (1219 mm). Grout shall be consolidated by mechanical vibration only, and shall be reconsolidated after excess moisture has been absorbed, but before plasticity is lost. The grouting of any section of a wall between control barriers shall be completed in one day, with no interruptions greater than one hour.

NOTE: *For special inspection requirements, see Section 2105A.3.*

8. **Stresses.** *All reinforced grouted masonry shall be so constructed that the unit stresses do not exceed those set forth in Sections 2107A.2.5 through 2107A.2.11.*

2104A.6.1.2 *Reinforced hollow-unit masonry.*

2104A.6.1.2.1 General. *Reinforced hollow-unit masonry is that type of construction made with hollow-masonry units in which cells are continuously filled with grout, and in which reinforcement is embedded. All cells shall be solidly filled with grout in reinforced hollow-unit masonry, except as provided in Section 2112A.1. Construction shall be one of the two following methods: The low-lift method where the maximum height of construction laid before grouting is 2 feet (610 mm), or the high-lift method where the full height of construction between horizontal cold joints is grouted in one operation. General requirements for construction shall be as follows:*

1. All reinforced hollow-unit masonry shall be built to preserve the unobstructed vertical continuity of the cells to be filled. All head joints shall be solidly filled with mortar for a distance in from the face of the wall or unit not less than the thickness of the longitudinal face shells.

2. Mortar shall be as specified in Section 2103A.

3. Walls and cross webs forming such cells to be filled shall be full bedded in mortar to prevent leakage of grout.

4. Bond shall be provided by lapping units in successive vertical courses. Where stack bond is used in reinforced hollow-unit masonry, the open-end type of unit shall be used with vertical reinforcement spaced a maximum of 16 inches (406 mm) on center.

5. Vertical cells to be filled shall have vertical alignment sufficient to maintain a clear unobstructed, continuous vertical cell measuring not less than 2 inches by 3 inches (51 mm by 76 mm), except the minimum cell dimension for high-lift grout shall be 3 inches (76 mm).

6. *At the time of laying, all masonry units shall be free of dust and dirt.*

7. *Grout shall be a workable mix suitable for placing without segregation and shall be thoroughly mixed. Grout shall be placed by pumping or an approved alternate method and shall be placed before initial set or hardening occurs. Grout shall be consolidated by mechanical vibration during placing and reconsolidated after excess moisture has been absorbed, but before workability is lost. The grouting of any section of a wall shall be completed in one day, with no interruptions greater than one hour.*

8. *All reinforcing and wire ties shall be embedded in the grout. The space between masonry unit surfaces and reinforcing shall be a minimum of one bar diameter.*

9. *Horizontal reinforcement shall be placed in bond beam units with a minimum grout cover of 1 inch (25 mm) above steel for each grout pour. The depth of the bond beam channel below the top of the unit shall be a minimum of 1 1/2 inches (38 mm) and the width shall be 3 inches (76 mm) minimum.*

2104A.6.1.2.2 Low-lift grouted construction. *Units shall be laid a maximum of 2 feet (610 mm) before grouting, and all overhanging mortar and mortar droppings shall be removed. Grouting shall follow each 2 feet (610 mm) of construction laid and shall be consolidated so as to completely fill all voids and embed all reinforcing steel. When grouting is stopped for one hour or longer, horizontal construction joints shall be formed by stopping the pour of grout not less than 1/2 inch (13 mm) or more than 2 inches (51 mm) below the top of the uppermost unit grouted. Horizontal steel shall be fully embedded in grout in an uninterrupted pour.*

2104A.6.1.2.3 High-lift grouted construction. *Where high-lift grouting is used, the method shall be approved by the enforcement agency. Cleanout openings shall be provided at the bottom of each pour of grout. The foundation or other horizontal construction joints shall be cleaned of all loose material and mortar droppings before each pour. The cleanouts shall be sealed before grouting. An approved admixture that reduces early water loss and produces an expansive action shall be used in the grout.*

2104A.6.1.2.4 Stresses. *All reinforced hollow-unit masonry shall be so constructed that the units stressed do not exceed those set forth in Sections 2107A.2.5 through 2107A.2.11.*

Vertical barriers of masonry may be built across the grout space. The grouting of any section of wall between barriers shall be completed in one day with no interruption longer than one hour.

NOTE: *See Section 2105A.3 for assumed masonry strength.*

2104A.6.2 Construction requirements. Reinforcement *and embedded items* shall be placed *and securely anchored against moving* prior to grouting. Bolts shall be accurately set with templates or by approved equivalent means and held in place to prevent dislocation during grouting.

Segregation of the grout materials and damage to the masonry shall be avoided during the grouting process.

Grout shall be consolidated by mechanical vibration during placement before loss of plasticity in a manner to fill the grout space. Grout pours greater than 12 inches (305 mm) in height shall be reconsolidated by mechanical vibration to minimize voids due to water loss. *Grout not mechanically vibrated shall be puddled.*

2104A.7 Aluminum Equipment. Grout shall not be handled nor pumped utilizing aluminum equipment unless it can be demonstrated with the materials and equipment to be used that there will be no deleterious effect on the strength of the grout.

2104A.8 Joint Reinforcement. Wire joint reinforcement used in the design as principal reinforcement in hollow-unit construction shall be continuous between supports unless splices are made by lapping:

1. Fifty-four wire diameters in a grouted cell, or

2. Seventy-five wire diameters in the mortared bed joint, or

3. In alternate bed joints of running bond masonry a distance not less than 54 diameters plus twice the spacing of the bed joints, or

4. As required by calculation and specific location in areas of minimum stress, such as points of inflection.

Side wires shall be deformed and shall conform to U.B.C. Standard 21-10, Part I, Joint Reinforcement for Masonry.

SECTION 2105A — QUALITY ASSURANCE

2105A.1 General. Quality assurance shall be provided to ensure that materials, construction and workmanship are in compliance with the plans and specifications, and the applicable requirements of this chapter. When required, inspection records shall be maintained and made available to the *enforcement agency.*

2105A.2 Scope. Quality assurance shall include, but is not limited to, assurance that:

1. Masonry units, reinforcement, cement, lime, aggregate and all other materials meet the requirements of the applicable standards of quality and that they are properly stored and prepared for use.

2. Mortar and grout are properly mixed using specified proportions of ingredients. The method of measuring materials for mortar and grout shall be such that proportions of materials are controlled.

3. Construction details, procedures and workmanship are in accordance with the plans and specifications.

4. Placement, splices and reinforcement sizes are in accordance with the provisions of this chapter and the plans and specifications.

2105A.3 Compliance with f'_m.

2105A.3.0 f'_m. *The specified compressive strength, f'_m, assumed in design shall be 1,500 psi (10.34 MPa) for all masonry construction using materials and details of construction required herein.*

> **EXCEPTION:** *Subject to the approval of the enforcement agency, higher values of f'_m may be used in the design of reinforced grouted masonry and reinforced hollow-unit masonry. The approval shall be based on test results submitted by the architect or engineer which demonstrate the ability of the proposed construction to meet prescribed performance criteria for strength and stiffness. The design shall assume that the reinforcement will be placed in a location that will produce the largest stresses within the tolerances allowed in Section 2104A.5 and shall take into account the mortar joint depth. In no case shall the f'_m assumed in design exceed 2,500 psi (17.24 MPa).*

Where an f'_m greater than 1,500 psi (10.34 MPa) is approved, the architect or structural engineer shall establish a method of quality control of the masonry construction acceptable to the enforcement agency which shall be described in the contract specifications. Compliance with the requirements for the specified compressive strength of masonry f'_m shall be in accordance with Section 2105A.3.2, 2105A.3.3, 2105A.3.4 or 2105A.3.5.

2105A.3.1 *Masonry core testing.* *Not less than two cores having a diameter of approximately two thirds the wall thickness shall be taken from each project. At least one core shall be taken from each building for each 5,000 square feet (465 m²) of floor area or fraction thereof. The architect or structural engineer in responsible charge of the project or his/her representative (inspector) shall select the areas for sampling. One half of the number of cores taken shall be tested in shear. The shear wall loadings shall test both joints between the grout core and the outside wythes of the masonry. Core samples shall not be soaked before testing. Materials and workmanship shall be such that for all* *masonry when tested in compression, cores shall show an ultimate strength at least equal to the f'_m assumed in design, but not less than 1,500 psi (10.34 MPa). When tested in shear, the unit shear on the cross section of the core shall not be less than*

$$2.5\sqrt{f'_m} \; psi$$

Shear testing apparatus shall be of a design approved by the enforcement agency. Visual examination of all cores shall be made to ascertain if the joints are filled.

The project inspector or testing agency shall inspect the coring of the masonry walls and shall prepare a report of coring operations for the testing laboratory files and mail one copy to the enforcement agency. Such reports shall include the total number of cores cut, the location, and the condition of all cores cut on each project, regardless of whether the core specimens failed during cutting operation. All cores shall be submitted to the laboratory for examination.

2105A.3.2 Masonry prism testing. The compressive strength of masonry determined in accordance with U.B.C. Standard 21-17 for each set of prisms shall equal or exceed f'_m. Compressive strength of prisms shall be based on tests at 28 days. Compressive strength at seven days or three days may be used provided a relationship between seven-day and three-day and 28-day strength has been established for the project prior to the start of construction. Verification by masonry prism testing shall meet the following:

1. A set of five masonry prisms shall be built and tested in accordance with U.B.C. Standard 21-17 prior to the start of construction. Materials used for the construction of the prisms shall be taken from those specified to be used in the project. Prisms shall be constructed under the observation of the engineer or special inspector or an approved agency and tested by an approved agency.

2. A set of three prisms shall be built and tested during construction in accordance with U.B.C. Standard 21-17 for each 5,000 square feet (465 m^2) of wall area, but not less than one set of three masonry prisms for the project.

2105A.3.3 Masonry prism test record. Compressive strength verification by masonry prism test records shall meet the following:

1. A masonry prism test record approved by the *enforcement agency* of at least 30 masonry prisms which were built and tested in accordance with U.B.C. Standard 21-17. Prisms shall have been constructed under the observation of an engineer or special inspector or an approved agency and shall have been tested by an approved agency.

2. Masonry prisms shall be representative of the corresponding construction.

3. The average compressive strength of the test record shall equal or exceed 1.33 f'_m.

4. * * * A set of three masonry prisms shall be built during construction in accordance with U.B.C. Standard 21-17 for each 5,000 square feet (465 m^2) of wall area, but not less than one set of three prisms for the project.

2105A.3.4 Unit strength method. Verification by the unit strength method shall meet the following:

1. *Masonry unit test. Test the* units prior to construction and test units during construction for each 5,000 square feet (465 m^2) of wall area for compressive strength to show compliance with the compressive strength required in Table 21A-D; and

> **EXCEPTION:** Prior to the start of construction, prism testing may be used in lieu of testing the unit strength. During construction, prism testing may also be used in lieu of testing the unit strength and the grout as required by Section 2105A.3.4, Item 4.

2. *Mortar and grout tests. At the beginning of all masonry work, at least one test sample of the mortar and grout shall be taken on three successive working days and at least at one-week intervals thereafter. The samples shall be continuously stored in moist air until tested. They shall meet the*

minimum strength requirement given in Sections 2103A.3 and 2103A.4 for mortar and grout, respectively. Additional samples shall be taken whenever any change in materials or job conditions occur, or whenever in the judgment of the architect, structural engineer or the enforcement agency such tests are necessary to determine the quality of the material.

Test specimens for mortar and grout shall be made as set forth in U.B.C. Standards 21-16 and 21-18. In making the mortar test specimens, the mortar shall be taken from the unit soon after spreading. After molding, the molds shall be carefully protected by a covering which shall be kept damp for at least 24 hours, after which the specimens shall be stored and tested as required for concrete cylinders.

In making grout test specimens, the masonry unit molds shall be broken away after the grout has taken its set, but before it has hardened. If an absorbent paper liner is used, the mold may be left in place until the specimen has hardened. The prisms shall be stored as required for concrete cylinders. They shall be tested in the vertical position.

2105A.3.5 Testing prisms from constructed masonry. When approved by the *enforcement agency*, acceptance of masonry which does not meet the requirements of Section 2105A.3.2, 2105A.3.3 or 2105A.3.4 shall be permitted to be based on tests of prisms cut from the masonry construction in accordance with the following:

1. A set of three masonry prisms that are at least 28 days old shall be saw cut from the masonry for each 5,000 square feet (465 m^2) of the wall area that is in question but not less than one set of three masonry prisms for the project. The length, width and height dimensions of the prisms shall comply with the requirements of U.B.C. Standard 21-17. Transporting, preparation and testing of prisms shall be in accordance with U.B.C. Standard 21-17.

2. The compressive strength of prisms shall be the value calculated in accordance with U.B.C. Standard 21-17, Section 21.1707.2, Item 2, except that the net cross-sectional area of the prism shall be based on the net mortar bedded area.

3. Compliance with the requirement for the specified compressive strength of masonry, f'_m, shall be considered satisfied provided the modified compressive strength equals or exceeds the specified f'_m. Additional testing of specimens cut from locations in question shall be permitted.

2105A.4 *Not adopted by the State of California.*

2105A.5 *Not adopted by the State of California.*

2105A.6 *Combination of Units. In walls or other structural members composed of different kinds or grades of units or materials, a full-scale test panel shall be constructed before the beginning of masonry work. The test panel will be cored and tested as approved by the enforcement agency to determine the compatibility of the materials (including bond strength between the materials). If the materials are not compatible, they will be rejected. The net thickness of any facing unit which is used to resist stress shall not be less than 1$^1\!/_2$ inches (38 mm).*

2105A.7 *Masonry Inspection. All structural masonry work shall be continuously inspected during laying and grouting by an inspector specially approved for that purpose by the enforcement agency. The inspector shall make test samples and perform such tests as are required and shall check the materials, details of construction, and construction procedures.*

The special masonry inspector shall furnish a verified report that, of his own personal knowledge, the work covered by the report has been performed, and materials used and installed in every material respect in compliance with the duly approved plans and specifications.

SECTION 2106A — GENERAL DESIGN REQUIREMENTS

2106A.1 General.

2106A.1.1 Scope. The design of masonry structures shall comply with the working stress design provisions of Section 2107A, or the strength design provisions of Section 2108A ✱ ✱ ✱ and with the provisions of this section. Unless otherwise stated, all calculations shall be made using or based on specified dimensions.

2106A.1.2 Plans. Plans submitted for approval shall describe the required design strengths of masonry materials and inspection requirements for which all parts of the structure were designed, and any load test requirements.

2106A.1.3 Design loads. See Chapter 16A for design loads.

2106A.1.4 Stack bond. In bearing and nonbearing walls, except veneer walls, if less than 75 percent of the units in any transverse vertical plane lap the ends of the units below a distance less than one half the height of the unit, or less than one fourth the length of the unit, the wall shall be considered laid in stack bond.

2106A.1.5 Multiwythe walls.

2106A.1.5.1 General. *All walls shall be reinforced as required by Section 2106A.1.12.4 and* all wythes bonded by grout *and when required,* tied together by corrosion-resistant wall ties or joint reinforcement conforming to the requirements of Section 2102A and as *follows:*

2106A.1.5.2 Wall ties. Wall ties shall be of sufficient length to engage all *the* wythes *to within 1 inch (25 mm) of the outside face.* The portion of the wall ties within the wythe shall be completely embedded in mortar or grout. The ends of the wall ties shall be bent to 90-degree angles with an extension not less than 2 inches (51 mm) long.

There shall be at least one $3/16$-inch-diameter (4.8 mm) wall tie for each *2 square feet (0.19 m^2)* of wall area.

Ties in alternate courses shall be staggered. The maximum vertical distance between ties shall not exceed 24 inches (610 mm).

Additional ties spaced not more than 36 inches (914 mm) apart shall be provided around openings within a distance of 12 inches (305 mm) from the edge of the opening.

Wall ties of different size and spacing that provide equivalent strength between wythes may be used.

2106A.1.5.3 *Not adopted by the State of California.*

2106A.1.5.4 Joint reinforcement. Prefabricated joint reinforcement for masonry walls shall have at least one cross wire of at least No. 9 gage steel for each 2 square feet (0.19 m^2) of wall area. The vertical spacing of the joint reinforcement shall not exceed 16 inches (406 mm). The longitudinal wires shall be thoroughly embedded in the bed joint mortar. The joint reinforcement shall engage all wythes.

2106A.1.6 Vertical support. Structural members providing vertical support of masonry shall provide a bearing surface on which the initial bed joint shall not be less than $1/4$ inch (6 mm) or more than 1 inch (25 mm) in thickness and shall be of noncombustible material. *Masonry shall not be supported by wood members, except as provided for in Sections 2316A.1 and 1631A.2.9, Item 5. Arches and lintels supporting masonry shall be of noncombustible material.*

2106A.1.7 Lateral support. Lateral support of masonry may be provided by cross walls, columns, pilasters, counterforts or buttresses where spanning horizontally or by floors, beams, girts or roofs where spanning vertically. *Where walls are supported laterally by vertical elements, the stiffness of each vertical element shall exceed that of the tributary area of the wall.*

The clear distance between lateral supports of a beam shall not exceed 32 times the least width of the compression area.

2106A.1.8 Protection of ties and joint reinforcement. A minimum of $5/8$-inch (16 mm) mortar cover shall be provided between ties or joint reinforcement and any exposed face. The thickness of grout or mortar between masonry units and joint reinforcement shall not be less than $^1/_4$ inch (6 mm), except that $^1/_4$ inch (6 mm) or smaller diameter reinforcement may be placed in bed joints which are at least twice the thickness of the reinforcement.

2106A.1.9 Pipes and conduits embedded in masonry. Pipes or conduit shall not be embedded in any masonry.

Placement of pipes or conduits in unfilled cores of hollow-unit masonry shall not be considered as embedment.

> **EXCEPTIONS:** 1. Rigid electric conduits may be embedded in structural masonry when their location has been detailed on the approved plan.
>
> 2. Any pipe or conduit may pass vertically or horizontally through any masonry by means of a sleeve at least large enough to pass any hub or coupling on the pipeline. Such sleeves shall not be placed closer than three diameters, center to center, nor shall they unduly impair the strength of construction.

2106A.1.10 Load tests. When a load test is required, the member or portion of the structure under consideration shall be subjected to a superimposed load equal to twice the design live load plus one half of the dead load. This load shall be left in position for a period of 24 hours before removal. If, during the test or upon removal of the load, the member or portion of the structure shows evidence of failure, such changes or modifications as are necessary to make the structure adequate for the rated capacity shall be made; or where approved, a lower rating shall be established. A flexural member shall be considered to have passed the test if the maximum deflection D at the end of the 24-hour period does not exceed the value of Formulas (6A-1) or (6A-2) and the beams and slabs show a recovery of at least 75 percent of the observed deflection within 24 hours after removal of the load.

$$D = \frac{l}{200} \qquad\qquad (6A\text{-}1)$$

$$D = \frac{l^2}{4,000t} \qquad\qquad (6A\text{-}2)$$

2106A.1.11 Reuse of masonry units. Masonry units may be reused when clean, whole and conforming to the other requirements of this section. All structural properties of masonry of reclaimed units shall be determined by approved test.

2106A.1.12 Special provisions in areas of seismic risk.

2106A.1.12.1 General. Masonry structures constructed ✳ ✳ ✳ shall be designed in accordance with the design requirements of this chapter and the special provisions for each seismic zone given in this section.

2106A.1.12.2 *Not adopted by the State of California.*

2106A.1.12.3 *Not adopted by the State of California.*

2106A.1.12.4 Special provisions for Seismic Zones 3 and 4. All masonry structures shall be *so* designed and constructed *that the unit stresses do not exceed those set forth in Section 2107A, and the* following additional requirements and limitations *are met.*

Wall reinforcement. *The total area of reinforcement in reinforced masonry walls shall not be less than 0.003 times the sectional area of the wall. Neither the horizontal nor the vertical reinforcement shall be less than one third of the total. Horizontal and vertical bars shall be spaced at not more than 24 inches (610 mm) center to center. The minimum reinforcing shall be No. 4, except that No. 3 bars may be used for ties and stirrups. Vertical wall steel shall have dowels of equal size and equal matched spacing in all footings.* Reinforcement shall be continuous around wall corners and through intersections. Only reinforcement which is continuous in the wall shall be considered in computing the minimum area of reinforcement. Reinforcement with splices conforming to Section 2107A.2.2.6 shall be considered as continuous reinforcement.

Horizontal reinforcement shall be provided in the top of footings, at the top of wall openings, at roof and floor levels, and at the top of parapet walls. For walls 12 inches (nominal) (305 mm) or more in thickness, reinforcing shall be equally divided into two layers, except where designed as retaining walls. Where reinforcement is added above the minimum requirements, such additional reinforcement need not be so divided.

In bearing walls of every type of reinforced masonry, there shall not be less than one No. 5 bar or two No. 4 bars on all sides of, and adjacent to, every opening which exceeds 24 inches (610 mm) in either direction, and such bars shall extend not less than 48 diameters, but in no case less than 24 inches (610 mm) beyond the corners of the opening. The bars required by this paragraph shall be in addition to the minimum reinforcement elsewhere required.

When the reinforcement in bearing walls is designed, placed and anchored in position as for columns, the allowable stresses shall be as for columns. The length of the wall to be considered effective shall not exceed the center-to-center distance between loads nor shall it exceed the width of the bearing plus four times the wall thickness.

1. **Column reinforcement.** The spacing of column ties shall *be as follows: not greater than 8 bar diameters, 24 tie diameters, or one half the least dimension of the column for* the full column height for columns stressed by tensile or compressive axial overturning forces due to the seismic loads of Part III of Chapter 16A; *and* for the tops and bottoms of all other columns for a distance of one sixth of the clear column height, but not less than *24 inches (610 mm)* or the maximum column dimension. Tie spacing for the remaining column height shall not exceed 16 bar diameters, 48 tie diameters, the least column dimension, nor 18 inches (457 mm). *Top tie shall be within 2 inches (51 mm) of the top of the column or of the bottom of the horizontal bar in the supported beam.*

Column ties shall terminate with a minimum 135-degree bend with a six-bar diameter, but not less than 4-inch (102 mm) extension that engages the longitudinal column reinforcement and projects into the interior of the column. Bends shall comply with Section 2107A.2.2.5, Item 3.

> **EXCEPTION:** Where the ties are placed in the horizontal bed joints, the hook may consist of a 90-degree bend having a radius of not less than four tie diameters plus an extension of 32 tie diameters.

2. **Shear walls.**

   2.1 **Reinforcement.** The portion of the reinforcement required to resist shear shall be uniformly distributed and shall be joint reinforcement, deformed bars or a combination thereof. The spacing of reinforcement in each direction shall not exceed *24 inches (610 mm) each way.*

   Joint reinforcement used in exterior walls and considered in the determination of the shear strength of the member shall be hot-dipped galvanized in accordance with U.B.C. Standard 21-10.

   Reinforcement required to resist in-plane shear shall be terminated with a standard hook as defined in Section 2107A.2.2.5 *which encircles the vertical reinforcing* or with an extension of proper embedment length beyond the reinforcement at the end of the wall section. The hook or extension may be turned up, down or horizontally. Provisions shall be made not to obstruct grout placement. Wall reinforcement terminating in columns or beams shall be fully anchored into these elements.

   2.2 **Bond.** Multiwythe grouted masonry shear walls shall be designed with consideration of the adhesion bond strength between the grout and masonry units. When bond strengths are not known from previous tests, the bond strength shall be determined by tests.

   2.3 **Wall reinforcement.** *Relocated above.*

   2.4 **Stack bond.** Reinforced hollow-unit stacked bond construction which is part of the seismic-resisting system shall use open-end units so that all head joints are made solid, shall use bond beam units to facilitate the flow of grout and shall be grouted solid.

3. **Type N mortar.** Type N mortar shall not be used as part of the vertical- or lateral-load-resisting system.

4. **Concrete abutting structural masonry.** Concrete abutting structural masonry, such as at starter courses or at wall intersections not designed as true separation joints, shall be roughened to a full amplitude of $^1/_{16}$ inch (1.6 mm) and shall be bonded to the masonry in accordance with the requirements of this chapter as if it were masonry. Unless keys or proper reinforcement is provided, vertical joints as specified in Section 2106A.1.4 shall be considered to be stack bond and the reinforcement as required for stack bond shall extend through the joint and be anchored into the concrete.

2106A.2 Working Stress Design and Strength Design Requirements for Unreinforced and Reinforced Masonry.

2106A.2.1 General. In addition to the requirements of Section 2106A.1, the design of masonry structures by the working stress design method and strength design method shall comply with the requirements of this section. Additionally, the design of reinforced masonry structures by these design methods shall comply with the requirements of Section 2106A.3

2106A.2.2 Specified compressive strength of masonry. The allowable stresses for the design of masonry shall be based on a value of f'_m selected for the construction.

Verification of the value of f'_m shall be based on compliance with Section 2105A.3. Unless otherwise specified, f'_m shall be based on 28-day tests. If other than a 28-day test age is used, the value of f'_m shall be as indicated in design drawings or specifications. Design drawings shall show the value of f'_m for which each part of the structure is designed.

2106A.2.3 Effective thickness.

2106A.2.3.1 Single-wythe walls. The effective thickness of single-wythe walls of either solid or hollow units is the specified thickness of the wall.

2106A.2.3.2 Multiwythe walls. The effective thickness of multiwythe walls is the specified thickness of the wall if the space between wythes is filled with mortar or grout.

2106A.2.3.3 *Walls and Piers.*

Thickness of Walls. For thickness limitations of walls as specified in this chapter, nominal thickness shall be used. Stresses shall be determined on the basis of the net thickness of the masonry, with consideration for reduction, such as raked joints.

The thickness of masonry walls shall be designed so that allowable maximum stresses specified in this chapter are not exceeded. Also, no masonry wall shall exceed the height or length-to-thickness ratio or the minimum thickness as specified in this chapter and as set forth in Table 21A-R, unless designed in accordance with Section 2108A.2.4.

Piers. Every pier or wall section which width is less than three times its thickness shall be designed and constructed as required for columns if such pier is a structural member. Every pier or wall section which width is between three and five times its thickness or less than one half the height of adjacent openings shall have all horizontal steel in the form of ties except that in walls 12 inches (305 mm) or less in thickness such steel may be in the form of hairpins.

2106A.2.3.4 Columns. The effective thickness for rectangular columns in the direction considered is the specified thickness. The effective thickness for nonrectangular columns is the thickness of the square column with the same moment of inertia about its axis as that about the axis considered in the actual column.

2106A.2.4 Effective height. The effective height of columns and walls is at least the clear height of members laterally supported at the top and bottom in a direction normal to the member axis considered. For members not supported at the top normal to the axis considered, the effective height is twice the height of the member above the support. Effective height less than clear height may be used if justified.

2106A.2.5 Effective area. The effective cross-sectional area shall be based on the minimum bedded area of hollow units, or the gross area of solid units plus any grouted area. ✳ ✳ ✳ Where bed joints are raked, the effective area shall be correspondingly reduced.

2106A.2.6 Effective width of intersecting walls. Where a shear wall is anchored to an intersecting wall or walls, the width of the overhanging flange formed by the intersected wall on either side of the shear wall, which may be assumed working with the shear wall for purposes of flexural stiffness calculations, shall not exceed six times the thickness of the intersected wall. Limits of the effective flange may be waived if justified. Only the effective area of the wall parallel to the shear forces may be assumed to carry horizontal shear.

2106A.2.7 Distribution of concentrated vertical loads in walls. The length of wall laid up in running bond which may be considered capable of working at the maximum allowable compressive stress to resist vertical concentrated loads shall not exceed the center-to-center distance between such loads, nor the width of bearing area plus four times the wall thickness. Concentrated vertical loads shall not be assumed to be distributed across continuous vertical mortar or control joints unless elements designed to distribute the concentrated vertical loads are employed. *Structural members framing into or supported by walls or columns shall be securely anchored. The end support of girders, beams or other concentrated loads on masonry shall have at least 3 inches (76 mm) in length upon solid bearing not less than 4 inches (102 mm) thick or upon metal bearing plate of adequate design and dimensions to distribute the loads safely on the wall or pier, or upon a continuous reinforced masonry member projecting not less than 3 inches (76 mm) from the face of the wall or other approved methods.*

Joists shall have bearing at least 3 inches (76 mm) in length upon solid masonry at least 2^1/$_2$ inches (64 mm) thick, or other provisions shall be made to distribute safely the loads on the wall or pier.

2106A.2.8 Loads on nonbearing walls. Masonry walls used as interior partitions or as exterior surfaces of a building which do not carry vertical loads imposed by other elements of the building shall be designed to carry their own weight plus any superimposed finish and lateral forces. Bonding or anchorage of nonbearing walls shall be adequate to support the walls and to transfer lateral forces to the supporting elements.

2106A.2.9 Vertical deflection. Elements supporting masonry shall be designed so that their vertical deflection will not exceed ^{1}/$_{600}$ of the clear span under total loads. Lintels shall bear on supporting masonry on each end such that allowable stresses in the supporting masonry are not exceeded. A minimum bearing length of 4 inches (102 mm) shall be provided for lintels bearing on masonry.

2106A.2.10 Structural continuity. Intersecting structural elements intended to act as a unit shall be anchored together to resist the design forces.

2106A.2.11 Walls intersecting with floors and roofs. Walls shall be anchored to all floors, roofs or other elements which provide lateral support for the wall. Where floors or roofs are designed to transmit horizontal forces to walls, the anchorage to such walls shall be designed to resist the horizontal force.

2106A.2.12 Modulus of elasticity of materials.

2106A.2.12.1 Modulus of elasticity of masonry. The moduli for masonry may be estimated as provided below. Actual values, where required, shall be established by test.

Modulus of elasticity of clay or shale unit masonry.

$$E_m = 750 f'_m, \text{ 3,000,000 psi (20.5 GPa) maximum} \qquad (6A\text{-}3)$$

Modulus of elasticity of concrete unit masonry.

$$E_m = 750 f'_m, \text{ 3,000,000 psi (20.5 GPa) maximum} \tag{6A-4}$$

2106A.2.12.2 Modulus of elasticity of steel.

$$E_s = 29,000,000 \text{ psi (200 GPa)} \tag{6A-5}$$

2106A.2.13 Shear modulus of masonry.

$$G = 0.4 \, E_m \tag{6A-6}$$

2106A.2.14 Placement of embedded anchor bolts.

2106A.2.14.1 General. Placement requirements for plate anchor bolts, headed anchor bolts and bent bar anchor bolts shall be determined in accordance with this subsection. The bent bar anchor bolt shall have a hook with a 90-degree bend with an inside diameter of three bolt diameters, plus an extension of $1^1/_2$ bolt diameters at the free end. Headed anchor bolts shall have a standard bolt head. Plate anchor bolts shall have a plate welded to the shank to provide anchorage equivalent to headed anchor bolts.

The effective embedment depth l_b for plate or headed anchor bolts shall be the length of embedment measured perpendicular from the surface of the masonry to the bearing surface of the plate or head of the anchorage, and l_b for bent bar anchors shall be the length of embedment measured perpendicular from the surface of the masonry to the bearing surface of the bent end minus one anchor bolt diameter. All bolts shall be grouted in place with at least 1 inch (25 mm) of grout between the bolt and the masonry *and shall be accurately set with templates.* $\overset{C}{\text{A}}$

2106A.2.14.2 Minimum edge distance. The minimum anchor bolt edge distance l_{be} measured from the edge of the masonry parallel with the anchor bolt to the surface of the anchor bolt shall be $1^1/_2$ inches (38 mm).

2106A.2.14.3 Minimum embedment depth. The minimum embedment depth of anchor bolts l_b shall be *eight* bolt diameters but not less than *4 inches (102 mm)*. $\overset{C}{\text{A}}$

2106A.2.14.4 Minimum spacing between bolts. The minimum center-to-center distance between anchor bolts shall be *eight* bolt diameters. $\overset{C}{\text{A}}$

2106A.2.15 *Not adopted by the State of California.*

2106A.3 Working Stress Design and Strength Design Requirements for Reinforced Masonry.

2106A.3.1 General. In addition to the requirements of Sections 2106A.1 and 2106A.2, the design of reinforced masonry structures by the working stress design method or the strength design method shall comply with the requirements of this section.

2106A.3.2 Plain bars. The use of plain bars larger than $^1/_4$ inch (6.4 mm) in diameter is not permitted.

2106A.3.3 Spacing of longitudinal reinforcement. The clear distance between parallel bars, except in columns, shall not be less than the nominal diameter of the bars or 1 inch (25 mm), except that bars in a splice may be in contact. This clear distance requirement applies to the clear distance between a contact splice and adjacent splices or bars.

The clear distance between the surface of a bar and any surface of a masonry unit shall not be less than $^1/_4$ inch (6 mm) for fine grout and $^1/_2$ inch (13 mm) for coarse grout. Cross webs of hollow units may be used as support for horizontal reinforcement.

2106A.3.4 Anchorage of flexural reinforcement. The tension or compression in any bar at any section shall be developed on each side of that section by the required development length. The development length of the bar may be achieved by a combination of an embedment length, anchorage or, for tension only, hooks.

Except at supports or at the free end of cantilevers, every reinforcing bar shall be extended beyond the point at which it is no longer needed to resist tensile stress for a distance equal to 12 bar diameters or the depth of the beam, whichever is greater. No flexural bar shall be terminated in a tensile zone unless at least one of the following conditions is satisfied:

1. The shear is not over one half that permitted, including allowance for shear reinforcement where provided.

2. Additional shear reinforcement in excess of that required is provided each way from the cutoff a distance equal to the depth of the beam. The shear reinforcement spacing shall not exceed $d/8r_b$.

3. The continuing bars provide double the area required for flexure at that point or double the perimeter required for reinforcing bond.

At least one third of the total reinforcement provided for negative moment at the support shall be extended beyond the extreme position of the point of inflection a distance sufficient to develop one half the allowable stress in the bar, not less than $^1/_{16}$ of the clear span, or the depth d of the member, whichever is greater.

Tensile reinforcement for negative moment in any span of a continuous restrained or cantilever beam, or in any member of a rigid frame, shall be adequately anchored by reinforcement bond, hooks or mechanical anchors in or through the supporting member.

At least one third of the required positive moment reinforcement in simple beams or at the freely supported end of continuous beams shall extend along the same face of the beam into the support at least 6 inches (153 mm). At least one fourth of the required positive moment reinforcement at the continuous end of continuous beams shall extend along the same face of the beam into the support at least 6 inches (153 mm).

Compression reinforcement in flexural members shall be anchored by ties or stirrups not less than $^3/_8$ inch (9.5 mm) in diameter, spaced not farther apart than 16 bar diameters or 48 tie diameters, whichever is less. Such ties or stirrups shall be used throughout the distance where compression reinforcement is required.

2106A.3.5 Anchorage of shear reinforcement. Single, separate bars used as shear reinforcement shall be anchored at each end by one of the following methods:

1. Hooking tightly around the longitudinal reinforcement through 180 degrees.

2. Embedment above or below the mid-depth of the beam on the compression side a distance sufficient to develop the stress in the bar for plain or deformed bars.

3. By a standard hook, as defined in Section 2107A.2.2.5, considered as developing 7,500 psi (52 MPa), plus embedment sufficient to develop the remainder of the stress to which the bar is subjected. The effective embedded length shall not be assumed to exceed the distance between the mid-depth of the beam and the tangent of the hook.

The ends of bars forming a single U or multiple U stirrup shall be anchored by one of the methods set forth in Items 1 through 3 above or shall be bent through an angle of at least 90 degrees tightly around a longitudinal reinforcing bar not less in diameter than the stirrup bar, and shall project beyond the bend at least 12 stirrup diameters.

The loops or closed ends of simple U or multiple U stirrups shall be anchored by bending around the longitudinal reinforcement through an angle of at least 90 degrees and project beyond the end of the bend at least 12 stirrup diameters.

2106A.3.6 Lateral ties. All longitudinal bars for columns shall be enclosed by lateral ties. Lateral support shall be provided to the longitudinal bars by the corner of a complete tie having an included angle of not more than 135 degrees or by a standard hook at the end of a tie. The corner bars shall have such support provided by a complete tie enclosing the longitudinal bars. Alternate longitudinal bars shall have such lateral support provided by ties and no bar shall be farther than 6 inches (153 mm) from such laterally supported bar.

Lateral ties and longitudinal bars shall be placed not less than $1^1/_2$ inches (38 mm) and not more than 5 inches (127 mm) from the surface of the column. Lateral ties *shall* be placed against the longitudinal bars. Spacing of ties shall *conform to 2106A.1.12.4, Item 1.*

C
A
C

Ties shall be at least No. 3 bars.

2106A.3.7 Column anchor bolt ties. Additional ties shall be provided around anchor bolts which are set in the top of columns. Such ties shall engage at least four bolts or, alternately, at least four vertical column bars or a combination of bolts and bars totaling at least four. Such ties shall be located within the top 5 inches (127 mm) of the column and shall provide a total of 0.4 square inch (260 mm^2) or more in cross-sectional area. The uppermost tie shall be within 2 inches (51 mm) of the top of the column.

2106A.3.8 Effective width *b* of compression area. In computing flexural stresses in walls where reinforcement occurs, the effective width assumed for running bond masonry shall not exceed six times the nominal wall thickness or the center-to-center distance between reinforcement. Where stack bond is used, the effective width shall not exceed three times the nominal wall thickness or the center-to-center distance between reinforcement or the length of one unit, unless solid grouted open-end units are used.

SECTION 2107A — WORKING STRESS DESIGN OF MASONRY

2107A.1 General.

2107A.1.1 Scope. The design of masonry structures using working stress design shall comply with the provisions of Section 2106A and this section. Stresses in clay or concrete masonry under service loads shall not exceed the values given in this section.

2107A.1.2 *Not adopted by the State of California.*

2107A.1.3 Minimum dimensions for masonry structures built in Seismic Zones 3 and 4.

2107A.1.3.1 Bearing walls. The nominal thickness of reinforced masonry bearing walls shall not be less than 6 inches (153 mm).

2107A.1.3.2 Columns. The least nominal dimension of a reinforced masonry column shall be 12 inches (305 mm). *No masonry column shall have an unsupported length greater than 20 times its least dimension.*

C
A
C

2107A.1.4 Design assumptions. The working stress design procedure is based on working stresses and linear stress-strain distribution assumptions with all stresses in the elastic range as follows:

1. Plane sections before bending remain plane after bending.

2. Stress is proportional to strain.

3. Masonry elements combine to form a homogenous member.

4. *Tensile forces are resisted only by the tensile reinforcement.*

5. *Reinforcement is completely surrounded by and bonded to the masonry materials so that they work together as a homogeneous material within the range of working stresses.*

6. *Masonry elements shall not be used as components for the design of rigid frames except as permitted in Section 2108A.2.6.*

2107A.1.5 Embedded anchor bolts.

2107A.1.5.1 General. Allowable loads for plate anchor bolts, headed anchor bolts and bent bar anchor bolts shall be determined in accordance with this section.

2107A.1.5.2 Tension. Allowable loads in tension shall be the lesser value selected from Tables 21A-E-1 and 21A-E-2 or shall be determined from the lesser of Formula (7A-1) or Formula (7A-2).

$$B_t = 0.5 \, A_p \sqrt{f'_m} \tag{7A-1}$$

For **SI:**
$$B_t = 0.042 \, A_p \sqrt{f'_m}$$

$$B_t = 0.2 \, A_b f_y \tag{7A-2}$$

The area A_p shall be the lesser of Formula (7A-3) or Formula (7A-4) and where the projected areas of adjacent anchor bolts overlap, A_p of each anchor bolt shall be reduced by one half of the overlapping area.

$$A_p = \pi l_b{}^2 \tag{7A-3}$$

$$A_p = \pi l_{b_e}{}^2 \tag{7A-4}$$

2107A.1.5.3 Shear. Allowable loads in shear shall be the value selected from Table 21A-F or shall be determined from the lesser of Formula (7A-5) or Formula (7A-6).

$$B_v = 350 \sqrt[4]{f'_m A_b} \tag{7A-5}$$

For **SI:**
$$B_v = 1070 \sqrt[4]{f'_m A_b}$$

$$B_v = 0.12 \, A_b f_y \tag{7A-6}$$

Where the anchor bolt edge distance l_{be} in the direction of load is less than 12 bolt diameters, the value of B_v in Formula (7A-5) shall be reduced by linear interpolation to zero at an l_{be} distance of

C
A $1^1/_2$ inches (38 mm) *and confining reinforcement consisting of not less than No. 3 hairpins, hooks or*
C *stirrups for end bolts and between horizontal reinforcing for other bolts shall be provided.* Where adjacent anchors are spaced closer than $8d_b$, the allowable shear of the adjacent anchors determined by Formula (7A-5) shall be reduced by linear interpolation to 0.75 times the allowable shear value

C at a center-to-center spacing of four bolt diameters. *For bolts in tops of columns, see Section*
A *2106A.3.7.*
C

2107A.1.5.4 Combined shear and tension. Anchor bolts subjected to combined shear and tension shall be designed in accordance with Formula (7A-7).

$$\frac{b_t}{B_t} + \frac{b_v}{B_v} \le 1.0 \tag{7A-7}$$

2107A.1.6 Compression in walls and columns.

2107A.1.6.1 Walls, axial loads. Stresses due to compressive forces applied at the centroid of wall may be computed by Formula (7A-8) assuming uniform distribution over the effective area.

$$f_a = P / A_e \tag{7A-8}$$

2107A.1.6.2 Columns, axial loads. Stresses due to compressive forces applied at the centroid of columns may be computed by Formula (7A-8) assuming uniform distribution over the effective area.

2107A.1.6.3 Columns, bending or combined bending and axial loads. Stresses in columns due to combined bending and axial loads shall satisfy the requirements of Section 2107A.2.7 where f_a/F_a is replaced by P/P_a. Columns subjected to bending shall meet all applicable requirements for flexural design.

2107A.1.7 Shear walls, design loads. When calculating shear or diagonal tension stresses, shear walls which resist seismic forces in Seismic Zones 3 and 4 shall be designed to resist 1.5 times the forces required by Section 1628A.

2107A.1.8 Design, composite construction.

2107A.1.8.1 General. The requirements of this section govern masonry in which at least one wythe has strength or composition characteristics different from the other wythe or wythes and is adequately bonded to act as a single structural element.

The following assumptions shall apply to the design of composite masonry:

1. Analysis shall be based on elastic transformed section of the net area.

2. The maximum computed stress in any portion of composite masonry shall not exceed the allowable stress for the material of that portion.

2107A.1.8.2 Determination of moduli of elasticity. The modulus of elasticity of each type of masonry in composite construction shall be measured by tests if the modular ratio of the respective types of masonry exceeds 2 to 1 as determined by Section 2106A.2.12.

2107A.1.8.3 Structural continuity.

2107A.1.8.3.1 Bonding of wythes. All wythes of composite masonry elements shall be tied together as specified in Section 2106A.1.5.2 as a minimum requirement. Additional ties or the combination of grout and metal ties shall be provided to transfer the calculated stress.

2107A.1.8.3.2 Material properties. The effect of dimensional changes of the various materials and different boundary conditions of various wythes shall be included in the design.

2107A.1.8.4 Design procedure, transformed sections. One material is chosen as the reference material, and the other materials are transformed to an equivalent area of the reference material by multiplying the area of the other materials by the respective ratios of the modulus of elasticity of the other materials to that of the reference material. Thickness of the transformed area and its distance perpendicular to a given bending axis remain unchanged. Effective height or length of the element remains unchanged.

2107A.1.9 Reuse of masonry units. The allowable working stresses for reused masonry units shall not exceed 50 percent of those permitted for new masonry units of the same properties.

2107A.2 Design of Reinforced Masonry.

2107A.2.1 Scope. The requirements of this section are in addition to the requirements of Sections 2106A and 2107A.1, and govern masonry in which reinforcement is used to resist forces.

Walls with openings used to resist lateral loads whose pier and beam elements are within the dimensional limits of Section 2108A.2.6.1.2 may be designed in accordance with Section 2108A.2.6. Walls used to resist lateral loads not meeting the dimensional limits of Section 2108A.2.6.1.2 may be designed as walls in accordance with this section or Section 2108A.2.5.

2107A.2.2 Reinforcement.

2107A.2.2.1 Maximum reinforcement size. The maximum size of reinforcement shall be No. 11 bars. Maximum reinforcement area in cells shall be 6 percent of the cell area without splices and 12 percent of the cell area with splices.

2107A.2.2.2 Cover. All reinforcing bars, except joint reinforcement, shall be completely embedded in mortar or grout and have a minimum cover, including the masonry unit, of at least $^3/_4$ inch (19 mm), $1^1/_2$ inches (38 mm) of cover when the masonry is exposed to weather and 2 inches (51 mm) of cover when the masonry is exposed to soil.

2107A.2.2.3 Development length. The required development length l_d for deformed bars or deformed wire shall be calculated by:

$$l_d = 0.002 \, d_b \, f_s \text{ for bars in tension} \tag{7A-9}$$

For SI: $\qquad l_d = 0.29 \, d_b \, f_s$ for bars in tension

$$l_d = 0.0015 \, d_b \, f_s \text{ for bars in compression} \tag{7A-10}$$

For SI: $\qquad l_d = 0.22 \, d_b \, f_s$ for bars in compression

Development length for smooth bars shall be twice the length determined by Formula (7A-9).

2107A.2.2.4 Reinforcement bond stress. Bond stress u in reinforcing bars shall not exceed the following:

Plain Bars	60 psi (413 kPa)
Deformed Bars	200 psi (1378 kPa)

2107A.2.2.5 Hooks.

1. The term "standard hook" shall mean one of the following:

    1.1 A 180-degree turn plus extension of at least four bar diameters, but not less than $2^1/_2$ inches (63 mm) at free end of bar.

    1.2 A 90-degree turn plus extension of at least 12 bar diameters at free end of bar.

    1.3 For stirrup and tie anchorage only, either a 90-degree or a 135-degree turn, plus an extension of at least six bar diameters, but not less than $2^1/_2$ inches (63 mm) at the free end of the bar.

2. Inside diameter of bend of the bars, other than for stirrups and ties, shall not be less than that set forth in Table 21A-G.

3. Inside diameter of bend for No. *4* or smaller stirrups and ties shall not be less than four bar diameters. Inside diameter of bend for *No. 5 or larger* stirrups and ties shall not be less than that given in Table 21A-G.

4. Hooks shall not be permitted in the tension portion of any beam, except at the ends of simple or cantilever beams or at the freely supported end of continuous or restrained beams.

5. Hooks shall not be assumed to carry a load which would produce a tensile stress in the bar greater than 7,500 psi (52 MPa).

6. Hooks shall not be considered effective in adding to the compressive resistance of bars.

7. Any mechanical device capable of developing the strength of the bar without damage to the masonry may be used in lieu of a hook. Data must be presented to show the adequacy of such devices.

2107A.2.2.6 Splices. *Splices may be made only at such points and in such a manner that the structural strength of the member will not be reduced.* The amount of lap of lapped splices shall be sufficient to transfer the allowable stress of the reinforcement as specified in Sections 2106A.3.4, 2107A.2.2.3 and 2107A.2.12. In no case shall the length of the *lap* splice be less than *36* bar diameters for compression and *48* bar diameters for tension.

Bars of size No. 8 and larger resisting tensile stresses shall be spliced by welding or by approved mechanical connectors. Welded or mechanical connections shall develop 125 percent of the specified yield strength of the bar in tension.

> **EXCEPTION:** For compression bars in columns that are not part of the seismic-resisting system and are not subject to flexure, only the compressive strength need be developed.

When adjacent splices in grouted masonry are separated by 3 inches (76 mm) or less, the required lap length shall be increased 30 percent.

> **EXCEPTION:** Where lap splices are staggered at least 24 bar diameters, no increase in lap length is required.

See Section 2107A.2.12 for lap splice increases.

2107A.2.3 Design assumptions. The following assumptions are in addition to those stated in Section 2107A.1.4:

1. Masonry carries no tensile stress.

2. Reinforcement is completely surrounded by and bonded to masonry material so that they work together as a homogenous material within the range of allowable working stresses.

2107A.2.4 Nonrectangular flexural elements. Flexural elements of nonrectangular cross section shall be designed in accordance with the assumptions given in Sections 2107A.1.4 and 2107A.2.3.

2107A.2.5 Allowable axial compressive stress and force. For members other than reinforced masonry columns, the allowable axial compressive stress F_a shall be determined as follows:

$$F_a = 0.25f'_m \left[1 - \left(\frac{h'}{140r} \right)^2 \right] \text{ for } h'/r \leq 99 \qquad (7A\text{-}11)$$

$$F_a = 0.25f'_m \left(\frac{70r}{h'} \right)^2 \text{ for } h'/r > 99 \qquad (7A\text{-}12)$$

For reinforced masonry columns, the allowable axial compressive force P_a shall be determined as follows:

$$P_a = [0.25f'_m A_e + 0.65A_s F_{sc}] \left[1 - \left(\frac{h'}{140r} \right)^2 \right] \text{ for } h'/r \leq 99 \qquad (7A\text{-}13)$$

$$P_a = [0.25f'_m A_e + 0.65A_s F_{sc}] \left(\frac{70r}{h'} \right)^2 \text{ for } h'/r > 99 \qquad (7A\text{-}14)$$

2107A.2.6 Allowable flexural compressive stress. The allowable flexural compressive stress F_b is:

$$F_b = 0.33 f'_m, \text{ 2,000 psi (13.8 MPa) maximum} \qquad (7A\text{-}15)$$

2107A.2.7 Combined compressive stresses, unity formula. Elements subjected to combined axial and flexural stresses shall be designed in accordance with accepted principles of mechanics or in accordance with Formula (7A-16):

$$\frac{f_a}{F_a} + \frac{f_b}{F_b} \leq 1 \qquad (7A\text{-}16)$$

2107A.2.8 Allowable shear stress in flexural members. Where no snear reinforcement is provided, the allowable shear stress F_v in flexural members is:

$$F_v = 1.0 \sqrt{f'_m}, \text{ 50 psi maximum} \qquad (7A\text{-}17)$$

For SI: $\qquad\qquad F_v = 0.083 \sqrt{f'_m}, \text{ 345 kPa maximum}$

EXCEPTION: For a distance of $^1/_{16}$ the clear span beyond the point of inflection, the maximum stress shall be 20 psi (140 kPa).

Where shear reinforcement designed to take entire shear force is provided, the allowable shear stress F_v in flexural members is:

$$F_v = 3.0 \sqrt{f'_m}, \text{ 150 psi maximum} \qquad (7A\text{-}18)$$

For SI: $\qquad F_v = 0.25 \sqrt{f'_m}, \text{ 1.0 MPa maximum}$

2107A.2.9 Allowable shear stress in shear walls. *Allowable shear stresses for reinforced masonry walls shall be as indicated below. When calculating shear or diagonal tension stresses, shear walls which resist seismic forces shall be designed to resist 1.5 times the forces required by Section 1627A.1.*

M is the design moment occurring simultaneously with shear load V at the section under consideration. Where in-plane flexural reinforcement is provided and masonry is used to resist all shear,

$$\text{For } M/Vd < 1, F_v = \tfrac{1}{3} \left(4 - \frac{M}{Vd}\right) \sqrt{f'_m}, \ \left(80 - 45\frac{M}{Vd}\right) \text{ maximum} \qquad (7A\text{-}19)$$

For SI: $\qquad F_v = \tfrac{1}{36} \left(4 - \frac{M}{Vd}\right) \sqrt{f'_m}, \ \left(80 - 45\frac{M}{Vd}\right) \text{ maximum}$

$$\text{For } M/Vd \geq 1, F_v = 1.0 \sqrt{f'_m}, \text{ 35 psi maximum} \qquad (7A\text{-}20)$$

For SI: $\qquad F_v = \tfrac{1}{12} \sqrt{f'_m}, \text{ 240 kPa maximum}$

Where shear reinforcement designed to take all the shear is provided, the allowable shear stress F_v in shear walls is:

$$\text{For } M/Vd < 1, F_v = \tfrac{1}{2} \left(4 - \frac{M}{Vd}\right) \sqrt{f'_m}, \ \left(120 - 45\frac{M}{Vd}\right) \text{ maximum} \qquad (7A\text{-}21)$$

For SI: $\quad \text{For } M/Vd < 1, F_v = \tfrac{1}{24} \left(4 - \frac{M}{Vd}\right) \sqrt{f'_m}, \ \left(120 - 45\frac{M}{Vd}\right) \text{ maximum}$

$$\text{For } M/Vd \geq 1, F_v = 1.5 \sqrt{f'_m}, \text{ 75 psi maximum} \qquad (7A\text{-}22)$$

For SI: $\qquad \text{For } M/Vd \geq 1, F_v = 0.12 \sqrt{f'_m}, \text{ 520 kPa maximum}$

2107A.2.10 Allowable bearing stress. When a member bears on the full area of a masonry element, the allowable bearing stress F_{br} is:

$$F_{br} = 0.26 f'_m \qquad (7A\text{-}23)$$

When a member bears on one third or less of a masonry element, the allowable bearing stress F_{br} is:

$$F_{br} = 0.38 f'_m \qquad (7A\text{-}24)$$

Formula (7A-24) applies only when the least dimension between the edges of the loaded and unloaded areas is a minimum of one fourth of the parallel side dimension of the loaded area. The allowable bearing stress on a reasonably concentric area greater than one third but less than the full area shall be interpolated between the values of Formulas (7A-23) and (7A-24).

2107A.2.11 Allowable stresses in reinforcement. The allowable stresses in reinforcement shall be as follows:

1. **Tensile stress, F_s.**

    1.1 Deformed bars,

    $$F_s = 0.5 f_y, \text{ 24,000 psi (165 MPa) maximum} \qquad (7A\text{-}25)$$

1.2 Wire reinforcement,

$$F_s = 0.5 f_y, \text{ 30,000 psi (207 MPa) maximum} \qquad \text{(7A-26)}$$

1.3 Ties, anchors and smooth bars,

$$F_s = 0.4 f_y, \text{ 20,000 psi (138 MPa) maximum} \qquad \text{(7A-27)}$$

2. **Compressive stress F_{sc}, F_s.**

2.1 Deformed bars in columns,

$$F_{sc} = 0.4 f_y, \text{ 24,000 psi (165 MPa) maximum} \qquad \text{(7A-28)}$$

2.2 Deformed bars in flexural members,

$$F_s = 0.5 f_y, \text{ 24,000 psi (165 MPa) maximum} \qquad \text{(7A-29)}$$

2.3 Deformed bars in shear walls which are confined by lateral ties throughout the distance where compression reinforcement is required and where such lateral ties are not less than $^1/_4$ inch in diameter and spaced not farther apart than 16 bar diameters or 48 tie diameters,

$$F_{sc} = 0.4 f_y, \text{ 24,000 psi (165 MPa) maximum} \qquad \text{(7A-30)}$$

2107A.2.12 Lap splice increases. In regions of moment where the design tensile stresses in the reinforcement are greater than 80 percent of the allowable steel tensile stress F_s, the lap length of splices shall be increased not less than 50 percent of the minimum required length. Other equivalent means of stress transfer to accomplish the same 50 percent increase may be used.

2107A.2.13 Reinforcement for columns. Columns shall be provided with reinforcement as specified in this subsection.

2107A.2.13.1 Vertical reinforcement. The area of vertical reinforcement shall not be less than $0.005 A_e$ and not more than $0.04 A_e$. At least four bars shall be provided. *The bars shall not be less than No. 4 or greater than No. 11. Welded splices shall be full butt welded.* The minimum clear distance between parallel bars in columns shall be two and one half times the bar diameter.

2107A.2.14 Compression in walls and columns.

2107A.2.14.1 General. Stresses due to compressive forces in walls and columns shall be calculated in accordance with Section 2107A.2.5.

2107A.2.14.2 Walls, bending or combined bending and axial loads. Stresses in walls due to combined bending and axial loads shall satisfy the requirements of Section 2107A.2.7 where f_a is given by Formula (7A-8). Walls subjected to bending with or without axial loads shall meet all applicable requirements for flexural design.

2107A.2.15 Flexural design, rectangular flexural elements. Rectangular flexural elements shall be designed in accordance with the following formulas or other methods based on the assumptions given in Sections 2107A.1.4, 2107A.2.3 and this section.

1. Compressive stress in the masonry:

$$f_b = \frac{M}{bd^2} \left(\frac{2}{jk} \right) \qquad \text{(7A-31)}$$

2. Tensile stress in the longitudinal reinforcement:

$$f_s = \frac{M}{A_s jd} \qquad \text{(7A-32)}$$

3. Design coefficients:

$$k = \sqrt{(n\rho)^2 + 2n\rho} - n\rho \qquad \text{(7A-33)}$$

or

$$k = \frac{1}{1 + \frac{f_s}{nf_b}} \qquad \text{(7A-34)}$$

$$j = 1 - \frac{k}{3} \qquad \text{(7A-35)}$$

2107A.2.16 Bond of flexural reinforcement. In flexural members in which tensile reinforcement is parallel to the compressive face, the bond stress shall be computed by the formula:

$$u = \frac{V}{\Sigma_o \, jd} \qquad \text{(7A-36)}$$

2107A.2.17 Shear in flexural members and shear walls. The shear stress in flexural members and shear walls shall be computed by:

$$f_v = \frac{V}{bjd} \qquad \text{(7A-37)}$$

For members of T or I section, b' shall be substituted for b. Where f_v as computed by Formula (7A-37) exceeds the allowable shear stress in masonry, F_v, web reinforcement shall be provided and designed to carry the total shear force. Both vertical and horizontal shear stresses shall be considered.

The area required for shear reinforcement placed perpendicular to the longitudinal reinforcement shall be computed by:

$$A_v = \frac{sV}{F_s d} \qquad \text{(7A-38)}$$

Where web reinforcement is required, it shall be so spaced that every 45-degree line extending from a point at $d/2$ of the beam to the longitudinal tension bars shall be crossed by at least one line of web reinforcement.

2107A.3 *Not adopted by the State of California.*

SECTION 2108A — STRENGTH DESIGN OF MASONRY

2108A.1 General.

2108A.1.1 General provisions. The design of reinforced hollow-unit clay and concrete masonry structures using strength design shall comply with the provisions of Section 2106A and this section.

> **EXCEPTION:** Two-wythe solid-unit masonry may be used under Sections 2108A.2.1 and 2108A.2.4.

2108A.1.2 Quality assurance provisions. Special inspection during construction shall be provided as set forth in Section *2105A.7.*

The value of f'_m shall be verified in accordance with Section 2105A.3.

2108A.1.3 Required strength. The required strength shall be determined as follows:

1. For earthquake loading, the load factors shall be:

$$U = 1.4 \, (D + L + E) \qquad \text{(8A-1)}$$

$$U = 0.9D \pm 1.4E \qquad \text{(8A-2)}$$

2. Required strength U to resist dead load D and live load L shall be at least equal to:

$$U = 1.4D + 1.7L \qquad (8A\text{-}3)$$

3. If resistance to structural effects of a specified wind load W are included in design, the following combinations of D, L and W shall be investigated to determine the greatest required strength U.

$$U = 0.75 (1.4D + 1.7L + 1.7W) \qquad (8A\text{-}4)$$

where load combinations shall include both full value and zero value of L to determine the more severe conditions, and

$$U = 0.9D + 1.3W \qquad (8A\text{-}5)$$

However, for any combination of D, L and W, required strength U shall not be less than the value determined from Formula (8A-3).

4. If resistance to earth pressure H is included in design, required strength U shall be at least equal to

$$U = 1.4D + 1.7L + 1.7H \qquad (8A\text{-}6)$$

except that where D or L reduces the effect of H, $0.9D$ shall be substituted for $1.4D$ and zero value of L shall be used to determine the greatest required strength U. For any combination of D, L and H, required strength U shall not be less than the value determined from Formula (8A-3).

5. If resistance to loadings due to weight and pressure of fluids with well-defined densities and controllable maximum heights F is included in design, such loadings shall have a load factor of 1.4 and be added to all loading combinations that include live load.

6. If resistance to impact effects is taken into account in design, such effects shall be included with live load, L.

7. Where structural effects T of differential settlement, creep, shrinkage or temperature change may be significant in design, required strength U shall be at least equal to:

$$U = 0.75 (1.4D + 1.4T + 1.7L) \qquad (8A\text{-}7)$$

but required strength U shall not be less than:

$$U = 1.4 (D + T) \qquad (8A\text{-}8)$$

2108A.1.4 Design strength. Design strength is the nominal strength, multiplied by the strength-reduction factor, ϕ, as specified in this section. Masonry members shall be proportioned such that the design strength exceeds the required strength.

2108A.1.4.1 Beams, piers and columns.

2108A.1.4.1.1 Flexure. Flexure with or without axial load, the value of ϕ shall be determined from Formula (8A-9):

$$\phi = 0.8 - \frac{P_u}{A_e f'_m} \qquad (8A\text{-}9)$$

and $0.60 \leq \phi \leq 0.80$

2108A.1.4.1.2 Shear. Shear: $\phi = 0.60$

2108A.1.4.2 Wall design for out-of-plane loads.

2108A.1.4.2.1 Walls with factored axial load of $0.04 f'_m$ or less.

Flexure: $\phi = 0.80$.

2108A.1.4.2.2 Walls with factored axial load greater than $0.04 f'_m$.

Axial load and axial load with flexure: $\phi = 0.80$.

Shear: $\phi = 0.60$.

2108A.1.4.3 Wall design for in-plane loads.

2108A.1.4.3.1 Axial load. Axial load and axial load with flexure: $\phi = 0.65$.

For walls with symmetrical reinforcement in which f_y does not exceed 60,000 psi (413 MPa), the value of ϕ may be increased linearly to 0.85 as the value of ϕP_n decreases from $0.10 f'_m A_e$ or $0.25 P_b$ to zero.

For solid grouted walls, the value of P_b may be calculated by Formula (8A-10)

$$P_b = 0.85 f'_m \, b a_b \tag{8A-10}$$

WHERE:

$$a_b = 0.85d \,\{e_{mu} / [e_{mu} + (f_y / E_s)]\} \tag{8A-11}$$

2108A.1.4.3.2 Shear. Shear: $\phi = 0.60$

The value of ϕ may be 0.80 for any shear wall when its nominal shear strength exceeds the shear corresponding to development of its nominal flexural strength for the factored-load combination.

2108A.1.4.4 Moment-resisting wall frames.

2108A.1.4.4.1 Flexure with or without axial load. The value of ϕ shall be as determined from Formula (8A-12); however, the value of ϕ shall not be less than 0.65 nor greater than 0.85.

$$\phi = 0.85 - 2\left(\frac{P_u}{A_n f'_m}\right) \tag{8A-12}$$

2108A.1.4.4.2 Shear. Shear: $\phi = 0.80$

2108A.1.4.5 Anchor. Anchor bolts: $\phi = 0.80$

2108A.1.4.6 Reinforcement.

2108A.1.4.6.1 Development. Development: $\phi = 0.80$

2108A.1.4.6.2 Splices. Splices: $\phi = 0.80$

2108A.1.5 Anchor bolts.

2108A.1.5.1 Required strength. The required strength of embedded anchor bolts shall be determined from factored loads as specified in Section 2108A.1.3.

2108A.1.5.2 Nominal anchor bolt strength. The nominal strength of anchor bolts times the strength-reduction factor shall equal or exceed the required strength.

The nominal tensile capacity of anchor bolts shall be determined from the lesser of Formula (8A-13) or (8A-14).

$$B_{tn} = 1.0 A_p \sqrt{f'_m} \tag{8A-13}$$

For SI:
$$B_{tn} = 0.084 A_p \sqrt{f'_m}$$

$$B_{tn} = 0.4 A_p f_y \tag{8A-14}$$

The area A_p shall be the lesser of Formula (8A-15) or (8A-16) and where the projected areas of adjacent anchor bolts overlap, the value of A_p of each anchor bolt shall be reduced by one half of the overlapping area.

$$A_p = \pi \, l_b{}^2 \qquad \text{(8A-15)}$$

$$A_p = \pi \, l_{be}{}^2 \qquad \text{(8A-16)}$$

The nominal shear capacity of anchor bolts shall be determined from the lesser of Formula (8A-17) or (8A-18).

$$B_{sn} = 900 \sqrt[4]{f'_m \, A_b} \qquad \text{(8A-17)}$$

For **SI:**
$$B_{sn} = 2750 \sqrt[4]{f'_m \, A_b}$$

$$B_{sn} = 0.25 A_s \, f_y \qquad \text{(8A-18)}$$

Where the anchor bolt edge distance, l_{be}, in the direction of load is less than 12 bolt diameters, the value of B_{tn} in Formula (8A-17) shall be reduced by linear interpolation to zero at an l_{be} distance of $1^1/_2$ inches (38 mm). Where adjacent anchor bolts are spaced closer than $8d_b$, the nominal shear strength of the adjacent anchors determined by Formula (8A-17) shall be reduced by linear interpolation to 0.75 times the nominal shear strength at a center-to-center spacing of four bolt diameters.

Anchor bolts subjected to combined shear and tension shall be designed in accordance with Formula (8A-19).

$$\frac{b_{tu}}{\phi B_{tn}} + \frac{b_{su}}{\phi B_{sn}} \le 1.0 \qquad \text{(8A-19)}$$

2108A.1.5.3 Anchor bolt placement. Anchor bolts shall be placed so as to meet the edge distance, embedment depth and spacing requirements of Sections 2106A.2.14.2, 2106A.2.14.3 and 2106A.2.14.4.

2108A.2 Reinforced Masonry.

2108A.2.1 General.

2108A.2.1.1 Scope. The requirements of this section are in addition to the requirements of Sections 2106A and 2108A.1 and govern masonry in which reinforcement is used to resist forces.

2108A.2.1.2 Design assumptions. The following assumptions apply:

Masonry carries no tensile stress greater than the modulus of rupture.

Reinforcement is completely surrounded by and bonded to masonry material so that they work together as a homogeneous material.

Nominal strength of singly reinforced masonry wall cross sections to combined flexure and axial load shall be based on applicable conditions of equilibrium and compatibility of strains. Strain in reinforcement and masonry walls shall be assumed to be directly proportional to the distance from the neutral axis.

Maximum usable strain, e_{mu}, at the extreme masonry compression fiber shall:

1. Be 0.003 for the design of beams, piers, columns and walls.

2. Not exceed 0.003 for moment-resisting wall frames, unless lateral reinforcement as defined in Section 2108A.2.6.2.6 is utilized.

Strain in reinforcement and masonry shall be assumed to be directly proportional to the distance from the neutral axis.

Stress in reinforcement below specified yield strength f_y for grade of reinforcement used shall be taken as E_s times steel strain. For strains greater than that corresponding to f_y, stress in reinforcement shall be considered independent of strain and equal to f_y.

Tensile strength of masonry walls shall be neglected in flexural calculations of strength, except when computing requirements for deflection.

Relationship between masonry compressive stress and masonry strain may be assumed to be rectangular as defined by the following:

Masonry stress of $0.85 f'_m$ shall be assumed uniformly distributed over an equivalent compression zone bounded by edges of the cross section and a straight line located parallel to the neutral axis at a distance $a = 0.85c$ from the fiber of maximum compressive strain. Distance c from fiber of maximum strain to the neutral axis shall be measured in a direction perpendicular to that axis.

2108A.2.2 Reinforcement requirements and details.

2108A.2.2.1 Maximum reinforcement. The maximum size of reinforcement shall be No. 9. The diameter of a bar shall not exceed one fourth the least dimension of a cell. No more than two bars shall be placed in a cell of a wall or a wall frame.

2108A.2.2.2 Placement. The placement of reinforcement shall comply with the following:

In columns and piers, the clear distance between vertical reinforcing bars shall not be less than one and one-half times the nominal bar diameter, nor less than $1^1/_2$ inches (38 mm).

2108A.2.2.3 Cover. All reinforcing bars shall be completely embedded in mortar or grout and shall have a cover of not less than $1^1/_2$ inches (38 mm) nor less than $2.5 d_b$.

2108A.2.2.4 Standard hooks. A standard hook shall be one of the following:

1. A 180-degree turn plus an extension of at least four bar diameters, but not less than $2^1/_2$ inches (63 mm) at the free end of the bar.

2. A 135-degree turn plus an extension of at least six bar diameters at the free end of the bar.

3. A 90-degree turn plus an extension of at least 12 bar diameters.

2108A.2.2.5 Minimum bend diameter for reinforcing bars. Diameter of bend measured on the inside of a bar other than for stirrups and ties in sizes No. 3 through No. 5 shall not be less than the values in Table 21A-G.

Inside diameter of bends for stirrups and ties shall not be less than $4d_b$ for No. 5 bars and smaller. For bars larger than No. 5, diameter of bend shall be in accordance with Table 21A-G.

2108A.2.2.6 Development. The calculated tension or compression reinforcement shall be developed in accordance with the following provisions:

The embedment length of reinforcement shall be determined by Formula (8A-20).

$$l_d = l_{de} / \phi \qquad \text{(8A-20)}$$

WHERE:

$$l_{de} = \frac{0.15 d_b{}^2 f_y}{K \sqrt{f'_m}} \leq 52 d_b \qquad \text{(8A-21)}$$

For **SI:**
$$l_{de} = \frac{1.8 d_b{}^2 f_y}{K \sqrt{f'_m}} \leq 52 d_b$$

K shall not exceed $3d_b$.

The minimum embedment length of reinforcement shall be 12 inches (305 mm).

2108A.2.2.7 Splices. Reinforcement splices shall comply with one of the following:

1. The minimum length of lap for bars shall be 12 inches (305 mm) or the length determined by Formula (8A-22).

$$l_d = l_{de}/\phi \qquad \text{(8A-22)}$$

Bars spliced by noncontact lap splices shall be spaced transversely not farther apart than one fifth the required length of lap nor more than 8 inches (203 mm).

2. A welded splice shall have the bars butted and welded to develop in tension 125 percent of the yield strength of the bar, f_y.

3. Mechanical splices shall have the bars connected to develop in tension or compression, as required, at least 125 percent of the yield strength of the bar, f_y.

2108A.2.3 Design of beams, piers and columns.

2108A.2.3.1 General. The requirements of this section are for the design of masonry beams, piers and columns.

The value of f'_m shall not be less than 1,500 psi (10.3 MPa). For computational purposes, the value of f'_m shall not exceed 4,000 psi (27.6 MPa).

2108A.2.3.2 Design assumptions.

Member design forces shall be based on an analysis which considers the relative stiffness of structural members. The calculation of lateral stiffness shall include the contribution of all piers, beams and columns.

The effects of cracking on member stiffness shall be considered. Unless stiffness values are obtained by a more comprehensive analysis, the effective moment of inertia shall be determined in accordance with Formula (8A-23).

$$I_e = \left(\frac{M_{cr}}{M_a}\right)I_{gt} + \left[1 - \left(\frac{M_{cr}}{M_a}\right)^3\right]I_{cr} \le I_g \qquad \text{(8A-23)}$$

The drift ratio of piers and columns shall satisfy the limits specified in Chapter 16A.

2108A.2.3.3 Balanced reinforcement ratio for compression limit state. Calculation of the balanced reinforcement ratio, ρ_b, shall be based on the following assumptions:

1. The distribution of strain across the section shall be assumed to vary linearly from the maximum usable strain, e_{mu}, at the extreme compression fiber of the element, to a yield strain of f_y/E_s at the extreme tension fiber of the element.

2. Compression forces shall be in equilibrium with the sum of tension forces in the reinforcement and the maximum axial load associated with a loading combination $1.0D + 1.0L + (1.4E$ or $1.3W)$.

3. The reinforcement shall be assumed to be uniformly distributed over the depth of the element and the balanced reinforcement ratio shall be calculated as the area of this reinforcement divided by the net area of the element.

4. All longitudinal reinforcement shall be included in calculating the balanced reinforcement ratio except that the contribution of compression reinforcement to resistance of compressive loads shall not be considered.

2108A.2.3.4 Required strength. Except as required by Sections 2108A.2.3.6 through 2108A.2.3.12, the required strength shall be determined in accordance with Section 2108A.1.3.

2108A.2.3.5 Design strength. Design strength provided by beam, pier or column cross sections in terms of axial force, shear and moment shall be computed as the nominal strength multiplied by the applicable strength-reduction factor, ϕ, specified in Section 2108A.1.4.

2108A.2.3.6 Nominal strength.

2108A.2.3.6.1 Nominal axial and flexural strength. The nominal axial strength, P_n, and the nominal flexural strength, M_n, of a cross section shall be determined in accordance with the design assumptions of Section 2108A.2.1.2 and 2108A.2.3.2.

The maximum nominal axial compressive strength shall be determined in accordance with Formula (8A-24).

$$P_n = 0.80[0.85f'_m(A_e - A_s) + f_y A_s] \qquad (8A\text{-}24)$$

2108A.2.3.6.2 Nominal shear strength. The nominal shear strength shall be determined in accordance with Formula (8A-25).

$$V_n = V_m + V_s \qquad (8A\text{-}25)$$

WHERE:

$$V_m = C_d A_e \sqrt{f'_m}, \ \ 63C_d A_e \text{ maximum} \qquad (8A\text{-}26)$$

For **SI:** $\qquad V_m = 0.083 \ C_d A_e \sqrt{f'_m}, \ \ 63C_d A_e \text{ maximum}$

and

$$V_s = A_e \rho_n f_y \qquad (8A\text{-}27)$$

1. The nominal shear strength shall not exceed the value given in Table 21A-J.

2. The value of V_m shall be assumed to be zero within any region subjected to net tension factored loads.

3. The value of V_m shall be assumed to be 25 psi (172 kPa) where M_u is greater than $0.7 M_n$. The required moment, M_u, for seismic design for comparison with the $0.7 M_n$ value of this section shall be based on an R_w of 3.

2108A.2.3.7 Reinforcement.

1. Where transverse reinforcement is required, the maximum spacing shall not exceed one half the depth of the member nor 48 inches (1219 mm).

2. Flexural reinforcement shall be uniformly distributed throughout the depth of the element.

3. Flexural elements subjected to load reversals shall be symmetrically reinforced.

4. The nominal moment strength at any section along a member shall not be less than one fourth of the maximum moment strength.

5. The flexural reinforcement ratio, ρ, shall not exceed $0.5 \ \rho_b$.

6. Lap splices shall comply with the provisions of Section 2108A.2.2.7.

7. Welded splices and mechanical splices which develop at least 125 percent of the specified yield strength of a bar may be used for splicing the reinforcement. Not more than two longitudinal bars shall be spliced at a section. The distance between splices of adjacent bars shall be at least 30 inches (762 mm) along the longitudinal axis.

8. Specified yield strength of reinforcement shall not exceed 60,000 psi (413 MPa). The actual yield strength based on mill tests shall not exceed 1.3 times the specified yield strength.

2108A.2.3.8 Seismic design provisions. The lateral seismic load resistance in any line or story level shall be provided by shear walls or wall frames, or a combination of shear walls and wall frames. Shear walls and wall frames shall provide at least 80 percent of the lateral stiffness in any line or story level.

> **EXCEPTION:** Where seismic loads are determined based on R_w not greater than three and where all joints satisfy the provisions of Section 2108A.2.6.2.9, the piers may be used to provide seismic load resistance.

2108A.2.3.9 Dimensional limits. Dimensions shall be in accordance with the following:

1. **Beams.**

1.1 The nominal width of a beam shall not be less than 6 inches (153 mm).

1.2 The clear distance between locations of lateral bracing of the compression side of the beam shall not exceed 32 times the least width of the compression area.

1.3 The nominal depth of a beam shall not be less than 8 inches (203 mm).

2. **Piers.**

2.1 The nominal width of a pier shall not be less than 6 inches (153 mm) and shall not exceed 16 inches (406 mm).

2.2 The distance between lateral supports of a pier shall not exceed 30 times the nominal width of the piers except as provided for in Section 2108A.2.3.9, Item 2.3.

2.3 When the distance between lateral supports of a pier exceeds 30 times the nominal width of the pier, the provisions of Section 2108A.2.4 shall be used for design.

2.4 The nominal length of a pier shall not be less than three times the nominal width of the pier. The nominal length of a pier shall not be greater than six times the nominal width of the pier. The clear height of a pier shall not exceed five times the nominal length of the pier.

EXCEPTION: The length of a pier may be equal to the width of the pier when the axial force at the location of maximum moment is less than $0.04 f'_m A_g$.

3. **Columns.**

3.1 The nominal width of a column shall not be less than 12 inches (305 mm).

3.2 The distance between lateral supports of a column shall not exceed *20* times the nominal width of the column.

3.3 The nominal length of a column shall not be less than 12 inches (305 mm) and not greater than three times the nominal width of the column.

2108A.2.3.10 Beams.

2108A.2.3.10.1 Scope. Members designed primarily to resist flexure shall comply with the requirements of this section. The factored axial compressive force on a beam shall not exceed $0.05 A_e f'_m$.

2108A.2.3.10.2 Longitudinal reinforcement.

1. The variation in the longitudinal reinforcing bars shall not be greater than one bar size. Not more than two bar sizes shall be used in a beam.

2. The nominal flexural strength of a beam shall not be less than 1.3 times the nominal cracking moment strength of the beam. The modulus of rupture, f_r, for this calculation shall be assumed to be 235 psi (1.6 MPa).

2108A.2.3.10.3 Transverse reinforcement. Transverse reinforcement shall be provided where V_u exceeds V_m. Required shear, V_u, shall include the effects of drift. The value of V_u shall be based on $3R_w/8$ times seismic drift. When transverse shear reinforcement is required, the following provisions shall apply:

1. Shear reinforcement shall be a single bar with a 180-degree hook at each end.

2. Shear reinforcement shall be hooked around the longitudinal reinforcement.

3. The minimum transverse shear reinforcement ratio shall be 0.0007.

4. The first transverse bar shall not be more than one fourth of the beam depth from the end of the beam.

2108A.2.3.10.4 Construction. Beams shall be solid grouted.

2108A.2.3.11 Piers.

2108A.2.3.11.1 Scope. Piers proportioned to resist flexure and shear in conjunction with axial load shall comply with the requirements of this section. The factored axial compression on the piers shall not exceed $0.3\,A_e\,f'_m$.

2108A.2.3.11.2 Longitudinal reinforcement. A pier subjected to in-plane stress reversals shall be longitudinally reinforced symmetrically on both sides of the neutral axis of the pier.

1. One bar shall be provided in the end cells.

2. The minimum longitudinal reinforcement ratio shall be 0.0007.

2108A.2.3.11.3 Transverse reinforcement. Transverse reinforcement shall be provided where V_u exceeds V_m. Required shear, V_u, shall include the effects of drift. The value of V_u shall be based on $3R_w/8$ times seismic drift. When transverse shear reinforcement is required, the following provisions shall apply:

1. Shear reinforcement shall be hooked around the extreme longitudinal bars with a 180-degree hook. Alternatively, at wall intersections, transverse reinforcement with a 90-degree standard hook around a vertical bar in the intersecting wall shall be permitted.

2. The minimum transverse reinforcement ratio shall be 0.0015.

2108A.2.3.12 Columns.

2108A.2.3.12.1 Scope. Columns shall comply with the requirements of this section.

2108A.2.3.12.2 Longitudinal reinforcement. Longitudinal reinforcement shall be a minimum of four bars, one in each corner of the column.

1. Maximum reinforcement area shall be $0.03\,A_e$.

2. Minimum reinforcement area shall be $0.005\,A_e$.

2108A.2.3.12.3 Lateral ties.

1. Lateral ties shall be provided in accordance with Section 2106A.3.6.

2. Minimum lateral reinforcement area shall be $0.0018\,A_g$.

2108A.2.3.12.4 Construction. Columns shall be solid grouted.

2108A.2.4 Wall design for out-of-plane loads.

2108A.2.4.1 General. The requirements of this section are for the design of walls for out-of-plane loads.

2108A.2.4.2 Maximum reinforcement. The reinforcement ratio shall not exceed $0.5\rho_b$.

2108A.2.4.3 Moment and deflection calculations. All moment and deflection calculations in Section 2108A.2.4 are based on simple support conditions top and bottom. Other support and fixity conditions, moments and deflections shall be calculated using established principles of mechanics.

2108A.2.4.4 Walls with axial load of $0.04f'_m$ or less. The procedures set forth in this section, which consider the slenderness of walls by representing effects of axial forces and deflection in calculation of moments, *may* be used when the vertical load stress at the location of maximum moment does not exceed $0.04f'_m$ as computed by Formula (8A-28). The value of f'_m shall not exceed *2,500* psi *(17.24 MPa)*.

$$\frac{P_w + P_f}{A_g} \leq 0.04\,f'_m \qquad (8A\text{-}28)$$

Walls shall have a minimum nominal thickness of *8 inches (203 mm)*. *The height-to-thickness ratio set forth in Section 2106A.2.3.3 and Table 21A-R shall not apply.*

Required moment and axial force shall be determined at the midheight of the wall and shall be used for design. The factored moment, M_u, at the midheight of the wall shall be determined by Formula (8A-29).

$$M_u = \frac{w_u h^2}{8} + P_{uf}\frac{e}{2} + P_u \Delta_u \qquad (8A-29)$$

WHERE:

Δ_u = deflection at midheight of wall due to factored loads

$$P_u = P_{uw} + P_{uf} \qquad (8A-30)$$

The design strength for out-of-plane wall loading shall be determined by Formula (8A-31).

$$M_u \leq \phi M_n \qquad (8A-31)$$

WHERE:

$$M_n = A_{se} f_y (d - a/2) \qquad (8A-32)$$

$$A_{se} = (A_s f_y + P_u) / f_y, \text{ effective area of steel} \qquad (8A-33)$$

$$a = (P_u + A_s f_y) / 0.85 f'_m b, \text{ depth of stress block due to factored loads} \qquad (8A-34)$$

2108A.2.4.5 Wall with axial load greater than 0.04f'_m. The procedures set forth in this section shall be used for the design of masonry walls when the vertical load stresses at the location of maximum moment exceed 0.04f'_m but are less than 0.2f'_m and the slenderness ratio h'/t does not exceed 30.

Design strength provided by the wall cross section in terms of axial force, shear and moment shall be computed as the nominal strength multiplied by the applicable strength-reduction factor, ϕ, specified in Section 2108A.1.4. Walls shall be proportioned such that the design strength exceeds the required strength.

The nominal shear strength shall be determined by Formula (8A-35).

$$V_n = 2A_{mv} \sqrt{f'_m} \qquad (8A-35)$$

For **SI:** $\qquad\qquad\qquad V_n = 0.166A_{mv} \sqrt{f'_m}$

2108A.2.4.6 Deflection design. The midheight deflection, Δ_s, under service lateral and vertical loads (without load factors) shall be limited by the relation:

$$\Delta_s = 0.005h \qquad (8A-36)$$

$P\Delta$ effects shall be included in deflection calculation. The midheight deflection shall be computed with the following formula:

$$\Delta_s = \frac{5 M_s h^2}{48 E_m I_g} \text{ for } M_{ser} \leq M_{cr} \qquad (8A-37)$$

$$\Delta_s = \frac{5 M_{cr} h^2}{48 E_m I_g} + \frac{5 (M_{ser} - M_{cr})h^2}{48 E_m I_{cr}} \text{ for } M_{cr} < M_{ser} < M_n \qquad (8A-38)$$

The cracking moment strength of the wall shall be determined from the formula:

$$M_{cr} = Sf_r \qquad (8A-39)$$

The modulus of rupture, f_r, shall be as follows:

1. For fully grouted hollow-unit masonry,

$$f_r = 4.0 \sqrt{f'_m}, \ 235 \text{ psi maximum} \tag{8A-40}$$

For SI: $\quad\quad\quad\quad\quad\quad f_r = 0.33 \sqrt{f'_m}, \ 1.6 \text{ MPa maximum}$

2. *Not adopted by the State of California.*

3. For two-wythe brick masonry,

$$f_r = 2.0 \sqrt{f'_m}, \ 125 \text{ psi maximum} \tag{8A-42}$$

For SI: $\quad\quad\quad\quad\quad\quad f_r = 0.166 \sqrt{f'_m}, \ 861 \text{ kPa maximum}$

2108A.2.5 Wall design for in-plane loads.

2108A.2.5.1 General. The requirements of this section are for the design of walls for in-plane loads.

The value of f'_m shall not be less than 1,500 psi (10.3 MPa) nor greater than 4,000 psi (27.6 MPa).

2108A.2.5.2 Reinforcement. Reinforcement shall be in accordance with the following:

1. Minimum reinforcement shall be provided in accordance with Section 2106A.1.12.4, Item 2.3, for all seismic areas using this method of analysis.

2. When the shear wall failure mode is in flexure, the nominal flexural strength of the shear wall shall be at least 1.8 times the cracking moment strength of a fully grouted wall ∗ ∗ ∗ from Formula (8A-39).

3. The amount of vertical reinforcement shall not be less than one half the horizontal reinforcement.

4. Spacing of horizontal reinforcement within the region defined in Section 2108A.2.5.5, Item 3, shall not exceed three times the nominal wall thickness nor 24 inches (610 mm).

2108A.2.5.3 Design strength. Design strength provided by the shear wall cross section in terms of axial force, shear and moment shall be computed as the nominal strength multiplied by the applicable strength-reduction factor, φ, specified in Section 2108A.1.4.3.

2108A.2.5.4 Axial strength. The nominal axial strength of the shear wall supporting axial loads only shall be calculated by Formula (8A-43).

$$P_o = 0.85 f'_m (A_e - A_s) + f_y A_s \tag{8A-43}$$

Axial design strength provided by the shear wall cross section shall satisfy Formula (8A-44).

$$P_u \leq 0.80 \ \phi \ P_o \tag{8A-44}$$

2108A.2.5.5 Shear strength. Shear strength shall be as follows:

1. The nominal shear strength shall be determined using either Item 2 or 3 below. Maximum nominal shear strength values are determined from Table 21A-J.

2. The nominal shear strength of the shear wall shall be determined from Formula (8A-45), except as provided in Item 3 below

$$V_n = V_m + V_s \tag{8A-45}$$

WHERE:

$$V_m = C_d A_{mv} \sqrt{f'_m} \tag{8A-46}$$

For SI: $\quad\quad\quad\quad\quad V_m = 0.083 \ C_d A_{mv} \sqrt{f'_m}$

and

$$V_s = A_{mv}\, \rho_n f_y \qquad\qquad (8A\text{-}47)$$

3. For a shear wall whose nominal shear strength exceeds the shear corresponding to development of its nominal flexural strength, two shear regions exist.

For all cross sections within the region defined by the base of the shear wall and a plane at a distance L_w above the base of the shear wall, the nominal shear strength shall be determined from Formula (8A-48).

$$V_n = A_{mv}\, \rho_n f_y \qquad\qquad (8A\text{-}48)$$

The required shear strength for this region shall be calculated at a distance $L_w/2$ above the base of the shear wall, but not to exceed one half story height.

For the other region, the nominal shear strength of the shear wall shall be determined from Formula (8A-45).

2108A.2.5.6 Boundary members. Boundary members shall be as follows:

1. Boundary members shall be provided at the boundaries of shear walls when the compressive strains in the wall exceed 0.0015. The strain shall be determined using factored forces and R_w equal to 1.5.

2. The minimum length of the boundary member shall be three times the thickness of the wall, but shall include all areas where the compressive strain per Section 2108A.2.6.2.7 is greater than 0.0015.

3. Lateral reinforcement shall be provided for the boundary elements. The lateral reinforcement shall be a minimum of No. 3 bars at a maximum of 8-inch (203 mm) spacing within the grouted core or equivalent confinement which can develop an ultimate compressive masonry strain of at least 0.006.

2108A.2.6 Design of moment-resisting wall frames.

2108A.2.6.1 General requirements.

2108A.2.6.1.1 Scope. The requirements of this section are for the design of fully grouted moment-resisting wall frames constructed of reinforced open-end hollow-unit concrete or hollow-unit clay masonry.

2108A.2.6.1.2 Dimensional limits. Dimensions shall be in accordance with the following.

Beams. Clear span for the beam shall not be less than two times its depth.

The nominal depth of the beam shall not be less than two units or 16 inches (406 mm), whichever is greater. The nominal beam depth to nominal beam width ratio shall not exceed 6.

The nominal width of the beam shall be the greater of 8 inches (203 mm) or $^1/_{26}$ of the clear span between pier faces.

Piers. The nominal depth of piers shall not exceed 96 inches (2438 mm). Nominal depth shall not be less than two full units or 32 inches (813 mm), whichever is greater.

The nominal width of piers shall not be less than the nominal width of the beam, nor less than 8 inches (203 mm) or $^1/_{14}$ of the clear height between beam faces, whichever is greater.

The clear height-to-depth ratio of piers shall not exceed 5.

2108A.2.6.1.3 Analysis. Member design forces shall be based on an analysis which considers the relative stiffness of pier and beam members, including the stiffening influence of joints.

The calculation of beam moment capacity for the determination of pier design shall include any contribution of floor slab reinforcement.

The out-of-plane drift ratio of all piers shall satisfy the drift-ratio limits specified in Section 1628A.8.2.

2108A.2.6.2 Design procedure.

2108A.2.6.2.1 Required strength. Except as required by Sections 2108A.2.6.2.7 and 2108A.2.6.2.8, the required strength shall be determined in accordance with Section 2108A.1.3.

2108A.2.6.2.2 Design strength. Design strength provided by frame member cross sections in terms of axial force, shear and moment shall be computed as the nominal strength multiplied by the applicable strength-reduction factor, ϕ, specified in Section 2108A.1.4.4.

Members shall be proportioned such that the design strength exceeds the required strength.

2108A.2.6.2.3 Design assumptions for nominal strength. The nominal strength of member cross sections shall be based on assumptions prescribed in Section 2108A.2.1.2.

The value of f'_m shall not be less than 1,500 psi (10.3 MPa) or greater than 4,000 psi (27.6 MPa).

2108A.2.6.2.4 Reinforcement. The nominal moment strength at any section along a member shall not be less than one fourth of the higher moment strength provided at the two ends of the member.

Lap splices shall be as defined in Section 2108A.2.2.7. The center of the lap splice shall be at the center of the member clear length.

Welded splices and mechanical connections conforming to Section 1912A.14.3, Items 1 through 4, may be used for splicing the reinforcement at any section provided not more than alternate longitudinal bars are spliced at a section, and the distance between splices of alternate bars is at least 24 inches (610 mm) along the longitudinal axis.

Reinforcement shall not have a specified yield strength greater than 60,000 psi (413 MPa). The actual yield strength based on mill tests shall not exceed the specified yield strength times 1.3.

2108A.2.6.2.5 Flexural members (beams). Requirements of this subsection apply to beams proportioned primarily to resist flexure as follows:

The axial compressive force on beams due to factored loads shall not exceed $0.10 \, A_n \, f'_m$.

1. **Longitudinal reinforcement.** At any section of a beam, each masonry unit through the beam depth shall contain longitudinal reinforcement.

The variation in the longitudinal reinforcement area between units at any section shall not be greater than 50 percent, except multiple No. 4 bars shall not be greater than 100 percent of the minimum area of longitudinal reinforcement contained by any one unit, except where splices occur.

Minimum reinforcement ratio calculated over the gross cross section shall be 0.002.

Maximum reinforcement ratio calculated over the gross cross section shall be $0.15 f'_m / f_y$.

2. **Transverse reinforcement.** Transverse reinforcement shall be hooked around top and bottom longitudinal bars with a standard 180-degree hook, as defined in Section 2108A.2.2.4, and shall be single pieces.

Within an end region extending one beam depth from pier faces and at any region at which beam flexural yielding may occur during seismic or wind loading, maximum spacing of transverse reinforcement shall not exceed one fourth the nominal depth of the beam.

The maximum spacing of transverse reinforcement shall not exceed one half the nominal depth of the beam.

Minimum reinforcement ratio shall be 0.0015.

The first transverse bar shall not be more than 4 inches (102 mm) from the face of the pier.

2108A.2.6.2.6 Members subjected to axial force and flexure.

The requirements set forth in this subsection apply to piers proportioned to resist flexure in conjunction with axial loads.

1. **Longitudinal reinforcement.** A minimum of four longitudinal bars shall be provided at all sections of every pier.

Flexural reinforcement shall be distributed across the member depth. Variation in reinforcement area between reinforced cells shall not exceed 50 percent.

Minimum reinforcement ratio calculated over the gross cross section shall be 0.002.

Maximum reinforcement ratio calculated over the gross cross section shall be $0.15f'_m/f_y$.

Maximum bar diameter shall be one eighth nominal width of the pier.

2. **Transverse reinforcement.** Transverse reinforcement shall be hooked around the extreme longitudinal bars with standard 180-degree hook as defined in Section 2108A.2.2.4.

Within an end region extending one pier depth from the end of the beam, and at any region at which flexural yielding may occur during seismic or wind loading, the maximum spacing of transverse reinforcement shall not exceed one fourth the nominal depth of the pier.

The maximum spacing of transverse reinforcement shall not exceed one half the nominal depth of the pier.

The minimum transverse reinforcement ratio shall be 0.0015.

3. **Lateral reinforcement.** Lateral reinforcement shall be provided to confine the grouted core when compressive strains due to axial and bending forces exceed 0.0015, corresponding to factored forces with R_w equal to 1.5. The unconfined portion of the cross section with strain exceeding 0.0015 shall be neglected in computing the nominal strength of the section.

The total cross-sectional area of rectangular tie reinforcement for the confined core shall not be less than:

$$A_{sh} = 0.09 s h_c f'_m / f_{yh} \qquad \text{(8A-49)}$$

Alternatively, equivalent confinement which can develop an ultimate compressive strain of at least 0.006 may be substituted for rectangular tie reinforcement.

2108A.2.6.2.7 Pier design forces. Pier nominal moment strength shall not be less than 1.6 times the pier moment corresponding to the development of beam plastic hinges, except at the foundation level.

Pier axial load based on the development of beam plastic hinges in accordance with the paragraph above and including factored dead and live loads shall not exceed $0.15 A_n f'_m$.

The drift ratio of piers shall satisfy the limits specified in Chapter 16A.

The effects of cracking on member stiffness shall be considered. Unless stiffness values are obtained by a more comprehensive analysis, the effective moment of inertia shall be determined in accordance with Formula (8A-50).

$$I_e = \left(\frac{M_{cr}}{M_a}\right) I_{gt} + \left[1 - \left(\frac{M_{cr}}{M_a}\right)^3\right] I_{cr} \le I_g \qquad \text{(8A-50)}$$

The base plastic hinge of the pier must form immediately adjacent to the level of lateral support provided at the base or foundation.

2108A.2.6.2.8 Shear design.

1. **General.** Beam and pier nominal shear strength shall not be less than 1.4 times the shears corresponding to the development of beam flexural yielding.

It shall be assumed in the calculation of member shear force that moments of opposite sign act at the joint faces and that the member is loaded with the tributary gravity load along its span.

2. **Vertical member shear strength.** The nominal shear strength shall be determined from Formula (8A-51):

$$V_n = V_m + V_s \tag{8A-51}$$

WHERE:

$$V_m = C_d A_{mv} \sqrt{f'_m} \tag{8A-52}$$

For **SI:** $\quad V_m = 0.083 C_d A_{mv} \sqrt{f'_m}$

and

$$V_s = A_{mv} \rho_n f_y \tag{8A-53}$$

The value of V_m shall be zero within an end region extending one pier depth from beam faces and at any region where pier flexural yielding may occur during seismic loading, and at piers subjected to net tension factored loads.

The nominal pier shear strength, V_n, shall not exceed the value determined from Table 21A-J.

3. **Beam shear strength.** The nominal shear strength shall be determined from Formula (8A-54),

WHERE:

$$V_m = 1.2 A_{mv} \sqrt{f'_m} \tag{8A-54}$$

For **SI:** $\quad V_m = 0.01 A_{mv} \sqrt{f'_m}$

The value of V_m shall be zero within an end region extending one beam depth from pier faces and at any region at which beam flexural yielding may occur during seismic loading.

The nominal beam shear strength, V_n, shall be determined from Formula (8A-55).

$$V_n \leq 4 A_{mv} \sqrt{f'_m} \tag{8-55}$$

For **SI:** $\quad V_n \leq 0.33 A_{mv} \sqrt{f'_m}$

2108A.2.6.2.9 Joints.

1. **General requirements.** Where reinforcing bars extend through a joint, the joint dimensions shall be proportioned such that

$$h_p > 4800 d_{bb} / \sqrt{f'_g} \tag{8A-56}$$

For **SI:** $\quad h_p > 57\ 827 d_{bb} / \sqrt{f'_g}$

and

$$h_b > 1800 d_{bp} / \sqrt{f'_g} \tag{8A-57}$$

For **SI:** $\quad h_b > 21\ 685 d_{bp} / \sqrt{f'_g}$

The grout strength shall not exceed 5,000 psi (34.4 MPa) for the purposes of Formulas (8A-56) and (8A-57).

Joint shear forces shall be calculated on the assumption that the stress in all flexural tension reinforcement of the beams at the pier faces is $1.4 f_y$.

Strength of joint shall be governed by the appropriate strength-reduction factors specified in Section 2108A.1.4.4.

Beam longitudinal reinforcement terminating in a pier shall be extended to the far face of the pier and anchored by a standard 90- or 180-degree hook, as defined in Section 2108A.2.2.4, bent back to the beam.

Pier longitudinal reinforcement terminating in a beam shall be extended to the far face of the beam and anchored by a standard 90- or 180-degree hook, as defined in Section 2108A.2.2.4, bent back to the beam.

2. **Transverse reinforcement.** Special horizontal joint shear reinforcement crossing a potential corner-to-corner diagonal joint shear crack, and anchored by standard hooks, as defined in Section 2108A.2.2.4, around the extreme pier reinforcing bars shall be provided such that

$$A_{jh} = 0.5 \, V_{jh} / f_y \qquad (8A\text{-}58)$$

Vertical shear forces may be considered to be carried by a combination of masonry shear-resisting mechanisms and truss mechanisms involving intermediate pier reinforcing bars.

3. **Shear strength.** The nominal horizontal shear strength of the joint shall not exceed $7\sqrt{f'_m}$ (For **SI:** $0.58\sqrt{f'_m}$) or 350 psi (2.4 MPa), whichever is less.

SECTION 2109A — *NOT ADOPTED BY THE STATE OF CALIFORNIA.*

SECTION 2110A — GLASS MASONRY

2110A.1 General. Masonry of glass blocks may be used in nonload-bearing exterior or interior walls and *shall conform to the requirements of Section 2113A. Stresses in glass block shall not be utilized.* Glass block may be solid or hollow and may contain inserts.

2110A.2 Mortar Joints. All mortar contact surfaces shall be treated to ensure adhesion between mortar and glass.

SECTION 2111A — CHIMNEYS, FIREPLACES AND BARBECUES

Chimneys, flues, fireplaces and barbecues and their connections carrying products of combustion shall be designed, anchored, supported and reinforced as set forth in Chapter 31 and any applicable provisions of this chapter.

SECTION 2112A — NONBEARING WALLS

2112A.1 General. All nonbearing masonry walls shall be reinforced as specified in Section 2106A.1.12.4. Fences and interior nonbearing nonshear walls may be of hollow-unit masonry construction grouted in cells containing vertical and horizontal reinforcement. Nonbearing walls may be used to carry a superimposed load of not more than 200 pounds per linear foot (2.92 kN/m).

1. Thickness. Every nonbearing masonry wall shall be so constructed and have a sufficient thickness to withstand all vertical loads and horizontal loads, but in no case shall the thickness of such walls be less than the values set forth in Table 21A-R.

Plaster shall not be considered as contributing to the thickness of a wall in computing the height-to-thickness ratio.

2. Anchorage. All nonbearing walls shall be anchored as required by Sections 1611A and 1631A.2.8. Suspended ceilings or other nonstructural elements shall not be used to provide anchorage for masonry walls.

SECTION 2113A — MASONRY SCREEN WALLS

2113A.1 General. Masonry units may be used in nonbearing decorative screen walls. Units may be laid up in panels with units on edge with the open pattern of the unit exposed in the completed wall.

1. **Horizontal Forces.** The panels shall be capable of spanning between supports to resist the horizontal forces specified in Chapter 16A. Wind loads shall be based on gross projected area of the block.

2. **Mortar Joints.** Horizontal and vertical joints shall not be less than $^1/_4$ inch (6 mm) thick. All joints shall be completely filled with mortar and shall be "shoved joint" work. The units of a panel shall be so arranged that either the horizontal or the vertical joint containing reinforcing is continuous without offset. This continuous joint shall be reinforced with a minimum of 0.03 square inch (19 mm²) of reinforcing steel. Reinforcement may be embedded in mortar.

3. **Reinforcing.** Joint reinforcing may be composed of two wires made with welded ladder or trussed wire cross ties. In calculating the resisting capacity of the system, compression and tension in the spaced wires may be utilized. Ladder wire reinforcing shall not be spliced and shall be the widest that the mortar joint will accommodate, allowing $^1/_2$ inch (13 mm) of mortar cover.

4. **Size of Panels.** The maximum size of panels shall be 144 square feet (13.4 m²), with the maximum dimension in either direction of 15 feet (4572 mm).

5. **Panel Support.** Each panel shall be supported on all edges by a structural member of concrete, masonry or steel. Supports at the top and ends of the panel shall be by means of confinement of the masonry by at least $^1/_2$ inch (13 mm) into and between the flanges of a steel channel. The space between the end of the panel and the web of the channel shall be filled with resilient material. The use of equivalent configuration in other steel section or in masonry or concrete is acceptable.

SECTION 2114A — USE OF EXISTING MASONRY

2114A.1 General. Existing masonry which does not meet the requirements for reinforced grouted masonry shall not be used for structural purposes.

1. **Method of Repair.** Pneumatically placed concrete (shotcrete) or concrete structural systems may be used to strengthen existing masonry which does not comply with the requirements for reinforced grouted masonry. Such masonry shall be relieved of stress by the shotcrete or concrete, except that it may be assumed to carry its own weight vertically.

2. **Pneumatically Placed Concrete (Shotcrete) or Concrete Method of Strengthening.**

2.1 Ribs or minor columns of shotcrete or concrete may be incorporated into existing masonry to support vertical loads and lateral forces normal to the wall. The width of the chases cut into the masonry to form the ribs shall be at least $1^1/_2$ times the depth. Concrete or shotcrete membranes between the ribs may be used to resist shear in the plane of the walls due to lateral loads. The thickness of the membrane shall not be less than $^1/_{24}$ of the clear distance between the ribs or minor columns, and in any case not less than that required by Chapter 19A.

2.2 A membrane will be required where ribs or minor columns are more than 5 feet (1524 mm) apart in the clear. Where less than 5 feet (1524 mm) apart in the clear, a membrane will be required unless the ribs extend to the far tier of masonry.

2.3 There shall be a rib or minor column within at least 2 feet (610 mm) of each vertical edge of each opening in the wall where there is a membrane, but where there is no membrane, there shall be a rib or minor column at each vertical edge of each opening.

2.4 If a shotcrete or concrete membrane is placed against only one side of a unit masonry wall, the shotcrete or concrete membrane shall be in contact with the far tier of the unit masonry at not less than one point in any rectangle 4 feet (1219 mm) wide and 3 feet (914 mm) high. The area of such contact shall not be less than 64 square inches (41,300 mm²).

EXCEPTIONS: 1. If continuous ribs or minor columns are provided not more than 5 feet (1524 mm) apart which extend through to the far tier, intermediate contact points may be omitted.

2. Where the membrane is in contact with header courses that extend to the far surface, where such header courses are spaced not farther apart than every sixth course and the headers are spaced not more than 12 inches (305 mm) apart, intermediate contact points may be omitted.

2.5 Where such walls are in contact with reinforced concrete or shotcrete columns, beams or walls, the concrete or shotcrete portion of such walls shall either be constructed integrally with the abutting columns, beams or walls, or keyways shall be provided to resist design forces.

2.6 Surfaces against which shotcrete is to be deposited shall be prepared in accordance with Section 1922A.1.

Surfaces against which concrete membranes are to be placed shall be prepared as prescribed for shotcrete and, in addition, joints at the header courses in brickwork shall be raked back $^1/_2$ inch (13 mm). The height of each pour and of each lift of forms shall not exceed nine times the thickness of the membrane.

2.7 Ribs supported laterally by shotcrete membranes shall be reinforced with not less than two No. 5 bars. Other ribs shall be reinforced with a minimum of four No. 5 bars. Ribs shall have No. 3 ties at 9 inches (229 mm) on centers and shall have one No. 6 dowel, top and bottom, where ribs frame into the existing concrete construction. The membrane shall be reinforced with not less than No. 3 bars at 18 inches (457 mm) on center each way. Dowels shall be provided to existing concrete construction where shear is transferred. Dowels shall be equivalent in area to the reinforcing, and spaced not more than 36 inches (914 mm) on center.

SECTION 2115A — TESTS AND INSPECTIONS

2115A.1 See Section 2105A.

TABLE 21A-A—MORTAR PROPORTIONS FOR UNIT MASONRY

MORTAR	TYPE	PROPORTIONS BY VOLUME (CEMENTITIOUS MATERIALS)								AGGREGATE MEASURED IN A DAMP, LOOSE CONDITION
		Portland Cement or Blended Cement	Masonry Cement[1] M	S	N	Mortar Cement[2] M	S	N	Hydrated Lime or Lime Putty	
Cement-lime	M	1	—	—	—	—	—	—	$^1/_4$	Not less than $2^1/_4$ and not more than 3 times the sum of the separate volumes of cementitious materials.
	S	1	—	—	—	—	—	—	over $^1/_4$ to $^1/_2$	
Mortar cement	M	1	—	—	—	—	—	1	—	
	M	—	—	—	—	1	—	—	—	
	S	$^1/_2$	—	—	—	—	—	1	—	
	S	—	—	—	—	—	1	—	—	
Masonry cement	M	1	—	—	1	—	—	—	—	
	M	—	1	—	—	—	—	—	—	
	S	$^1/_2$	—	—	1	—	—	—	—	
	S	—	—	1	—	—	—	—	—	

[1]Masonry cement conforming to the requirements of U.B.C. Standard 21-11.

[2]Mortar cement conforming to the requirements of U.B.C. Standard 21-14.

TABLE 21A-B—GROUT PROPORTIONS BY VOLUME[1]

TYPE	PARTS BY VOLUME OF PORTLAND CEMENT OR BLENDED CEMENT	PARTS BY VOLUME OF HYDRATED LIME OR LIME PUTTY	AGGREGATE MEASURED IN A DAMP, LOOSE CONDITION	
			Fine	Coarse
Fine grout	1	0 to $^1/_{10}$	$2^1/_4$ to 3 times the sum of the volumes of the cementitious materials	
Coarse grout	1	0 to $^1/_{10}$	$2^1/_4$ to 3 times the sum of the volumes of the cementitious materials	1 to 2 times the sum of the volumes of the cementitious materials

[1]Grout shall attain a minimum compressive strength at 28 days of 2,000 psi (13.8 MPa). The *enforcement agency* may require a compressive field strength test of grout made in accordance with U.B.C. Standard 21-18.

C
A

TABLE 21A-C—GROUTING LIMITATIONS

GROUT TYPE	GROUT POUR MAXIMUM HEIGHT (feet)[1]	MINIMUM DIMENSIONS OF THE TOTAL CLEAR AREAS WITHIN GROUT SPACES AND CELLS[2,3]	
		× 25.4 for mm	
	× 304.8 for mm	Multiwythe Masonry	Hollow-unit Masonry
Fine	1	$^3/_4$	$1^1/_2$ x 2
Fine	5	$1^1/_2$	$1^1/_2$ x 2
Fine	8	$1^1/_2$	$1^1/_2$ x 3
Fine	12	$1^1/_2$	$1^3/_4$ x 3
Fine	24	2	3 x 3
Coarse	1	$1^1/_2$	$1^1/_2$ x 3
Coarse	5	2	$2^1/_2$ x 3
Coarse	8	2	3 x 3
Coarse	12	$2^1/_2$	3 x 3
Coarse	24	3	3 x 4

[1]See also Section 2104A.6.
[2]The actual grout space or grout cell dimensions must be larger than the sum of the following items: (1) The required minimum dimensions of total clear areas in Table 21A-G; (2) The width of any mortar projections within the space; and (3) The horizontal projections of the diameters of the horizontal reinforcing bars within a cross section of the grout space or cell.
[3]The minimum dimensions of the total clear areas shall be made up of one or more open areas, with at least one area being $^3/_4$ inch (19 mm) or greater in width.

TABLE 21A-D—SPECIFIED COMPRESSIVE STRENGTH OF MASONRY, f'_m (psi) BASED ON SPECIFYING THE COMPRESSIVE STRENGTH OF MASONRY UNITS

COMPRESSIVE STRENGTH OF CLAY MASONRY UNITS[1, 2] (psi)	SPECIFIED COMPRESSIVE STRENGTH OF MASONRY, f'_m
	Type M or S Mortar[3] (psi)
× 6.89 for kPa	
14,000 or more	5,300
12,000	4,700
10,000	4,000
8,000	3,350
6,000	2,700
4,000	2,000

COMPRESSIVE STRENGTH OF CONCRETE MASONRY UNITS[2, 4] (psi)	SPECIFIED COMPRESSIVE STRENGTH OF MASONRY, f'_m
	Type M or S Mortar[3] (psi)
× 6.89 for kPa	
4,800 or more	3,000
3,750	2,500
2,800	2,000
1,900	1,500
1,250	1,000

[1]Compressive strength of solid clay masonry units is based on gross area. Compressive strength of hollow clay masonry units is based on minimum net area. Values may be interpolated. When hollow clay masonry units are grouted, the grout shall conform to the proportions in Table 21A-B.
[2]Assumed assemblage. The specified compressive strength of masonry f'_m is based on gross area strength when using solid units or solid grouted masonry and net area strength when using ungrouted hollow units.
[3]Mortar for unit masonry, proportion specification, as specified in Table 21A-B. These values apply to portland cement-lime mortars without added air-entraining materials.
[4]Values may be interpolated. In grouted concrete masonry, the compressive strength of grout shall be equal to or greater than the compressive strength of the concrete masonry units.

TABLE 21A-E-1—ALLOWABLE TENSION, B_t, FOR EMBEDDED ANCHOR BOLTS FOR CLAY AND CONCRETE MASONRY, pounds[1,2,3]

f'_m (psi)	EMBEDMENT LENGTH, l_b, or EDGE DISTANCE, l_{be} (Inches)						
	2	3	4	5	6	8	10
× 6.89 for kPa	× 25.4 for mm × 4.45 for N						
1,500	240	550	970	1,520	2,190	3,890	6,080
1,800	270	600	1,070	1,670	2,400	4,260	6,660
2,000	280	630	1,120	1,760	2,520	4,500	7,020
2,500	310	710	1,260	1,960	2,830	5,030	7,850
3,000	340	770	1,380	2,150	3,100	5,510	8,600
4,000	400	890	1,590	2,480	3,580	6,360	9,930
5,000	440	1,000	1,780	2,780	4,000	7,110	11,100
6,000	480	1,090	1,950	3,040	4,380	7,790	12,200

[1]The allowable tension values in Table 21A-E-1 are based on compressive strength of masonry assemblages. Where yield strength of anchor bolt steel governs, the allowable tension in pounds is given in Table 21A-E-2.
[2]Values are for bolts of at least A 307 quality. Bolts shall be those specified in Section 2106A.2.14.1.
[3]Values shown are for work with ✳ ✳ ✳ special inspection.

TABLE 21A-E-2—ALLOWABLE TENSION, B_t, FOR EMBEDDED ANCHOR BOLTS FOR CLAY AND CONCRETE MASONRY, pounds[1,2]

BENT BAR ANCHOR BOLT DIAMETER (inches)							
× 25.4 for mm							
1/4	3/8	1/2	5/8	3/4	7/8	1	1 1/8
× 4.45 for N							
350	790	1,410	2,210	3,180	4,330	5,650	7,160

[1]Values are for bolts of at least A 307 quality. Bolts shall be those specified in Section 2106A.2.14.1.
[2]Values shown are for work with ✳ ✳ ✳ special inspection.

TABLE 21A-F—ALLOWABLE SHEAR, B_v, FOR EMBEDDED ANCHOR BOLTS FOR CLAY AND CONCRETE MASONRY, pounds[1,2]

f'_m (psi)	BENT BAR ANCHOR BOLT DIAMETER (inches)						
	× 25.4 for mm						
	3/8	1/2	5/8	3/4	7/8	1	1 1/8
	× 4.45 for N						
1,500	480	850	1,330	1,780	1,920	2,050	2,170
1,800	480	850	1,330	1,860	2,010	2,150	2,280
2,000	480	850	1,330	1,900	2,060	2,200	2,340
2,500	480	850	1,330	1,900	2,180	2,330	2,470
3,000	480	850	1,330	1,900	2,280	2,440	2,590
4,000	480	850	1,330	1,900	2,450	2,620	2,780
5,000	480	850	1,330	1,900	2,590	2,770	2,940
6,000	480	850	1,330	1,900	2,600	2,900	3,080

[1]Values are for bolts of at least A 307 quality. Bolts shall be those specified in Section 2106A.2.14.1.
[2]Values shown are for work with ✳ ✳ ✳ special inspection.

TABLE 21A-G—MINIMUM DIAMETERS OF BEND

BAR SIZE	MINIMUM DIAMETER
No. 3 through No. 8	6 bar diameters
No. 9 through No. 11	8 bar diameters

TABLE 21A-H-1—RADIUS OF GYRATION[1] FOR CONCRETE MASONRY UNITS[2]

Grout Spacing (inches)	Nominal Width of Wall (inches)				
	× 25.4 for mm				
× 25.4 for mm	4	6	8	10	12
Solid grouted	1.04	1.62	2.19	2.77	3.34

[1]For single-wythe masonry or for an individual wythe of a cavity wall.

$$r = \sqrt{I/A_e}$$

[2]The radius of gyration shall be based on the specified dimensions of the masonry units or shall be in accordance with the values shown which are based on the minimum dimensions of hollow concrete masonry unit face shells and webs in accordance with U.B.C. Standard 21-4 for two cell units.

TABLE 21A-H-2—RADIUS OF GYRATION[1] FOR CLAY MASONRY UNIT LENGTH, 16 INCHES[2]

Grout Spacing (inches)	Nominal Width of Wall (inches)				
× 25.4 for mm	× 25.4 for mm				
× 25.4 for mm	4	6	8	10	12
Solid grouted	1.06	1.64	2.23	2.81	3.39

[1]For single-wythe masonry or for an individual wythe of a cavity wall.

$$r = \sqrt{I/A_e}$$

[2]The radius of gyration shall be based on the specified dimensions of the masonry units or shall be in accordance with the values shown which are based on the minimum dimensions of hollow clay masonry face shells and webs in accordance with U.B.C. Standard 21-1 for two cell units.

TABLE 21A-H-3—RADIUS OF GYRATION[1] FOR CLAY MASONRY UNIT LENGTH, 12 INCHES[2]

Grout Spacing (inches)	Nominal Width of Wall (inches)				
× 25.4 for mm	× 25.4 for mm				
× 25.4 for mm	4	6	8	10	12
Solid grouted	1.06	1.65	2.24	2.82	3.41

[1]For single-wythe masonry or for an individual wythe of a cavity wall.

$$r = \sqrt{I/A_e}$$

[2]The radius of gyration shall be based on the specified dimensions of the masonry units or shall be in accordance with the values shown which are based on the minimum dimensions of hollow clay masonry face shells and webs in accordance with U.B.C. Standard 21-1 for two cell units.

TABLE 21A-I—ALLOWABLE FLEXURAL TENSION (psi)

TABLE 21-I NOT ADOPTED BY THE STATE OF CALIFORNIA.

TABLE 21A-J—MAXIMUM NOMINAL SHEAR STRENGTH VALUES[1,2]

M/Vd	V_n MAXIMUM
≤ 0.25	$6.0 \sqrt{f'_m} = 380$ maximum ($322 \sqrt{f'_m} = 1691$ maximum)
≥ 1.00	$4.0 \sqrt{f'_m} = 250$ maximum ($214 \sqrt{f'_m} = 1113$ maximum)

[1]M is the maximum bending moment that occurs simultaneously with the shear load V at the section under consideration. Interpolation may be by straight line for M/Vd values between 0.25 and 1.00.

[2]V_n is in pounds (N), and f'_m is in pounds per square inches (kPa).

TABLE 21A-K—NOMINAL SHEAR STRENGTH COEFFICIENT

M/Vd[1]	C_d
≤ 0.25	2.4
≥ 1.00	1.2

[1]M is the maximum bending moment that occurs simultaneously with the shear load V at the section under consideration. Interpolation may be by straight line for M/Vd values between 0.25 and 1.00.

TABLE 21A-L—SHEAR WALL SPACING REQUIREMENTS FOR EMPIRICAL DESIGN OF MASONRY

TABLE 21-L NOT ADOPTED BY THE STATE OF CALIFORNIA.

TABLE 21A-M—ALLOWABLE COMPRESSIVE STRESSES FOR EMPIRICAL DESIGN OF MASONRY

TABLE 21-M NOT ADOPTED BY THE STATE OF CALIFORNIA.

TABLE 21A-N—ALLOWABLE SHEAR ON BOLTS FOR EMPIRICALLY DESIGNED MASONRY EXCEPT UNBURNED CLAY UNITS

TABLE 21-N NOT ADOPTED BY THE STATE OF CALIFORNIA.

TABLE 21A-O—WALL LATERAL SUPPORT REQUIREMENTS FOR EMPIRICAL DESIGN OF MASONRY

TABLE 21-O NOT ADOPTED BY THE STATE OF CALIFORNIA.

TABLE 21A-P—THICKNESS OF FOUNDATION WALLS FOR EMPIRICAL DESIGN OF MASONRY

TABLE 21-P NOT ADOPTED BY THE STATE OF CALIFORNIA.

TABLE 21A-Q—ALLOWABLE SHEAR ON BOLTS FOR MASONRY OF UNBURNED CLAY UNITS

TABLE 21-Q NOT ADOPTED BY THE STATE OF CALIFORNIA.

TABLE 21A-R—MINIMUM THICKNESS OF MASONRY WALLS

TYPE OF MASONRY	MAXIMUM RATIO UNSUPPORTED HEIGHT OR LENGTH TO THICKNESS[1,2]	NOMINAL MINIMUM THICKNESS (inches)
BEARING OR SHEAR WALLS:		
1. Stone masonry	*14*	*16*
2. Reinforced grouted masonry	*25*	*6*
3. Reinforced hollow-unit masonry	*25*	*6*
NONBEARING WALLS:		
4. Exterior reinforced walls	*30*	*6*
5. Interior partitions reinforced	*36*	*4*

[1]*In determining the height or length-to-thickness ratio of a cantilevered wall, the dimension to be used shall be twice the dimension of the end of the wall from the lateral support.*

[2]*Cantilevered walls not part of a building and not carrying applied vertical loads need not meet these minimum requirements but their design must comply with stress and overturning requirements.*

Chapter 14A *[DSA/SS, OSHPD 1]*
EXTERIOR WALL COVERINGS

NOTE: This state chapter is applicable to hospitals, skilled nursing facilities, intermediate-care facilities, public schools and state-owned or state-leased essential services buildings regulated by the Office of Statewide Health Planning and Development and the Division of the State Architect.

EXCEPTION: Single-story, Type V skilled nursing or intermediate-care facilities utilizing wood-frame or light-steel-frame construction which shall comply with U.B.C. Chapter 14 and any applicable amendments therein.

SECTION 1401A — GENERAL

1401A.1 Applicability. Exterior wall coverings for the building shall provide weather protection for the building at its exterior boundaries.

Exterior wall covering shall be in accordance with this chapter and as specified by the applicable provisions elsewhere in this code. For additional provisions see Chapter 19A for concrete, Chapter 20A for lightweight metals, Chapter 21A for masonry, Chapter 22A for steel, Chapter 23A for wood, Chapter 25A for gypsum wallboard and plaster and Chapter 26 for plastics. Also, see the following:

SECTION	SUBJECT
601.5.4	Walls fronting on streets
602.1	Materials in Type I construction
603.1	Materials in Type II construction
604.3.1	Exterior walls in Type III construction
605.3.1	Exterior walls in Type IV construction
606.1	Materials in Type V construction

1401A.2 Standards. The standards listed below labeled a "U.B.C. standard" are also listed in Chapter 35, Part II, and are part of this code.

1. U.B.C. Standard 14-1, Kraft Waterproof Building Paper

2. U.B.C. Standard 14-2, Vinyl Siding

SECTION 1402A — WEATHER PROTECTION

1402A.1 Weather-resistive Barriers. All weather-exposed surfaces shall have a weather-resistive barrier to protect the interior wall covering. Such barrier shall be equal to that provided for in U.B.C. Standard 14-1 for kraft waterproof building paper or asphalt-saturated rag felt. This standard is listed in Chapter 35, Part II, and is a part of this code. Building paper and felt shall be free from holes and breaks other than those created by fasteners and construction system due to attaching of the building paper, and shall be applied over studs or sheathing of all exterior walls. Such felt or paper shall be applied horizontally, with the upper layer lapped over the lower layer not less than 2 inches (51 mm). Where vertical joints occur, felt or paper shall be lapped not less than 6 inches (152 mm).

Weather-protected barrier may be omitted in the following cases:

1. When exterior covering is of approved weatherproof panels.

2. In back-plastered construction.

3. When there is no human occupancy.

4. Over water-repellent panel sheathing.

5. Under approved paperbacked metal or wire fabric lath.

6. Behind lath and portland cement plaster applied to the underside of roof and eave projections.

1402A.2 Flashing and Counterflashing. Exterior openings exposed to the weather shall be flashed in such a manner as to make them weatherproof.

All parapets shall be provided with coping of approved materials. All flashing, counterflashing and coping, when of metal, shall have a minimum thickness of 0.019 inches (0.48 mm) (No. 26 galvanized sheet metal gage) corrosion-resistant metal.

1402A.3 Waterproofing Weather-exposed Areas. Balconies, landings, exterior stairways, occupied roofs and similar surfaces exposed to the weather and sealed underneath shall be waterproofed and sloped a minimum of $^1/_4$ unit vertical in 12 units horizontal (2% slope) for drainage.

1402A.4 Dampproofing Foundation Walls. Unless otherwise approved by the building official, foundation walls enclosing a basement below finished grade shall be dampproofed outside by approved methods and materials.

SECTION 1403A — VENEER

1403A.1 Scope.

1403A.1.1 General. All veneer and its application shall conform to the requirements of this code. Wainscots not exceeding 4 feet (1219 mm) in height measured above the adjacent ground elevation for exterior veneer or the finish floor elevation for interior veneer may be exempted from the provisions of this chapter if approved by the *enforcement agency.*

> NOTE: See Section 1405A for inspection requirements.

1403A.1.2 Limitations. Exterior veneer shall not be attached to wood-frame construction at a point more than *25 feet (7620 mm)* in height above the *adjacent ground elevation except when approved by the enforcement agency considering* special construction is designed to provide for differential movement.

Where wood-frame construction provides lateral support for veneer, studs or similar vertical load-supporting members shall be continuous between the foundation and the top of the veneer.

1403A.2 Definitions. For the purpose of this chapter, certain terms are defined as follows:

BACKING as used in this chapter is the surface or assembly to which veneer is attached.

VENEER is nonstructural facing of brick, concrete, stone, tile, metal, plastic or other similar approved material attached to a backing for the purpose of ornamentation, protection or insulation.

Adhered Veneer is veneer secured and supported through adhesion to an approved bonding material applied over an approved backing.

Anchored Veneer is veneer secured to and supported by approved connectors attached to an approved backing.

Exterior Veneer is veneer applied to weather-exposed surfaces as defined in Section 224.

Interior Veneer is veneer applied to surfaces other than weather-exposed surfaces as defined in Section 224.

1403A.3 Materials. Materials used in the application of veneer shall conform to the applicable requirements for such materials as set forth elsewhere in this code.

For masonry units and mortar, see Chapter 21A.

For precast concrete units, see Chapter 19A.

For portland cement plaster, see Chapter 25A.

Anchors, supports and ties shall be noncombustible and corrosion resistant.

When the terms "corrosion resistant" or "noncorrosive" are used in this chapter they shall mean having a corrosion resistance equal to or greater than a hot-dipped galvanized coating of 1.5 ounces of zinc per square foot (458 g/m^2) of surface area. When an element is required to be corrosion resistant or noncorrosive, all of its parts, such as screws, nails, wire, dowels, bolts, nuts, washers, shims, anchors, ties and attachments, shall be corrosion resistant.

Grout shall conform to the requirements of Section 2103A.4.2 for fine grout. C
A

1403A.4 Design.

1403A.4.1 General. The design of all veneer shall comply with the requirements of Chapter 16A and this section.

Veneer shall support no load other than its own weight and the vertical dead load of veneer above.

Surfaces to which veneer is attached shall be designed to support the additional vertical and lateral loads imposed by the veneer.

Consideration shall be given for differential movement of supports, including that caused by temperature changes, shrinkage, creep and deflection.

In no case shall veneer be considered as part of the wall in computing strength or deflection nor shall it be considered a part of the required thickness of the wall.

Veneer shall be anchored in a manner which will not allow relative movement between the veneer and the wall.

1403A.4.2 Adhered veneer. With the exception of ceramic tile, adhered veneer and its backing shall be designed to have a bond to the supporting element sufficient to withstand a shearing stress of 50 psi (345 kPa).

1403A.4.3 Anchored veneer. Anchored veneer and its attachments shall be designed to resist a horizontal force equal to at least twice the weight of the veneer.

1403A.4.4 Adhered veneer or anchored veneer shall not be used on overhead horizontal surfaces. C
A

1403A.5 Adhered Veneer.

1403A.5.1 Permitted backing. Backing shall be continuous and may be of any material permitted by this code. It shall have surfaces prepared to secure and support the imposed loads of veneer.

Exterior veneer, including its backing, shall provide a weatherproof covering.

For additional backing requirements, see Section 1402A.

1403A.5.2 Area limitations. The height and length of veneered areas shall be unlimited except as required to control expansion and contraction and as limited by Section 1403A.1.2.

1403A.5.3 Unit size limitations. Veneer units shall not exceed 36 inches (914 mm) in the greatest dimension or more than 720 square inches (0.46 m^2) in total area and shall weigh not more than *10 pounds per square foot (psf) (48.8 kg/m^2) unless approved by the enforcement agency.*

Units of tile, masonry, stone or terra cotta which exceed $^5/_8$ inch (16 mm) in thickness shall be applied as for anchored veneer where used over exit ways or more than 20 feet (6096 mm) in height above adjacent ground elevation.

> **EXCEPTION:** Veneer units weighing less than 3 psf (14.6 kg/m^2) shall not be limited in dimension or area.

1403A.5.4 Application. In lieu of the design required by Sections 1403A.4.1 and 1403A.4.2, adhered veneer may be applied by one of the following application methods:

1. A paste of neat portland cement shall be brushed on the backing and the back of the veneer unit. Type S mortar then shall be applied to the backing and the veneer unit. Sufficient mortar shall

be used to create a slight excess to be forced out the edges of the units. The units shall be tapped into place so as to completely fill the space between the units and the backing. The resulting thickness of mortar in back of the units shall not be less than $1/2$ inch (13 mm) or more than $11/4$ inches (32 mm).

2. Units of tile, masonry, stone or terra cotta, not over 1 inch (25 mm) in thickness shall be restricted to 81 square inches (52 258 mm^2) in area unless the back side of each unit is ground or box screeded to true up any deviations from plane. These units and glass mosaic units of tile not over 2 inches by 2 inches by $3/8$ inch (51 mm by 51 mm by 9.5 mm) in size may be adhered by means of portland cement. Backing may be of masonry, concrete or portland cement plaster on metal lath. Metal lath shall be fastened to the supports in accordance with the requirements of Chapter 25A. Mortar as described in Table 14A-A shall be applied to the backing as a setting bed. The setting bed shall be a minimum of $3/8$ inch (10 mm) thick and a maximum of $3/4$ inch (19 mm) thick. A paste of neat portland cement or one-half portland cement and one-half graded sand shall be applied to the back of the exterior veneer units and to the setting bed and the veneer pressed and tapped into place to provide complete coverage between the mortar bed and veneer unit. A cement mortar shall be used to point the veneer.

1403A.5.5 Ceramic tile. Portland cement mortars for installing ceramic tile on walls, floors and ceilings shall be as set forth in Table 14A-A.

__1403A.5.6 Bond strength and tests.__ Veneer shall develop a bond to the supporting element of sufficient strength to provide a working shear stress of 50 psi (345 kPa).

Not less than two shear tests shall be performed for the adhered veneer between the units and the supporting element. At least one shear test shall be performed at each building for each 5,000 square feet (465 m^2) of floor area or fraction thereof.

The bond strength as determined by the tests shall have a minimum shear strength of 100 psi (690 kPa).

1403A.6 Anchored Veneer.

1403A.6.1 Permitted backing. Backing may be of any material permitted by this code. Exterior veneer including its backing shall provide a weatherproof covering.

1403A.6.2 Height and support limitations. Anchored veneers shall be supported on footings, foundations or other noncombustible support except as provided under Section 2316A.

In Seismic Zones 2, 3 and 4 the weight of all anchored veneers installed on structures more than 30 feet (9144 mm) in height above the noncombustible foundation or support shall be supported by noncombustible, corrosion-resistant structural framing. The structural framing shall have horizontal supports spaced not more than 12 feet (3658 mm) vertically above the initial 30-foot (9144 mm) height. The vertical spacing between horizontal supports may be increased when special design techniques, approved by the *enforcement agency*, are used in the construction.

Noncombustible, noncorrosive lintels and noncombustible supports shall be provided over all openings where the veneer unit is not self-spanning. The deflections of all structural lintels and horizontal supports required by this subsection shall not exceed $1/600$ of the span under full load of the veneer.

1403A.6.3 Area limitations. The area and length of anchored veneer walls shall be unlimited, except as required to control expansion and contraction and by Section 1403A.1.2.

1403A.6.4 Application.

1403A.6.4.1 General. In lieu of the design required by Sections 1403A.4.1 and 1403A.4.3, anchored veneer may be applied *by any one of the methods specified below:*

1403A.6.4.2 Masonry and stone units [5 inches (127 mm) maximum in thickness]. Masonry and stone veneer not exceeding 5 inches (127 mm) in thickness may be anchored directly to structural masonry, concrete or studs in one of the following manners:

All masonry and stone veneer shall be placed 1 inch (25 mm) clear of its backing.

1. Anchor ties shall be corrosion resistant, and if made of sheet metal, shall have a minimum *size of No. 16 gage by 1 inch (25 mm) and may either be attached to the backing by screws as the veneer is laid, or may be inserted in anchor slots. Anchor ties shall be adjusted for height to lay in horizontal joints and* shall be spaced so as to support not more than 2 square feet (0.19 m²) of wall area but shall not be more than 24 inches (610 mm) on center horizontally. In Seismic Zones 3 and 4, wall ties shall have a lip or hook on the extended leg that will engage or enclose a horizontal joint reinforcement wire having a diameter of 0.148 inch (3.76 mm) (No. 9 B.W. gage) or equivalent. The joint reinforcement shall be continuous with *4-inch (102 mm) lap splices permitted and shall be located in the middle third of the width of the masonry units.*

When applied over stud construction, the studs shall be spaced a maximum of 16 inches (406 mm) on center and approved paper shall first be applied over the sheathing or wires between studs except as otherwise provided in Section 1402A. *The 1-inch (25 mm) space between veneer and its backing shall be solidly filled with a fine grout as specified in Section 2103A.4.2. Grout shall be poured in lifts not exceeding 16 inches (406 mm) in height.*

As an alternate, an air space of at least 1 inch (25 mm) may be maintained between the backing and the veneer in which case spot bedding at all ties shall be of cement mortar.

2. *If spot bedding of cement mortar is not used to entirely surround the ties and to provide a fill bed to solid backing, anchorage shall be No. 14 gage galvanized anchors in No. 22 gage galvanized anchor slots of a type which will not open under direct load. Slots shall be secured to the studs with 10d common hot-dipped galvanized nails at 12-inch-maximum (305 mm) spacing, or by 1-inch-wide (25 mm) No. 14 gage galvanized angle ties fastened to each stud with a 2-inch (51 mm), No. 10 galvanized screw placed not more than ¹/₄ inch (6.4 mm) above the extended leg of the angle tie. Anchors shall be capable of taking tension or compression without end play.*

1403A.6.4.3 Stone units [10 inches (254 mm) maximum in thickness]. Stone veneer units not exceeding 10 inches (254 mm) in thickness may be anchored directly to structural masonry or concrete *in the following manner:*

∗ ∗ ∗ Anchor ties shall not be less than 0.109 inch (2.77 mm) (No. 12 B.W. gage) galvanized wire, or approved equal, formed as an exposed eye and extending not less than ¹/₂ inch (13 mm) beyond the face of the backing. The legs of the loops shall not be less than 6 inches (152 mm) in length bent at right angles and laid in the masonry mortar joint and spaced so that the eyes or loops are 12 inches (254 mm) maximum on center in both directions. There shall be provided not less than a 0.109 inch (2.77 mm) (No. 12 B.W. gage) galvanized wire tie, or approved equal, threaded through the exposed loops for every 2 square feet (0.19 m²) of stone veneer. This tie shall be a loop having legs not less than 15 inches (381 mm) in length so bent that it will lie in the stone veneer mortar joint. The last 2 inches (51 mm) of each wire leg shall have a right angle bend. One inch (25 mm) of cement grout shall be placed between the backing and the stone veneer.

1403A.6.4.4 Slab-type units [2 inches (51 mm) maximum in thickness]. *All veneer units of marble, travertine, granite or other stone units of slab form shall be 2 inches (51 mm) maximum in thickness and 1¹/₄ inches (32 mm) in minimum thickness for exterior and ³/₄ inch (19 mm) minimum thickness for interior use and shall be set a minimum of 1 inch (25 mm) clear of the backing, and the space shall be solidly filled with fine grout or each piece shall be set rigidly against spot bedding pads. Spot bedding pads shall be of cement mortar not less than 6 inches (152 mm) in diameter and shall be located at each anchor tie and over the back face of the veneer at a maximum spacing of 18 inches (457 mm) on center. The spot bedding at ties shall entirely surround the ties.*

Ties shall engage drilled eyes of corrosion-resistant metal dowels of ¹/₈-inch (3.2 mm) diameter or more penetrating at least ¹/₂ inch (13 mm) into the edge of the veneer located in the middle third of the edge of the units spaced a maximum of 18 inches (457 mm) apart around the periphery of each unit with not less than four ties per veneer unit. Units shall not exceed 20 square feet (1.9 m²) in area.

If the dowels are not tightfitting, the holes may be drilled not more than ¹/₁₆ inch (1.6 mm) larger in diameter than the dowel with the hole countersunk to a diameter and depth equal to twice the diameter of the dowel in order to provide a tight-fitting key of cement mortar at the dowel locations when the mortar in the joint has set.

All veneer ties shall be corrosion-resistant metal capable of resisting in tension or compression a force equal to two times the weight of the attached veneer.

If made of sheet metal, veneer ties shall have a minimum size of No. 16 gage by 1 inch (25 mm) or, if made of wire, not smaller in diameter than 0.148 inch (3.76 mm) (No. 9 B.W. gage) wire.

Anchored products which do not comply with these requirements shall be submitted to the enforcement agency for evaluation and approval.

1403A.6.4.5 Unit size limitations. *The thickness of anchored unit masonry and stone veneer, including grout, shall not be less than 3 inches (76 mm).*

SECTION 1404A — VINYL SIDING

1404A.1 Vinyl siding conforming to the requirements of this section and complying with U.B.C. Standard 14-2 may be installed on exterior walls of buildings of Type V construction located in areas where the wind speed specified in Figure 16-1 does not exceed 80 miles per hour (129 km/h) and the building height is less than 40 feet (12 192 mm) in Exposure C. If construction is located in areas where wind speed exceeds 80 miles per hour (129 km/h), or building heights are in excess of 40 feet (12 192 mm), data indicating compliance with Chapter 16 must be submitted. Vinyl siding shall be secured to the building so as to provide weather protection for the exterior walls of the building.

1404A.2 Application. The siding shall be applied over sheathing or materials listed in Section 2320A. Siding shall be applied to conform with the weather-resistive barrier requirements in Section 1402A.1. Siding and accessories shall be installed in accordance with approved manufacturer's instructions.

Nails used to fasten the siding and accessories shall have a minimum ³/₈-inch (9.5 mm) head diameter and 0.120-inch (3.05 mm) shank diameter. The nails shall be corrosion resistant and shall be long enough to penetrate the studs or nailing strip at least ³/₄ inch (19 mm). Where the siding is installed horizontally, the fastener spacing shall not exceed 16 inches (406 mm) horizontally and 12 inches (305 mm) vertically. Where the siding is installed vertically, the fastener spacing shall not exceed 12 inches (305 mm) horizontally and 12 inches (305 mm) vertically.

SECTION 1405A — INSPECTION

1405A.1 *All veneer shall be continuously inspected during application by an inspector specially approved for that purpose by the enforcement agency.*

TABLE 14A-A—CERAMIC TILE SETTING MORTARS

COAT		VOLUME TYPE 1 PORTLAND CEMENT	VOLUME TYPE S HYDRATED LIME	VOLUME U.B.C. STD. 24-21 SAND		MAXIMUM THICKNESS OF COAT (inches)	MINIMUM INTERVAL BETWEEN COATS (hours)
				Dry	Damp	× 25.4 for mm	
1. Walls and ceilings over 10 sq. ft. (0.93 m²)	Scratch	1	$^1/_2$	4	5	$^3/_8$	24
		1	0	3	4	$^3/_8$	24
	Float or leveling	1	$^1/_2$	4	5	$^3/_4$	24
		1	1	6	7	$^3/_4$	24
2. Walls and ceilings 10 sq. ft. (0.93 m²) or less	Scratch and float	1	$^1/_2$	$2^1/_2$	3	$^3/_8$ $^3/_4$	24
3. Floors	Setting bed	1	0	5	6	$1^1/_4$	—
		1	$^1/_{10}$	5	6	$1^1/_4$	—

(Text continues on page 1–219.)

TABLE 14A-A
CA-UBC 1-218.7

Chapter 31

SPECIAL CONSTRUCTION

SECTION 3101 — SCOPE

The provisions of this chapter shall apply to special construction described herein.

SECTION 3102 — CHIMNEYS, FIREPLACES AND BARBECUES

3102.1 Scope. Chimneys, flues, fireplaces and barbecues, and their connections, carrying products of combustion shall conform to the requirements of this section.

[For DSA/SS, OSHPD 1] The quality, design, fabrication and erection of all masonry or concrete chimneys, flues, fireplaces and barbecues shall comply with the requirements set forth in this chapter.

> *NOTE: See Sections 1928A and 2105A for testing and inspection.*

3102.2 Definitions.

BARBECUE is a stationary open hearth or brazier, either fuel fired or electric, used for food preparation.

CHIMNEY is a hollow shaft containing one or more passageways, vertical or nearly so, for conveying products of combustion to the outside atmosphere.

CHIMNEY CLASSIFICATIONS:

Chimney, High-heat Industrial Appliance-type, is a factory-built, masonry or metal chimney suitable for removing the products of combustion from fuel-burning high-heat appliances producing combustion gases in excess of 2,000°F. (1093°C.) measured at the appliance flue outlet.

Chimney, Low-heat Industrial Appliance-type, is a factory-built, masonry or metal chimney suitable for removing the products of combustion from fuel-burning low-heat appliances producing combustion gases not in excess of 1,000°F. (538°C.) under normal operating conditions but capable of producing combustion gases of 1,400°F. (760°C.) during intermittent forced firing for periods up to one hour. All temperatures are measured at the appliance flue outlet.

Chimney, Medium-heat Industrial Appliance-type, is a factory-built, masonry or metal chimney suitable for removing the products of combustion from fuel-burning medium-heat appliances producing combustion gases not in excess of 2,000°F. (1093°C.) measured at the appliance flue outlet.

Chimney, Residential Appliance-type, is a factory-built or masonry chimney suitable for removing products of combustion from residential-type appliances producing combustion gases not in excess of 1,000°F. (538°C.) measured at the appliance flue outlet.

CHIMNEY CONNECTOR is the pipe or breeching which connects a fuel-burning appliance to a chimney. (See Chapter 9, Mechanical Code.)

CHIMNEY, FACTORY-BUILT, is a chimney manufactured at a location other than the building site and composed of listed factory-built components assembled in accordance with the terms of the listing to form the completed chimney.

CHIMNEY LINER is a lining material of fireclay tile or approved fireclay refractory brick. For guideline standard on fireclay refractory brick see Sections 3502 and 3503, ASTM C 27, Fireclay Refractories.

FIREBRICK is a refractory brick.

FIREPLACE is a hearth and fire chamber or similar prepared place in which a fire may be made and which is built in conjunction with a chimney.

Factory-built Fireplace is a listed assembly of a fire chamber, its chimney and related factory-made parts designed for unit assembly without requiring field construction. Factory-built fireplaces are not dependent on mortar-filled joints for continued safe use.

Masonry Fireplace is a hearth and fire chamber of solid masonry units such as bricks, stones, masonry units, or reinforced concrete provided with a suitable chimney.

MASONRY CHIMNEY is a chimney of masonry units, bricks, stones or listed masonry chimney units lined with approved flue liners. For the purpose of this chapter, masonry chimneys shall include reinforced concrete chimneys.

3102.3 Chimneys, General.

3102.3.1 Chimney support. Chimneys shall be designed, anchored, supported and reinforced as required in this chapter and applicable provisions of Chapters 16, 18, 19, 21 and 22 of this code. A chimney shall not support any structural load other than its own weight unless designed as a supporting member.

3102.3.2 Construction. Each chimney shall be so constructed as to safely convey flue gases not exceeding the maximum temperatures for the type of construction as set forth in Table 31-B and shall be capable of producing a draft at the appliance not less than that required for safe operation.

3102.3.3 Clearance. Clearance to combustible material shall be as required by Table 31-B.

3102.3.4 Lining. When required by Table 31-B, chimneys shall be lined with fireclay flue tile, firebrick, molded refractory units or other approved lining not less than $^5/_8$ inch (15.9 mm) thick as set forth in Table 31-B. Chimney liners shall be carefully bedded in approved mortar with close-fitting joints left smooth on the inside.

3102.3.5 Area. Chimney passageways shall not be smaller in area than the vent connection on the appliance attached thereto or not less than that set forth in Table 31-A, unless engineering methods approved by the building official have been used to design the system.

3102.3.6 Height and termination. Every chimney shall extend above the roof and the highest elevation of any part of a building as shown in Table 31-B. For altitudes over 2,000 feet (610 m), the building official shall be consulted in determining the height of the chimney.

3102.3.7 Cleanouts. Cleanout openings shall be provided within 6 inches (152 mm) of the base of every masonry chimney.

3102.3.8 Spark arrester. Where determined necessary by the building official due to local climatic conditions or where sparks escaping from the chimney would create a hazard, chimneys attached to any appliance or fireplace that burns solid fuel shall be equipped with an approved spark arrester. The net free area of the spark arrester shall not be less than four times the net free area of the outlet of the chimney. The spark arrester screen shall have heat and corrosion resistance equivalent to (0.109 inch (2.77 mm) (No. 12 B.W. gage) wire, 0.042 inch (1.07 mm) (No. 19 B.W. gage) galvanized wire or 0.022 inch (0.56 mm) (No. 24 B.W. gage) stainless steel. Openings shall not permit the passage of spheres having a diameter larger than $^1/_2$ inch (12.7 mm) and shall not block the passage of spheres having a diameter of less than $^3/_8$ inch (9.5 mm).

Chimneys used with fireplaces or heating appliances in which solid or liquid fuel is used shall be provided with a spark arrester as required in the Fire Code.

> **EXCEPTION:** Chimneys which are located more than 200 feet (60 960 mm) from any mountainous, brush-covered or forest-covered land or land covered with flammable material and are not attached to a structure having less than a Class C roof covering, as set forth in Chapter 15.

3102.3.8a [For SFM] Spark arrester. *All chimneys attached to any appliance or fireplace that burns solid fuel shall be equipped with an approved spark arrester. The net free area of the spark arrester shall not be less than four times the net free area of the outlet of the chimney. The spark arrester screen shall have heat and corrosion resistance equivalent to 12 gage wire, 19 gage galvanized wire or 24 gage stainless steel. Openings shall not permit the passage of spheres having a diameter larger than $^1/_2$ inch (12.7 mm) and shall not block the passage of spheres having a diameter of less than $^3/_8$ inch (9.5 mm).*

3102.4 Masonry Chimneys.

3102.4.1 Design. Masonry chimneys shall be designed and constructed to comply with Sections 3102.3.2 and 3102.4.2.

3102.4.2 Walls. Walls of masonry chimneys shall be constructed as set forth in Table 31-B.

3102.4.3 Reinforcing and seismic anchorage. Unless a specific design is provided, every masonry or concrete chimney in Seismic Zones 2, 3 and 4 shall be reinforced with not less than four No. 4 steel reinforcing bars conforming to the provisions of Chapter 19 or 21 of this code. The bars shall extend the full height of the chimney and shall be spliced in accordance with the applicable requirements of Chapter 19 or 21. In masonry chimneys the vertical bars shall have a minimum cover of $^1/_2$ inch (13 mm) of grout or mortar tempered to a pouring consistency. The bars shall be tied horizontally at 18-inch (457 mm) intervals with not less than $^1/_4$-inch-diameter (6.4 mm) steel ties. The slope of the inclined portion of the offset in vertical bars shall not exceed 2 units vertical in 1 unit horizontal (200% slope). Two ties shall also be placed at each bend in vertical bars. Where the width of the chimney exceeds 40 inches (1016 mm), two additional No. 4 vertical bars shall be provided for each additional flue incorporated in the chimney or for each additional 40 inches (1016 mm) in width or fraction thereof.

In Seismic Zones 2, 3 and 4, all masonry and concrete chimneys shall be anchored at each floor or ceiling line more than 6 feet (1829 mm) above grade, except when constructed completely within the exterior walls of the building. Anchorage shall consist of two $^3/_{16}$-inch by 1-inch (4.8 mm by 25 mm) steel straps cast at least 12 inches (305 mm) into the chimney with a 180-degree bend with a 6-inch (152 mm) extension around the vertical reinforcing bars in the outer face of the chimney.

Each strap shall be fastened to the structural framework of the building with two $^1/_2$-inch-diameter (12.7 mm) bolts per strap. Where the joists do not head into the chimney, the anchor strap shall be connected to 2-inch by 4-inch (51 mm by 102 mm) ties crossing a minimum of four joists. The ties shall be connected to each joist with two 16d nails. As an alternative to the 2-inch by 4-inch (51 mm by 102 mm) ties, each anchor strap shall be connected to the structural framework by two $^1/_2$-inch-diameter (12.7 mm) bolts in an approved manner.

3102.4.3.1 [For DSA/SS, OSHPD 1 & 2] Reinforcing and seismic anchorage. *Every element of a masonry or concrete chimney, flue, fireplace or barbecue that extends 6 feet (1829 mm) or more above grade or that is part of a building shall be designed and constructed in accordance with Chapters 16, 18, 19, 21 and 22.*

3102.4.4 Chimney offset. Masonry chimneys may be offset at a slope of not more than 4 inches in 24 inches (102 mm in 610 mm), but not more than one third of the dimension of the chimney, in the direction of the offset. The slope of the transition from the fireplace to the chimney shall not exceed 2 units vertical in 1 unit horizontal (200% slope).

3102.4.5 Change in size or shape. Changes in the size or shape of a masonry chimney, where the chimney passes through the roof, shall not be made within a distance of 6 inches (152 mm) above or below the roof joists or rafters.

3102.4.6 Separation of masonry chimney passageways. Two or more flues in a chimney shall be separated by masonry not less than 4 inches (102 mm) thick bonded into the masonry wall of the chimney.

3102.4.7 Inlets. Every inlet to any masonry chimney shall enter the side thereof and shall be of not less than $^1/_8$-inch-thick (3.2 mm) metal or $^5/_8$-inch-thick (16 mm) refractory material. Where there is no other opening below the inlet other than the cleanout, a masonry plug shall be constructed in the chimney not more than 16 inches (406 mm) below the inlet and the cleanout shall be located where it is accessible above the plug. If the plug is located less than 6 inches (152 mm) below the inlet, the inlet may serve as the cleanout.

3102.5 Factory-built Chimneys and Fireplaces. *[For SFM, DSA/SS, OSHPD 1 & 2] All factory-built chimneys shall be designed and constructed in accordance with Chapters 16, 18, 19, 21 and 22 or shall be as specifically approved by the enforcement agency.*

C L
A L
C L
A L
C L

3102.5.1 General. Factory-built chimneys and factory-built fireplaces shall be listed and shall be installed in accordance with the terms of their listings and the manufacturer's instructions as specified in the Mechanical Code.

3102.5.2 Hearth extensions. Hearth extensions of listed factory-built fireplaces shall conform to the conditions of listing and the manufacturer's installation instructions.

3102.5.3 Multiple venting in vertical shafts. Factory-built chimneys utilized with listed factory-built fireplaces may be used in a common vertical shaft having the required fire-resistance rating.

3102.6 Metal Chimneys. Metal chimneys shall be constructed and installed to meet the requirements of the *[for SFM, DSA/SS, OSHPD 1 & 2] California* Mechanical Code.

C
A

Metal chimneys shall be anchored at each floor and roof with two $1^1/_2$-inch by $^1/_8$-inch (38 mm by 3.2 mm) metal straps looped around the outside of the chimney installation and nailed with not less than six 8d nails per strap at each joist.

3102.7 Masonry and Concrete Fireplaces and Barbecues.

3102.7.1 General. Masonry fireplaces, barbecues, smoke chambers and fireplace chimneys shall be of masonry or reinforced concrete and shall conform to the requirements of this section.

3102.7.2 Support. *[For DSA/SS, OSHPD 1 & 2] All designs shall be shown on the approved plans and specifications.* Masonry fireplaces shall be supported on foundations designed as specified in Chapters 16, 18 and 21.

C
A
C

When an approved design is not provided, foundations for masonry and concrete fireplaces shall not be less than 12 inches (305 mm) thick, extend not less than 6 inches (152 mm) outside the fireplace wall and project below the natural ground surface in accordance with the depth of foundations set forth in Table 18-I-D.

3102.7.3 Fireplace walls. Masonry walls of fireplaces shall not be less than 8 inches (203 mm) in thickness. Walls of fireboxes shall not be less than 10 inches (254 mm) in thickness, except that where a lining of firebrick is used, such walls shall not be less than a total of 8 inches (203 mm) in thickness. The firebox shall not be less than 20 inches (508 mm) in depth. Joints in firebrick shall not exceed $^1/_4$ inch (6 mm).

> **EXCEPTION:** For Rumford fireplaces, the depth may be reduced to 12 inches (305 mm) when:
>
> 1. The depth is at least one third the width of the fireplace opening.
>
> 2. The throat is at least 12 inches (305 mm) above the lintel and is at least $^1/_{20}$ of the cross-sectional area of the fireplace opening.

3102.7.4 Hoods. Metal hoods used as part of a fireplace or barbecue shall not be less than 0.036 inch (0.92 mm) (No. 19 carbon sheet steel gage) copper, galvanized steel or other equivalent corrosion-resistant ferrous metal with all seams and connections of smokeproof unsoldered constructions. The hoods shall be sloped at an angle of 45 degrees or less from the vertical and shall extend horizontally at least 6 inches (152 mm) beyond the limits of the firebox. Metal hoods shall be kept a minimum of 18 inches (457 mm) from combustible materials unless approved for reduced clearances.

3102.7.5 Metal heat circulators. Approved metal heat circulators may be installed in fireplaces.

3102.7.6 Smoke chamber. Front and side walls shall not be less than 8 inches (203 mm) in thickness. Smoke chamber back walls shall not be less than 6 inches (152 mm) in thickness.

3102.7.7 Chimneys. Chimneys for fireplaces shall be constructed as specified in Sections 3102.3, 3102.4 and 3102.5 for residential-type appliances.

3102.7.8 Clearance to combustible material. Combustible materials shall not be placed within 2 inches (51 mm) of fireplace, smoke chamber or chimney walls. Combustible material shall not be placed within 6 inches (152 mm) of the fireplace opening. No such combustible material within 12 inches (305 mm) of the fireplace opening shall project more than $1/8$ inch (3 mm) for each 1-inch (25 mm) clearance from such opening.

No part of metal hoods used as part of a fireplace or barbecue shall be less than 18 inches (457 mm) from combustible material. This clearance may be reduced to the minimum requirements specified in the Mechanical Code.

3102.7.9 Areas of flues, throats and dampers. The net cross-sectional area of the flue and of the throat between the firebox and the smoke chamber of a fireplace shall not be less than as set forth in Table 31-A. Metal dampers equivalent to not less than 0.097 inch (2.46 mm) (No. 12 carbon sheet metal gage) steel shall be installed. When fully opened, damper openings shall not be less than 90 percent of the required flue area.

3102.7.10 Lintel. Masonry over the fireplace opening shall be supported by a noncombustible lintel.

3102.7.11 Hearth. Masonry fireplaces shall be provided with a brick, concrete, stone or other approved noncombustible hearth slab. This slab shall not be less than 4 inches (102 mm) thick and shall be supported by noncombustible materials or reinforced to carry its own weight and all imposed loads. Combustible forms and centering shall be removed.

3102.7.12 Hearth extensions. Hearths shall extend at least 16 inches (406 mm) from the front of, and at least 8 inches (203 mm) beyond each side of, the fireplace opening. Where the fireplace opening is 6 square feet (0.56 m²) or larger, the hearth extension shall extend at least 20 inches (508 mm) in front of, and at least 12 inches (305 mm) beyond each side of, the fireplace opening.

Except for fireplaces which open to the exterior of the building, the hearth slab shall be readily distinguishable from the surrounding or adjacent floor.

3102.7.13 Fire blocking. Fire blocking between chimneys and combustible construction shall meet the requirements specified in Section 708.

SECTION 3103 — TEMPORARY BUILDINGS OR STRUCTURES

Temporary buildings or structures such as reviewing stands and other miscellaneous structures, sheds, canopies or fences used for the protection of the public around and in conjunction with construction work may be erected by special permit from the building official for a limited period of time. Such buildings or structures need not comply with the type of construction or fire-resistive time periods required by this code. Temporary buildings or structures shall be completely removed upon the expiration of the time limit stated in the permit.

TABLE 31-A—MINIMUM PASSAGEWAY AREAS FOR MASONRY CHIMNEYS[1]

Type of Masonry Chimney	MINIMUM CROSS-SECTIONAL AREA		
	× 645 for mm²		
	Tile Lined		Lined with Firebrick or Unlined
	Round	Square or Rectangle	
1. Residential	50 sq. in.	50 sq. in.	85 sq. in.
2. Fireplace[2]	$^1/_{12}$ of opening Minimum 50 sq. in.	$^1/_{10}$ of opening Minimum 64 sq. in.	$^1/_8$ of opening Minimum 100 sq. in.
3. Low heat	50 sq. in.	57 sq. in.	135 sq. in.

4. Incinerator
Apartment type
   1 opening
   2 to 6 openings
   7 to 14 openings
   15 or more openings
	196 sq. in.
	324 sq. in.
	484 sq. in.
	484 sq. in. plus 10 sq. in. for each additional opening
			Not applicable

[1] Areas for medium- and high-heat chimneys shall be determined using accepted engineering methods and as approved by the building official.

[2] Where fireplaces open on more than one side, the fireplace opening shall be measured along the greatest dimension.

NOTE: For altitudes over 2,000 feet (610 m) above sea level, the building official shall be consulted in determining the area of the passageway.

TABLE 31-A
CA-UBC 1-376.1

1994 UBC - 1995 CA-UBC - FIREPLACES

221

THE AMERICAN SOCIETY FOR TESTING AND MATERIALS

ASTM, founded in 1898, is a developer and publisher of technical information designed to promote the understanding and development of technology and to ensure the quality of commodities, services and safety of products.

ASTM's primary mission is to develop voluntary full consensus standards for materials, products, systems and services. It provides a forum for producers, users, ultimate consumers and those having a general interest (representatives of government and academia) to meet on common ground to write standards that best meet their needs.

The Society operates through more than 132 main technical committees. These committees function in prescribed fields under regulations that ensure balanced representation among producers, consumers and general interest participants.

The Society currently has 33,000 active members of which approximately 22,000 serve as technical experts on committees.

Membership in the society is open to all concerned with the fields of work in which ASTM is active. Additional information may be obtained from ASTM, Member Service, American Society for Testing and Materials, 100 Barr Harbor Drive, West Conshohocken, PA 19428-2959.

These Standards are reprinted with permission of the American Society for Testing and Materials.

Standard Specification for
Masonry Joint Reinforcement[1]

This standard is issued under the fixed designation A 951; the number immediately following the designation indicates the year of original adoption or, in the case of revision, the year of last revision. A number in parentheses indicates the year of last reapproval. A superscript epsilon (ε) indicates an editorial change since the last revision or reapproval.

1. Scope

1.1 This specification covers masonry joint reinforcement fabricated from cold drawn steel wire. Joint reinforcement consists of longitudinal wires welded to cross wires.

1.2 The values stated in inch-pound units are to be regarded as the standard. The values given in parentheses are for information only.

2. Referenced Documents

2.1 *ASTM Standards:*
A 82 Specification for Steel Wire, Plain, for Concrete Reinforcement[2]
A 153 Specification for Zinc Coating (Hot-Dipped) on Iron and Steel Hardware[3]
A 185 Specification for Steel Welded Wire Fabric, Plain, for Concrete Reinforcement[2]
A 496 Specification for Steel Wire, Deformed, for Concrete Reinforcement[2]
A 580 Specification for Stainless Steel Wire[4]
A 641 Specification for Zinc-Coated (Galvanized) Carbon Steel Wire[3]

3. Ordering Information

3.1 Orders for material to this specification should include the following information:
3.1.1 Quantity (linear feet) (metres),
3.1.2 Type (truss, ladder),
3.1.3 Width (nominal thickness of masonry wall),
3.1.4 Wire size and wire specification,
3.1.5 Finish (see Section 6),
3.1.6 Packaging (see Section 12), and
3.1.7 ASTM designation number and year of issue.

NOTE 1—A typical ordering description is as follows:
10 000 ft joint reinforcement, No. 8 standard (W1.7) truss, wire to ASTM A 82 –_____, hot-dip galvanized to ASTM A 153 –_____, conforming to ASTM A 951 –_____.

4. Materials and Manufacture

4.1 Wire used in the manufacture of masonry joint reinforcement shall conform to Specification A 82 or A 580, Type 304, except as modified herein.

4.2 Masonry joint reinforcement shall be assembled by automatic machines or by other suitable mechanical means

that will assure accurate spacing and alignment of all members of the finished product.

4.3 Longitudinal and cross wires shall be securely connected at every intersection by an electric-resistance welding process that includes both fusion welding together with applied pressure to join the materials.

4.4 Longitudinal wires shall be deformed. One set of two deformations shall occur around the perimeter of the wire at a maximum spacing of 0.7 times the diameter of the wire but not less than eight sets per inch (25 mm) of length.

NOTE 2—Wire used for joint reinforcement is knurled to form deformations and as such it does not come under the deformation requirements in Specification A 496.

5. Mechanical Requirements

5.1 *Tensile Properties*—Wire used in the fabrication of masonry joint reinforcement shall conform to the requirements of Table 1 based on nominal area of the wire.

5.2 *Tension Tests:*
5.2.1 Tension tests shall be made on individual wires cut from the finished product and tested either across or between the welds. No less than 50 % shall be across welds.

5.2.2 Tension tests across a weld shall have the welded joint located approximately at the center of the wire being tested.

5.2.3 Tensile strength shall be the average of four test values determined by dividing the maximum test load by the specified nominal cross-sectional area of the wire.

5.3 *Reduction of Area*—The ruptured section of the tensile specimen is measured to determine this property. The measurement shall be made only when rupture has occurred at a sufficient distance from the center of the weld to permit an accurate measurement of the fractured section. Additional testing is permitted when a suitable ruptured section is not obtained from the initial test. The wire shall meet the minimum reduction of area requirements of Table 1.

5.4 *Weld Shear Strength*—The least weld shear strength in pounds-force shall be not less than 25 000 multiplied by the specified nominal area of the larger wire in square inches (or in Newtons shall not be less than 172 multiplied by the nominal area in square millimetres).

NOTE 3—Since industry practice is to use butt welds in the manufacture of joint reinforcement, the weld shear strength in pounds-force is prescribed as 25 000 times the area of the larger wire rather than 35 000 times the area of the larger wire.

5.5 *Weld Shear Strength Tests:*

TABLE 1 Tension Test Requirements

Tensile strength, min, ksi (MPa)	80 (550)
Yield strength, min, ksi (MPa)	70 (485)
Reduction of area, min, %	30

---

[1] This specification is under the jurisdiction of ASTM Committee A-1 on Steel, Stainless Steel, and Related Alloys and is the direct responsibility of Subcommittee A01.05 on Steel Reinforcement.
Current edition approved March 10, 1996. Published May 1996.
[2] *Annual Book of ASTM Standards*, Vol 01.04.
[3] *Annual Book of ASTM Standards*, Vol 01.06.
[4] *Annual Book of ASTM Standards*, Vol 01.03.

5.5.1 Test specimens shall be obtained from the finished product by cutting a section of longitudinal wire that includes one weld.

5.5.2 Weld shear strength tests shall be conducted using a fixture of such design as to prevent rotation of the cross wire. The cross wire shall be placed in the anvil of the testing device which is secured in the tensile machine and the load then applied to the longitudinal wire.

5.5.3 Weld shear strength shall be the average test load in pounds (Newtons) of four test specimens selected at random.

5.6 *Bend Tests:*

5.6.1 Test specimens shall be obtained from the fabricated product by cutting a section of longitudinal wire without welds.

5.6.2 The test specimens shall be bent at room temperature through 180 degrees around a pin, the diameter of which is equal to the nominal diameter of the specified wire.

5.6.3 The specimen shall not break nor shall there be visible cracks of the base metal on the outside diameter of the bend.

6. Other Requirements

6.1 When corrosion protection of joint reinforcement is provided, it shall be in accordance with one of the following:

6.1.1 *Mill Galvanized*—Zinc coated, in accordance with the hot-dip method of Specification A 641, with a minimum of 0.1 oz of zinc per square foot (30 g/m²) of surface area. The coating may be applied before fabrication.

6.1.2 *Hot-Dip Galvanized*—Zinc coated, by the hot-dip method, in accordance with Specification A 153, Class B average of 1.50 oz of zinc per square foot (458 g/m²)). The coating shall be applied after fabrication.

7. Dimensions and Tolerances

7.1 *Longitudinal Wires*—The minimum size of longitudinal wires shall be W1.1 (11 gage).

7.2 *Cross Wires*—The minimum size of cross wires shall be W1.1 (11 gage). Cross wires shall not project beyond the outside longitudinal wires by more than ⅛ in. (3 mm).

7.3 *Width*—The width of joint reinforcement is defined as the out-to-out distance of outside longitudinal wires. Width shall not vary by more than ⅛ in. (3 mm) from the manufacturer's specified standard dimension. Width shall be measured as follows:

7.3.1 *Ladder Type Joint Reinforcement*—At opposite weld points, and

7.3.2 *Truss Type Joint Reinforcement*—From a weld point on one longitudinal wire perpendicular to a point on the opposite wire between adjacent weld points.

7.4 *Length*—The length of pieces of joint reinforcement shall not vary by more than ½ in. (13 mm) from the specified length.

7.5 *Dimensions:*

7.5.1 The required dimensions shall be measured on three samples of joint reinforcement prior to galvanizing or on three samples from which the galvanizing has been removed.

7.5.2 Measure the diameter of both longitudinal and cross wires to the nearest 0.001 in. (0.03 mm).

7.5.3 Measure the gaps between the ends of deformations around the circumference of the wire to the nearest 0.001 in. (0.03 mm).

8. Number of Tests

8.1 *Number of Tests*—One set of each test described herein shall be performed for each 300 000 linear feet (91 500 m) of joint reinforcement, but not less than one set each week.

9. Inspection

9.1 The inspector representing the purchaser shall have free entry at all times, while work on the contract of the purchaser is being performed, to all parts of the manufacturer's works that concern the manufacture of the material ordered. The manufacturer shall afford the inspector all reasonable facilities to assure that the material is being furnished in accordance with this specification.

9.2 Except for yield strength, all tests and inspections shall be made at the place of manufacture prior to shipment, unless otherwise specified. Such tests shall be conducted so as not to interfere unnecessarily with the operation of the works.

9.3 If the purchaser considers it desirable to determine compliance with the yield strength requirements of Specification A 82, yield strength tests may be made in a recognized laboratory, or their representative may make the test at the mill, if such tests do not interfere unnecessarily with the mill operations.

10. Rejection and Rehearing

10.1 Material that does not meet the requirements of this specification may be rejected. Unless otherwise specified, any rejection shall be reported to the manufacturer within five days from the time of selection of test specimens.

10.2 In case a specimen fails to meet the tension or bend test, the material shall not be rejected until two additional specimens taken from other wires in the same bundle have been tested. The material shall be considered as meeting the specification in respect to any prescribed tensile property, provided the tested average for the three specimens, including the specimen originally tested, is equal to or exceeds the required minimum for the particular property in question, and further provided that none of the three specimens develops less than 80 % of the required minimum for the tensile property in question. The material shall be considered as meeting this specification in respect to bend test requirements provided both additional specimens satisfactorily pass the prescribed bend test.

10.3 Any material that shows injurious imperfections subsequent to its acceptance at the manufacturer's works may be rejected and the manufacturer shall be promptly notified.

10.4 Welded joints shall withstand normal shipping and handling without becoming broken, but the presence of broken welds, regardless of cause, shall not constitute cause for rejection unless the number of broken welds per bundle exceeds 1 % of the total number of joints in a bundle.

10.5 In the event of rejection because of failure to meet the weld shear requirements, four additional specimens shall be taken from four different bundles and tested in accordance with 5.5. If the average of all the weld shear tests performed does not meet the requirement, the material shall be rejected.

10.6 In the event of rejection because of failure to meet

the requirements for dimensions, the amount of material rejected shall be limited to those bundles which fail to meet this specification.

10.7 Rust, surface seams, or surface irregularities will not be cause for rejection provided the minimum dimensions, cross-sectional area, and tensile properties of a hand wire brushed test specimen are not less than the requirements of this specification.

10.8 *Rehearing*—Rejected materials shall be preserved for a period of at least two weeks from the date of inspection, during which time the manufacturer may make claim for a rehearing and retesting.

11. Certification

11.1 If outside inspection is waived, a manufacturer's certification that the material has been tested in accordance with and meets the requirements of this specification shall be the basis of acceptance of the material. The certification shall include the specification number, year-date of issue, and revision letter, if any.

12. Packaging and Marking

12.1 Joint reinforcement shall be assembled in bundles containing 250 to 500 linear feet and securely fastened together.

12.2 Each bundle shall have attached thereto a suitable tag bearing the name of the manufacturer, description of the material, ASTM A 951, and other such information as may be specified by the purchaser.

13. Keywords

13.1 cross wires; deformed; galvanized wire; hot-dipped galvanized; joint reinforcement; ladder type; longitudinal wires; tensile properties; truss type; weld shear strength

APPENDIX

(Nonmandatory Information)

X1. WELD SHEAR TESTING

X1.1 *Scope*—This appendix provides information leading to a better understanding of the purpose and significance of the weld shear strength provisions of this specification.

X1.2 *Background*—Joint reinforcement has been used in the masonry industry since 1940. For most of the period since then, its manufacture has been limited to a relatively small group of producers and users who simply referred to "manufacturers' recommendations" as the standard of quality and acceptance. With the adoption of a new consensus standard for the design of masonry, it became clear that a standard for the manufacture of joint reinforcement was needed. In developing this standard it was decided to use a format similar to that used for the ASTM Standard for Welded Wire Fabric, Plain, for Concrete Reinforcement, Specification A 185, since many people had the notion that joint reinforcement was used in a manner similar to wire mesh. A significant difference between wire mesh and joint reinforcement arose when an attempt was made to fashion the requirements for weld shear strength after those in Specification A 185.

X1.3 *Manufacturing Differences*—Welded wire mesh is manufactured with lap welds while almost all of the manufacturers of joint reinforcement use butt welds so that the total thickness of material at a weld is as small as possible.

This is important since there is not much room to install joint reinforcement in conventional mortar bed joints. In addition, virtually all of the wire mesh manufactured is square or rectangular with intersecting wires perpendicular to each other. This is not the case with joint reinforcement where the majority of product is produced with a "truss" configuration. Compounding the difference is the fact that the angle of intersection varies for each different width of product produced since the pitch between welds is a constant 16 in.

X1.4 *Weld Tests*—Because the shape of truss type joint reinforcement is so variable, the established test method used for wire mesh does not apply and reference to it has caused problems. It was decided that the best way to handle the testing was to omit any description of the test setup and to simply state the minimum acceptable weld strength, recognizing that this is the method used by manufacturers to ensure that they are producing "satisfactory welds." For ladder type joint reinforcement, manufacturers use a setup similar to that used for wire mesh (see Fig. X1.1). For truss type, the method used is to attach the opposite jaws of a testing machine to the longitudinal and diagonal wires as shown in Fig. X1.2. The strength of the weld used to connect longitudinal and cross wires of joint reinforcement is not critical.

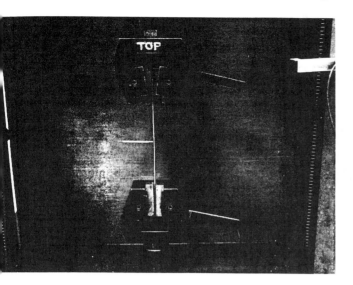

a   b

FIG. X1.1   Test Set-up for Ladder-Type Joint Reinforcement

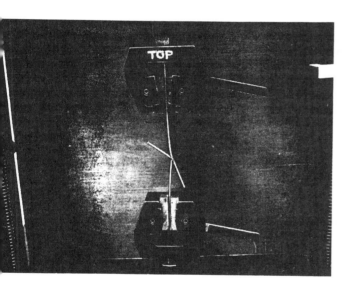
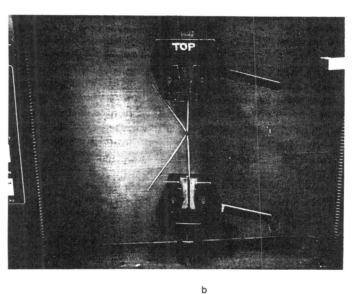

a   b

FIG. X1.2   Test Set-up for Truss-Type Joint Reinforcement

Standard Specification for
Building Brick (Solid Masonry Units Made From Clay or Shale)[1]

This standard is issued under the fixed designation C 62; the number immediately following the designation indicates the year of original adoption or, in the case of revision, the year of last revision. A number in parentheses indicates the year of last reapproval. A superscript epsilon (ε) indicates an editorial change since the last revision or reapproval.

This standard has been approved for use by agencies of the Department of Defense. Consult the DoD Index of Specifications and Standards for the specific year of issue which has been adopted by the Department of Defense.

1. Scope

1.1 This specification covers brick intended for both structural and nonstructural masonry where external appearance is not a requirement. The brick are prismatic units available in a variety of sizes, shapes, textures, and colors. The specification does not cover brick intended for use as facing units or where surface appearance is a requirement, (see Specification C 216). This specification does not cover brick intended for use as paving brick (see Specification C 902).

1.2 The property requirements of this standard apply at the time of purchase. The use of results from testing of brick extracted from masonry structures for determining conformance or non-conformance to the property requirements (Section 3) of this standard is beyond the scope of this standard.

1.3 Brick are manufactured from clay, shale, or similar naturally occurring earthy substances and subjected to a heat treatment at elevated temperatures (firing). The heat treatment must develop sufficient fired bond between the particulate constituents to provide the strength and durability requirements of this specification. (See firing, fired bond, and incipient fusion in Terminology C 43.)

1.4 Brick are shaped during manufacture by molding, pressing, or extrusion, and the shaping method may be used to describe the brick (see Terminology C 43).

1.5 The values stated in inch-pound units are to be regarded as the standard. The values given in parentheses are for information only.

1.6 The text of this standard references notes and footnotes which provide explanatory material. These notes and footnotes (excluding those in tables and figures) shall not be considered as requirements of the standard.

2. Referenced Documents

2.1 *ASTM Standards:*
C 43 Terminology of Structural Clay Products[2]
C 67 Test Methods for Sampling and Testing Brick and Structural Clay Tile[2]
C 216 Specification for Facing Brick (Solid Masonry Units Made from Clay or Shale)[2]

C 902 Specification for Pedestrian and Light Traffic Paving Brick[2]
E 835/E 835M Guide for Modular Coordination of Clay and Concrete Masonry Units[3]

3. Grades

3.1 Grades classify brick according to their resistance to damage by freezing when wet, as defined in Note 1. Three grades are covered and the grade requirements are shown in Table 1.

3.1.1 *Grade SW*—Brick intended for use where high and uniform resistance to damage caused by cyclic freezing is desired and where the brick may be frozen when saturated with water.

3.1.2 *Grade MW*—Brick intended for use where moderate resistance to cyclic freezing damage is permissible or where the brick may be damp but not saturated with water when freezing occurs.

3.1.3 *Grade NW*—Brick with little resistance to cyclic freezing damage but which are acceptable for applications protected from water absorption and freezing.

NOTE 1—The word "saturated," with respect to this standard, refers to the condition of a brick that has absorbed water to an amount equal to that resulting from submersion in room temperature water for 24 h.

4. Physical Properties

4.1 *Appearance*—If brick are required to have a particular color, texture, finish, uniformity, or limits on cracks, warpage or other imperfections detracting from the appearance they are purchased under Specification C 216.

4.2 *Durability*—When Grade is not specified, the requirements for Grade SW shall govern. Unless otherwise specified by the purchaser, brick of Grade SW or MW shall be accepted instead of Grade NW; and Grade SW instead of Grade MW.

4.2.1 *Physical Property Requirements*—The brick shall conform to the physical requirements for the Grade specified as prescribed in Table 1. For the compressive strength requirements in Table 1, test the unit with the compressive force perpendicular to the bed surface of the unit, with the unit in the stretcher position.

4.2.2 *Absorption Alternate*—The saturation coefficient requirement does not apply, provided the cold water absorption of any single unit of a random sample of five brick does not exceed 8 %.

---

[1] This specification is under the jurisdiction of ASTM Committee C-15 on Manufactured Masonry Units and is the direct responsibility of Subcommittee C15.02 on Clay Brick and Structural Clay Tile.
Current edition approved July 10, 1996. Published September 1996. Originally published as C 62 – 27. Last previous edition C 62 – 95a.
[2] *Annual Book of ASTM Standards*, Vol 04.05.

[3] *Annual Book of ASTM Standards*, Vol 04.07.

TABLE 1 Physical Requirements

Designation	Minimum Compressive Strength gross area, psi (MPa)		Maximum Water Absorption by 5-h Boiling, %		Maximum Saturation Coefficient[A]	
	Average of 5 Brick	Individual	Average of 5 Brick	Individual	Average of 5 Brick	Individual
Grade SW	3000 (20.7)	2500 (17.2)	17.0	20.0	0.78	0.80
Grade MW	2500 (17.2)	2200 (15.2)	22.0	25.0	0.88	0.90
Grade NW	1500 (10.3)	1250 (8.6)	no limit	no limit	no limit	no limit

[A] The saturation coefficient is the ratio of absorption by 24-h submersion in cold water to that after 5-h submersion in boiling water.

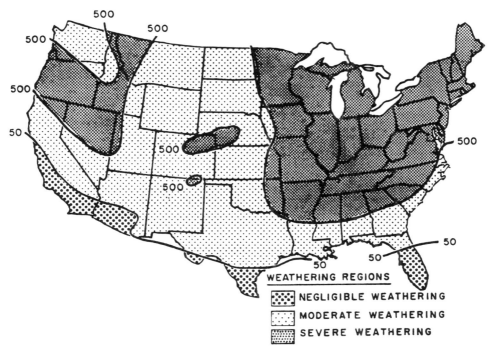

FIG. 1 Weathering Indexes in the United States

4.2.3 *Freezing and Thawing Alternative*—The requirements for 5 h boiling water absorption and saturation coefficient do not apply, provided a sample of five brick, meeting the strength requirements of Table 1, passes the freezing and thawing test as described in the Rating Section of the Freezing and Thawing test procedures of Test Methods C 67:

4.2.3.1 *Grade SW—Weight Loss Requirement*—Not greater than 0.5 % loss in dry weight of any individual unit.

NOTE 2—The 50 cycle freezing and thawing test is specified as an alternative only when brick do not conform to either Table 1 requirements for maximum water absorption and saturation coefficient, or to the requirements of the Absorption Alternate in Section 4.2.2.

4.2.4 *Waiver of Absorption and Saturation Coefficient Requirements*—If the brick are intended for use exposed to weather where the weathering index is less than 50 (see Fig. 1), and unless otherwise specified, the requirements given in Table 1 for 5 h boiling water absorption and for saturation coefficient shall not apply but the minimum average compressive strength requirement of 2500 psi (17.2 MPa) shall apply.

NOTE 3—*Weathering Index* The effect of weathering on brick is related to the weathering index, which for any locality is the product of

the average annual number of *freezing cycle days* and the average annual *winter rainfall* in inches defined as follows:[4]

A Freezing Cycle Day is any day during which the air temperature passes either above or below 32°F (0°C). The average number of freezing cycle days in a year may be taken to equal the difference between the mean number of days during which the minimum temperature was 32°F or below and the mean number of days during which the maximum temperature was 32°F or below.

Winter Rainfall is the sum, in inches, of the mean monthly corrected precipitation (rainfall) occurring during the period between and including the normal date of the first killing frost in the fall and the normal date of the last killing frost in the spring. The winter rainfall for any period is equal to the total precipitation less one tenth of the total fall of snow, sleet, and hail. Rainfall for a portion of a month is prorated.

Figure 1 indicates general areas of the United States in which brick masonry is subject to severe, moderate, and negligible weathering. The severe weathering region has a weathering index greater than 500. The moderate weathering region has a weathering index of 50 to 500. The negligible weathering region has a weathering index of less than 50.

The use of Grade MW brick in a wall area above grade is structurally adequate in the severe weathering region, but Grade SW would provide a higher and more uniform degree of resistance to frost action. The degree of durability called for by Grade SW is not necessary for use in

---

[4] Data needed to determine the weathering index for any locality may be found or estimated from the tables of Local Climatological Data, published by the Weather Bureau, U. S. Department of Commerce.

TABLE 1 - 4.2.4   ASTM C 62   229

wall areas above grade in the moderate weathering region. Grade MW brick performs satisfactorily in wall areas above grade in the no-weathering region, where the average compressive strength of the units is at least 2500 psi (17.2 MPa). Grade SW brick should be used in any region when the units are in contact with the ground, in horizontal surfaces, or in any position where they are likely to be permeated with water.

The recommended correlation between grade of building brick, weathering index, and exposure is found in Table 2. The specifier can use these recommendations or use the grade descriptions and physical requirements along with use exposure and local climatological conditions to select grade.

4.3 *Strength*—When brick are required having strengths greater than prescribed by this specification, the purchaser shall specify minimum strength.

4.4 *Initial Rate of Absorption (IRA)*—Test results for initial rate of absorption (IRA) shall be determined in accordance with the IRA (Suction) (Laboratory Test) of Test Methods C 67 and shall be furnished at the request of the specifier or purchaser. IRA is not a qualifying condition or property of units in this specification. This property is measured in order to assist in mortar selection and material handling in the construction process. See Note 4.

NOTE 4—*Initial Rate of Absorption (Suction)*—Both laboratory and field investigation have shown that strong and watertight joints between mortar and masonry units are not achieved by ordinary construction methods when the units as laid have excessive initial rates of absorption. Mortar that has stiffened somewhat because of loss of excessive mixing water to a unit may not make complete and intimate contact with the second unit, resulting in poor adhesion, incomplete bond, and water-permeable joints of low strength. The IRA of the units is determined by the oven-dried procedure described in the IRA (Suction) (Laboratory Test) of Test Methods C 67. IRA in the field depends on the moisture content of the masonry unit and is determined in accordance with the IRA (Suction)—Field Test of Test Methods C 67. Units having average field IRA exceeding 30 g/min per 30 in² (30 g/min 194 cm²) should have their IRA reduced below 30 g/min per 30 in² prior to laying. They may be wetted immediately before they are laid, but it is preferable to wet them thoroughly 3 to 24 h prior to their use so as to allow time for moisture to become distributed throughout the unit.

5. Size and Coring

5.1 *Size*—The size of brick shall be as specified by the purchaser. The maximum permissible variation in dimensions of individual units shall not exceed those given in Table 3.

NOTE 5—For a list of modular sizes see Guide E 835/E 835M. Sizes listed in this standard are not produced in all parts of the United States. Brick names denoting sizes may be regional and, therefore, may not be included in all reference books. Purchasers should ascertain the size of brick available in their locality and should specify accordingly, stating the desired dimensions (width by height by length).

5.2 *Coring*—Unless otherwise specified in the invitation for bids, brick shall be either solid or cored at the option of

TABLE 2 Grade Recommendations for Face Exposures

Exposure	Weathering Index		
	Less than 50	50 to 500	500 and greater
In vertical surfaces:			
In contact with earth	MW	SW	SW
Not in contact with earth	MW	SW	SW
In other than vertical surfaces:			
In contact with earth	SW	SW	SW
Not in contact with earth	MW	SW	SW

TABLE 3 Permissible Variations in Dimensions

Specified dimension, in. (mm)	Maximum Permissible Variations from Specified Dimension, plus or minus, in. (mm)
Up to 3 (76), incl	³⁄₃₂ (2.4)
Over 3 to 4 (76 to 102), incl	⅛ (3.2)
Over 4 to 6 (102 to 152), incl	³⁄₁₆ (4.8)
Over 6 to 8 (152 to 203), incl	¼ (6.4)
Over 8 to 12 (203 to 305), incl	⁵⁄₁₆ (7.9)
Over 12 to 16 (305 to 406), incl	⅜ (9.5)

the seller. The net cross-sectional area of cored brick in any plane parallel to the surface containing the cores shall be at least 75 % of the gross cross-sectional area measured in the same plane. No part of any hole shall be less than ¾ in. (19.1 mm) from any edge of the brick.

5.3 *Frogging*—Unless otherwise specified in the invitation for bids, one bearing face of each brick may have a recess or panel frog and deep frogs. The recess or panel frog shall not exceed ⅜ in. (9.5 mm) in depth and no part of the recess or panel frog shall be less than ¾ in. (19.1 mm) from any edge of the brick. In brick containing deep frogs, frogs deeper than ⅜ in. (9.5 mm), any cross-section through the deep frogs parallel to the surface containing the deep frogs shall conform to the requirements of 4.2.

6. Sampling and Testing

6.1 For purpose of tests, brick that are representative of the commercial product shall be selected by a competent person appointed by the purchaser, the place or places of selection to be designated when the purchase order is placed. The manufacturer or the seller shall furnish specimens for tests without charge.

6.2 The brick shall be sampled and tested in accordance with Test Methods C 67.

NOTE 6—Unless otherwise specified in the purchase order, the cost of tests is typically borne as follows: If the results of the test show that the brick do not conform to the requirements of this specification, the cost is typically borne by the seller. If the results of the tests show that the brick do conform to the requirements of this specification, the cost is typically borne by the purchaser.

7. Visual Inspection

7.1 The brick, as delivered to the site, shall, by visual inspection, conform to the requirements specified by the purchaser or to the sample or samples approved as the standard of comparison and to the samples passing the tests for physical requirements. Minor indentations or surface cracks incidental to the usual method of manufacture, or the chipping resulting from the customary methods of handling in shipment and delivery, shall not be deemed grounds for rejection.

7.2 The brick shall be free of defects, deficiencies, and surface treatments, including coatings, that would interfere with the proper setting of the brick or significantly impair the strength or performance of the construction.

7.3 Unless otherwise agreed upon between the purchaser and the seller, a delivery of brick is permitted to contain not more than 5 % broken brick.

8. Keywords

8.1 building brick; clay; fired masonry units; masonry

construction; physical properties; shale; solid brick

Designation: C 67 – 96

American Association State Highway and Transportation
Officials Standard
AASHTO No.: T 32-70

Standard Test Methods for
Sampling and Testing Brick and Structural Clay Tile[1]

This standard is issued under the fixed designation C 67; the number immediately following the designation indicates the year of original adoption or, in the case of revision, the year of last revision. A number in parentheses indicates the year of last reapproval. A superscript epsilon (ϵ) indicates an editorial change since the last revision or reapproval.

This standard has been approved for use by agencies of the Department of Defense. Consult the DoD Index of Specifications and Standards for the specific year of issue which has been adopted by the Department of Defense.

1. Scope

1.1 These test methods cover procedures for the sampling and testing of brick and structural clay tile. Although not necessarily applicable to all types of units, tests include modulus of rupture, compressive strength, absorption, saturation coefficient, effect of freezing and thawing, efflorescence, initial rate of absorption and determination of weight, size, warpage, length change, and void area. (Additional methods of test pertinent to ceramic glazed facing tile are included in Specification C 126.)

1.2 The text of this standard references notes and footnotes which provide explanatory material. These notes and footnotes (excluding those in tables and figures) shall not be considered as requirements of the standard.

1.3 The values stated in inch-pound units are to be regarded as the standard. The values given in parentheses are for information only.

1.4 *This standard does not purport to address all of the safety concerns, if any, associated with its use. It is the responsibility of the user of this standard to establish appropriate safety and health practices and determine the applicability of regulatory limitations prior to use.*

2. Referenced Documents

2.1 *ASTM Standards:*
C 43 Terminology of Structural Clay Products[2]
C 126 Specification for Ceramic Glazed Structural Clay Facing Tile, Facing Brick, and Solid Masonry Units[2]
C 150 Specification for Portland Cement[3]
E 4 Practices for Force Verification of Testing Machines[4]
E 6 Terminology Relating to Methods of Mechanical Testing[4]

3. Terminology

3.1 *Definitions*—Terminology E 6 and Terminology C 43 shall be considered as applying to the terms used in these test methods.

4. Sampling

4.1 *Selection of Test Specimens*—For the purpose of these tests, full-size brick, tile, or solid masonry units shall be selected by the purchaser or by his authorized representative. Specimens shall be representative of the whole lot of units from which they are selected and shall include specimens representative of the complete range of colors, textures and sizes in the shipment and shall be free of dirt, mud, mortar, or other foreign materials unassociated with the manufacturing process.

4.2 *Number of Specimens:*

4.2.1 *Brick*—For the modulus of rupture, compressive strength, abrasion resistance, and absorption determinations, at least ten individual brick shall be selected for lots of 1 000 000 brick or fraction thereof. For larger lots, five additional specimens shall be selected from each additional 500 000 brick or fraction thereof. Additional specimens are taken at the discretion of the purchaser.

4.2.2 *Structural Clay Tile*—For the weight determination and for compressive strength and absorption tests, at least five tile shall be selected from each lot of 250 tons (226.8 Mg) or fraction thereof. For larger lots, five additional specimens shall be tested for each 500 tons (453.6 Mg) or fraction thereof. In no case shall less than five tile be taken. Additional specimens are taken at the discretion of the purchaser.

4.3 *Identification*—Each specimen shall be marked so that it may be identified at any time. Markings shall cover not more than 5 % of the superficial area of the specimen.

4.4 *Weight Determination:*

4.4.1 *Drying*—Dry the test specimens in a ventilated oven at 230 to 239°F (110 to 115°C) for not less than 24 h and until two successive weighings at intervals of 2 h show an increment of loss not greater than 0.2 % of the last previously determined weight of the specimen.

4.4.2 *Cooling*—After drying, cool the specimens in a drying room maintained at a temperature of 75 ± 15°F (24 ± 8°C), with a relative humidity between 30 and 70 %. Store the units free from drafts, unstacked, with separate placement, for a period of at least 4 h. Do not use specimens noticeably warm to the touch for any test requiring dry units.

4.4.2.1 An alternative method of cooling the specimens to approximate room temperature shall be permitted as follows: Store units, unstacked, with separate placement, in a ventilated room for a period of 4 h, with a current of air from an electric fan passing over them for a period of at least 2 h.

4.4.3 *Calculations and Report:*

4.4.3.1 Calculate the weight per unit area of a specimen by dividing the total weight in pounds by the average area in square feet of the two faces of the unit as normally laid in a wall.

4.4.3.2 Report results separately for each unit with the

---

[1] These test methods are under the jurisdiction of Committee C-15 on Manufactured Masonry Units and is the direct responsibility of Subcommittee C15.02 on Clay Brick and Structural Clay Tile.

Current edition approved Aug. 10, 1996. Published September 1996. Originally published as C 67 – 37 T and C 112 – 34 T. Last previous edition C 67 – 94.

[2] *Annual Book of ASTM Standards*, Vol 04.05.
[3] *Annual Book of ASTM Standards*, Vol 04.01.
[4] *Annual Book of ASTM Standards*, Vol 03.01.

ASTM C 67

average for five units or more.

4.5 *Removal of Silicone Coatings from Brick Units*—The silicone coatings intended to be removed by this process are any of the various polymeric organic silicone compounds used for water-resistant coatings of brick units. Heat the brick at 950 ± 50°F (510 ± 10°C) in an oxidizing atmosphere for a period of not less than 3 h. The rate of heating and cooling shall not exceed 300°F (149°C) per h.

5. Modulus of Rupture (Flexure Test)

5.1 *Test Specimens*—The test specimens shall consist of whole dry full-size units (see 4.4.1). Five such specimens shall be tested.

5.2 *Procedure:*

5.2.1 Support the test specimen flatwise unless specified and reported otherwise (that is, apply the load in the direction of the depth of the unit) on a span approximately 1 in. (25.4 mm) less than the basic unit length and loaded at midspan. If the specimens have recesses (panels or depressions) place them so that such recesses are on the compression side. Apply the load to the upper surface of the specimen through a steel bearing plate ¼ in. (6.35 mm) in thickness and 1½ in. (38.10 mm) in width and of a length at least equal to the width of the specimen.

5.2.2 Make sure the supports for the test specimen are free to rotate in the longitudinal and transverse directions of the test specimen and adjust them so that they will exert no force in these directions.

5.2.3 *Speed of Testing*—The rate of loading shall not exceed 2000 lbf (8896 N)/min. but this requirement is considered as being met if the speed of the moving head of the testing machine immediately prior to application of the load is not more than 0.05 in. (1.27 mm)/min.

5.3 *Calculation and Report:*

5.3.1 Calculate the modulus of rupture of each specimen as follows:

$$S = 3W(l/2 - x)/bd^2 \qquad (1)$$

where:

S = modulus of rupture of the specimen at the plane of failure, lb/in.^{2} (Pa),

W = maximum load indicated by the testing machine, lbf (N),

l = distance between the supports, in. (mm),

b = net width, (face to face minus voids), of the specimen at the plane of failure, in. (mm),

d = depth, (bed surface to bed surface), of the specimen at the plane of failure, in. (mm), and

x = average distance from the midspan of the specimen to the plane of failure measured in the direction of the span along the centerline of the bed surface subjected to tension, in. (mm).

5.3.2 Report the average of the modulus of rupture determinations of all the specimens tested as the modulus of rupture of the lot.

6. Compressive Strength

6.1 *Test Specimens:*

6.1.1 *Brick*—The test specimens shall consist of dry half brick (see 4.4.1), the full height and width of the unit, with a length equal to one half the full length of the unit ±1 in.

(25.4 mm), except as described below. If the test specimen, described above, exceeds the testing machine capacity, the test specimens shall consist of dry pieces of brick, the full height and width of the unit, with a length not less than one quarter of the full length of the unit, and with a gross cross-sectional area perpendicular to bearing not less than 14 in.^{2} (90.3 cm^2). Test specimens shall be obtained by any method that will produce, without shattering or cracking, a specimen with approximately plane and parallel ends. Five specimens shall be tested.

6.1.2 *Structural Clay Tile*—Test five dry tile specimens in a bearing bed length equal to the width ± 1 in. (25.4 mm); or test full-size units.

6.2 *Capping Test Specimens:*

6.2.1 All specimens shall be dry and cool within the meaning of 4.4.1 and 4.4.2 before any portion of the capping procedure is carried out.

6.2.2 If the surface which will become bearing surfaces during the compression test are recessed or paneled, fill the depressions with a mortar composed of 1 part by weight of quick-hardening cement conforming to the requirements for Type III cement of Specification C 150, and 2 parts by weight of sand. Age the specimens at least 48 h before capping them. Where the recess exceeds ½ in. (12.7 mm), use a brick or tile slab section or metal plate as a core fill. Cap the test specimens using one of the two procedures described in 6.2.3 and 6.2.4.

6.2.3 *Gypsum Capping*—Coat the two opposite bearing surfaces of each specimen with shellac and allow to dry thoroughly. Bed one of the dry shellacked surfaces of the specimen in a thin coat of neat paste of calcined gypsum (plaster of paris) that has been spread on an oiled nonabsorbent plate, such as glass or machined metal. The casting surface plate shall be plane within 0.003 in. (0.076 mm) in 16 in. (406.4 mm) and sufficiently rigid; and so supported that it will not be measurably deflected during the capping operation. Lightly coat it with oil or other suitable material. Repeat this procedure with the other shellacked surface. Take care that the opposite bearing surfaces so formed will be approximately parallel and perpendicular to the vertical axis of the specimen and the thickness of the caps will be approximately the same and not exceeding ⅛ in. (3.18 mm). Age the caps at least 24 h before testing the specimens.

NOTE 1—A rapid-setting industrial type gypsum, such as Hydrocal or Hydrostone, is frequently used for capping.

6.2.4 *Sulfur-Filler Capping*—Use a mixture containing 40 to 60 weight % sulfur, the remainder being ground fire clay or other suitable inert material passing a No. 100 (150-µm) sieve with or without plasticizer. The casting surface plate requirements shall be as described in 6.2.3. Place four 1-in. (25.4-mm) square steel bars on the surface plate to form a rectangular mold approximately ½ in. (12.7 mm) greater in either inside dimension than the specimen. Heat the sulfur mixture in a thermostatically controlled heating pot to a temperature sufficient to maintain fluidity for a reasonable period of time after contact with the surface being capped. Take care to prevent overheating, and stir the liquid in the pot just before use. Fill the mold to a depth of ¼ in. (6.35 mm) with molten sulfur material. Place the surface of the unit to be capped quickly in the liquid, and hold the specimen so that its vertical axis is at right angles to the capping surface. The thickness of the caps shall be approximately the same. Allow the unit to remain undisturbed until solidification is complete. Allow the caps to cool for a minimum of 2 h before testing the specimens.

6.3 *Procedure:*

6.3.1 Test brick specimens flatwise (that is, the load shall be applied in the direction of the depth of the brick). Test structural clay tile specimens in a position such that the load is applied in the same direction as in service. Center the specimens under the spherical upper bearing within 1/16 in. (1.59 mm).

6.3.2 The testing machine shall conform to the requirements of Practices E 4.

6.3.3 The upper bearing shall be a spherically seated, hardened metal block firmly attached at the center of the upper head of the machine. The center of the sphere shall lie at the center of the surface of the block in contact with the specimen. The block shall be closely held in its spherical seat, but shall be free to turn in any direction, and its perimeter shall have at least 1/4 in. (6.35 mm) clearance from the head to allow for specimens whose bearing surfaces are not exactly parallel. The diameter of the bearing surface shall be at least 5 in. (127.00 mm). Use a hardened metal bearing block beneath the specimen to minimize wear of the lower platen of the machine. The bearing block surfaces intended for contact with the specimen shall have a hardness not less than HRC60 (HB 620). These surfaces shall not depart from plane surfaces by more than 0.001 in. (0.03 mm). When the bearing area of the spherical bearing block is not sufficient to cover the area of the specimen, place a steel plate with surfaces machined to true planes within ± 0.001 in. (0.03 mm), and with a thickness equal to at least one third of the distance from the edge of the spherical bearing to the most distant corner between the spherical bearing block and the capped specimen.

6.3.4 *Speed of Testing*—Apply the load, up to one half of the expected maximum load, at any convenient rate, after which, adjust the controls of the machine so that the remaining load is applied at a uniform rate in not less than 1 nor more than 2 min.

6.4 *Calculation and Report:*

6.4.1 Calculate the compressive strength of each specimen as follows:

$$\text{Compressive strength, } C = W/A \qquad (2)$$

where:

C = compressive strength of the specimen, lb/in.^{2} (or kg/cm^2) (or Pa·10^4)

W = maximum load, lbf, (or kgf) (or N), indicated by the testing machine, and

A = average of the gross areas of the upper and lower bearing surfaces of the specimen, in.^{2} (or cm^2).

NOTE 2—When compressive strength is to be based on net area (example: clay floor tile), substitute for A in the above formula the net area, in.^{2} (or cm^2), of the fired clay in the section of minimum area perpendicular to the direction of the load.

7. Absorption

7.1 *Accuracy of Weighings:*

7.1.1 *Brick*—The scale or balance used shall have a capacity of not less than 2000 g, and shall be sensitive to within 0.5 g.

7.1.2 *Tile*—The balance used shall be sensitive to within 0.2 % of the weight of the smallest specimen tested.

7.2 *Test Specimens:*

7.2.1 *Brick*—The test specimens shall consist of half brick conforming to the requirements of 6.1.1. Five specimens shall be tested.

7.2.2 *Tile*—The specimens for the absorption test shall consist of five tile or three representative pieces from each of these five tile. If small pieces are used, take two from the shell and one from an interior web, the weight of each piece being not less than 227 g. The specimens shall have had their rough edges or loose particles ground off and, if taken from tile that have been subjected to compressive strength tests, specimens shall be free of cracks due to failure in compression.

7.3 *5-h and 24-h Submersion Tests:*

7.3.1 *Procedure:*

7.3.1.1 Dry and cool the test specimens in accordance with 4.4.1 and 4.4.2 and weigh each one.

7.3.1.2 *Saturation*—Submerge the dry, cooled specimen, without preliminary partial immersion, in clean water (soft, distilled or rain water) at 60 to 86°F (15.5 to 30°C) for the specified time. Remove the specimen, wipe off the surface water with a damp cloth and weigh the specimen. Complete weighing of each specimen within 5 min after removing the specimen from the bath.

7.3.2 *Calculation and Report:*

7.3.2.1 Calculate the absorption of each specimen as follows:

$$\text{Absorption, \%} = 100(W_s - W_d)/W_d \qquad (3)$$

where:

W_d = dry weight of the specimen, and

W_s = saturated weight of the specimen after submersion in cold water.

7.3.2.2 Report the average absorption of all the specimens tested as the absorption of the lot.

7.4 *1-h, 2-h, and 5-h Boiling Tests:*

7.4.1 *Test Specimens*—The test specimens shall be the same five specimens used in the 5-h or 24-h cold-water submersion test where required and shall be used in the state of saturation existing at the completion of that test.

7.4.2 *Procedure:*

7.4.2.1 Return the specimen that has been subjected to the cold-water submersion to the bath, and subject it to the boiling test as described in 7.4.2.2.

7.4.2.2 Submerge the specimen in clean water (soft, distilled or rain water) at 60 to 86°F (15.5 to 30°C) in such a manner that water circulates freely on all sides of the specimen. Heat the water to boiling, within 1 h, boil continuously for specified time, and then allow to cool to 60 to 86°F (15.5 to 30°C) by natural loss of heat. Remove the specimen, wipe off the surface water with a damp cloth, and weigh the specimen. Complete weighing of each specimen within 5 min after removing the specimen from the bath.

7.4.2.3 If the tank is equipped with a drain so that water at 60 to 86°F (15.5 to 30°C) passes through the tank continuously and at such a rate that a complete change of water takes place in not more than 2 min, make weighings at the end of 1 h.

7.4.3 Calculation and Report:

7.4.3.1 Calculate the absorption of each specimen as follows:

$$\text{Absorption, \%} = 100(W_b - W_d)/W_d \qquad (4)$$

where:
W_d = dry weight of the specimen, and
W_b = saturated weight of the specimen after submersion in boiling water.

7.4.3.2 Report the average absorption of all the specimens tested as the absorption of the lot.

7.4.4 *Saturation Coefficient*—Calculate the saturation coefficient of each specimen as follows:

$$\text{Saturation coefficient} = W_{s2} - W_d/W_{b5} - W_d \qquad (5)$$

where:
W_d = dry weight of the specimen,
W_{s2} = saturated weight of the specimen after 24-h submersion in cold water, and
W_{b5} = saturated weight of the specimen after 5-h submersion in boiling water.

8. Freezing and Thawing

8.1 *Apparatus:*

8.1.1 *Compressor, Freezing Chamber, and Circulator* of such design and capacity that the temperature of the air in the freezing chamber will not exceed 16°F (−9°C) 1 h after introducing the maximum charge of units, initially at a temperature not exceeding 90°F (32°C).

8.1.2 *Trays and Containers*, shallow, metal, having an inside depth of $1\frac{1}{2} \pm \frac{1}{2}$ in. (38.1 ± 12.7 mm), and of suitable strength and size so that the tray with a charge of frozen units can be removed from the freezing chamber by one man.

8.1.3 *Balance*, having a capacity of not less than 2000 g and sensitive to 0.5 g.

8.1.4 *Drying Oven* that provides a free circulation of air through the oven and is capable of maintaining a temperature between 230 and 239°F (110 and 115°C).

8.1.5 *Thawing Tank* of such dimensions as to permit complete submersion of the specimens in their trays. Adequate means shall be provided so that the water in the tank may be kept at a temperature of 75 ± 10°F (24 ± 5.5°C).

8.1.6 *Drying Room*, maintained at a temperature of 75 ± 15°F (24 ± 8°C), with a relative humidity between 30 and 70 %, and free from drafts.

8.2 *Test Specimens:*

8.2.1 *Brick*—The test specimens shall consist of half brick with approximately plane and parallel ends. If necessary, the rough ends may be smoothed by trimming off a thin section with a masonry saw. The specimens shall be free from shattering or unsoundness, visually observed, resulting from the flexure or from the absorption tests. Additionally, prepare specimens by removing all loosely adhering particles, sand or edge shards from the surface or cores. Test five specimens.

8.2.2 *Structural Clay Tile*—The test specimens shall consist of five tile or of a cell not less than 4 in. (101.6 mm) in length sawed from each of the five tile.

8.3 *Procedure:*

8.3.1 Dry and cool the test specimens as prescribed in 4.4.1 and 4.4.2 and weigh and record the dry weight of each.

8.3.2 Carefully examine each specimen for cracks. A crack is defined as a fissure or separation visible to a person with normal vision from a distance of one foot under an illumination of not less than 50 fc. Mark each crack its full length with an indelible felt marking pen.

8.3.3 Submerge the test specimens in the water of the thawing tank for 4 ± ½ h.

8.3.4 Remove the specimens from the thawing tank and stand them in the freezing trays with one of their head faces down. Head face is defined as the end surfaces of a whole rectangular brick (which have the smallest area). A space of at least ½ in. (12.7 mm) shall separate the specimens as placed in the tray. Pour sufficient water into the trays so that each specimen stands in ½ in. depth of water and then place the trays and their contents in the freezing chamber for 20 ± 1 h.

8.3.5 Remove the trays from the freezing chamber after 20 ± 1 h and totally immerse them and their contents in the water of the thawing tank for 4 ± ½ h.

8.3.6 Freeze the test specimens by the procedure in 8.3.4 one cycle each day of the normal work week. Following the 4 ± ½ h thawing after the last freeze-thaw cycle of the normal work week, remove the specimens from the trays and store them for 44 ± 1 h in the drying room. Do not stack or pile units. Provide a space of at least 1 in. (25.4 mm) between all specimens. Following this period of air drying, inspect the specimens, submerge them in the water of the thawing tank for 4 ± ½ h, and again subject them to a normal week of freezing and thawing cycles in accordance with 8.3.4 and 8.3.5. If a laboratory has personnel available for testing 7 days a week, the requirement for storing the specimens for 44 ± 1 h in the drying room following the 4 ± ½ h thawing after the last freezing cycle of the week may be waived. The specimens may then be subjected to 50 cycles of freezing and thawing on 50 consecutive days. When a normal 5-day work week is interrupted, put specimens into a drying cycle which may extend past the 44 ± 1 h drying time outlined in the procedures of this section.

8.3.7 Continue the alternations of drying and submersion in water for 4 ± ½ h, followed by 5 cycles of freezing and thawing or the number of cycles needed to complete a normal work week, until a total of 50 cycles of freezing and thawing has been completed. Stop the test if the test specimen has been broken or appears to have lost more than 3 % of its original weight as judged by visual inspection.

8.3.8 After completion of 50 cycles, or when the test specimen has been withdrawn from test as a result of disintegration, dry and weigh the specimen as prescribed in 8.3.1.

8.4 *Calculations, Examination, Rating and Report:*

8.4.1 *Calculation*—Calculate the loss in weight as a percentage of the original weight of the dried specimen.

8.4.2 *Examination*—Reexamine the surface of the specimens for cracks (see 8.3.2) and record the presence of any new cracks developed during the freezing-thawing testing procedure. Measure and record the length of the new cracks.

8.4.3 *Rating*—A specimen is considered to fail the freezing and thawing test under any one of three circumstances:

8.4.3.1 *Weight Loss*—A weight loss of greater than 0.5 %.

8.4.3.2 *Breakage*—The specimen separates into two or more significant pieces, or

8.4.3.3 *Cracking*—A specimen develops a crack during the freezing and thawing procedure that exceeds in length the minimum dimension of the specimen.

If none of the above circumstances occur, the specimens are considered to pass the freezing and thawing test.

8.4.4 *Report*—The report shall state whether the sample passed or failed the test. Any failures shall include the rating and the reason for classification as a failure and the number of cycles causing failure in the event failure occurs prior to 50 cycles.

9. Initial Rate of Absorption (Suction) (Laboratory Test)

9.1 *Apparatus:*

9.1.1 *Trays or Containers*—Watertight trays or containers, having an inside depth of not less than ½ in. (12.7 mm), and of such length and width that an area of not less than 300 in.² (1935.5 cm.²) of water surface is provided. The bottom of the tray shall provide a plane, horizontal upper surface, when suitably supported, so that an area not less than 8 in. (203.2 mm) in length by 6 in. (152.4 mm) in width will be level when tested by a spirit level.

9.1.2 *Supports for Brick*—Two noncorrodible metal supports consisting of bars between 5 and 6 in. (127.00 and 152.5 mm) in length, having triangular, half-round, or rectangular cross sections such that the thickness (height) will be approximately ¼ in. (6.35 mm). The thickness of the two bars shall agree within 0.001 in. (0.03 mm) and, if the bars are rectangular in cross section, their width shall not exceed ⁵⁄₁₆ in. (1.94 mm).

9.1.3 *Means for Maintaining Constant Water Level*—Suitable means for controlling the water level above the upper surface of the supports for the brick within ± 0.01 in. (0.25 mm) (see Note 3), including means for adding water to the tray at a rate corresponding to the rate of removal by the brick undergoing test (see Note 4). For use in checking the adequacy of the method of controlling the rate of flow of the added water, a reference brick or half brick shall be provided whose displacement in ⅛ in. (3.18 mm) of water corresponds to the brick or half brick to be tested within ± 2.5 %. Completely submerge the reference brick in water for not less than 3 h preceding its use.

NOTE 3—A suitable means for obtaining accuracy in control of the water level may be provided by attaching to the end of one of the bars two stiff metal wires that project upward and return, terminating in points; one of which is ⅛ − 0.01 in. (3.18 − 0.25 mm) and the other ⅛ + 0.01 in. (3.18 + 0.25 mm) above the upper surface or edge of the bar. Such precise adjustment is obtainable by the use of depth plates or a micrometer microscope. When the water level with respect to the upper surface or edge of the bar is adjusted so that the lower point dimples the water surface when viewed by reflected light and the upper point is not in contact with the water, the water level is within the limits specified. Any other suitable means for fixing and maintaining a constant depth of immersion may be used if equivalent accuracy is obtained. As an example of such other suitable means, there may be mentioned the use of rigid supports movable with respect to the water level.

NOTE 4—A rubber tube leading from a siphon or gravity feed and closed by a spring clip will provide a suitable manual control. The so-called "chicken-feed" devices as a rule lack sensitivity and do not operate with the very small changes in water level permissible in this test.

9.1.4 *Balance*, having a capacity of not less than 3000 g, and sensitive to 0.5 g.

9.1.5 *Drying Oven*, conforming to the requirements of 8.1.4.

9.1.6 *Constant-Temperature Room*, maintained at a temperature of 70 ± 2.5°F (21 ± 1.4°C).

9.1.7 *Timing Device*—A suitable timing device, preferably a stop watch or stop clock, which shall indicate a time of 1 min to the nearest 1 s.

9.2 *Test Specimens*, consisting of whole brick. Five specimens shall be tested.

9.3 *Procedure:*

9.3.1 Dry and cool the test specimens in accordance with one of the following procedures.

9.3.1.1 *Oven-dried Procedure*—Dry and cool the test specimens in accordance with 4.4.1 and 4.4.2.

9.3.1.2 *Ambient Air-dried Procedure*—Store units unstacked, with separate placement in a ventilated room maintained at a temperature of 75 ± 15°F (24 ± 8°C) with a relative humidity between 30 % and 70 % for a period of 4 h, with a current of air from an electric fan passing over them for a period of at least 2 h. Continue until two successive weighings at intervals of 2 h show an increment of loss not greater than 0.2 % of the last previously determined weight of the specimen.

9.3.2 Measure to the nearest 0.05 in. (1.27 mm) the length and width of the flatwise surface of the test specimen of rectangular units or determine the area of other shapes to similar accuracy that will be in contact with the water. Weigh the specimen to the nearest 0.5 g.

9.3.3 Adjust the position of the tray for the absorption test so that the upper surface of its bottom will be level when tested by a spirit level, and set the saturated reference brick (9.1.3) in place on top of the supports. Add water until the water level is ⅛ ± 0.01 in. (3.18 ± 0.25 mm) above the top of the supports. When testing tile with scored bed surfaces, the depth of water level is ⅛ ± 0.01 in. plus the depth of scores.

9.3.4 After removal of the reference brick, set the test brick in place flatwise, counting zero time as the moment of contact of the brick with the water. During the period of contact (1 min ± 1 s) keep the water level within the prescribed limits by adding water as required. At the end of 1 min ± 1 s, lift the brick from contact with the water, wipe off the surface water with a damp cloth, and reweigh the brick to the nearest 0.5 g. Wiping shall be completed within 10 s of removal from contact with the water, and weighing shall be completed within 2 min.

NOTE 5—Place the brick in contact with the water quickly, but without splashing. Set the brick in position with a rocking motion to avoid the entrapping of air on its under surface. Test brick with frogs or depressions in one flatwise surface with the frog or depression uppermost.

9.4 *Calculation and Report:*

9.4.1 The difference in weight in grams between the initial and final weighings is the weight in grams of water absorbed by the brick during 1-min contact with the water. If the area of its flatwise surface (length times width) does not differ more than ± 0.75 in.² (4.84 cm²) (±2.5 %) from 30 in.² (193.55 cm²), report the gain in weight in grams as the initial rate of absorption in 1 min.

9.4.2 If the area of its flatwise surface differs more than ±

0.75 in.² (4.84 cm²) (±2.5 %) from 30 in.² (193.55 cm²), calculate the equivalent gain in weight from 30 in.² (193.55 cm²) as follows:

$$X = 30\,W/LB \quad \text{(metric } X = 193.55\,W/LB) \qquad (6)$$

where:

X = gain in weight corrected to basis of 30 in.² (193.55 cm²) flatwise area,
W = actual gain in weight of specimen, g,
L = length of specimen, in., (cm), and
B = width of specimen, in., (cm).

9.4.3 Report the corrected gain in weight, X, as the initial rate of absorption in 1 min.

9.4.4 If the test specimen is a cored brick, calculate the net area and substitute for LB in the equation given in 9.4.2. Report the corrected gain in weight as the initial rate of absorption in 1 min.

9.4.5 If specimen is non-prismatic, calculate the net area by suitable geometric means and substitute for LB in the equation given in 9.4.2.

9.4.6 Report the method of drying as oven-dried (in accordance with 9.3.1.1) or ambient air-dried (in acordance with 9.3.1.2).

10. Efflorescence

10.1 *Apparatus:*

10.1.1 *Trays and Containers*—Watertight shallow pans or trays made of corrosion-resistant metal or other material that will not provide soluble salts when in contact with distilled water containing leachings from brick. The pan shall be of such dimensions that it will provide not less than a 1-in. (25.4-mm) depth of water. Unless the pan provides an area such that the total volume of water is large in comparison with the amount evaporated each day, suitable apparatus shall be provided for keeping a constant level of water in the pan.

10.1.2 *Drying Room*, conforming to the requirements of 8.1.6.

10.1.3 *Drying Oven*, conforming to the requirements of 8.1.4.

10.2 *Test Specimens:*

10.2.1 The sample shall consist of ten full-size brick.

10.2.2 The ten specimens shall be sorted into five pairs so that both specimens of each pair will have the same appearance as nearly as possible.

10.3 *Preparation of Specimens*—Remove by brushing any adhering dirt that might be mistaken for efflorescence. Dry the specimens and cool them as prescribed in 4.4.1 and 4.4.2.

10.4 *Procedure:*

10.4.1 Set one specimen from each of the five pairs, on end, partially immersed in distilled water to a depth of approximately 1 in. (25.4 mm) for 7 days in the drying room. When several specimens are tested in the same container, separate the individual specimens by a spacing of at least 2 in. (50.8 mm).

NOTE 6—Do not test specimens from different sources simultaneously in the same container, because specimens with a considerable content of soluble salts may contaminate salt-free specimens.
NOTE 7—Empty and clean the pans or trays after each test.

10.4.2 Store the second specimen from each of the five pairs in the drying room without contact with water.

10.4.3 At the end of 7 days, inspect the first set of specimens and then dry both sets in the drying oven for 24 h.

10.5 *Examination and Rating*—After drying, examine and compare each pair of specimens, observing the top and all four faces of each specimen from a distance of 10 ft. (3 m) under an illumination of not less than 50 footcandles (538.2 lm/m²) by an observer with normal vision. If under these conditions no difference is noted, report the rating as "not effloresced." If a perceptible difference due to efflorescence is noted under these conditions, report the rating as "effloresced." Record the appearance and distribution of the efflorescence.

11. Measurement of Size

11.1 *Apparatus*—Either a 1-ft (or metric) steel rule, graduated in ¹⁄₃₂-in. (or 1-mm) divisions, or a gage or caliper having a scale ranging from 1 to 12 in. (25 to 300 mm), and having parallel jaws, shall be used for measuring the individual units. Steel rules or calipers of corresponding accuracy and size required shall be used for measurement of larger brick, solid masonry units and tile.

11.2 *Test Specimens*—Measure ten dry full-size units. These units shall be representative of the shipment and shall include the extremes of color range and size as determined by visual inspection of the shipment. (The same samples may be used for determining efflorescence and other properties.)

11.3 *Individual Measurements of Width, Length, and Height*—Measure the width across both ends and both beds from the midpoints of the edges bounding the faces. Record these four measurements to the nearest ¹⁄₃₂ in. (1 mm) and record the average to the nearest ¹⁄₆₄ in. (0.5 mm) as the width. Measure the length along both beds and along both faces from the midpoints of the edges bounding the ends. Record these four measurements to the nearest ¹⁄₃₂ in. (1 mm) and record the average to the nearest ¹⁄₆₄ in. (0.5 mm) as the length. Measure the height across both faces and both ends from the midpoints of the edges bounding the beds. Record these four measurements to the nearest ¹⁄₃₂ in. (1 mm) and record the average to the nearest ¹⁄₆₄ in. (0.5 mm) as the height. Use the apparatus described in 11.1. Retest by the same method when required.

12. Measurement of Warpage

12.1 *Apparatus:*

12.1.1 *Steel Straightedge:*

12.1.2 *Rule or Measuring Wedge*—A steel rule graduated from one end in ¹⁄₃₂-in. (or 1-mm) divisions, or alternatively, a steel measuring wedge 2.5 in. (60 mm) in length by 0.5 in. (12.5 mm) in width by 0.5 in. (12.5 mm) in thickness at one end and tapered, starting at a line 0.5 in. (12.5 mm) from one end, to zero thickness at the other end. The wedge shall be graduated in ¹⁄₃₂-in. (or 1-mm) divisions and numbered to show the thickness of the wedge between the base, *AB*, and the slope, *AC*, Fig. 1.

12.1.3 *Flat Surface*, of steel or glass, not less than 12 by 12 in. (305 by 305 mm) and plane to within 0.001 in. (0.025 mm).

12.2 *Sampling*—Use the sample of ten units selected for determination of size.

12.3 *Preparation of Samples*—Test the specimens as re-

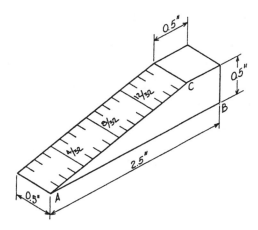

FIG. 1 Measuring Wedge

ceived, except remove any adhering dirt by brushing.

12.4 *Procedure:*

12.4.1 *Concave Surfaces*—Where the warpage to be measured is of a surface and is concave, place the straightedge lengthwise or diagonally along the surface to be measured, selecting the location that gives the greatest departure from straightness. Select the greatest distance from the unit surface to the straightedge. Using the steel rule or wedge, measure this distance to the nearest 1/32 in. (1 mm), and record as the concave warpage of the surface.

12.4.2 *Concave Edges*—Where the warpage to be measured is of an edge and is concave, place the straightedge between the ends of the concave edge to be measured. Select the greatest distance from the unit edge to the straightedge. Using the steel rule or wedge, measure this distance to the nearest 1/32 in. (1 mm), and record as the concave warpage of the edge.

12.4.3 *Convex Surfaces*—When the warpage to be measured is of a surface and is convex, place the unit with the convex surface in contact with a plane surface and with the corners approximately equidistant from the plane surface. Using the steel rule or wedge, measure the distance to the nearest 1/32 in. (1 mm) of each of the four corners from the plane surface. Record the average of the four measurements as the convex warpage of the unit.

12.4.4 *Convex Edges*—Where the warpage to be measured is of an edge and is convex, place the straightedge between the ends of the convex edge. Select the greatest distance from the unit edge to the straightedge. Using the steel rule or wedge, measure this distance to the nearest 1/32 in. (1 mm) and record as the convex warpage of the edge.

13. Measurement of Length Change

13.1 *Apparatus*—A dial micrometer or other suitable measuring device graduated to read in 0.0001-in. (or 0.001-mm) increments, mounted on a stand suitable for holding the specimen in such a manner that reproducible results can be obtained, shall be used for measuring specimen length. Provisions shall be made to permit changing the position of the dial micrometer on its mounting rod so as to accommodate large variations in specimen size. The base of the stand and the tip of the dial micrometer shall have a conical depression to accept a 1/4-in. (6.35-mm) steel ball. A suitable

reference instrument shall be provided for checking the measuring device.

13.2 *Preparation of Specimen*—Remove the ends of deeply textured specimens to the depth of the texture by cutting perpendicular to the length and parallel to each other. Drill a hole in each end of the specimen with a 1/4-in. (6.35-mm) carbide drill. Drill these holes at the intersection of the two diagonals from the corners. Place 1/4-in. (6.35-mm) steel balls in these depressions by cementing in place with a calcium aluminate cement. Any equivalent method for establishing the reference length is permissible.

13.3 *Procedure*—Mark the specimen for identification and measure to the nearest 0.0001 in. (or 0.001 mm) in a controlled environment and make subsequent measurements in the same controlled environment, ± 2°F and ± 5 % relative humidity. Record the temperature and relative humidity. Apply a reference mark to the specimen for orientation in the measuring device. Check the measuring device with the reference instrument before each series of measurements.

14. Initial Rate of Absorption (Suction)—Field Test

14.1 *Scope*—This test method is intended to serve as a volumetric means of determining the initial rate of absorption (IRA) of any size brick when weighing determination, described in Section 9 of this standard, is impractical. This test method is applicable to assess the need for wetting the brick. This test method is performed on specimens taken from the field with no modification of moisture content, therefore, the IRA determined by this test method may differ from the IRA determined by the laboratory test method in Section 9, which requires drying the specimens.

14.2 *Apparatus:*

14.2.1 *Absorption Test Pan*—A watertight, rectangular pan, constructed of noncorroding material, with a flat, rigid bottom and inside depth of about 1½ in. (38.1 mm). The inside length and width of the pan shall exceed the length and width of the tested brick by a minimum of 3 in. (76.2 mm) but not more than 5 in. (127.0 mm).

14.2.2 *Brick Supports*—Two noncorroding rectangular bars, 1/4 in. (6.4 mm) in height and width and 1 in. (25.4 mm) shorter than the inside width of the pan in length. The brick supports can be placed on the bottom of the pan just before the test or permanently affixed to the bottom of the pan. The space between the supports should be about 4 in. (101.6 mm) shorter than the length of the tested brick. A device indicating the desired water level can be permanently

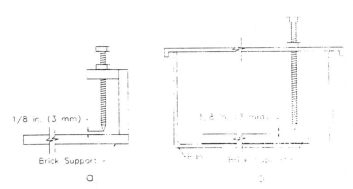

FIG. 2 Water Level Indicators

FIGURE 1 - FIGURE 2

attached to the end of one of the brick supports or suspended from the top of the pan (see Figs. 2a and b). Any other device of equivalent accuracy for controlling the required water level, 1/8 in. (3.2 mm) above the brick supports, can be used in place of that depicted in Fig. 2.

14.2.3 *Timing Device*—A suitable timing device that shall indicate a time of 1 min to the nearest 1 s.

14.2.4 *Squeeze Bottle*—A plastic squeeze bottle, 100 mL capacity.

14.2.5 *Graduated Cylinder*—A plastic or glass graduated measuring cylinder, 100 mL capacity.

14.3 *Test Specimens*—Select six whole brick in accordance with the requirements of Paragraph 4.1.

14.4 *Procedure:*

14.4.1 Completely immerse one brick specimen in a container of water for 2 h.

14.4.2 Measure to the nearest 1/16 in. (1.6 mm) the length and width of the five remaining specimens at the surface that will be in contact with water. If the test specimens are cored, determine the area of the cores at the same surface.

14.4.3 Pre-wet and drain the absorption pan and place it on a flat, level surface.

14.4.4 Remove the pre-wetted specimen from the container, shake off the surface water, and place the specimen on brick supports in the pan. Pour water into the pan until the water reaches a level 1/8 in. (3.2 mm) above the brick supports. (If using a pointed level water indicator, pour water into the pan until the water makes a minimum contact (dimpling effect).) Remove the pre-wetted brick, and tilt the brick sharply so that one corner serves as a drip point for clinging surface water to return to the pan. A gentle shake of the brick may be necessary to make the last drop fall. Put the pre-wetted brick back into the container of water.

14.4.5 Using the graduated cylinder, fill the squeeze bottle with exactly 100 mL of water.

14.4.6 Set the first test specimen squarely on the brick supports, counting zero time as the moment the brick contacts the water. At the end of 1 min ± 1 s lift the test specimen from water and tilt the brick sharply so that one corner serves as a drip point for clinging surface water to return to the pan. A gentle shake of the brick may be necessary to make the last drop fall.

14.4.6.1 Continue setting the remaining test specimens into the pan in the same way until all five specimens are tested. During the test add water to the pan, using the squeeze bottle, to keep the water level approximately constant at the 1/8 in. depth. Refill the squeeze bottle with 100 mL of water when empty, recording each refill.

14.4.6.2 After the last specimen is tested, place the pre-wetted brick back in the pan and restore the original level with water from the squeeze bottle.

NOTE 8—Place the brick in contact with the water quickly, but without splashing. Set the brick in position with a rocking motion to avoid the entrapping of air on its under surface. Test brick with frogs or depressions in one flatwise surface with the frog or depression uppermost.

14.4.7 Using the graduated cylinder, measure the volume of water remaining in the squeeze bottle.

14.5 *Calculation and Report:*

14.5.1 The number of refills plus the first full bottle, times 100 mL, minus the volume of water remaining in the squeeze bottle, is the total measured volume of water in millilitres absorbed by the five specimens.

$$V_t = 100 (n + 1) - V_r \qquad (7)$$

where:
V_t = total measured volume of water absorbed by all tested specimens, mL,
n = the number of squeeze bottle refills, and
V_r = the volume of water remaining in the squeeze bottle, mL.

14.5.2 When the average net surface area in contact with water of a single specimen (sum of net surface areas divided by the number of specimens) differs by ±0.75 in.^{2} (4.84 cm^2) or less from 30 in.^{2} (193.5 cm^2), report the total measured absorbed volume of water divided by five, the number of tested specimens, as the IRA (Field) in g/min/30 in.^{2}

$$IRA\ (Field) = \frac{V_t}{5} \qquad (8)$$

14.5.3 If the average net surface area in contact with water differs by more than ±0.75 in.^{2} (4.84 cm^2) from 30 in.^{2} (193.5 cm^2), calculate the equivalent volume in 1 min for 30 in.^{2} (193.5 cm^2) of surface as follows:

$$V_c = \frac{30\ V_t}{A_n} \left(metric\ V_c = \frac{193.5\ V_t}{A_n} \right) \qquad (9)$$

where:
V_c = average volume of absorbed water by a specimen, corrected to basis of 30 in.^{2} (193.5 cm^2) of surface, mL, and
A_n = sum of net surface areas in contact with water of all tested specimens, in.^{2} (cm^2).

14.5.4 *Report*—Report the corrected volume (V_c) as the IRA (Field) in g/l min/30 in.^{2}

14.6 *Precision and Bias*—Insufficient data is currently available for a precision and bias statement.

15. Measurement of Void Area in Cored Units

15.1 *Apparatus:*

15.1.1 *Steel Rule or Calipers*—As described in 11.1.

15.1.2 *Graduated Cylinder*—A glass cylinder with a capacity of 500 mL and graduated in 1-mL increments.

15.1.3 *Paper*—A sheet of smooth, hard-finish paper not less than 24 by 24 in. (610 by 610 mm).

15.1.4 *Sand*—500 mL of clean, dry sand.

15.1.5 *Steel Straightedge.*

15.1.6 *Flat Surface*—A level, flat, smooth, clean dry surface.

15.1.7 *Brush*—A soft-bristle brush.

15.1.8 *Neoprene Mat*—24 by 24 in. (610 by 610 mm) open-cell neoprene sponge 1/4 in. (6.4 mm) in thickness.

15.2 *Test Specimens*—Use of a sample of ten units selected as described for the determination of size (The samples taken for the determination of size may be used).

15.3 *Preparation of Samples*—Test the specimens as received, except remove any adhering dirt by brushing.

15.4 *Procedure:*

15.4.1 Measure and record the length, width, and depth of the unit as described for the determination of size.

15.4.2 Place the unit to be tested bed down (cores vertical) on the sheet of paper that has been spread over the neoprene mat on the flat surface.

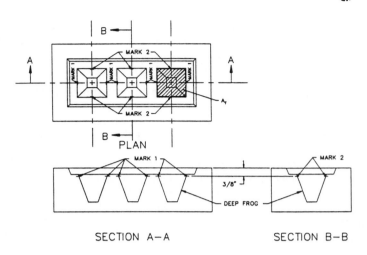

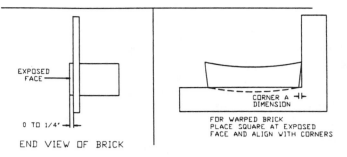

FIG. 4 Location of Carpenter's Square

FIG. 3 Deep Frogged Units

15.4.3 Fill the cores with sand, allowing the sand to fall naturally. Do not work the sand into the cores. Using the steel straightedge, bring the level of the sand in the cores down to the top of the unit. With the brush, remove all excess sand from the top of the unit and from the paper sheet.

15.4.4 Lifting the unit up, allow all of the sand in the cores to fall on the sheet of paper.

15.4.5 Transfer the sand from the sheet of paper to the graduated cylinder allowing the sand to fall naturally. Do not shake or vibrate the cylinder. Level the sand in the cylinder. Read and record the sand level to the nearest 1 mL.

15.5 *Calculation and Report:*

15.5.1 Determine the percentage of void as follows:

$$\% \text{ Void area} = \frac{V_s}{V_u} \times \frac{1}{16.4} \times 100 \qquad (10)$$

where:

V_s = amount of sand recorded in 15.4.5, mL, and
V_u = length × width × depth recorded in 15.4.1, in.[3]

15.5.2 Report the results of the equation in 15.5.1 as the units percentage of void area.

16. Measurement of Void Area In Deep Frogged Units

NOTE 9—The area measured corresponds to a section located ⅜ in. (9.5 mm) distant from the voided bed of the units.

16.1 *Apparatus:*

16.1.1 *Steel Rule or Gage or Calipers (inside and outside)*—as described in 11.1.

16.1.2 *Steel Straightedge.*

16.1.3 *Marking Pen or Scribe.*

16.2 *Test Specimens*—Use a sample of 10 units selected as described for the determination of size. (The samples taken for the determination of size may be used.)

16.3 *Preparation of Sample*—Test the specimens as received except remove any adhering dirt by brushing.

16.4 *Procedure:*

16.4.1 Measure the length along both faces and the width along both ends at a distance of ⅜ in. (9.5 mm) down from the bed containing the deep frogs. Record the measurements to the nearest ¹⁄₃₂ in. (1 mm). Record the average of the two length measurements to the nearest ¹⁄₃₂ in. (1 mm) as the

length of the unit and the average of the two width measurements to the nearest ¹⁄₃₂ in. (1 mm) as the width of the unit.

16.4.2 With the steel straightedge parallel to the length of the unit and centered over the deep frog or frogs, inscribe a mark on both faces of the frog ⅜ in. (9.5 mm) below the underside of the steel straightedge (mark 1 on Fig. 3). With the steel straightedge parallel to the width of the unit and centered over the deep frog, inscribe a mark on both faces of each frog ⅜ in. (9.5 mm) below the underside of the steel straightedge (mark 2 on Fig. 3).

16.4.3 Measure and record to the nearest ¹⁄₃₂ in. (1 mm) the distance between the inscribed marks on a line parallel to the length of the unit for each frog, and measure and record to the nearest ¹⁄₃₂ in. (1 mm) the distance between the inscribed marks on a line parallel to the width of the unit for each frog.

16.5 *Calculations and Report:*

16.5.1 Using the recorded length and width measurements calculate the gross area of the unit (A_u) in the plane of the unit ⅜ in. (9.5 mm) down from the frogged bed.

16.5.2 Using the distance between the inscribed marks calculate the inside area of each deep frog (A_f) in the plane of the unit ⅜ in. (9.5 mm) down from the frogged bed (see Fig. 3).

16.5.3 Determine the percentage of void as follows:

$$\% \text{ Void area} = \frac{\Sigma A_f \times 100}{A_u} \qquad (11)$$

where:

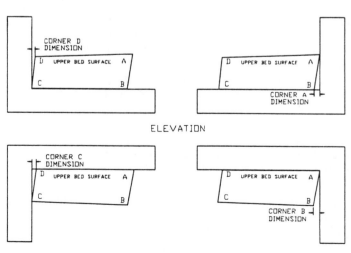

FIG. 5 Out-of-Square Measurements

C 67

ΣA_f = sum of the inside area of the deep frogs
A_u = gross area of unit

16.5.4 Report the results of the equation in 16.5.3 as the unit's percentage of void area.

17. Measurement of Out of Square

17.1 *Apparatus:*

17.1.1 *Steel Rule or Calipers*, as described in 11.1.

17.1.2 *Steel Carpenter's Square.*

17.2 *Procedure:*

17.2.1 Place one leg of a carpenter's square adjacent to the length of the unit when laid as a stretcher. Align the leg of the square parallel to the length of the unit by having the corners of the face of the unit in contact with the leg of the square. Locate the square parallel to and at or within ¼ in. (6.4 mm) of the face to be exposed. See Fig. 4.

17.2.2 Measure the deviation due to the departure from the 90° angle at each corner of the exposed face of the unit. Record the measurement to the nearest ¹⁄₃₂ in. (0.8 mm) for each corner. See Fig. 5.

18. Keywords

18.1 absorption; compressive strength; efflorescence; freezing and thawing; initial rate of absorption; length change; modulus of rupture; out-of-square; sampling; size; void area; warpage

Standard Specification for
Loadbearing Concrete Masonry Units[1]

This standard is issued under the fixed designation C 90; the number immediately following the designation indicates the year of original adoption or, in the case of revision, the year of last revision. A number in parentheses indicates the year of last reapproval. A superscript epsilon (ε) indicates an editorial change since the last revision or reapproval.

This standard has been approved for use by agencies of the Department of Defense. Consult the DoD Index of Specifications and Standards for the specific year of issue which has been adopted by the Department of Defense.

1. Scope

1.1 This specification covers hollow and solid (see 5.4 and 5.5) concrete masonry units made from portland cement, water, and mineral aggregates with or without the inclusion of other materials. There are three classes of concrete masonry units: (1) normal weight, (2) medium weight, and (3) lightweight. There are two types of concrete masonry units: (1) Type I, moisture-controlled, and (2) Type II, nonmoisture-controlled. These units are suitable for both loadbearing and nonloadbearing applications.

1.2 Concrete masonry units covered by this specification are made from lightweight or normal weight aggregates, or both.

1.3 The values stated in inch-pound units are to be regarded as the standard. The values given in parentheses are for information only.

NOTE 1—When particular features are desired such as surface textures for appearance or bond, finish, color, or particular properties such as weight classification, higher compressive strength, fire resistance, thermal performance or acoustical performance, these features should be specified separately by the purchaser. Local suppliers should be consulted as to the availability of units having the desired features.

2. Referenced Documents

2.1 *ASTM Standards:*
C 33 Specification for Concrete Aggregates[2]
C 140 Test Methods of Sampling and Testing Concrete Masonry Units[3]
C 150 Specification for Portland Cement[4]
C 331 Specification for Lightweight Aggregates for Concrete Masonry Units[2]
C 426 Test Method for Drying Shrinkage of Concrete Masonry Units[3]
C 595/C 595M Specification for Blended Hydraulic Cements[4]
C 618 Specification for Coal Fly Ash and Raw or Calcined Natural Pozzolan for Use as a Mineral Admixture in Concrete[2]

C 989 Specification for Ground Granulated Blast-Furnace Slag for Use in Concrete and Mortars[2]
E 72 Methods for Conducting Strength Tests of Panels for Building Construction[5]
E 447 Test Methods for Compressive Strength of Masonry Prisms[3]
E 519 Test Method for Diagonal Tension (Shear) in Masonry Assemblages[3]

3. Classification

3.1 *Types*—Two types of concrete masonry units are covered as follows:

3.1.1 *Type I, Moisture-Controlled Units*—Units designated as Type I shall conform to the requirements of this specification.

3.1.2 *Type II, Nonmoisture-Controlled Units*—Units designated as Type II shall conform to the requirements of this specification, except the requirements of Table 1.

4. Materials

4.1 *Cementitious Materials*—Materials shall conform to the following applicable specifications:

4.1.1 *Portland Cement*—Specification C 150.

4.1.2 *Modified Portland Cement*—Portland cement conforming to Specification C 150, modified as follows:

4.1.2.1 *Limestone*—Calcium carbonate, with a minimum 85 % $CaCO_3$ content, may be added to the cement, provided these requirements of Specification C 150 as modified are met:

(1) *Limitation on Insoluble Residue*—1.5 %.

(2) *Limitation on Air Content of Mortar*—Volume percent, 22 % max.

(3) *Limitation on Loss on Ignition*—7 %.

4.1.3 *Blended Cements*—Specification C 595/C 595M.

4.1.4 *Pozzolans*—Specification C 618.

4.1.5 *Blast Furnace Slag Cement*—Specification C 989.

4.2 *Aggregates*—Aggregates shall conform to the following specifications, except that grading requirements shall not necessarily apply:

4.2.1 *Normal Weight Aggregates*—Specification C 33.

4.2.2 *Lightweight Aggregates*—Specification C 331.

4.3 *Other Constituents*—Air-entraining agents, coloring pigments, integral water repellents, finely ground silica, and other constituents shall be previously established as suitable for use in concrete masonry units and shall conform to applicable ASTM standards or shall be shown by test or experience not to be detrimental to the durability of the

---

[1] This specification is under the jurisdiction of ASTM Committee C-15 on Manufactured Masonry Units and is the direct responsibility of Subcommittee C15.03 on Concrete Masonry Units and Related Units.
Current edition approved June 10, 1996. Published August 1996. Originally published as C 90 – 31T. Last previous edition C 90 – 96.
[2] *Annual Book of ASTM Standards*, Vol 04.02.
[3] *Annual Book of ASTM Standards*, Vol 04.05.
[4] *Annual Book of ASTM Standards*, Vol 04.01.

[5] *Annual Book of ASTM Standards*, Vol 04.07.

TABLE 1 Moisture Content Requirements for Type I Units

| Total Linear Drying Shrinkage, % | Moisture Content, max, % of Total Absorption (Average of 3 Units) | | |
| | Humidity[A] Conditions at Job Site or Point of Use | | |
	Humid[B]	Intermediate[C]	Arid[D]
Less than 0.03	45	40	35
0.03 to less than 0.045	40	35	30
0.045 to 0.065, max	35	30	25

[A] See Appendix X1 for map of mean annual relative humidity.
[B] Mean annual relative humidity above 75 %.
[C] Mean annual relative humidity 50 to 75 %.
[D] Mean annual relative humidity less than 50 %.

TABLE 2 Minimum Thickness of Face Shells and Webs

| Nominal Width (W) of Units, in. (mm) | Face Shell Thickness (FST), min, in. (mm)[A] | Web Thickness (WT) | |
		Webs[A] min, in. (mm)	Equivalent Web Thickness, min, in./linear ft[B,C] (mm/linear m)
3 (76.2) and 4 (102)	¾ (19)	¾ (19)	1⅝ (136)
6 (152)	1 (25)[D]	1 (25)	2¼ (188)
8 (203)	1¼ (32)[D]	1 (25)	2¼ (188)
10 (254)	1⅜ (35)[D] 1¼ (32)[D,E]	1⅛ (29)	2½ (209)
12 (305)	1½ (38) 1¼ (32)[D,E]	1⅛ (29)	2½ (209)

[A] Average of measurements on 3 units taken at the thinnest point when measured as described in Test Methods C 140. When this standard is used for split face units, a maximum of 10 % of a split face shell area may have thickness less than those shown, but not less than ¾ inch (19.1 mm). When the units are solid grouted the 10 % limit does not apply.
[B] Average of measurements on 3 units taken at the thinnest point when measured as described in Test Methods C 140. The minimum web thickness for units with webs closer than 1 in. (25.4 mm) apart shall be ¾ in. (19.1 mm).
[C] Sum of the measured thicknesses of all webs in the unit, multiplied by 12 and divided by the length of the unit. Equivalent web thickness does not apply to the portion of the unit to be filled with grout. The length of that portion shall be deducted from the overall length of the unit for the calculation of the equivalent web thickness.
[D] For solid grouted masonry construction, minimum face shell thickness shall be not less than ⅝ in. (16 mm).
[E] This face shell thickness (FST) is applicable where allowable design load is reduced in proportion to the reduction in thickness from basic face shell thicknesses shown, except that allowable design loads on solid grouted units shall not be reduced.

concrete masonry units or any material customarily used in masonry construction.

5. Physical Requirements

5.1 At the time of delivery to the purchaser, all units shall conform to the physical requirements prescribed in Tables 2 and 3.

5.2 At the time of delivery to the purchaser, Type I units shall conform to the requirements prescribed in Table 1.

5.3 At the time of delivery to the purchaser, the linear shrinkage of Type II units shall not exceed 0.065 %.

NOTE 2—The purchaser is the public body or authority, association, corporation, partnership, or individual entering into a contract or agreement to purchase or install, or both, concrete masonry units. The time of delivery to the purchaser is FOB plant when the purchaser or the purchaser's agent transports the concrete masonry units, or at the time unloaded at the worksite if the manufacturer or the manufacturer's agent transports the concrete masonry units.

5.4 Hollow Units:

5.4.1 Face shell thickness (FST) and web thickness (WT) shall conform to the requirements prescribed in Table 2.

NOTE 3—Web thickness (WT) not conforming to the requirements prescribed in Table 2 may be approved, provided equivalent structural capability has been established when tested in accordance with the applicable provisions of Methods E 72, Test Method E 519, Test Methods E 447, or other applicable tests and the appropriate design criteria developed is in accordance with applicable building codes.

5.5 Solid Units:

5.5.1 The net cross-sectional area of solid units in every plane parallel to the bearing surface shall be not less than 75 % of the gross cross-sectional area measured in the same plane.

5.6 End Flanges:

5.6.1 For units having end flanges, the thickness of each flange shall not be less than the minimum face shell thickness.

NOTE 4—Flanges beveled at the ends for mortarless head joint applications that will be filled with grout are exempt from this requirement. Flanges which are specially shaped for mortarless head joint applications which have been shown by testing or field experience to provide equivalent performance are exempt from this requirement.

6. Permissible Variations in Dimensions

6.1 Overall dimensions for width, height, and length shall differ by not more than ±⅛ in. (3.2 mm) from the specified standard dimensions.

6.2 Permissible variations in dimensions for architectural features such as scores, dummy joints, flutes, and ribs shall be ⅟₁₆ in. (1.6 mm) from the specified standard dimensions. These requirements do not apply to split faces.

NOTE 5—Standard dimensions of units are the manufacturer's designated dimension. Nominal dimensions of units are equal to the standard dimensions plus the thickness of one mortar joint.

7. Finish and Appearance

7.1 All units shall be sound and free of cracks or other defects that would interfere with the proper placement of the unit or would significantly impair the strength or permanence of the construction. Minor cracks incidental to the usual method of manufacture or minor chipping resulting from customary methods of handling in shipment and delivery are not grounds for rejection.

7.2 Where units are to be used in exposed wall construc-

TABLE 3 Strength and Absorption Requirements

| Compressive Strength,[A] min, psi (MPa) | | Water Absorption, max, lb/ft³ (kg/m³) (Average of 3 Units) | | |
| Average Net Area | | Weight Classification—Oven-Dry Weight of Concrete, lb/ft³ (kg/m³) | | |
Average of 3 Units	Individual Unit	Lightweight, less than 105 (1680)	Medium Weight, 105 to less than 125 (1680–2000)	Normal Weight, 125 (2000) or more
1900 (13.1)	1700 (11.7)	18 (288)	15 (240)	13 (208)

[A] Higher compressive strengths may be specified where required by design. Consult with local suppliers to determine availability of units of higher compressive strength.

TABLE 1 - TABLE 3   ASTM C 90   243

tion, the face or faces that are to be exposed shall not show chips or cracks, not otherwise permitted, or other imperfections when viewed from a distance of not less than 20 ft (6.1 m) under diffused lighting.

7.2.1 Five percent of a shipment containing chips not larger than 1 in. (25.4 mm) in any dimension, or cracks not wider than 0.02 in. (0.5 mm) and not longer than 25 % of the nominal height of the unit is permitted.

7.3 The color and texture of units shall be specified by the purchaser. The finished surfaces that will be exposed in place shall conform to an approved sample consisting of not less than four units, representing the range of texture and color permitted.

8. Sampling and Testing

8.1 The purchaser or authorized representative shall be accorded proper facilities to inspect and sample the units at the place of manufacture from the lots ready for delivery.

8.2 Sample and test units in accordance with Test Methods C 140.

8.3 Total linear drying shrinkage shall be based on tests of concrete masonry units made with the same materials, concrete mix design, manufacturing process, and curing method, conducted in accordance with Test Method C 426 and not more than 24 months prior to delivery.

9. Rejection

9.1 If the samples tested from a shipment fail to conform to the specified requirements, the manufacturer shall be permitted to remove units from the shipment, and new specimens shall be selected by the purchaser from the retained lot and tested at the expense of the manufacturer. If the second set of specimens fails to conform to the specified requirements, the entire lot shall be rejected.

NOTE 6—Unless otherwise specified in the purchase order, the cost of tests is typically borne as follows: (1) if the results of the tests show that the units do not conform to the requirements of this specification, the cost is typically borne by the seller; (2) if the results of the tests show that the units conform to the specification requirements, the cost is typically borne by the purchaser.

10. Keywords

10.1 absorption; climatic map; concrete masonry units; equivalent web thickness; face shell; flange; lightweight; linear shrinkage; loadbearing; medium weight; moisture-controlled; normal weight; webs

APPENDIX

(Nonmandatory Information)

X1. CLIMATIC MAP

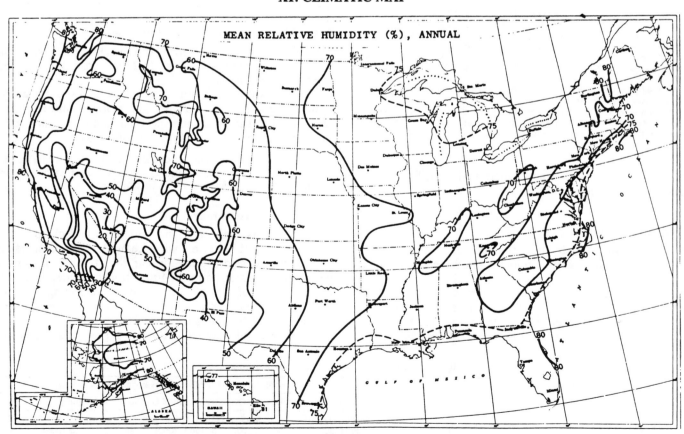

NOTE—Based on 1:30 AM and PM and 7:30 AM and PM Eastern Standard Time, observations for 20 years or more through 1964.

FIG. X1.1 Selected Climatic Maps of the United States

X2. WATER PENETRATION RESISTANCE

X2.1 Exterior walls are often subjected to moisture penetration from one or more sources. For example, basement walls may be exposed to water from saturated soil. Above-grade exterior walls are usually exposed to wind-driven rain. To prevent water penetration, proper detailing, construction, flashing, and drainage should be provided. Proper water penetration resistant treatments should be applied to the walls. While it is not within the scope of Specification C 90 to include information on resistance to water penetration, such information and guidelines are available from other organizations.

AMERICAN SOCIETY FOR TESTING AND MATERIALS
100 Barr Harbor Dr., West Conshohocken, PA 19428
Reprinted from the Annual Book of ASTM Standards. Copyright AST
If not listed in the current combined index, will appear in the next editi

Standard Test Methods of
Sampling and Testing Concrete Masonry Units[1]

This standard is issued under the fixed designation C 140; the number immediately following the designation indicates the year of original adoption or, in the case of revision, the year of last revision. A number in parentheses indicates the year of last reapproval. A superscript epsilon (ϵ) indicates an editorial change since the last revision or reapproval.

This standard has been approved for use by agencies of the Department of Defense. Consult the DoD Index of Specifications and Standards for the specific year of issue which has been approved by the Department of Defense.

1. Scope*

1.1 These test methods cover the sampling and testing of concrete masonry units for dimensions, compressive strength, absorption, unit weight (density), and moisture content. Flexural load testing and ballast weight determination of concrete roof pavers are also covered.

NOTE 1—The testing laboratory performing these test methods should be evaluated in accordance with Practice C 1093.

1.2 The values stated in inch-pound units are to be regarded as the standard. The values given in parentheses are for information only.

1.3 *This standard does not purport to address all of the safety concerns, if any, associated with its use. It is the responsibility of the user of this standard to establish appropriate safety and health practices and determine the applicability of regulatory limitations prior to use.*

2. Referenced Documents

2.1 *ASTM Standards:*
C 143 Test Method for Slump of Hydraulic Cement Concrete[2]
C 1093 Practice for Accreditation of Testing Agencies for Unit Masonry[3]
E 4 Practices for Force Verification of Testing Machines[4]
E 6 Terminology Relating to Methods of Mechanical Testing[4]

3. Terminology

3.1 *Definitions*—For definitions of terms listed in these test methods, refer to Terminology E 6.

4. Sampling

4.1 *Selection of Test Specimens:*
4.1.1 For purposes of test, full-size concrete masonry units shall be selected by the purchaser or authorized representative. The selected specimens shall be of similar configuration and dimensions. Specimens shall be representative of the whole lot of units from which they are selected. If test specimens are selected at the work site, units for moistu content tests shall be sampled upon delivery to the purchas and placed in a sealed container until the received weigh (W_r) is determined in accordance with 4.3.2.

4.1.2 The term "lot" refers to any number of concre masonry units of any configuration or dimension manufac tured by the producer using the same materials, concrete mi design, manufacturing process, and curing method.

4.2 *Number of Specimens:*
4.2.1 For the compressive strength, absorption, un weight (density), and moisture content determinations, si units shall be selected from each lot of 10 000 units c fraction thereof and 12 units from each lot of more tha 10 000 and less than 100 000 units. For lots of more tha 100 000 units, six units shall be selected from each 50 00 units or fraction thereof contained in the lot. Additiona specimens may be taken at the discretion of the purchaser.

4.3 *Identification:*
4.3.1 Mark each specimen so that it may be identified a any time. Markings shall cover not more than 5 % of th superficial area of the specimen.

4.3.2 Weigh units for moisture content tests immediatel after sampling and marking and record as W_r (receive weight).

5. Measurement of Dimensions

5.1 *Apparatus:*
5.1.1 Measure overall dimensions with a steel scale grad uated in $\frac{1}{32}$-in. (1-mm) divisions. Face shell and we thicknesses shall be measured with a caliper rule graduated i $\frac{1}{64}$-in. (0.4-mm) divisions and having parallel jaws not les than $\frac{1}{2}$ in. (12.7 mm) nor more than 1 in. (25.4 mm) i length.

5.2 *Specimens*—Three full-size units shall be measure for width, height, and length, and minimum thicknesses o face shells and webs.

NOTE 2—The same specimens may be used in other tests.

5.3 *Measurements:*
5.3.1 For each unit, measure and record the width (W across the top and bottom bearing surfaces at mid-length height (H) at mid-length of each face, and length (L) a mid-height of each face.

5.3.2 For each unit, measure face shell thicknesses (FST and web thicknesses (WT) at the thinnest point of each such element $\frac{1}{2}$ in. (12.7 mm) above the mortar-bed plane and record to the nearest division of the scale or caliper. Where

[1] These test methods are under the jurisdiction of ASTM Committee C-15 on Manufactured Masonry Units and are the direct responsibility of Subcommittee C15.03 on Concrete Masonry Units and Related Units.
Current edition approved Nov. 10, 1996. Published February 1997. Originally published as C 140 – 38 T. Last previous edition C 140 – 96a.
[2] *Annual Book of ASTM Standards*, Vol 04.02.
[3] *Annual Book of ASTM Standards*, Vol 04.05.
[4] *Annual Book of ASTM Standards*, Vol 03.01.

* A Summary of Changes section appears at the end of these test methods.

the thinnest point of opposite face shells differ in thickness by less than ⅛ in. (3.2 mm), average their measurements to determine the minimum face shell thickness for that unit. Disregard sash grooves, dummy joints, and similar details in the measurements.

6. Compressive Strength

6.1 *Apparatus:*

6.1.1 *Testing Machine*—The testing machine shall conform to the requirements prescribed in Practices E 4. The machine shall be equipped with two steel bearing blocks (Note 3), one of which is a spherically seated block that will transmit load to the upper surface of the masonry specimen, and the other a plane rigid block on which the specimen will rest. When the bearing area of the steel blocks is not sufficient to cover the bearing area of the masonry specimen, steel bearing plates meeting the requirements of 6.1.2 shall be placed between the bearing blocks and the capped specimen after the centroid of the masonry bearing surface has been aligned with the center of thrust of the bearing blocks (see 6.4.1).

6.1.2 *Steel Bearing Blocks and Plates*—The surfaces of the steel bearing blocks and plates shall not depart from a plane by more than 0.001 in. (0.025 mm) in any 6-in. (152.4-mm) dimension. The center of the sphere of the spherically seated upper bearing block shall coincide with the center of its bearing face. If a bearing plate is used, the center of the sphere of the spherically seated bearing block shall lie on a line passing vertically through the centroid of the specimen bearing face. The spherically seated block shall be held closely in its seat but shall be free to turn in any direction. The diameter of the face of the bearing blocks shall be at least 6 in. (152.4 mm). When the bearing area of the spherical bearing block is not sufficient to cover the area of the specimen, use a single-thickness steel plate with a thickness equal to at least one half of the distance from the edge of the spherical bearings to the most distant corner of the specimen. In no case shall the plate thickness be less than 1 in. (25.4 mm). The length and width of the steel plate shall be at least ¼ in. (6 mm) greater than the length and width of the specimen.

NOTE 3—It is desirable that the bearing faces of blocks and plates used for compression testing of concrete masonry have a Rockwell hardness of not less than HRC 60 (BHN 620).

6.2 *Test Specimens:*

6.2.1 Of the six units sampled, three shall be tested in compression. After delivery to the laboratory, store (unstacked and separated by not less than 0.5 in. (13 mm) on all sides) continuously in air at a temperature of 75 ± 15°F (24 ± 8°C) and a relative humidity of less than 80 % for not less than 48 h. Alternatively, if compression results are required sooner, store units unstacked in the same environment described above with a current of air from an electric fan passing over them for a period of not less than 4 h. Continued until two successive weighings at intervals of 2 h show an increment of loss of not greater than 0.2 % of the previous determined weight of the specimen and until no moisture or dampness is visible on any surface of the unit. Specimens shall not be subjected to oven-drying. Specimens shall be full sized units except as modified in 6.2.2 through 6.2.5.

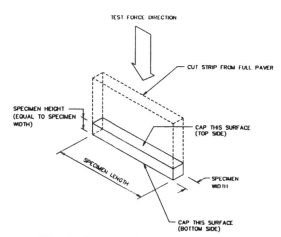

FIG. 1 Compressive Strength Test Setup

NOTE 4—In this test method, net area (other than certain solid units, see 9.4) is determined from specimens other than those subjected to compression testing. The compressive strength method is based on the assumption that units used for determining net volume (absorption specimens) have the same net volume as units used for compression testing. Sampled split face units, which have irregular surfaces, should be divided at the time they are sampled from the lot, such that the absorption test specimens have a net volume that is visually representative and a weight that is representative of the compression test specimens.

6.2.2 Unsupported projections having a length greater than the thickness of the projection shall be removed by saw-cutting. For units with recessed webs, the face shell projecting above the web shall be removed by saw-cutting to provide a full bearing surface over the net cross section of the unit. Where the resulting unit height would be reduced by more than one-third of the original unit height, the unit shall be coupon tested in accordance with 6.2.4.

6.2.3 When compression testing full-sized units that are too large for the test machine's bearing block and platens or are beyond the load capacity of the test machine, saw-cut the units to properly size them to conform to the capabilities of the testing machine. The resulting specimen shall have no face shell projections or irregular webs and shall be fully enclosed in a four-sided cell or cells. The compressive strength of the segment shall be considered to be the compressive strength of the whole unit.

6.2.4 When compression testing units of unusual size and shape (such as, but not limited to, bond beam units, open end units, and pilaster units), the specimens shall be sawed to remove any face shell projections. The resulting specimen shall be a cell or cells containing four sides that will ensure a 100 % bearing surface. Where saw-cutting will not result in an enclosed four-sided unit, the specimen shall be a coupon cut from a face shell of each unit. The coupon size shall have a height to thickness ratio of 2 to 1 before capping and a length to thickness ratio of 4 to 1. The coupon shall be cut from the unit such that the coupon height dimension is in the same direction as the unit height dimension. The compressive strength of the coupon shall be the net area compressive strength of the whole unit.

6.2.5 For concrete roof paver compressive strength tests, cut three test specimens from three whole paver units. Each specimen shall consist of a strip of paver with specimen height equal to specimen width. Where a unit contains supporting ribs, obtain specimens by cutting perpendicular to the direction of the ribs so as to avoid inclusion of bevelled or recessed surfaces at top or bottom edges (see Fig. 1).

6.2.6 For segmental retaining wall unit compressive strength tests, tested specimens shall be not less than 75 % solid and have a height to thickness ratio of not less than 1:1 and not more than 2:1. Saw-cutting, if necessary to obtain the required test specimen, shall be performed in accordance with 6.2.3 and 6.2.7.

6.2.7 Sawing shall be performed in an accurate, competent manner subjecting the specimen to as little saw vibration as possible. Use a diamond saw blade of proper hardness. If the specimen is wetted during sawing, allow the specimen to dry to equilibrium with laboratory air conditions before testing using the procedures outlined in 6.2.1.

6.2.8 If compression test specimens have been saw-cut from full-size units in accordance with the provisions of 6.2.2 through 6.2.5 and the net area of the compression test specimens can not be determined by 9.4.1, saw-cut an additional three units to the dimensions and configuration of the three compression test specimens. The average net area for the saw-cut compression specimens shall be taken as the average net area of the additional three saw-cut units calculated as required in 9.4. Calculated net volumes of saw-cut specimens shall not be used in calculating equivalent thickness.

6.3 *Capping Test Specimens:*

6.3.1 Cap bearing surfaces of units by one of the methods in 6.3.2 or 6.3.3.

6.3.2 *Sulfur and Granular Materials*—Spread evenly on a nonabsorbent capping surface that has been lightly coated with oil (Note 5) or sprayed with a TFE-fluorocarbon coating. Use proprietary or laboratory prepared mixtures of 40 to 60 % sulfur by weight, the remainder being ground fire clay or other suitable inert material passing a No. 100 (150-μm) sieve with or without a plasticizer. Heat the sulfur mixture in a thermostatically controlled heating pot to a temperature sufficient to maintain fluidity after contact with the capping surface. Take care to prevent overheating, and stir the liquid in the pot just before use. The capping surface shall be plane within 0.003 in. (0.08 mm) in 16 in. (406.4 mm) and shall be sufficiently rigid and supported so as not to be measurably deflected during the capping operation. Place four 1-in. (25-mm) square steel bars on the *capping* surface plate to form a rectangular mold approximately ½ in. (12.7 mm) greater in either inside dimension than the masonry unit. Fill the mold to a depth of ¼ in. (6.4 mm) with molten sulfur material. Bring the surface of the unit to be capped quickly into contact with the liquid, and insert the specimen, holding it so that its axis is at right angles to the surface of the capping liquid. Allow the unit to remain undisturbed until solidification is complete. Allow the caps to cool for a minimum of 2 h before testing the specimens. Patching of caps shall not be permitted. Remove imperfect caps and replace with new ones.

NOTE 5—The use of oil on capping plates may be omitted if it is found that plate and unit can be separated without damaging the cap.

6.3.3 *Gypsum Cement Capping*—Spread evenly on a nonabsorbent capping surface that has been lightly coated with oil (Note 5) or sprayed with a TFE-fluorocarbon coating, a neat paste of special high-strength gypsum cement

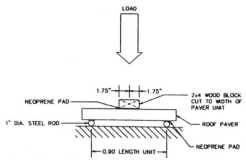

FIG. 2 Flexural Strength Test Setup

(Note 6) and water. Such gypsum cement, when gaged with water at the capping consistency, shall have a compressive strength at a 2-h age of not less than 3500 psi (24.1 MPa) when tested as 2-in. (50.8-mm) cubes. The casting surface plate shall conform to the requirements described in 6.3.2. Bring the surface of the unit to be capped into contact with the capping paste; firmly press down the specimen with a single motion, holding it so that its axis is at right angles to the capping surface. The average thickness of the cap shall not exceed ⅛ in. (3.2 mm). Patching of caps shall not be permitted. Remove imperfect caps and replace with new ones. Age the caps for at least 2 h before testing the specimens.

NOTE 6—The following two gypsum cements are considered to be in this classification: Hydrostone and Hydrocal white gypsum cement. Other cements should not be used unless shown by test to meet the strength requirement.

6.4 *Procedure:*

6.4.1 *Position of Specimens*—Test specimens with the centroid of their bearing surfaces aligned vertically with the center of thrust of the spherically seated steel bearing block of the testing machine (Note 7). Except for special units intended for use with their cores in a horizontal direction, test all hollow concrete masonry units with their cores in a vertical direction. Test masonry units that are 100 % solid and special hollow units intended for use with their hollow cores in a horizontal direction in the same direction as in service.

NOTE 7—For those masonry units that are symmetrical about an axis, the location of that axis can be determined geometrically by dividing the dimension perpendicular to that axis (but in the same plane) by two. For those masonry units that are nonsymmetrical about an axis, the location of that axis can be determined by balancing the masonry unit on a knife edge or a metal rod placed parallel to that axis. If a metal rod is used, the rod shall be straight, cylindrical (able to roll freely on a flat surface), have a diameter of not less than ¼ in. (6.4 mm) and not more than ¾ in. (19.1 mm), and its length shall be sufficient to extend past each end of the specimen when placed upon it. The metal rod shall be placed on a smooth, flat, level surface. One determined, the centroidal axis shall be marked on the end of the unit using a pencil or marker having a marking width of not greater than 0.05 in. (1.3 mm). A tamping rod used for consolidation of concrete and grout for slump tests performed in accordance with Test Method C 143 is often used as a balancing rod.

6.4.2 *Moisture Condition of Specimens*—At the time the specimens are tested, they shall be free of visible moisture or dampness.

6.4.3 *Speed of Testing*—Apply the load up to one half of the expected maximum load at any convenient rate, after which adjust the controls of the machine as required to give a uniform rate of travel of the moving head such that the remaining load is applied in not less than 1 nor more than 2 min.

6.4.4 *Maximum Load*—Record the maximum compressive load in pounds (newtons) as P_{max}.

7. Flexural Load on Concrete Roof Pavers

7.1 Three full-size units shall be tested.

7.2 *Capping*—Units with wearing (top) surfaces containing recesses or other irregularities shall have such recesses capped flush with the uppermost surface by either the sulfur and granular method or the gypsum cement casting method in accordance with 6.3.

7.3 *Testing*—The testing arrangement shall be as shown in Fig. 2. The load from the upper bearing block of the testing machine shall be applied through the centroid of the concrete roof paver by the bearing assembly illustrated. The flexural length of the paver units is taken as the end to end plan dimension of the units.

8. Absorption

8.1 *Apparatus*—The balance used shall be sensitive to within 0.5 % of the weight of the smallest specimen tested.

8.2 *Test Specimens*—Three full-size units that have been marked, weighed, and recorded in accordance with 4.3.2 shall be used.

8.3 *Procedure:*

8.3.1 *Saturation*—Immerse the test specimens in water at a temperature of 60 to 80°F (15.6 to 26.7°C) for 24 h. Weigh the specimens while suspended by a metal wire and completely submerged in water and record W_i (immersed weight). Remove from the water and allow to drain for 1 min by placing them on a ⅜-in. (9.5-mm) or coarser wire mesh, removing visible surface water with a damp cloth; weigh and record as W_s (saturated weight).

8.3.2 *Drying*—Subsequent to saturation, dry all specimens in a ventilated oven at 212 to 239°F (100 to 115°C) for not less than 24 h and until two successive weighings at intervals of 2 h show an increment of loss not greater than 0.2 % of the last previously determined weight of the specimen. Record weight of dried specimens as W_d (oven-dry weight).

9. Calculations

9.1 *Absorption*—Calculate absorption as follows:

$$\text{Absorption, lb/ft}^3 = [(W_s - W_d)/(W_s - W_i)] \times 62.4$$
$$\text{Absorption, kg/m}^3 = [(W_s - W_d)/(W_s - W_i)] \times 1000$$
$$\text{Absorption, \%} = [(W_s - W_d)/W_d] \times 100 \quad (1)$$

where:
W_s = saturated weight of unit, lb (kg) (see 8.3.1),
W_i = immersed weight of unit, lb (kg) (see 8.3.1), and
W_d = oven-dry weight of unit, lb (kg) (see 8.3.2).

9.2 *Moisture Content*—Calculate the moisture content as follows:

$$\text{Moisture Content, \% of total absorption} = [(W_r - W_d)/(W_s - W_d)] \times 100 \quad (2)$$

where:
W_r = received weight of unit, lb (kg) (see 4.3.2),
W_d = oven-dry weight of unit, lb (kg) (see 8.3.2), and

W_s = saturated weight of unit, lb (kg) (see 8.3.1).

9.3 *Density*—Calculate oven-dry density as follows:

$$\text{Density }(D), \text{lb/ft}^3 = [W_d/(W_s - W_i)] \times 62.4$$
$$\text{Density }(D), \text{kg/m}^3 = [W_d/(W_s - W_i)] \times 1000 \quad (3)$$

where:
W_d = oven-dry weight of unit, lb (kg) (see 8.3.2),
W_s = saturated weight of unit, lb (kg) (see 8.3.1), and
W_i = immersed weight of unit, lb (kg) (see 8.3.1).

9.4 *Average Net Area*—Calculate average net area as follows:

$$\text{Net Volume }(V_n), \text{ft}^3 = W_d/D = (W_s - W_i)/62.4$$
$$\text{Net Volume }(V_n), \text{mm}^3 = W_d/D = (W_s - W_i) \times 10^4$$
$$\text{Average Net Area }(A_n), \text{in.}^2 = (V_n \times 1728)/H$$
$$\text{Average Net Area }(A_n), \text{mm}^2 = V_n/H \quad (4)$$

where:
V_n = net volume of unit, ft³ (mm³),
W_d = oven-dry weight of unit, lb (kg) (see 8.3.2),
D = oven-dry density of unit, lb/ft³ (kg/m³) (see 9.3),
W_s = saturated weight of unit, lb (kg) (see 8.3.1),
W_i = immersed weight of unit, lb (kg) (see 8.3.1),
A_n = average net area of unit, in.² (mm²), and
H = average height of unit, in. (mm) (see 5.3.2).

9.4.1 Except for irregularly shaped units, such as those with split surfaces, calculate the net area of coupons and those units whose net cross-sectional area in every plane parallel to the bearing surface is the gross cross-sectional area measured in the same plane, as follows:

$$\text{Net Area }(A_n), \text{in.}^2 \text{ (mm}^2) = L \times W \quad (5)$$

where:
A_n = net area of the coupon or unit, in.² (mm²),
L = average length of the coupon or unit, in. (mm) (see 5.3.1), and
W = average width of the coupon or unit, in. (mm) (see 5.3.1)

9.5 *Gross Area*—Calculate gross area as follows:

$$\text{Gross Area }(A_g), \text{in.}^2 \text{ (mm}^2) = L \times W \quad (6)$$

where:
A_g = gross area of the unit, in.² (mm²),
L = average length of the unit, in. (mm) (see 5.3.2), and
W = average width of the unit, in. (mm) (see 5.3.2).
The gross cross-sectional area of a unit is the total area of a section perpendicular to the direction of the load, including areas within cells and reentrant spaces, unless these spaces are to be occupied in the masonry by portions of adjacent masonry.

9.6 *Compressive Strength:*

9.6.1 *Net Area Compressive Strength*—Calculate the net area compressive strength of the unit as follows:

$$\text{Net Area Compressive Strength, psi (MPa)} = P_{max}/A_n \quad (7)$$

where:
P_{max} = maximum compressive load, lb (N) (see 6.4.3), and
A_n = average net area of unit, in.² (mm²) (see 9.4).

9.6.2 *Gross Area Compressive Strength*—Calculate the gross area compressive strength of the unit as follows:

$$\text{Gross Area Compressive Strength, psi (MPa)} = P_{max}/A_g \quad (8)$$

where:

TABLE 1 Height to Thickness Correction Factors for Segmental Retaining Wall Unit Compression Tests

h/t[A]	1.0	1.1	1.2	1.3	1.4	1.5	1.6	1.7	1.8	1.9	2.0
correction factor	0.85	0.88	0.90	0.92	0.94	0.95	0.96	0.97	0.98	0.99	1.00

[A] h/t = ratio of measured specimen height to least measured lateral dimension.

P_{max} = maximum compressive load, lb (N) (see 6.4.3), and
A_g = gross area of unit, in.^{2} (mm^2) (see 9.5).

9.6.3 *Corrected Net Area Compressive Strength for Segmental Retaining Wall Unit Test Specimens*—Multiply the calculated net area compressive strength of each specimen by the height to thickness ratio correction factor in Table 1.

9.7 *Equivalent Web Thickness*—Equivalent web thickness of each unit (in inches per linear foot of specimen) is equal to the sum of the measured thicknesses of all webs in the unit multiplied by 12 and divided by the length of the unit.

NOTE 8—Equivalent web thickness does not apply to the portion of the unit to be filled with grout. The length of that portion should be deducted from the overall length of the unit.

9.8 *Equivalent Thickness*—Equivalent thickness for concrete masonry is defined as the average thickness of solid material in the unit and is calculated as follows:

$$T_e, \text{in.} = [V_n/(L \times H)] \times 1728 \qquad (9)$$
$$T_e, \text{mm} = [V_n/(L \times H)]$$

where:
T_e = equivalent thickness, in. (mm),
V_n = average net volume of full-size units, ft^3 (mm^3) (see 9.4),
L = average length of full-size units, in. (mm) (see 5.3.2), and
H = average height of full-size units, in. (mm) (see 5.3.2).

9.9 *Ballast Weight*—For concrete roof pavers, calculate ballast weight as follows:

$$W_b \text{ (lb/ft}^2) = \frac{W_d}{A_g} \times 144$$
$$\qquad (10)$$
$$W_b \text{ (kg/m}^2) = \frac{W_d}{A_g} \times 10^6$$

where:
W_b = ballast weight, lb/ft^2 (kg/m^2),
W_d = oven-dry weight of unit, lb (kg) (see 8.3.2), and
A_g = gross area of unit, in.^{2} (mm^2) (see 9.5).

10. Report

10.1 A complete report shall include the following:

10.1.1 The net area compressive strength to the nearest 10 psi (0.1 MPa) separately for each specimen and as the average for three specimens as determined by 9.6.1.

10.1.1.1 For segmental retaining wall units, report the tested net area compressive strength to the nearest 10 psi (0.1 MPa), the height to thickness ratio, and the corrected net area compressive strength separately for each specimen as determined by 9.6. Also, report the average corrected net area compressive strength for the set of three specimens.

10.1.2 The absorption and density results separately for each unit and as the average for the three units as determined by 9.1 and 9.3.

10.1.3 The average width, height, and length of each specimen as determined by 5.3.2.

10.1.4 The minimum face shell thickness as an average of the minimum face shell thicknesses recorded for each of three specimens as determined by 5.3.2.

10.1.5 The minimum web thickness as an average of the minimum web thicknesses recorded for each of three specimens as determined by 5.3.2.

10.1.6 The equivalent web thickness as an average for three specimens as determined by 9.7.

10.1.7 The equivalent thickness as an average for three specimens as determined by 9.8 when required.

10.1.8 The moisture content as an average for three specimens as determined by 9.2 when required.

10.1.9 *Flexural Strength of Concrete Roof Pavers*—Report the flexural load required to fail a unit separately and as an average for three units.

11. Keywords

11.1 absorption; compressive strength; concrete masonry units; density; equivalent thickness; equivalent web thickness; face shell; moisture content; roof paver; web thickness; webs

APPENDIX

(Nonmandatory Information)

XI. WORKSHEET AND TEST REPORT

ASTM C 140 Worksheet

Lab Proj. No.: _____
Date Received: _____

Client: _____
Address: _____
_____
_____

Testing Agency: _____
Address: _____
_____
_____

Job No./Description: _____

Sampling Party: _____

Unit Designation/Description:
_____
_____
_____

Specified Overall Dimensions: Width (in.) _____
Height (in.) _____
Length (in.) _____

Compression Units
(Determine the following information for each of the three units to be tested in compression.)

	Unit #1	Unit #2	Unit #3	
Width (W) @ Top	_____	_____	_____	in.
@ Bottom	_____	_____	_____	in.
Height (H) @ Face 1	_____	_____	_____	in.
@ Face 2	_____	_____	_____	in.
Length (L) @ Face 1	_____	_____	_____	in.
@ Face 2	_____	_____	_____	in.
Faceshell Thickness (FST) @ Face 1	_____	_____	_____	in.
@ Face 2	_____	_____	_____	in.
Web Thickness (WT) @ Web 1	_____	_____	_____	in.
@ Web 2	_____	_____	_____	in.
@ Web 3	_____	_____	_____	in. Measurements...
@ Web 4	_____	_____	_____	in. Date _____ By _____
Received Weight (W_R)	_____	_____	_____	lb Date _____ By _____
Max. Compressive Load (P_{MAX})	_____	_____	_____	lb Date _____ By _____

Absorption Units
(Determine the following information for each of the three units to be immersed in water for absorption testing.)

	Unit #4	Unit #5	Unit #6	
Width (W) @ Top	_____	_____	_____	in.
@ Bottom	_____	_____	_____	in.
Height (H) @ Face 1	_____	_____	_____	in.
@ Face 2	_____	_____	_____	in.
Length (L) @ Face 1	_____	_____	_____	in. Measurements...
@ Face 2	_____	_____	_____	in. Date _____ By _____
Received Weight (W_R)	_____	_____	_____	lb Date _____ By _____
Immersed Weight (W_I)	_____	_____	_____	lb Date _____ By _____
Saturated Weight (W_s)	_____	_____	_____	lb Date _____ By _____
Oven-Dry Weight (W_D)	_____	_____	_____	lb Date _____ By _____

FIG. X1.1 Worksheet

ASTM C 140

ASTM C 140 Test Report

Job No.: _____
Report Date: _____

Client: _____
Address: _____
_____
_____

Testing Agency: _____
Address: _____
_____
_____

Job No./Description: _____

Sampling Party: _____

Unit Specification: ASTM C_____

Unit Designation/Description:
_____
_____
_____

Unit Configuration:

(sketch)

Summary of Test Results

	Required Values	Tested Values
Net Area Compressive Strength	****	----- psi
Gross Area Compressive Strength	****	----- psi
Density	****	----- pcf
Absorption	****	----- pcf
Minimum Faceshell Thickness	****	----- in.
Minimum Web Thickness	****	----- in.
Equivalent Web Thickness	****	----- in.
Equivalent Thickness	****	----- in.
Max. Variation from Specified Dimensions	****	----- in.
Net Cross-Sectional Area	****	----- in.^{2}
Gross Cross-Sectional Area	****	----- in.^{2}
Percent Solid	****	----- %
Moisture Content	****	----- %

Individual Unit Test Results

	Avg. Width in.	Avg. Height in.	Avg. Length in.	Received Weight lb	Max. Load lb	Cross-Sectional Area Gross in.^{2}	Net in.^{2}	Compressive Strength Gross psi	Net psi
Unit #1	-----	-----	-----	-----	-----	-----	-----	-----	-----
Unit #2	-----	-----	-----	-----	-----	-----	-----	-----	-----
Unit #3	-----	-----	-----	-----	-----	-----	-----	-----	-----
Average	-----	-----	-----	-----	-----	-----	-----	-----	-----

	Avg. Width in.	Avg. Height in.	Avg. Length in.	Received Weight lb	Absorp. pcf	Density pcf	Gross Volume ft^3	Net Volume ft^3	Percent Solid %
Unit #4	-----	-----	-----	-----	-----	-----	-----	-----	-----
Unit #5	-----	-----	-----	-----	-----	-----	-----	-----	-----
Unit #6	-----	-----	-----	-----	-----	-----	-----	-----	-----
Average	-----	-----	-----	-----	-----	-----	-----	-----	-----

Signature of Lab Director
Name of Lab Director
Title of Lab Director

FIG. X1.2 Test Report

SUMMARY OF CHANGES

Committee C-15 has identified the location of selected changes to this standard since the C 140-96a edition that may impact the use of this standard.

(*1*) Inserted new Note 1 suggesting laboratory accreditation in accordance with Practice C 1093 and renumbered subsequent notes.

(*2*) Inserted Practice C 1093 into Section 2, Referenced Documents.

(*3*) Revised drying requirements for wet saw-cut specimens in 6.2.7.

(*4*) Revised precision requirements for reporting compressive strength values in 10.1.1 and 10.1.1.1.

Standard Specification for
Facing Brick (Solid Masonry Units Made from Clay or Shale)[1]

This standard is issued under the fixed designation C 216; the number immediately following the designation indicates the year of original adoption or, in the case of revision, the year of last revision. A number in parentheses indicates the year of last reapproval. A superscript epsilon (ε) indicates an editorial change since the last revision or reapproval.

This standard has been approved for use by agencies of the Department of Defense. Consult the DoD Index of Specifications and Standards for the specific year of issue which has been adopted by the Department of Defense.

1. Scope

1.1 This specification covers brick intended for use in masonry and supplying structural or facing components, or both, to the structure.

1.2 The property requirements of this standard apply at the time of purchase. The use of results from testing of brick extracted from masonry structures for determining conformance or nonconformance to the property requirements (Section 5) of this standard is beyond the scope of this standard.

1.3 The brick are prismatic units available in a variety of sizes, textures, colors, and shapes. This specification is not intended to provide specifications for paving brick (see Specification C 902).

1.4 Brick are manufactured from clay, shale, or similar naturally occurring earthy substances and subjected to a heat treatment at elevated temperatures (firing). The heat treatment must develop a fired bond between the particulate constituents to provide the strength and durability requirements of this specification (see firing, fired bond, and incipient fusion in Terminology C 43).

1.5 Brick may be shaped during manufacture by molding, pressing, or extrusion, and the shaping method may be used to describe the brick.

1.6 Three types of brick in each of two grades are covered.

1.7 The values stated in inch-pound units are to be regarded as the standard. The values given in parentheses are for information only.

2. Referenced Documents

2.1 *ASTM Standards:*
C 43 Terminology of Structural Clay Products[2]
C 67 Test Methods of Sampling and Testing Brick and Structural Clay Tile[2]
C 902 Specification for Pedestrian and Light Traffic Paving Brick[2]
E 835/E 835M Guide for Modular Coordination of Clay and Concrete Masonry Units[3]

3. Grades

3.1 Grades classify brick according to their resistance to damage by freezing when wet, as defined in Note 1. Two grades of facing brick are covered and the requirements are shown in Table 1.

3.1.1 *Grade SW*—Brick intended for use where high and uniform resistance to damage caused by cyclic freezing is desired and where the brick may be frozen when saturated with water.

3.1.2 *Grade MW*—Brick which may be used where moderate resistance to cyclic freezing damage is permissible or where the brick may be damp but not saturated with water when freezing occurs.

NOTE 1—The word "saturated," with respect to this standard, refers to the condition of a brick that has absorbed water to an amount equal to that resulting from submersion in room temperature water for 24 h.

4. Types

4.1 Three types of facing brick are covered:

4.1.1 *Type FBS*—Brick for general use in masonry.

4.1.2 *Type FBX*—Brick for general use in masonry where a higher degree of precision and lower permissible variation in size than permitted for Type FBS is required.

4.1.3 *Type FBA*—Brick for general use in masonry selected to produce characteristic architectural effects resulting from nonuniformity in size and texture of the individual units.

4.2 When the type is not specified, the requirements for Type FBS shall govern.

5. Physical Properties

5.1 *Durability*—When Grade is not specified, the requirements for Grade SW shall govern. Unless otherwise specified by the purchaser, brick of Grade SW shall be accepted instead of Grade MW.

5.1.1 *Physical Property Requirements*—The brick shall conform to the physical requirements for the Grade specified as prescribed in Table 1. For the compressive strength requirements in Table 1, test the unit with the compressive force perpendicular to the bed surface of the unit, with the unit in the stretcher position.

5.1.2 *Absorption Alternate*—The saturation coefficient requirement does not apply, provided the cold water absorption of any single unit of a random sample of five brick does not exceed 8 %.

5.1.3 *Freezing and Thawing Alternative*—The requirements for 5 h boiling water absorption and saturation coefficient do not apply, provided a sample of five brick, meeting the strength requirements of Table 1, passes the freezing and thawing test as described in the Rating Section

[1] This specification is under the jurisdiction of ASTM Committee C-15 on Manufactured Masonry Units and is the direct responsibility of Subcommittee C15.02 on Clay Brick and Structural Clay Tile.
Current edition approved June 15, 1995. Published August 1995. Originally published as C 216 – 47T. Last previous edition C 216 – 95.
[2] *Annual Book of ASTM Standards*, Vol 04.05.
[3] *Annual Book of ASTM Standards*, Vol 04.07.

TABLE 1 Physical Requirements

Designation	Minimum Compressive Strength psi, (MPa) gross area		Maximum Water Absorption by 5-h Boiling, %		Maximum Saturation Coefficient[A]	
	Average of 5 brick	Individual	Average of 5 brick	Individual	Average of 5 brick	Individual
Grade SW	3000 (20.7)	2500 (17.2)	17.0	20.0	0.78	0.80
Grade MW	2500 (17.2)	2200 (15.2)	22.0	25.0	0.88	0.90

[A] The saturation coefficient is the ratio of absorption by 24-h submersion in cold water to that after 5-h submersion in boiling water.

TABLE 2 Grade Recommendations for Face Exposures

Exposure	Weathering Index (Explanatory Note 2)	
	Less than 50	50 and greater
vertical surfaces:		
In contact with earth	MW	SW
Not in contact with earth	MW	SW
other than vertical surfaces:		
In contact with earth	SW	SW
Not in contact with earth	MW	SW

f the Freezing and Thawing test procedures of Test Methods
67:

5.1.3.1 *Grade SW—Weight Loss Requirement*—Not reater than 0.5 % loss in dry weight of any individual unit.

NOTE 2—The 50 cycle freezing and thawing test is specified as an ternative only when brick do not conform to either Table 1 require- ents for maximum water absorption and saturation coefficient, or to e requirements of the Absorption Alternate in Section 5.1.2.

5.1.4 *Waiver of Absorption and Saturation Coefficient equirements*—If the brick are intended for use exposed to eather where the weathering index is less than 50 (see Fig.), and unless otherwise specified, the requirements given in able 1 for 5-h boiling water absorption and for saturation oefficient shall not apply, but the minimum average com-

pressive strength requirement of 2500 psi (17.2 MPa) shall apply.

NOTE 3—The effect of weathering on brick is related to the weath- ering index, which for any locality is the product of the average annual number of *freezing cycle days* and the average annual *winter rainfall* in inches (millimetres), defined as follows.[4]

A Freezing Cycle Day is any day during which the air temperature passes either above or below 32°F (0°C). The average number of freezing cycle days in a year may be taken to equal the difference between the mean number of days during which the minimum temperature was 32°F or below, and the mean number of days during which the maximum temperature was 32°F or below.

Winter Rainfall is the sum, in inches (millimetres), of the mean monthly corrected precipitation (rainfall) occurring during the period between and including the normal date of the first killing frost in the fall and the normal date of the last killing frost in the spring. The winter rainfall for any period is equal to the total precipitation less one tenth of the total fall of snow, sleet, and hail. Rainfall for a portion of a month is prorated.

Fig. 1 indicates general areas of the United States in which brick masonry is subject to severe, moderate, and negligible weathering. The severe weathering region has a weathering index greater than 500. The moderate weathering region has a weathering index of 50 to 500. The

---

[4] Data needed to determine the weathering for any locality may be found or estimated from tables of Local Climatological Data—Annual Summary with Comparative Data available from the National Oceanic and Atmospheric Admin- istration.

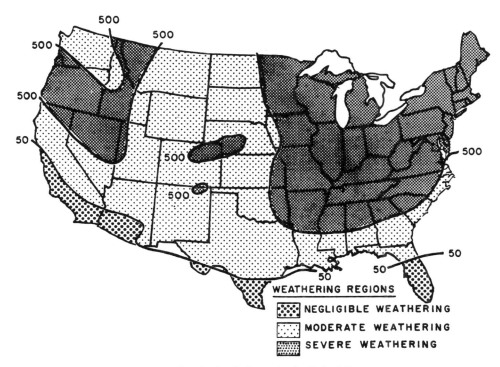

WEATHERING REGIONS
▨ NEGLIGIBLE WEATHERING
▫ MODERATE WEATHERING
▨ SEVERE WEATHERING

FIG. 1 Weathering Indexes in the United States

negligible weathering region has a weathering index of less than 50. The index for geographic locations near the 50 and 500-in. cycle lines should be determined by analysis of weather bureau local climatological summaries, with due regard to the effect of microclimatic conditions, especially altitude.

The recommended correlation between grade of facing brick, weathering index, and exposure is found in Table 2. The specifier may use these recommendations or use the grade descriptions and physical requirements along with use exposure and local climatological conditions to select grade.

5.2 *Strength*—When brick are required having strengths greater than prescribed by this specification, the purchaser shall specify the desired minimum compressive strength.

5.3 *Initial Rate of Absorption (IRA)*—Test results for IRA shall be determined in accordance with the IRA (Suction) (Laboratory Test) of Test Methods C 67 and shall be furnished at the request of the specifier or purchaser. IRA is not a qualifying condition or property of units in this specification. This property is measured in order to assist in mortar selection and material handling in the construction process. See Note 4.

NOTE 4—*Initial Rate of Absorption (Suction)*—Both laboratory and field investigation have shown that strong and watertight joints between mortar and masonry units are not achieved by ordinary construction methods when the units as laid have excessive initial rates of absorption. Mortar that has stiffened somewhat because of excessive loss of mixing water to a unit may not make complete and intimate contact with the second unit, resulting in poor adhesion, incomplete bond, and water-permeable joints of low strength. IRA of the units is determined by the oven-dried procedure described in the IRA (Suction) (Laboratory Test) of Test Methods C 67. IRA in the field depends on the moisture content of the masonry unit and is determined in accordance with the IRA (Suction)—Field Test of Test Methods C 67. Units having average field IRA exceeding 30 g/min·30 in.^{2} (30 g/min·194 cm^2) should have their IRA reduced below 30 g/min·30 in.^{2} prior to laying. They may be wetted immediately before they are laid, but it is preferable to wet them thoroughly 3 to 24 h prior to their use so as to allow time for moisture to become distributed throughout the unit.

6. Efflorescence

6.1 When the brick are tested in accordance with Test Methods C 67, the rating for efflorescence shall be: "not effloresced."

7. Material and Finish

7.1 Colors and textures produced by application of inorganic coatings to the faces of the brick are permissible with the consent of the purchaser, provided that evidence is furnished of the durability of the coatings. Brick that are colored by flashing or textured by sanding, where the sand does not form a continuous coating, are not considered as surface-colored brick for the purpose of this specification.

NOTE 5—When surface colored brick, other than sanded or flashed, are specified for *exterior* use, the purchaser should require that data be submitted showing that after 50 cycles of freezing thawing there is no observable difference in the applied finish when viewed from a distance of 10 ft (3.0 m) under an illumination of not less than 50 ft-candles (538 lx) by an observer with normal vision.

Service records of the performance of the particular coated brick in exterior locations may be accepted in place of the freezing and thawing test, upon consent of the purchaser.

7.2 The brick shall be free of defects, deficiencies, and surface treatments, including coatings, that would interfere with the proper setting of the brick or significantly impair the strength or performance of the construction.

7.3 If any post-firing coatings or surface treatments are applied by the manufacturer, the manufacturer shall report the type and extent of these coatings or surface treatments in all certificates of compliance with this specification.

7.4 The face or faces that will be exposed in place shall be free of chips that exceed the limits given in Table 3. The aggregate length of chips shall not exceed 10 % of the perimeter of the face of the brick.

NOTE 6—Of all the units that will be exposed in place, a small percentage of the units may have chips that are larger in size than those chips allowed for the majority of the units. This special allowed percentage, listed in the second column from the left of Table 3 ranges up to 5 % for FBX, up to 10 % for FBS Smooth, and up to 15 % for FBS Rough. The remainder of the units that will be exposed in place, listed in the fifth column from the left, must conform to the chip sizes listed in the sixth and seventh columns from the left.

Example: Type FBS Smooth units will conform to the requirements of Table 3 if not more than 10 % of the units have edge chips greater than ¼ in. (6.4 mm) but less than 5/16 in. (7.9 mm) or corner chips greater than 3/8 in. (9.5 mm) but less than ½ in. (12.7 mm) and the remainder of the units, in this maximum case 90 % (100 % − 10 %) do not have edge chips greater than ¼ in. (6.4 mm) in from the edge nor corner chips greater than 3/8 in. (9.5 mm) in from the corner.

7.4.1 Other than chips, the face or faces shall be free of cracks or other imperfections detracting from the appearance of the designated sample when viewed from a distance of 15 ft (4.6 m) for Type FBX and a distance of 20 ft (6.1 m) for Types FBS and FBA.

7.5 Unless otherwise agreed upon between the purchaser and the seller, a delivery of brick may contain not more than 5 % brick, including broken brick, that do not meet the requirements for chippage and tolerances.

7.6 After brick are placed in usage the manufacturer or his agent shall not be held responsible for compliance of brick

TABLE 3   Maximum Permissible Extent of Chippage From the Edges and Corners of Finished Face or Faces onto the Surface

Type	Percentage Allowed[A]	Chippage in in. (mm) in from		Percentage Allowed[A]	Chippage in in. (mm) in from	
		Edge	Corner		Edge	Corner
FBX	5 % or less	⅛–¼ (3.2–6.4)	¼–⅜ (6.4–9.5)	95 to 100 %	0–⅛ (0–3.2)	0–¼ (0–6.4)
FBS[B] (Smooth)	10 % or less	¼–5/16 (6.4–7.9)	⅜–½ (9.5–12.7)	90 to 100 %	0–¼ (0–6.4)	0–⅜ (0–9.5)
FBS[C] (Rough)	15 % or less	5/16–7/16 (7.9–11.1)	½–¾ (12.7–19.1)	85 to 100 %	0–5/16 (0–7.9)	0–½ (0–12.7)
FBA	to meet the designated sample or as specified by the purchaser, but not more restrictive than Type FBS (rough)					

[A] Percentage of exposed brick allowed in the wall with chips measured the listed dimensions in from an edge or corner.
[B] Smooth texture is the unbroken natural die finish.
[C] Rough texture is the finish produced when the face is sanded, combed, scratched, or scarified or the die skin on the face is entirely broken by mechanical means such as wire-cutting or wire-brushing.

TABLE 4   Tolerances on Dimensions

| Specified Dimension or Average Brick Size in Job Lot Sample, in. (mm) | Maximum Permissible Variation, in. (mm) plus or minus from: | | | | |
| | Column A (for Specified Dimension) | | Column B (for Average Brick Size in Job Lot Sample)[A] | | |
	Type FBX	Type FBS	Type FBX	Type FBS Smooth[B]	Type FBS Rough[C]
3 (76) and under	¹⁄₁₆ (1.6)	³⁄₃₂ (2.4)	¹⁄₁₆ (1.6)	¹⁄₁₆ (1.6)	³⁄₃₂ (2.4)
Over 3–4 (76 to 102), incl	³⁄₃₂ (2.4)	¹⁄₈ (3.2)	¹⁄₁₆ (1.6)	³⁄₃₂ (2.4)	¹⁄₈ (3.2)
Over 4–6 (102 to 152), incl	¹⁄₈ (3.2)	³⁄₁₆ (4.8)	³⁄₃₂ (2.4)	³⁄₃₂ (2.4)	³⁄₁₆ (4.8)
Over 6–8 (152 to 203), incl	⁵⁄₃₂ (4.0)	¹⁄₄ (6.4)	³⁄₃₂ (2.4)	¹⁄₈ (3.2)	¹⁄₄ (6.4)
Over 8–12 (203 to 305), incl	⁷⁄₃₂ (5.6)	⁵⁄₁₆ (7.9)	¹⁄₈ (3.2)	³⁄₁₆ (4.8)	⁵⁄₁₆ (7.9)
Over 12–16 (305 to 406), incl	⁹⁄₃₂ (7.1)	³⁄₈ (9.5)	³⁄₁₆ (4.8)	¹⁄₄ (6.4)	³⁄₈ (9.5)

[A] Lot size shall be determined by agreement between purchaser and seller. If not specified, lot size shall be understood to include all brick of one size and color in the job order.
[B] Type FBS Smooth units have relatively fine texture and smooth edges, including wire cut surfaces. These definitions relate to dimensional tolerances only.
[C] Type FBS Rough units have textured, rounded, or tumbled edges or faces. These definitions apply to dimensional tolerances only.

TABLE 5   Tolerances on Distortion

| Maximum Dimension, in. (mm) | Maximum Permissible Distortion, in. (mm) | |
	Type FBX	Type FBS
8 (203) and under	¹⁄₁₆ (1.6)	³⁄₃₂ (2.4)
Over 8–12 (203 to 305), incl	³⁄₃₂ (2.4)	¹⁄₈ (3.2)
Over 12–16 (305 to 406), incl	¹⁄₈ (3.2)	⁵⁄₃₂ (4.0)

with the requirements of this specification for chippage and dimensional tolerances.

8. Texture and Color

8.1 The color, color range, and texture should be specified by the purchaser. Unless otherwise specified by the purchaser, at least one end of the majority of the individual brick shall have the same general texture and general color tone as the approved sample. The texture of the finished surfaces that will be exposed when in place shall conform to an approved sample consisting of not less than four stretcher brick, each representing the texture desired. The color range shall be indicated by the approved sample.

8.2 Where brick with other than one finished face and one finished end are required (brick with two finished faces or ends, or other types), all such special brick shall be explicitly specified by the purchaser.

NOTE 7—The manufacturer should be consulted for the availability of specialty units suitable for the intended purpose.

9. Size

9.1 *Size*—The size of brick shall be as specified by the purchaser (see Note 8). In a sample of ten brick selected to represent the extreme range of sizes of brick to be supplied, no brick shall depart from the specified size by more than the individual tolerance for the type specified as prescribed in Table 4, Column A. The average size of the ten brick sample shall be determined, and no brick in the job lot (delivered brick) shall vary from this average size by more than the individual tolerance for the type specified as prescribed in Table 4, Column B. No individual brick in the job lot shall fall outside of the dimensional tolerances of Table 4, Column A. Tolerances on dimensions for Type FBA shall be as specified by the purchaser, but not more restrictive than FBS.

NOTE 8—For a list of modular sizes, see Guide E 835/E 835M. Sizes listed in this standard are not produced in all parts of the United States. Brick names denoting sizes may be regional and, therefore, may not be included in all reference books. Purchasers should ascertain the sizes of brick available in their locality and should specify accordingly, stating the desired dimensions (width by height by length).

9.2 *Warpage*—Tolerances for distortion or warpage of surfaces or edges intended to be exposed in use of individual brick from a plane surface and from a straight line, respectively, shall not exceed the maximum for the type specified as prescribed in Table 5. Tolerances on distortion for Type FBA shall be as specified by the purchaser.

9.3 *Out-of-Square*—The maximum permitted dimension for out-of-square of the exposed face of the brick is ¹⁄₈ in. (3.2 mm) for Type FBS brick and ³⁄₃₂ in. (2.4 mm) for Type FBX brick. Tolerances on out-of-square for Type FBA brick shall be specified by the purchaser.

NOTE 9—Linear dimensions and flat surfaces of specially shaped brick shall meet the requirements for size and warpage, respectively, of the specified type. Tolerances for size and warpage of nonlinear dimensions and surfaces, and out-of-square shall be determined by agreement with the manufacturer.

10. Coring and Frogging

10.1 *Coring*—Unless otherwise specified in the invitation for bids, brick may or may not be cored at the option of the seller. The net cross-sectional area of cored brick in any plane parallel to the surface containing the cores shall be at least 75 % of the gross cross-sectional area measured in the same plane. No part of any hole shall be less than ¾ in. (19.1 mm) from any edge of the brick.

10.2 *Frogging*—Unless otherwise specified in the invitation for bids, one bearing face of each brick may have a recess or panel frog and deep frogs. The recess or panel frog shall not exceed ³⁄₈ in. (9.5 mm) in depth and no part of the recess or panel frog shall be less than ¾ in. (19.1 mm) from any edge of the brick. In brick containing deep frogs, frogs deeper than ³⁄₈ in. (9.5 mm), any cross-section through the deep frogs parallel to the surface containing the deep frogs shall conform to the requirements of 10.1.

11. Sampling and Testing

11.1 For purposes of tests, brick that are representative of the commercial product shall be selected by a competent person appointed by the purchaser, the place or places of

TABLE 4 - 11.1   ASTM C 216   257

selection to be designated when the purchase order is placed. The sample or samples shall include specimens representative of the complete range of colors and sizes of the brick supplied or to be supplied. The manufacturer or the seller shall furnish specimens for tests without charge.

11.2 The brick shall be sampled and tested in accordance with Test Methods C 67.

NOTE 10—Unless otherwise specified in the purchase order, the cost of tests is typically borne as follows: If the results of the tests show that the brick do not conform to the requirements of this specification, the cost is typically borne by the seller. If the results of the tests show that the brick do conform to the requirements of this specification, the cost is typically borne by the purchaser.

12. Keywords

12.1 appearance requirements; clay; facing brick; fired masonry units; masonry construction; physical properties; shale; solid brick

Standard Specification for
Mortar for Unit Masonry[1]

This standard is issued under the fixed designation C 270; the number immediately following the designation indicates the year of original adoption or, in the case of revision, the year of last revision. A number in parentheses indicates the year of last reapproval. A superscript epsilon (ϵ) indicates an editorial change since the last revision or reapproval.

This standard has been approved for use by agencies of the Department of Defense. Consult the DoD Index of Specifications and Standards for the specific year of issue which has been adopted by the Department of Defense.

1. Scope

1.1 This specification covers mortars for use in the construction of non-reinforced and reinforced unit masonry structures. Four types of mortar are covered in each of two alternative specifications: (1) proportion specifications and (2) property specifications.

1.2 The proportion or property specifications shall govern as specified.

1.3 When neither proportion or property specifications are specified, the proportion specifications shall govern, unless data are presented to and accepted by the specifier to show that mortar meets the requirements of the property specifications.

1.4 The text of this standard references notes and footnotes which provide explanatory material. These notes and footnotes (excluding those in tables and figures) shall not be considered as requirements of the standard.

1.5 The following safety hazards caveat pertains only to the test methods section of this specification: *This standard does not purport to address all of the safety concerns, if any, associated with its use. It is the responsibility of the user of this standard to establish appropriate safety and health practices and determine the applicability of regulatory limitations prior to use.*

2. Referenced Documents

2.1 *ASTM Standards:*
C 5 Specification for Quicklime for Structural Purposes[2]
C 91 Specification for Masonry Cement[2]
C 109 Test Method for Compressive Strength of Hydraulic Cement Mortars (Using 2-in. or 50-mm Cube Specimens)[2]
C 144 Specification for Aggregate for Masonry Mortar[3]
C 150 Specification for Portland Cement[2]
C 207 Specification for Hydrated Lime for Masonry Purposes[2]
C 305 Practice for Mechanical Mixing of Hydraulic Cement Pastes and Mortars of Plastic Consistency[2]
C 511 Specification for Moist Cabinets, Moist Rooms, and Water Storage Tanks Used in the Testing of Hydraulic Cements and Concretes[2]

C 595/C 595M Specification for Blended Hydraulic Cements[2]
C 780 Test Method for Preconstruction and Construction Evaluation of Mortars for Plain and Reinforced Unit Masonry[3]
E 514 Test Method for Water Penetration and Leakage Through Masonry[3]
E 518 Test Methods for Flexural Bond Strength of Masonry[3]

2.2 *International Masonry All Weather Council:*[4]
Recommended Practices and Guide Specifications for Cold Weather Masonry Construction; Section 04200, Article 3 of the Guide Specifications, Sixth Edition, July 1977

3. Requirements

3.1 *Proportion Specifications*—Mortar conforming to the proportion specifications shall consist of a mixture of cementitious material, aggregate, and water, all conforming to the requirements of Section 4 and the proportion specifications' requirements of Table 1. See Appendixes X1 or X3 for a guide for selecting masonry mortars.

3.1.1 Unless otherwise stated, either a cement/lime mortar or a masonry cement mortar is permitted. Mortar of known higher strength shall not be indiscriminately substituted where a mortar type of anticipated lower strength is specified.

3.2 *Property Specifications*—Mortar conformance to the property specifications shall be established by tests of laboratory prepared mortar in accordance with Sections 5 and 6.2. The laboratory prepared mortar shall consist of a mixture of cementitious material, aggregate, and water, all conforming to the requirements of Section 4 and the properties of the laboratory prepared mortar shall conform to the requirements of Table 2. See Appendix X1 for a guide for selecting masonry mortars.

3.2.1 No change shall be made in the laboratory established proportions for mortar accepted under the property specifications, except for the quantity of mixing water. Materials with different physical characteristics shall not be utilized in the mortar used in the work unless compliance with the requirements of the property specifications is re-established.

NOTE 1—The required properties of the mortar in Table 2 are for laboratory prepared mortar mixed with a quantity of water to produce a

---

[1] This specification is under the jurisdiction of ASTM Committee C-12 on Mortars for Unit Masonry and is the direct responsibility of Subcommittee C12.03 on Specifications for Mortar.
Current edition approved Aug. 10, 1996. Published September 1996. Originally published as C 270 – 51 T. Last previous edition C 270 – 96.
[2] *Annual Book of ASTM Standards*, Vol 04.01.
[3] *Annual Book of ASTM Standards*, Vol 04.05.

[4] Available from the International Masonry All Weather Council, 823 Fifteenth Street, N. W., Washington, DC 20005.

TABLE 1 Proportion Specification Requirements

NOTE—Two air-entraining materials shall not be combined in mortar.

Mortar	Type	Proportions by Volume (Cementitious Materials)					Aggregate Ratio (Measured in Damp, Loose Conditions)
		Portland Cement or Blended Cement	Masonry Cement			Hydrated Lime or Lime Putty	
			M	S	N		
Cement-lime	M	1	...	...	...	¼	
	S	1	...	...	...	over ¼ to ½	
	N	1	...	...	...	over ½ to 1¼	
	O	1	...	...	...	over 1¼ to 2½	
Masonry cement	M	1	...	...	1	...	Not less than 2¼ and not more than 3 times the sum of the separate volumes of cementitious materials.
	M	...	1	...	...	...	
	S	½	...	...	1	...	
	S	...	...	1	...	...	
	N	...	...	...	1	...	
	O	...	...	...	1	...	

flow of 110 ± 5 %. This quantity of water is not sufficient to produce a mortar with a workable consistency suitable for laying masonry units in the field. Mortar for use in the field must be mixed with the maximum amount of water, consistent with workability, in order to provide sufficient water to satisfy the initial rate of absorption (suction) of the masonry units. The properties of laboratory prepared mortar at a flow of 110 ± 5, as required by this specification, are intended to approximate the flow and properties of field prepared mortar after it has been placed in use and the suction of the masonry units has been satisfied. The properties of field prepared mortar mixed with the greater quantity of water, prior to being placed in contact with the masonry units, will differ from the property requirements in Table 2. Therefore, the property requirements in Table 2 cannot be used as requirements for quality control of field prepared mortar. Test Method C 780 may be used for this purpose.

NOTE 2—The physical properties of plastic and hardened mortar complying with the proportion specification (3.1) may differ from the physical properties of mortar of the same type complying with the property specification (3.2).

4. Materials

4.1 Materials used as ingredients in the mortar shall conform to the requirements specified in 4.1.1 to 4.1.4.

4.1.1 *Cementitious Materials*—Cementitious materials shall conform to the following ASTM specifications:

4.1.1.1 *Portland Cement*—Types I, IA, II, IIA, III, or IIIA of Specification C 150.

4.1.1.2 *Blended Hydraulic Cements*—Types IS, IS-A, IP, IP-A, I(PM) or I(PM)-A of Specification C 595/C 595M.

4.1.1.3 *Slag Cement (for Use in Property Specifications Only)*—Types S or SA of Specification C 595/C 595M.

4.1.1.4 *Masonry Cement*—See Specification C 91.

4.1.1.5 *Quicklime*—See Specification C 5.

4.1.1.6 *Hydrated Lime*—Specification C 207, Types S or SA. Types N or NA limes are permitted if shown by test or performance record to be not detrimental to the soundness of the mortar.

4.1.2 *Aggregates*—See Specification C 144.

4.1.3 *Water*—Water shall be clean and free of amounts of oils, acids, alkalies, salts, organic materials, or other substances that are deleterious to mortar or any metal in the wall.

4.1.4 *Admixtures*—Admixtures such as coloring pigments, air-entraining agents, accelerators, retarders, water-repellent agents, antifreeze compounds, and other admixtures shall not be added to mortar unless specified. Calcium chloride, when explicitly provided for in the contract documents, is permitted to be used as an accelerator in amounts not exceeding 2 % by weight of the portland cement content or 1 % by weight of the masonry cement content, or both, of the mortar.

NOTE 3—If calcium chloride is allowed, it should be used with caution as it may have a detrimental effect on metals and on some wall finishes.

NOTE 4—Air content of non-air-entrained portland cement-lime mortar is generally less than 8 %.

5. Test Methods

5.1 *Proportions of Materials for Test Specimens*—Laboratory mixed mortar used for determining conformance to this property specification shall contain construction materials in proportions indicated in project specifications. Measure materials by weight for laboratory mixed batches. Convert proportions, by volume, to proportions, by weight, using a batch factor calculated as follows:

TABLE 2 Property Specification Requirements[A]

Mortar	Type	Average Compressive Strength at 28 Days, Min. psi (MPa)	Water Retention, min, %	Air Content, max, %[B]	Aggregate Ratio (Measured in Damp, Loose Conditions)
Cement-lime	M	2500 (17.2)	75	12	
	S	1800 (12.4)	75	12	
	N	750 (5.2)	75	14[C]	
	O	350 (2.4)	75	14[C]	Not less than 2¼ and not more than 3½ times the sum of the separate volumes of cementitious materials.
Masonry Cement	M	2500 (17.2)	75	18	
	S	1800 (12.4)	75	18	
	N	750 (5.2)	75	20[D]	
	O	350 (2.4)	75	20[D]	

[A] Laboratory prepared mortar only (see Note 1).
[B] See Note 4.
[C] When structural reinforcement is incorporated in cement-lime mortar, the maximum air content shall be 12 %.
[D] When structural reinforcement is incorporated in masonry cement mortar, the maximum air content shall be 18 %.

Batch factor = 1440/(80 times total sand volume proportion)

NOTE 5—See Appendix X4 for examples of material proportioning.

5.1.1 Oven dry and cool to room temperature all sand for laboratory mixed mortars. Sand weight shall be 1440 g for each individual batch of mortar prepared. Add water to obtain flow of 110 ± 5 %. A test batch provides sufficient mortar for completing the water retention test and fabricating three 2-in. cubes for the compressive strength test.

5.2 *Mixing of Mortars*—Mix the mortar in accordance with Practice C 305.

5.3 *Water Retention*—Determine water retention in accordance with Specification C 91, except that the laboratory-mixed mortar shall be of the materials and proportions to be used in the construction.

5.4 *Compressive Strength*—Determine compressive strength in accordance with Test Method C 109. The mortar shall be composed of materials and proportions that are to be used in the construction with mixing water to produce a flow of 110 ± 5.

5.4.1 *Specimen Storage*—Keep mortar cubes for compressive strength tests in the molds on plane plates in a moist room or a cabinet meeting the requirements of Specification C 511, from 48 to 52 h in such a manner that the upper surfaces shall be exposed to the moist air. Remove mortar specimens from the molds and place in a moist cabinet or moist room until tested.

5.5 *Air Content*—Determine air content in accordance with Specification C 91 *except* that the laboratory mixed mortar is to be of the materials and proportions to be used in the construction. Calculate the air content to the nearest 0.1 % as follows:

$$D = \frac{(W_1 + W_2 + W_3 + W_4 + V_w)}{\frac{W_1}{P_1} + \frac{W_2}{P_2} + \frac{W_3}{P_3} + \frac{W_4}{P_4} + V_w}$$

$$A = 100 - \frac{W_m}{4D}$$

where:

D = density of air-free mortar, g/cm^3,
W_1 = weight of portland cement, g,
W_2 = weight of hydrated lime, g,
W_3 = weight of masonry cement, g,
W_4 = weight of sand, g,
V_w = millilitres of water used,
P_1 = density of portland cement, g/cm^3,
P_2 = density of hydrated lime, g/cm^3,
P_3 = density of masonry cement, g/cm^3,
P_4 = density of sand, g/cm^3,
A = volume of air, %, and
W_m = weight of 400 mL of mortar, g.

6. Construction Practices

6.1 *Storage of Materials*—Cementitious materials and aggregates shall be stored in such a manner as to prevent deterioration or intrusion of foreign material.

6.2 *Measurement of Materials*—The method of measuring materials for the mortar used in construction shall be such that the specified proportions of the mortar materials are controlled and accurately maintained.

NOTE 6—The weights per cubic foot of the materials are considered to be as follows:

Material	Weight, lb/ft^3 (kg/m^3)
Portland cement	94(1505)
Blended cement	Weight printed on bag
Masonry cement	Weight printed on bag
Hydrated lime	40(640)
Lime putty[A]	80(1280)
Sand, damp and loose[B]	80 lb (1280 kg) of dry sand

[A] All quicklime should be slaked in accordance with the manufacturer's directions. All quicklime putty, except pulverized quicklime putty, should be sieved through a No. 20 (850-μm) sieve and allowed to cool until it has reached a temperature of 80°F (26.7°C). Quicklime putty should weigh at least 80 lb/ft^3 (1280 kg/m^3). Putty that weighs less than this may be used in the proportion specifications, if the required quantity of extra putty is added to meet the minimum weight requirement.

[B] For the purposes of this specification, a weight of 80 lb of oven-dried sand shall be used. This is, in most cases, equivalent to one cubic foot of loose, damp sand.

6.3 *Mixing Mortars*—All cementitious materials and aggregate shall be mixed between 3 and 5 min in a mechanical batch mixer with the maximum amount of water to produce a workable consistency. Hand mixing of the mortar is permitted with the written approval of the specifier outlining hand mixing procedures.

NOTE 7—These mixing water requirements differ from those in test methods in Section 5.

6.4 *Tempering Mortars*—Mortars that have stiffened shall be re-tempered by adding water as frequently as needed to restore the required consistency. No mortars shall be used beyond 2½ h after mixing.

6.5 *Climatic Conditions*—Unless superseded by other contractual relationships or the requirements of local building codes, cold weather masonry construction relating to mortar shall comply with the International Masonry All-Weather Council's "Guide Specification for Cold Weather Masonry Construction, Section 04200, Article 3."

NOTE 8—*Limitations*—Mortar type should be correlated with the particular masonry unit to be used because certain mortars are more compatible with certain masonry units.

The specifier should evaluate the interaction of the mortar type and masonry unit specified, that is, masonry units having a high initial rate of absorption will have greater compatibility with mortar of high-water retentivity.

7. Specification Limitations

7.1 Specification C 270 is *not* a specification to determine mortar strengths through field testing.

7.2 Test Method C 780 is acceptable for preconstruction and construction evaluation of mortars for plain and reinforced unit masonry.

7.3 *Tests of Hardened Mortars*—There is no ASTM method for determining the conformance or nonconformance of a mortar to Specification C 270 by tests on hardened mortar samples removed from a structure, but such standards are under development.

NOTE 9—Where necessary, testing of a wall or a masonry prism from the wall is generally more desirable than attempting to test individual components.

NOTE 10—The cost of tests to show initial compliance are typically borne by the seller. The party initiating a change of materials typically bear the cost for recompliance.

Unless otherwise specified, the cost of other tests are typically borne as follows:

If the results of the tests show that the mortar does not conform to the requirements of the specification, the costs are typically borne by the seller.

If the results of the tests show that the mortar does conform to the requirements of the specification, the costs are typically borne by the purchaser.

8. Keywords

8.1 air content; compressive strength; masonry; masonry cement; mortar; portland cement-lime; water retention

APPENDIXES

(Nonmandatory Information)

X1. SELECTION AND USE OF MORTAR FOR UNIT MASONRY

X1.1 *Scope*—This appendix provides information to allow a more knowledgeable decision in the selection of mortar for a specific use.

X1.2 *Significance and Use*—Masonry mortar is a versatile material capable of satisfying a variety of diverse requirements. The relatively small portion of mortar in masonry significantly influences the total performance. There is no single mortar mix that satisfies all situations. Only an understanding of mortar materials and their properties, singly and collectively, will enable selection of a mortar that will perform satisfactorily for each specific endeavor.

X1.3 *Function:*

X1.3.1 The primary purpose of mortar in masonry is to bond masonry units into an assemblage which acts as an integral element having desired functional performance characteristics. Mortar influences the structural properties of the assemblage while adding to its water resistance.

X1.3.2 Because portland cement concretes and masonry mortars contain some of the same principal ingredients, it is often erroneously assumed that good concrete practice is also good mortar practice. Realistically, mortars differ from concrete in working consistencies, in methods of placement and in the curing environment. Masonry mortar is commonly used to bind masonry units into a single structural element, while concrete is usually a structural element in itself.

X1.3.3 A major distinction between the two materials is illustrated by the manner in which they are handled during construction. Concrete is usually placed in nonabsorbent metal or wooden forms or otherwise treated so that most of the water will be retained. Mortar is usually placed between absorbent masonry units, and as soon as contact is made the mortar loses water to the units. Compressive strength is a prime consideration in concrete, but it is only one of several important factors in mortar.

X1.4 *Properties:*

X1.4.1 Masonry mortars have two distinct, important sets of properties, those of plastic mortars and those of hardened mortars. Plastic properties determine a mortar's construction suitability, which in turn relate to the properties of the hardened mortar and, hence, of finished structural elements. Properties of plastic mortars that help determine their construction suitability include workability and water retentivity. Properties of hardened mortars that help determine the performance of the finished masonry include bond, durability, elasticity, and compressive strength.

X1.4.2 Many properties of mortar are not quantitatively definable in precise terms because of a lack of measurement standards. For this and other reasons there are no mortar standards wholly based upon performance, thus the continued use of the traditional prescription specification in most situations.

X1.4.3 It is recommended that Test Method C 780 and assemblage testing be considered with proper interpretation to aid in determining the field suitability of a given masonry mortar for an intended use.

X1.5 Plastic Mortars

X1.5.1 *Workability*—Workability is the most important property of plastic mortar. Workable mortar can be spread easily with a trowel into the separations and crevices of the masonry unit. Workable mortar also supports the weight of masonry units when placed and facilitates alignment. It adheres to vertical masonry surfaces and readily extrudes from the mortar joints when the mason applies pressure to bring the unit into alignment. Workability is a combination of several properties, including plasticity, consistency, cohesion, and adhesion, which have defied exact laboratory measurement. The mason can best assess workability by observing the response of the mortar to the trowel.

X1.5.2 Workability is the result of a ball bearing affect of aggregate particles lubricated by the cementing paste. Although largely determined by aggregate grading, material proportions and air content, the final adjustment to workability depends on water content. This can be, and usually is, regulated on the mortar board near the working face of the masonry. The capacity of a masonry mortar to retain satisfactory workability under the influence of masonry unit suction and evaporation rate depends on the water retentivity and setting characteristics of the mortar. Good workability is essential for maximum bond with masonry units.

X1.5.3 *Flow*—Initial flow is a laboratory measured property of mortar that indicates the percent increase in diameter of the base of a truncated cone of mortar when it is placed on a flow table and mechanically raised ½ in. (12.7 mm) and dropped 25 times in 15 s. Flow after suction is another laboratory property which is determined by the same test, but performed on a mortar sample which has had some water removed by a specific applied vacuum. Water retentivity is the ratio of flow after suction to initial flow, expressed in percent.

X1.5.3.1 Construction mortar normally requires a greater flow value than laboratory mortar, and consequently possesses a greater water content. Mortar standards commonly require a minimum water retention of 75 %, based on an initial flow of only 105 to 115 %. Construction mortars normally have initial flows, although infrequently measured, in the range of 130 to 150 % (50–60 mm by cone penetration, as outlined in the annex of Test Method C 780) in order to produce a workability satisfactory to the mason. The lower initial flow requirements for laboratory mortars were arbitrarily set because the low flow mortars more closely indicated the mortar compressive strength in the masonry. This is because most masonry units will remove some water from the mortar once contact is made. While there may be some discernible relationship between bond and compressive strength of mortar, the relationship between mortar flow and tensile bond strength is apparent. For most mortars, and with minor exceptions for all but very low suction masonry units, bond strength increases as flow increases to where detectable bleeding begins. Bleeding is defined as migration of free water through the mortar to its surface.

X1.5.4 *Water Retentivity*—Water retentivity is a measure of the ability of a mortar under suction to retain its mixing water. This mortar property gives the mason time to place and adjust a masonry unit without the mortar stiffening. Water retentivity is increased through higher lime or air content, addition of sand fines within allowable gradation limits, or use of water retaining materials.

X1.5.5 *Stiffening Characteristics*—Hardening of plastic mortar relates to the setting characteristics of the mortar, as indicated by resistance to deformation. Initial set as measured in the laboratory for cementitious materials indicates extent of hydration or setting characteristics of neat cement pastes. Too rapid stiffening of the mortar before use is harmful. Mortar in masonry stiffens through loss of water and hardens through normal setting of cement. This transformation may be accelerated by heat or retarded by cold. A consistent rate of stiffening assists the mason in tooling joints.

X1.6 *Hardened Mortars:*

X1.6.1 *Bond*—Bond is probably the most important single physical property of hardened mortar. It is also the most inconstant and unpredictable. Bond actually has three facets; strength, extent and durability. Because many variables affect bond, it is difficult to devise a single laboratory test for each of these categories that will consistently yield reproducible results and which will approximate construction results. These variables include air content and cohesiveness of mortar, elapsed time between spreading mortar and laying masonry unit, suction of masonry unit, water retentivity of mortar, pressure applied to masonry joint during placement and tooling, texture of masonry unit's bedded surfaces, and curing conditions.

X1.6.1.1 The test method for flexural bond strength of masonry as prescribed in Test Methods E 518 is presently the most common method for evaluating this property of mortar. Test Methods E 518 consists of loading to failure a stack-bond, mortar and unit masonry prism, tested as a simple beam. Test Methods E 518 replaced a crossed-brick couplet test. Research on new test methods is currently underway. Presently the bend wrench method of test is under scrutiny as an alternative to Test Methods E 518.

X1.6.1.2 Extent of bond may be observed under the microscope. Lack of extent of bond, where severe, may be measured indirectly by testing for relative movement of

water through the masonry at the unit-mortar interface, such as prescribed in Test Method E 514. This laboratory test method consists of subjecting a sample wall to a through-the-wall pressure differential and applying water to the high pressure side. Time, location and rate of leakage must be observed and interpreted.

X1.6.1.3 The tensile and compressive strength of mortar far exceeds the bond strength between the mortar and the masonry unit. Mortar joints, therefore, are subject to bond failures at lower tensile or shear stress levels. A lack of bond at the interface of mortar and masonry unit may lead to moisture penetration through those areas. Complete and intimate contact between mortar and masonry unit is essential for good bond. This can best be achieved through use of mortar having proper composition and good workability, and being properly placed.

X1.6.1.4 In general, the tensile bond strength of laboratory mortars increase with an increase in cement content. Because of mortar workability, it has been found that Type S mortar generally results with the maximum tensile bond strength that can practically be achieved in the field.

X1.6.2 *Extensibility and Plastic Flow*—Extensibility is maximum unit tensile strain at rupture. It reflects the maximum elongation possible under tensile forces. Low strength mortars, which have lower moduli of elasticity, exhibit greater plastic flow than their high moduli counterparts at equal paste to aggregate ratios. For this reason, mortars with higher strength than necessary should not be used. Plastic flow or creep will impart flexibility to the masonry, permitting slight movement without apparent joint opening.

X1.6.3 *Compressive Strength*—The compressive strength of mortar is sometimes used as a principal criterion for selecting mortar type, since compressive strength is relatively easy to measure, and it commonly relates to some other properties, such as tensile strength and absorption of the mortar.

X1.6.3.1 The compressive strength of mortar depends largely upon the cement content and the water-cement ratio. The accepted laboratory means for measuring compressive strength is to test 2 in. (50.8 mm) cubes of mortar. Because the referenced test in this specification is relatively simple, and because it gives consistent, reproducible results, compressive strength is considered a basis for assessing the compatibility of mortar ingredients. Field testing compressive strength of mortar is accomplished with Test Method C 780 using either 2 in. (50.8 mm) cubes or small cylindrical specimens of mortar.

X1.6.3.2 Perhaps because of the previously noted confusion regarding mortar and concrete, the importance of compressive strength of mortar is over-emphasized. Compressive strength should not be the sole criterion for mortar selection. Bond strength is generally more important, as is good workability and water retentivity, both of which are required for maximum bond. Flexural strength is also important because it measures the ability of a mortar to resist cracking. Often overlooked is the size/shape of mortar joints in that the ultimate compressive load carrying capacity of a typical ⅜ in. (9.5 mm) bed joint will probably be well over twice the value obtained when the mortar is tested as a 2 in. (50.8 mm) cube. Mortars should typically be weaker than the masonry units, so that any cracks will occur in the mortar joints where they can more easily be repaired.

X1.6.3.3 Compressive strength of mortar increases with an increase in cement content and decreases with an increase in lime, sand, water or air content. Retempering is associated with a decrease in mortar compressive strength. The amount of the reduction increases with water addition and time between mixing and retempering. It is frequently desirable to sacrifice some compressive strength of the mortar in favor of improved bond, consequently retempering within reasonable time limits is recommended to improve bond.

X1.6.4 *Durability*—The durability of relatively dry masonry which resists water penetration is not a serious problem. The coupling of mortars with certain masonry units, and design without exposure considerations, can lead to unit or mortar durability problems. It is generally conceded that masonry walls, heated on one side, will stand many years before requiring maintenance, an indication of mortar's potential longevity. Parapets, masonry paving, retaining walls, and other masonry exposed to freezing while saturated represent extreme exposures and thus require a more durable mortar.

X1.6.4.1 Mortar, when tested in the laboratory for durability, is subjected to repeated cycles of freezing and thawing. Unless a masonry assemblage is allowed to become nearly saturated, there is little danger of substantial damage due to freezing. Properly entrained air in masonry mortar generally increases its resistance to freeze-thaw damage where extreme exposure (such as repeated cycles of freezing and thawing while saturated with water) exists. Air content within the specification limits for mortar, however, may be above the minimum amount required for durability. Durability is adversely affected by oversanded or overtempered mortars as well as use of highly absorbent masonry units.

X1.7 *Composition and Its Effect on Properties:*

X1.7.1 Essentially, mortars contain cementitious materials, aggregate and water. Sometimes admixtures are used also.

X1.7.2 Each of the principal constituents of mortar makes a definite contribution to its performance. Portland cement contributes to strength and durability. Lime, in its hydroxide state, provides workability, water retentivity, and elasticity. Both portland cement and lime contribute to bond strength. Instead of portland cement-lime combinations, masonry cement is frequently used. Sand acts as a filler and enables the unset mortar to retain its shape and thickness under the weight of subsequent courses of masonry. Water is the mixing agent which gives fluidity and causes cement hydration to take place.

X1.7.3 Mortar should be composed of materials which will produce the best combination of mortar properties for the intended service conditions.

X1.7.4 *Cementitious Materials Based on Hydration*—Portland cement, a hydraulic cement, is the principal cementitious ingredient in most masonry mortars. Portland cement contributes strength to masonry mortar, particularly early strength, which is essential for speed of construction. Straight portland cement mortars are not used because they lack plasticity, have low water retentivity, and are harsh and less workable than portland cement-lime or masonry cement mortars.

X1.7.4.1 Masonry cement is a proprietary product usually containing portland cement and fines, such as ground limestone or other materials in various proportions, plus additives such as air entraining and water repellency agents.

X1.7.5 *Cementitious Materials Based on Carbonation*—Hydrated lime contributes to workability, water retentivity, and elasticity. Lime mortars carbonate gradually under the influence of carbon dioxide in the air, a process slowed by cold, wet weather. Because of this, complete hardening occurs very slowly over a long period of time. This allows healing, the recementing of small hairline cracks.

X1.7.5.1 Lime goes into solution when water is present and migrates through the masonry where it can be deposited in cracks and crevices as water evaporates. This could also cause some leaching, especially at early ages. Successive deposits may eventually fill the cracks. Such autogenous healing will tend to reduce water permeance.

X1.7.5.2 Portland cement will produce approximately 25 percent of its weight in calcium hydroxide at complete hydration. This calcium hydroxide performs the same as lime during carbonation, solubilizing, and redepositing.

X1.7.6 *Aggregates*—Aggregates for mortar consist of natural or manufactured sand and are the largest volume and weight constituent of the mortar. Sand acts as an inert filler, providing economy, workability and reduced shrinkage, while influencing compressive strength. An increase in sand content increases the setting time of a masonry mortar, but reduces potential cracking due to shrinkage of the mortar joint. The special or standard sand required for certain laboratory mortar tests may produce quite different test results from sand that is used in the construction mortar.

X1.7.6.1 Well graded aggregate reduces separation of materials in plastic mortar, which reduces bleeding and improves workability. Sands deficient in fines produce harsh mortars, while sands with excessive fines produce weak mortars and increase shrinkage. High lime or high air content mortars can carry more sand, even with poorly graded aggregates, and still provide adequate workability.

X1.7.6.2 Field sands deficient in fines can result in the cementitious material acting as fines. Excess fines in the sand, however, is more common and can result in oversanding, since workability is not substantially affected by such excess.

X1.7.6.3 Unfortunately, aggregates are frequently selected on the basis of availability and cost rather than grading. Mortar properties are not seriously affected by some variation in grading, but quality is improved by more attention to aggregate selection. Often gradation can be easily and sometimes inexpensively altered by adding fine or coarse sands. Frequently the most feasible method requires proportioning the mortar mix to suit the available sand within permissible aggregate ratio tolerances, rather than requiring sand to meet a particular gradation.

X1.7.7 *Water*—Water performs three functions. It contributes to workability, hydrates cement, and facilitates carbonation of lime. The amount of water needed depends primarily on the ingredients of the mortar. Water should be clean and free from injurious amounts of any substances that may be deleterious to mortar or metal in the masonry. Usually, potable water is acceptable.

X1.7.7.1 Water content is possibly the most misunderstood aspect of masonry mortar, probably due to the confusion between mortar and concrete requirements. Water requirement for mortar is quite different from that for concrete where a low water/cement ratio is desirable. Mortars should contain the maximum amount of water consistent with optimum workability. Mortar should also be retempered to replace water lost by evaporation.

X1.7.8 *Admixtures*—Admixtures for masonry mortars are available in a wide variety and affect the properties of fresh or hardened mortar physically or chemically. Some chemical additions are essential in the manufacture of basic mortar materials. The inclusion of an additive is also necessary for the production of ready mixed mortars. Undoubtedly there are also some special situations where the use of admixtures may be advantageous when added at the job site mixer. In general, however, such use of admixtures is not recommended. Careful selection of the mortar mix, use of quality materials, and good practice will usually result in sound masonry. Improprieties cannot be corrected by admixtures, some of which are definitely harmful.

X1.7.8.1 Admixtures are usually commercially prepared products and their compositions are not generally disclosed. Admixtures are functionally classified as agents promoting air entrainment, water retentivity, workability, accelerated set, and so on. Limited data are available regarding the effect of proprietary admixtures on mortar bond, compressive strength, or water permeance of masonry. Field experience indicates that detrimental results have frequently occurred. For these reasons, admixtures should be used in the field only after it has been established by laboratory test under conditions duplicating their intended use, and experience, that they improve the masonry.

X1.7.8.2 Use of an air entraining admixture, along with the limits on air content in a field mortar, still continues to create controversy. Most masonry cements, all Type "A" portland cements and all Type "A" limes incorporate air entraining additions during their manufacture to provide required minimum as well as maximum levels of air in a laboratory mortar. Such materials should never be combined, nor should admixtures which increase the entrained air content of the mortar be added in the field, except under the most special of circumstances.

X1.7.8.3 The uncontrolled use of air entraining agents should be prohibited. At high air levels, a definite inverse relationship exists between air content and tensile bond strength of mortar as measured in the laboratory. In general, any increase in air content is accompanied by a decrease in bond as well as compressive strength. Data on masonry grouts indicate that lower bond strength between grout and reinforcing steel is associated with high air content. Most highly air entrained mortar systems can utilize higher sand contents without losing workability, which could be detrimental to the masonry if excessive sand were used. The use of any mortar containing air entraining materials, where resulting levels of air are high or unknown, should be based on a knowledge of local performance or on laboratory tests of mortar and masonry assemblages.

X1.7.8.4 Air can be removed from plastic mortar containing air entraining material by use of a defoamer, although its use in the field is strongly discouraged.

X1.7.8.5 Color can be added to mortar using selected aggregates or inorganic pigments. Inorganic pigments should be of mineral oxide composition and should not exceed 10 % of the weight of portland cement, with carbon black limited to 2 %, to avoid excessive strength reduction of the mortar. Pigments should be carefully chosen and used in the smallest amount that will produce the desired color. To minimize variations from batch to batch it is advisable to purchase cementitious materials to which coloring has been added at the plant or to use preweighed individual packets of coloring compounds for each batch of mortar, and to mix the mortar in batches large enough to permit accurate batching. Mortar mixing procedures should remain constant for color consistency.

X1.8 *Kinds of Mortars:*

X1.8.1 *History*—History records that burned gypsum and sand mortars were used in Egypt at least as early as 2690 B.C. Later in ancient Greece and Rome, mortars were produced from various materials such as burned lime, volcanic tuff, and sand. When the first settlements appeared in North America, a relatively weak product was still being made from lime and sand. The common use of portland cement in mortar began in the early part of the twentieth century and led to greatly strengthened mortar, either when portland cement was used alone or in combination with lime. Modern mortar is still made from portland cement and hydrated lime, in addition to mortars made from masonry cement.

X1.8.2 *Portland Cement-Hydrated Lime*—Cement-lime mortars have a wide range of properties. At one extreme, a straight portland cement and sand mortar would have high compressive strength and low water retentivity. A wall containing such a mortar would be strong but vulnerable to cracking and rain penetration. At the other extreme, a straight lime and sand mortar would have low compressive strength and high water retentivity. A wall containing such a mortar would have lower strength, particularly early strength, but greater resistance to cracking and rain penetration. Between the two extremes, various combinations of cement and lime provide a balance with a wide variety of properties, the high strength and early setting characteristics of cement modified by the excellent workability and water retentivity of lime. Selective proportions are found in this specification.

X1.8.3 *Masonry Cement*—Masonry cement mortars generally have excellent workability. Microscopic bubbles of entrained air contribute to the ball bearing action and provide a part of this workability. Freeze-thaw durability of masonry cement mortars in the laboratory is outstanding. Three types of masonry cement are recognized by Specification C 91. These masonry cements are formulated to produce mortars conforming to either the proportion or the property specifications of this specification. Such masonry cements provide the total cementitious material in a single bag to which sand and water are added at the mixer. A consistent appearance of mortar made from masonry cements should be easier to obtain because all the cementitious ingredients are proportioned, and ground or blended together before being packaged.

X1.8.4 *Portland Cement-Masonry Cement*—The addition of portland cement to Type N masonry cement mortars also allow qualification as Types M and S Mortars in this specification.

X1.8.5 *Prebatched or Premixed*—Recently, prebatched or premixed mortars have been made readily available in two options. One is a wet, ready mixed combination of hydrated lime or lime putty, sand, and water delivered to the construction project, and when mixed with cement and additional water is ready for use. The other is dry, packaged mortar mixtures requiring only the addition of water and mixing. Special attention should be given to the dry system, in that resulting mortars may have to be mixed for a longer period of time to overcome the water affinity of oven dry sand and subsequent workability loss in the mortar. The use of ready mixed mortar is also on the increase. These are mixtures consisting of cementitious materials, aggregates, and admixtures, batched and mixed at a central location, and delivered to the construction project with suitable workability characteristics for a period in excess of 2½ h after mixing. Systems utilizing continuous batching of mortar are also available.

X1.9 *Related Items That Have an Effect on Properties:*

X1.9.1 The factors influencing the successful conclusion of any project with the desired performance characteristics are the design, material, procedure and craftsmanship selected and used.

X1.9.2 The supervision, inspecting and testing necessary for compliance with requirements should be appropriate and predetermined.

X1.9.3 *Masonry Units*—Masonry units are absorptive by nature, with the result that water is extracted from the mortar as soon as the masonry unit and the mortar come into contact. The amount of water removal and its consequences effect the strength of the mortar, the properties of the boundary between the mortar and the masonry units, and thus the strength, as well as other properties, of the masonry assemblage.

X1.9.3.1 The suction exerted by the masonry unit is a very important external factor which affects the fresh mortar and initiates the development of bond. Masonry units vary widely in initial rate of absorption (suction). It is therefore necessary that the mortar chosen have properties that will provide compatibility with the properties of the masonry unit being used, as well as environmental conditions that exist during construction and the construction practices peculiar to the job.

X1.9.3.2 Mortar generally bonds best to masonry units having moderate initial rates of absorption (IRA), from 5 to 25 g/min · 30 in.² (194 cm²), at the time of laying. More than adequate bond can be obtained, however, with many units having IRA's less than or greater than these values.

X1.9.3.3 The extraction of too much or too little of the available water in the mortar tends to reduce the bond between the masonry unit and the mortar. A loss of too much water from the mortar can be caused by low water retentivity mortar, high suction masonry units, or dry, windy conditions. When this occurs, the mortar is incapable of forming a complete bond when the next unit is placed. Where lowering the suction by prewetting the units is not proper or possible, the time lapse between spreading the mortar and laying of a masonry unit should be kept to a minimum. When a very low suction masonry unit is used, the unit tends to float and bond is difficult to accomplish. There is no available means of increasing the suction of a low

suction masonry unit, and thus the time lapse between spreading the mortar and placing the unit may have to be increased.

X1.9.3.4 Mortars having higher water retentivity are desirable for use in summer or with masonry units having high suction. Mortars having lower water retentivity are desirable for use in winter or with masonry units having low suction.

X1.9.3.5 Shrinkage or swelling of the masonry unit or mortar once contact has been achieved affects the quality of the mortar joint. Protection should be provided to prevent excessive wetting, drying, heating or cooling, until the mortar has at least achieved final set.

X1.9.3.6 Mortar bond is less to surfaces having an unbroken die skin or sanded finish than it is to roughened surfaces such as a wire cut or textured finish.

X1.9.4 *Construction Practice*—Careful attention to good practice on the construction site is essential to achieve quality. Cementitious materials and aggregate should be protected from rain and ground moisture and air borne contaminants.

X1.9.4.1 Proper batching procedures include use of a known volume container (such as a one cubic foot batching box) for measuring sand. When necessary, sand quantities should be adjusted to provide for bulking of the sand. Shovel measuring cannot be expected to produce mortar of consistent quality. Alternatively, a combination volumetric measure calibration of a mixer followed by full bag cementitious additions and shovel additions of sand to achieve the same volume of mortar in the mixer with subsequent batches, should prove adequate.

X1.9.4.2 Good mixing results can be obtained where about three-fourths of the required water, one-half of the sand, and all of the cementitious materials are briefly mixed together. The balance of the sand is then charged and the remaining water added. The mixer should be charged to its full design capacity for each batch and completely emptied before charging the next batch.

X1.9.4.3 Mixing time in a paddle mixer should usually be a minimum of 3 and a maximum of 5 min. after the last mixing water has been added, to insure homogeneity and workability of the mortar. Overmixing results in changing the air content of the mortar. Worn paddles and rubber scrapers will greatly influence the mixing efficiency. Concern for quality suggests use of an automatic timer on the mixing machine. Mixing time should not be determined by the demand of the working force.

X1.9.4.4 Since all mortar is not used immediately after mixing, evaporation may require the addition of water, retempering the mortar, to restore its original consistency. The addition of water to mortar within specified time limits should not be prohibited. Although compressive strength of the mortar is reduced slightly by retempering, bond strength is usually increased. For this reason, retempering should be required to replace water lost by evaporation. Because retempering is harmful only after mortar has begun to set, all site prepared mortar should be placed in final position as soon as possible, but always within 2½ h after the original mixing, or the mortar discarded.

X1.9.4.5 Weather conditions also should be considered when selecting mortar. During warm, dry, windy, summer weather, mortar must have a high water retentivity to

TABLE X1.1 Guide for the Selection of Masonry Mortars[A]

Location	Building Segment	Mortar Type Recommended	Mortar Type Alternative
Exterior, above grade	load-bearing wall non-load bearing wall parapet wall	N O[B] N	S or M N or S S
Exterior, at or below grade	foundation wall, retaining wall, manholes, sewers, pavements, walks, and patios	S[C]	M or N[C]
Interior	load-bearing wall	N	S or M
Interior or Exterior	non-bearing partitions tuck pointing	O see X3	N see X3

[A] This table does not provide for many specialized mortar uses, such as chimney, reinforced masonry, and acid-resistant mortars.

[B] Type O mortar is recommended for use where the masonry is unlikely to be frozen when saturated, or unlikely to be subjected to high winds or other significant lateral loads. Type N or S mortar should be used in other cases.

[C] Masonry exposed to weather in a nominally horizontal surface is extremely vulnerable to weathering. Mortar for such masonry should be selected with due caution.

minimize the effect of water lost by evaporation. In winter, a lower water retentivity has merit because it facilitates water loss from the mortar to the units prior to a freeze. To minimize the risk of reduced bond in cold weather, the masonry units being used as well as the surface on which the mortar is placed should both be brought to a temperature at least above 32°F (0°C) before any work commences.[5]

X1.9.5 *Workmanship*—Workmanship has a substantial effect on strength and extent of bond. The time lapse between spreading mortar and placing masonry units should be kept to a minimum because the flow will be reduced through suction of the unit on which it is first placed. This time lapse should normally not exceed one minute. Reduce this time lapse for hot, dry and windy conditions, or with use of highly absorptive masonry units. If excessive time elapses before a unit is placed on the mortar, bond will be reduced. Elimination of deep furrows in horizontal bed joints and providing full head joints are essential. Any metal embedded in mortar should be completely surrounded by mortar.

X1.9.5.1 Once the mortar between adjacent units has begun to stiffen, tapping or otherwise attempting to move masonry units is highly detrimental to bond and should be prohibited. The movement breaks the bond between the mortar and the masonry unit, and the mortar will not be sufficiently plastic to re-establish adherence to the masonry unit.

X1.9.5.2 Tooling of the mortar joint should be done when its surface is thumb-print hard utilizing a jointer having a diameter slightly larger than the mortar joint width. Joint configurations other than concave can result in increased water permeance of the masonry assemblage. Striking joints with the same degree of hardness produces uniform joint appearance. Finishing is not only for appearance, but to seal the interface between mortar and masonry unit, while densifying the surface of the mortar joint.

X1.9.5.3 The benefits of the finishing operation should be protected from improper cleaning of the masonry. Use of strong chemical or harsh physical methods of cleaning may be detrimental to the mortar. Colored mortars are especially susceptible to damage from such cleaning. Most chemicals used in cleaning attack the cementitious materials within the mortar system, as well as enlarge cracks between mortar and masonry unit.

X1.9.5.4 With very rapid drying under hot, dry and windy conditions, very light wetting of the in-place masonry, such as fog spray, can improve its quality. Curing of mortar by the addition of considerable water to the masonry assemblage, however, could prove to be more detrimental than curing of mortar by retention of water in the system from its construction. The addition of excess moisture might saturate the masonry, creating movements which decrease the adhesion between mortar and masonry unit.

X1.10 *Summary:*

X1.10.1 No one combination of ingredients provides a mortar possessing an optimum in all desirable properties. Factors that improve one property generally do so at the expense of others. Testing of mortars in the laboratory by this specification's referenced methods, and in the field by Test Method C 780 is beneficial. Some physical properties of mortar, however, are of equal or greater significance to masonry performance than those properties commonly specified. When selecting a mortar, evaluate all properties, and then select the mortar providing the best compromise for the particular requirements.

X1.10.2 Bond is probably the most important single property of a conventional mortar. Many variables affect bond. To obtain optimum bond, use a mortar with properties that are compatible with the masonry units to be used. To increase tensile bond strength in general, increase the cement content of the mortar (see X1.6.1.4); keep air content of the mortar to a minimum; use mortars having high water retentivity; mix mortar to the water content compatible with workability; allow retempering of the mortar; use masonry units having moderate initial rates of absorption when laid (see X1.9.3.2); bond mortar to a rough surface rather than to a die skin surface; minimize time between spreading mortar and placing masonry units; apply pressure in forming the mortar joint; and do not subsequently disturb laid units.

X1.10.3 Table X1.1 is a general guide for the selection of mortar type for various masonry wall construction. Selection of mortar type should also be based on the type of masonry units to be used as well as the applicable building code and engineering practice standard requirements, such as allowable design stresses, and lateral support.

---

[5] For more inclusive suggestions, see "Recommended Practices for Cold Weather Masonry Construction" available from the International Masonry Industry All-Weather Council.

X2. EFFLORESCENCE

X2.1 Efflorescence is a crystalline deposit, usually white, of water soluble salts on the surface of masonry. The principal objection to efflorescence is the appearance of the salts and the nuisance of their removal. Under certain circumstances, particularly when exterior coatings are present, salts can be deposited below the surface of the masonry units. When this cryptoflorescence occurs, the force of crystallization can cause disintegration of the masonry.

X2.2 A combination of circumstances is necessary for the formation of efflorescence. First, there must be a source of soluble salts. Second, there must be moisture present to pick up the soluble salts and carry them to the surface. Third, evaporation or hydrostatic pressure must cause the solution to migrate. If any one of these conditions is eliminated, efflorescence will not occur.

X2.3 Salts may be found in the masonry units, mortar components, admixtures or other secondary sources. Water-soluble salts that appear in chemical analyses as only a few tenths of one percent are sufficient to cause efflorescence when leached out and concentrated on the surface. The amount and character of the deposits vary according to the nature of the soluble materials and the atmospheric conditions. A test for the efflorescence of individual masonry units is contained within ASTM standards. Efflorescence can occur with any C 270 mortar when moisture migration occurs. There is no ASTM test method that will predict the potential for efflorescence of mortar. Further, there is no ASTM test method to evaluate the efflorescence potential of combined masonry materials.

X2.4 The probability of efflorescence in masonry as related directly to materials may be reduced by the restrictive selection of materials. Masonry units with a rating of "not effloresced" are the least likely to contribute towards efflorescence. The potential for efflorescence decreases as the alkali content of cement decreases. Admixtures should not be used in the field. Washed sand and clean, potable water should be used.

X2.5 Moisture can enter masonry in a number of ways. Attention must be paid to the design and installation of flashing, vapor barriers, coping and caulking to minimize penetration of rainwater into the masonry. During construction, masonry materials and unfinished walls should be protected from rain and construction applied water. Full bed and head joints, along with a compacting finish on a concave mortar joint, will reduce water penetration. Condensation occurring within the masonry is a further source of water.

X2.6 Although selection of masonry construction materials having a minimum of soluble salts is desirable, the prevention of moisture migration through the wall holds the greatest potential in minimizing efflorescence. Design of masonry using the principle of pressure equalization between the outside and a void space within the wall will greatly reduce the chances of water penetration and subsequently efflorescence.

X2.7 Removal of efflorescence from the face of the masonry can frequently be achieved by dry brushing. Since many salts are highly soluble in water, they will disappear of their own accord under normal weathering processes. Some salts, however, may require harsh physical or even chemical treatment, if they are to be removed.

X3. TUCK POINTING MORTAR

X3.1 *General:*

X3.1.1 Tuck pointing mortars are replacement mortars used at or near the surface of the masonry wall to restore integrity or improve appearance. Mortars made without portland cement may require special considerations in selecting tuck pointing mortars.

X3.1.2 If the entire wall is not to be tuck pointed, the color and texture should closely match those of the original mortar. An exact match is virtually impossible to achieve.

X3.2 *Materials:*

X3.2.1 Use cementitious materials that conform to the requirements of this specification (C 270).

X3.2.2 Use sand that conforms to the requirements of this specification (C 270). Sand may be selected to have color, size, and gradation similar to that of the original mortar, if color and texture are important.

X3.3 *Selection Guide*—Use tuck pointing mortar of the same or weaker composition as the original mortar.

X3.4 *Materials*—Mortar shall be specified as one of the following:

X3.4.1 The proportion specification of C 270, Type _____.

X3.4.2 *Type K*—One part portland cement and 2½ to 4 parts hydrated lime. Aggregate Ratio of 2¼ to 3 times sum of volume of cement and lime.

NOTE X3.1—Type K mortar proportions were referenced in this specification (C 270) prior to 1982.

X3.5 *Mixing:*

X3.5.1 Dry mix all solid materials.

X3.5.2 Add sufficient water to produce a damp mix that will retain its shape when pressed into a ball by hand. Mix from 3 to 7 min., preferably with a mechanical mixer.

X3.5.3 Let mortar stand for not less than 1 h nor more than 1½ h for prehydration.

X3.5.4 Add sufficient water to bring the mortar to the proper consistency for tuck pointing, somewhat drier than mortar used for laying the units.

X3.5.5 Use the mortar within 2½ h of its initial mixing. Permit tempering of the mortar within this time interval.

TABLE X3.1   Guide for Selection of Tuck Pointing Mortar[A]

Location or Service	Mortar Type	
	Recommended	Alternate
interior	O	K,N
exterior, above grade exposed on one side, unlikely to be frozen when saturated, not subject to high wind or other significant lateral load	O	N,K
exterior, other than above	N	O

[A] In some applications structural concerns may dictate the use of mortars other than those recommended. This table is not applicable to pavement applications.

X4. EXAMPLES OF MATERIAL PROPORTIONING FOR TEST BATCHES OF MORTAR

X4.1 *Example A*—A mortar consisting of one part portland cement, 1¼ parts lime, and 6¾ parts of sand [A] is to be tested. The weights of the materials used in the mortar are calculated as follows:

Batch factor = 1440/(80 × 6.75) = 2.67

Weight of portland cement = 1 × 94 × 2.67 = 251
Weight of lime   = 1¼ × 40 × 2.67 = 133
Weight of sand   = 6¾ × 80 × 2.67 = 1440

	Portland Cement	Lime	Sand
Proportions by volume	1	1¼	6¾
Unit weight (lb/ft³)	94	40	80
Batch factor	2.67	2.67	2.67
Weight of material [B] (in g)	251	133	1440

_____

[A] Total sand content is calculated as: (1 volume part of portland cement plus 1¼ volume parts of hydrated lime) times three = 6¾ parts of sand.
[B] Weight of material = volume proportion times unit weight times batch factor.

X4.2 *Example B*—A mortar consisting of one part masonry cement, three parts sand [C] is to be tested. The weights of the materials used in the mortar are calculated as follows:

Batch factor = 1440/(80 × 3) = 6.00

Weight of masonry cement = 1 × 70 × 6.00 = 420
Weight of sand   = 3 × 80 × 6.00 = 1440

	Masonry Cement	Sand
Proportions by volume	1	3
Unit weight (lb/ft³) (Weight printed on bag for masonry cement)	70	80
Batch factor	6.00	6.00
Weight of material [B] (in g)	420	1440

_____

[C] Total sand content is calculated as: (1 volume part of masonry cement) times three = 3 parts of sand.

Standard Specification for
Hollow Brick (Hollow Masonry Units Made From Clay or Shale)[1]

This standard is issued under the fixed designation C 652; the number immediately following the designation indicates the year of original adoption or, in the case of revision, the year of last revision. A number in parentheses indicates the year of last reapproval. A superscript epsilon (ε) indicates an editorial change since the last revision or reapproval.

This standard has been approved for use by agencies of the Department of Defense. Consult the DoD Index of Specifications and Standards for the specific year of issue which has been adopted by the Department of Defense.

1. Scope

1.1 This specification covers hollow building brick and hollow facing brick made from clay, shale, fire clay, or mixtures thereof, and fired to incipient fusion. Four types of hollow brick in each of two grades and two classes are covered. In this specification the term hollow brick shall be understood to mean hollow clay masonry units whose net cross-sectional area (solid area) in any plane parallel to the surface, containing the cores, cells, or deep frogs, is less than 75 % of its gross cross sectional area measured in the same plane (see 3.3). This specification does not cover brick intended for use as paving brick (see Specification C 902).

1.2 The property requirements of this standard apply at the time of purchase. The use of results from testing of brick extracted from masonry structures for determining conformance or nonconformance to the property requirements (Section 5) of this standard is beyond the scope of this standard.

1.3 Brick covered by this specification are manufactured from clay, shale, or similar naturally occurring substances and subjected to a heat treatment at elevated temperatures (firing). The heat treatment shall develop sufficient fired bond between the particulate constituents to provide the strength and durability requirements of this specification. (See "firing" and "firing bond" in Terminology C 43.)

1.4 Hollow brick differ from unglazed structural clay tile (Specifications C 34 and C 212) and solid brick (Specifications C 62 and C 216). Hollow brick require greater shell and web thicknesses and higher minimum compressive strength than structural clay tile, but permit greater void area and lesser distance from exposed edge to core hole than solid brick. Therefore, environmental and structural performance may be different in elements constructed of hollow brick from those constructed of structural clay tile or solid brick.

1.5 The text of this standard references notes and footnotes which provide explanatory material. These notes and footnotes (excluding those in tables and figures) shall not be considered as requirements of the standard.

1.6 The values stated in inch-pound units are to be regarded as the standard. The values given in parentheses are for information only.

2. Referenced Documents

2.1 *ASTM Standards:*
C 34 Specification for Structural Clay Load-Bearing Wall Tile[2]
C 43 Terminology of Structural Clay Products[2]
C 62 Specification for Building Brick (Solid Masonry Units Made from Clay or Shale)[2]
C 67 Test Methods of Sampling and Testing Brick and Structural Clay Tile[2]
C 212 Specification for Structural Clay Facing Tile[2]
C 216 Specification for Facing Brick (Solid Masonry Units Made from Clay or Shale)[2]
C 902 Specification for Pedestrian and Light Traffic Paving Brick[2]
E 835/E 835M Guide for Modular Coordination of Clay and Concrete Masonry Units[3]

3. Classification

3.1 *Grades*—Two grades of hollow brick are covered:
3.1.1 *Grade SW*—Hollow brick intended for use where a high and uniform degree of resistance to frost action and disintegration by weathering is desired and the exposure is such that the hollow brick may be frozen when permeated with water.

3.1.2 *Grade MW*—Hollow brick intended for use where a moderate and somewhat nonuniform degree of resistance to frost action is permissible or where they are unlikely to be permeated with water when exposed to temperatures below freezing.

3.2 *Types*—Four types of hollow brick are covered:
3.2.1 *Type HBS*—Hollow brick for general use in masonry.

3.2.2 *Type HBX*—Hollow brick for general use in masonry where a higher degree of precision and lower permissible variation in size than permitted for Type HBS is required.

3.2.3 *Type HBA*—Hollow brick for general use in masonry selected to produce characteristic architectural effects resulting from nonuniformity in size and texture of the individual units.

3.2.4 *Type HBB*—Hollow brick for general use in masonry where a particular color, texture, finish, uniformity, or limits on cracks, warpage, or other imperfections detracting from the appearance are not a consideration.

---

[1] This specification is under the jurisdiction of ASTM Committee C-15 on Manufactured Masonry Units and is the direct responsibility of Subcommittee C15.02 on Clay Brick and Structural Clay Tile.
Current edition approved Jan. 10, 1997. Published March 1997. Originally published as C 652 – 70. Last previous edition C 652 – 95a.

[2] *Annual Book of ASTM Standards*, Vol 04.05.
[3] *Annual Book of ASTM Standards*, Vol 04.07.

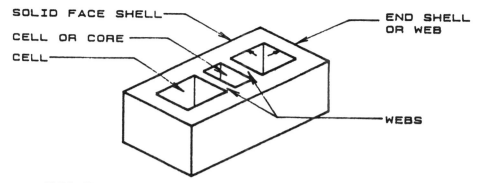

SOLID FACE SHELL — — END SHELL OR WEB

CELL OR CORE

CELL

— WEBS

SOLID SHELL HOLLOW BRICK UNITS

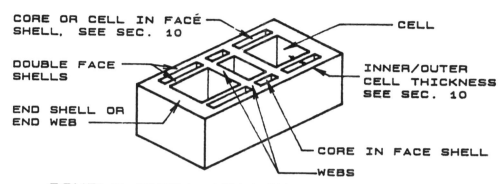

CORE OR CELL IN FACE SHELL, SEE SEC. 10 — CELL

DOUBLE FACE SHELLS

END SHELL OR END WEB — INNER/OUTER CELL THICKNESS SEE SEC. 10

— CORE IN FACE SHELL

— WEBS

DOUBLE SHELL HOLLOW BRICK UNITS

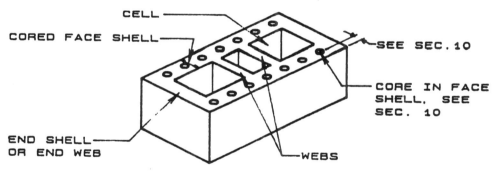

CELL

CORED FACE SHELL — SEE SEC. 10

CORE IN FACE SHELL, SEE SEC. 10

END SHELL OR END WEB

— WEBS

CORED SHELL HOLLOW BRICK UNITS

FOR MINIMUM DIMENSIONS SEE SECTION 10 AND TABLE 5

FIG. 1 Hollow Brick Units

3.2.5 When the type is not specified, the requirements for Type HBS shall govern.

3.3 *Class*—Two classes of hollow brick are covered:

3.3.1 *Class H40V*—Hollow brick intended for use where void areas or hollow spaces greater than 25 %, but not greater than 40 %, of the gross cross-sectional area of the unit measured in any plane parallel to the surface containing the cores, cells, or deep frogs are desired. The void spaces, the web thicknesses, and the shell thicknesses shall comply with the requirements of Section 10 on Hollow Spaces.

3.3.2 *Class H60V*—Hollow brick intended for use where larger void areas are desired. The sum of these void areas shall be greater than 40 %, but not greater than 60 %, of the gross cross-sectional area of the unit measured in any plane parallel to the surface containing the cores, cells or deep

frogs. The void spaces, the web thicknesses, and the shell thicknesses shall comply with the requirements of Section 10 on Hollow Spaces and to the minimum requirements contained in Table 1 (see Fig 1).

3.3.3 When the class is not specified, the requirements of class H40V shall govern.

4. Materials and Manufacture

4.1 The body of all hollow brick shall be of clay, shale, fire clay, or mixtures of these materials, with or without admixtures, burned to meet the requirements of this specification. Any coloring or other materials added to the clay shall be suitable ceramic materials and shall be well distributed throughout the body. Unless otherwise specified by the purchaser, surface coloring shall not be applied to any brick

FIGURE 1 - SEC. 4.1   ASTM C 652   271

TABLE 1 Class H60V—Hollow Brick Minimum Thickness of Face Shells and Webs, in. (mm)

Nominal Width of Units	Face Shell Thicknesses		End Shells or End Webs
	Solid	Cored or Double Shell	
3 and 4 (76 and 101)	¾ (19.05)	...	¾ (19.05)
6 (152)	1 (25.4)	1½ (38)	1 (25.4)
8 (203)	1¼ (32)	1½ (38)	1 (25.4)
10 (254)	1⅜ (35)	1⅝ (41)	1⅛ (29.5)
12 (306)	1½ (38)	2 (50)	1⅛ (29.5)

TABLE 2 Physical Requirements

Desig-nation	Compressive Strength gross area, min, psi (MPa)		Water Absorption by 5-h Boiling, max, %		Saturation Coefficient, max	
	Average of 5 brick	Individual	Average of 5 brick	Indi-vidual	Average of 5 brick	Indi-vidual
Grade SW	3000 (20.7)	2500 (17.2)	17.0	20.0	0.78	0.80
Grade MW	2500 (17.2)	2200 (15.2)	22.0	25.0	0.88	0.90

TABLE 3 Tolerances on Dimensions, in. (mm)

Specified Dimensions	Permissible Variation, max	
	Type HBX	Type HBS and HBB
3 (76) and under	±1/16 (1.58)	±3/32 (2.38)
Over 3 to 4 (102), incl	±3/32 (2.38)	±1/8 (3.18)
Over 4 to 6 (152), incl	±1/8 (3.18)	±3/16 (4.76)
Over 6 to 8 (204), incl	±5/32 (3.97)	±1/4 (6.35)
Over 8 to 12 (306), incl	±7/32 (5.56)	±5/16 (7.94)
Over 12 to 16 (408), incl	±9/32 (7.14)	±3/8 (9.52)

TABLE 4 Tolerances on Distortion, in. (mm)

Dimension, max	Permissible Distortion, max	
	Type HBX	Type HBS
8 (204) and under	1/16 (1.58)	3/32 (2.38)
Over 8 to 12, (306), incl	3/32 (2.38)	1/8 (3.18)
Over 12 to 16 (408), incl	1/8 (3.18)	5/32 (3.97)

other than by sanding or flashing.

5. Physical Properties

5.1 *Durability*—When Grade is not specified, the requirements for Grade SW shall govern. Unless otherwise specified by the purchaser, brick of Grade SW shall be accepted instead of Grade MW.

5.1.1 *Physical Property Requirements*—The brick shall conform to the physical requirements for the Grade specified as prescribed in Table 2. For the compressive strength requirements in Table 2, test the unit with the compressive force perpendicular to the bed surface of the unit, with the unit in the stretcher position.

5.1.2 *Absorption Alternate*—The saturation coefficient requirement does not apply, provided the cold water absorption of any single unit of a random sample of five brick does not exceed 8 %.

5.1.3 *Freezing and Thawing Alternative*—The requirements for 5 h boiling water absorption and saturation coefficient do not apply, provided a sample of five brick, meeting the strength requirements of Table 2, passes the freezing and thawing test as described in the Rating Section of the Freezing and Thawing test procedures of Test Methods C 67.

5.1.3.1 *Grade SW—Weight Loss Requirement*—Not greater than 0.5 % loss in dry weight of any individual unit.

NOTE 1—The 50 cycle freezing and thawing test is specified as an alternative only when brick do not conform to either Table 2 requirements for maximum water absorption and saturation coefficient, or to the requirements of the Absorption Alternate in Section 5.1.2.

5.2 *Strength*—When hollow brick are required having strengths greater than prescribed in Table 2, the purchaser shall specify the desired minimum compressive strength.

5.3 *Initial Rate of Absorption (IRA)*—Test results for IRA shall be determined in accordance with Section 9 of Test Methods C 67 and shall be furnished at the request of the specifier or purchaser. IRA is not a qualifying condition or property of units in this specification. This property is measured in order to assist in mortar selection and material handling in the construction process. See Note 2.

NOTE 2—*Initial Rate of Absorption (Suction)*—Both laboratory and field investigation have shown that strong and watertight joints between mortar and masonry units are not achieved by ordinary construction methods when the units as laid have excessive initial rates of absorption. Mortar that has stiffened somewhat because of excessive loss of mixing water to a unit may not make complete and intimate contact with the second unit, resulting in poor adhesion, incomplete bond, and water-permeable joints of low strength. IRA of the units is determined by the oven-dried procedure described in the IRA (Suction) (Laboratory Test) of Test Methods C 67. IRA in the field depends on the moisture content of the masonry unit and is determined in accordance with the IRA (Suction)—Field Test of Test Methods C 67. Units having average field IRA exceeding 30 g/min per 30 in.² (30 g/min 194 cm²) should have their IRA reduced below 30 g/min per 30 in.² prior to laying. They may be wetted immediately before they are laid, but it is preferable to wet them thoroughly 3 to 24 h prior to their use so as to allow time for moisture to become distributed throughout the unit.

6. Efflorescence

6.1 When the hollow brick are tested in accordance with Test Methods C 67, the rating for efflorescence shall be: "not effloresced."

7. Dimensions and Permissible Variations

7.1 *Size*—The size of hollow brick shall be as specified by the purchaser. In a sample of ten hollow brick selected to include the extreme range of color and sizes to be supplied, no hollow brick shall depart from the specified size by more than the individual tolerance for the type specified as prescribed in Table 3. Tolerances on dimensions for Type HBA shall be as specified by the purchaser, but not more restrictive than HBS and HBB.

NOTE 3—For a list of modular sizes see Guide E 835/E 835M. Sizes listed in this standard are not produced in all parts of the United States. Brick names denoting sizes may be regional and therefore may not be included in all reference books. Purchasers should ascertain the size of brick available in their locality and should specify accordingly, stating the desired dimensions (width by height by length).

7.2 *Warpage*—Tolerances for distortion or warpage of surfaces or edges intended to be exposed in use of individual hollow brick from a plane surface and from a straight line, respectively, shall not exceed the maximum for the type specified as prescribed in Table 4. Tolerances on distortion for Type HBA shall be as specified by the purchaser.

8. Workmanship, Finish, and Appearance

8.1 The face or faces that will be exposed in place shall be

TABLE 5 Maximum Permissible Range of Chippage That Extends from the Edges and Corners of the Finished Face or Faces Onto the Surface

Type	Percentage Allowed[A]	Chippage in in. (mm) in from		Percentage Allowed[A]	Chippage in in. (mm) in from	
		Edge	Corner		Edge	Corner
HBX	5 % or less	⅛ to ¼ (3.18 to 6.35)	¼ to ⅜ (6.35 to 9.52)	95 to 100 %	0 to ⅛ (0 to 3.18)	0 to ¼ (0 to 0.35)
HBS[B] (smooth)	10 % or less	¼ to ⁵⁄₁₆ (6.35 to 7.94)	⅜ to ½ (9.52 to 12.7)	90 to 100 %	0 to ¼ (0 to 6.35)	0 to ⅜ (0 to 9.52)
HBS[C] (rough)	15 % or less	⁵⁄₁₆ to ⁷⁄₁₆ (7.94 to 11.11)	½ to ¾ (12.7 to 19.05)	85 to 100 %	0 to ⁵⁄₁₆ (0 to 7.94)	0 to ½ (0 to 12.7)
HBA and HBB	to meet the designated sample or as specified by the purchaser, but not more restrictive than HBS (rough)					

[A] The allowable percentage of brick that will be exposed in the wall having the allowed maximum size chips measured the listed maximum dimensions in from an edge or corner.

[B] Smooth texture is unbroken natural die finish.

[C] Rough texture is the finish produced when the face is sanded, combed, scratched, or scarified or the die skin on the face is entirely broken by mechanical means such as wire-cutting or wire brushing.

free of chips that exceed the limits of Table 5. The limits apply to the types as specified. The aggregate length of chips shall not exceed 10 % of the perimeter of the exposed face or faces of the hollow brick.

8.2 *Other than Chips*—The face or faces shall be free of other imperfections detracting from the appearance of a sample wall when viewed from a distance of 15 ft (4.6 m) for Type HBX and a distance of 20 ft (6.1 m) for Types HBS and HBA.

NOTE 4—Of all the units that will be exposed in place, a small percentage of the units may have chips that are larger in size than those chips allowed for the majority of the units. This special allowed percentage, listed in the second column from the left of Table 5, ranges up to 5 % for HBX, up to 10 % for HBS (smooth), and up to 15 % for HBS (rough). The remainder of the units that will be exposed in place, listed in the fifth column from the left, must conform to the chip sizes listed in the sixth and seventh columns from the left.

Example—Type HBS (smooth) units will conform to the requirements of Table 5 if not more than 10 % of the units have edge chips greater than ¼ in. (6.4 mm) but less than ⁵⁄₁₆ in. (7.9 mm), or corner chips greater than ⅜ in. (9.5 mm) but less than ½ in. (2.7 mm) and the remainder of the units, in this maximum case 90 % (100 % − 10 %), do not have edge chips greater than ¼ in. (6.4 mm) in from the edge nor corner chips greater than ⅜ in. (9.5 mm) in from the corner.

8.3 The brick shall be free of defects, deficiencies, and surface treatments, including coatings, that interfere with the proper setting of the brick or significantly impair the strength or performance of the construction.

8.4 Unless otherwise agreed upon by the purchaser and the seller, a delivery of brick may contain not more than 5 % brick, including broken brick, that do not meet the requirements for chippage and tolerances.

8.5 After brick are placed in usage the manufacturer or the manufacturer's agent shall not be held responsible for compliance of brick with the requirements of this specification for chippage and dimensional tolerances.

9. Texture and Color

9.1 If brick having a particular color, color range, or texture are desired, these features shall be specified separately by the purchaser. Unless otherwise specified by the purchaser, at least one end of the majority of the individual hollow brick shall have the same general texture and general color tone as the approved sample. The texture of the finished surfaces that will be exposed when in place shall conform to an approved sample consisting of not less than four stretcher hollow brick, each representing the texture desired. The color range shall be indicated by the approved sample.

9.2 Where brick with other than one finished face and one finished end are required (brick with two finished faces or ends, or other types), all such special brick shall be explicitly specified by the purchaser.

NOTE 5—The manufacturer should be consulted for the availability of specialty units suitable for the intended purpose.

10. Hollow Spaces

10.1 *Cores*—The distance of any core (void space having a gross cross-sectional area equal to or less than 1½ in.² (9.68 cm²)) from exposed edges shall be not less than ⅝ in. (16 mm), except for cored-shell hollow brick.

10.1.1 Cored-shell hollow brick shall have a minimum shell thickness of 1½ in. Cores greater than 1 in.² (6.45 cm²) in cored shells shall be not less than ½ in. (13 mm) from any edge. Cores not greater than 1 in.² in shells cored not more than 35 %, shall be not less than ⅜ in. (10 mm) from any edge.

10.2 *Cells*—The distance of cells (void space having a gross cross-sectional area greater than 1½ in.²) from any exposed edge of the unit shall not be less than ¾ in. (19 mm), except for double-shell hollow brick.

10.2.1 Double-shell hollow brick with inner and outer shells not less than ½ in. (13 mm) are permitted to have cells not greater than ⅝ in. (16 mm) in width nor 5 in. (127 mm) in length between the inner and outer shell.

10.3 *Webs*—The thickness of webs between cells shall not be less than ½ in. (13 mm), ⅜ in. (9.5 mm) between cells and cores, or ¼ in. (6 mm) between cores.

10.4 *Unexposed Edges*—The distance of voids from unexposed edges, which are recessed not less than ½ in. (13 mm), shall be not less than ½ in.

10.5 *Frogging*—Unless otherwise specified in the invitation for bids, one bearing face of each brick is permitted to have a recess or panel frog and deep frogs. The recess or panel frog shall not exceed ⅜ in. (9.5 mm) in depth and no part of the recess or panel frog shall be less than ⅝ in. (15.9 mm) from any edge of the brick. In brick containing deep frogs, frogs deeper than ⅜ in. (9.5 mm), any cross-section through the deep frogs parallel to the bearing surface shall conform to other requirements of this specification for hollow spaces and void area.

TABLE 5 - SEC. 10.5   ASTM C 652   273

ASTM C 652

11. Sampling and Testing

11.1 For purposes of tests, brick that are representative of the commercial product shall be selected by a competent person appointed by the purchaser, the place or places of selection to be designated when the purchase order is placed. The sample or samples shall include specimens representative of the complete range of colors and sizes of the brick supplied or to be supplied. The manufacturer or the seller shall furnish specimens for test without charge.

11.2 Sample and test the brick in accordance with Test Methods C 67.

Note 6—Unless otherwise specified in the purchase order, the cost of tests is typically borne as follows: If the results of the tests show that the brick do not conform to the requirements of this specification, the cost is typically borne by the seller. If the results of the tests show that the brick do conform to the requirements of this specification, the cost is typically borne by the purchaser.

12. Keywords

12.1 appearance requirements; clay; fired masonry unit; hollow brick; hollow building brick; hollow facing brick; masonry construction; physical properties; shale

Standard Specification for
Pedestrian and Light Traffic Paving Brick[1]

This standard is issued under the fixed designation C 902; the number immediately following the designation indicates the year of original adoption or, in the case of revision, the year of last revision. A number in parentheses indicates the year of last reapproval. A superscript epsilon (ε) indicates an editorial change since the last revision or reapproval.

1. Scope

1.1 This specification covers brick intended for use as paving material to support pedestrian and light vehicular traffic. The units are designed for use in such places as patios, walkways, floors, plazas, and driveways. The units are not intended to support heavy vehicular traffic or for applications covered by Specification C 410.

1.2 The property requirements of this standard apply at the time of purchase. The use of results from testing of brick extracted from masonry structures for determining conformance or non-conformance to the property requirements (Section 4) of this standard is beyond the scope of this standard.

1.3 Brick are manufactured from clay, shale, or similar naturally occurring earthy substances and subjected to a heat treatment at elevated temperatures (firing). The heat treatment must develop sufficient fired bond between the particulate constituents to provide the strength and durability requirement of this specification (see firing, fired bond and incipient fusion in Terminology C 43).

1.4 The brick are available in a variety of sizes, colors, and shapes. They are available in three classes according to exposure environment and three types according to type of traffic exposure.

1.5 The values stated in inch-pound units are to be regarded as the standard.

2. Referenced Documents

2.1 *ASTM Standards:*
C 43 Terminology of Structural Clay Products[2]
C 67 Test Methods of Sampling and Testing Brick and Structural Clay Tile[2]
C 88 Test Method for Soundness of Aggregates by Use of Sodium Sulfate or Magnesium Sulfate[3]
C 410 Specification for Industrial Floor Brick[2]
C 418 Test Method for Abrasion Resistance of Concrete by Sandblasting[3]

3. Classification

3.1 Light traffic paving brick are classified according to the severity of their use-environment. Two types of environment are considered: (*1*) weather and (*2*) traffic:
3.1.1 *Weather:*

3.1.1.1 *Class SX*—Brick intended for use where the brick may be frozen while saturated with water.

3.1.1.2 *Class MX*—Brick intended for exterior use where resistance to freezing is not a factor.

3.1.1.3 *Class NX*—Brick not intended for exterior use but which may be acceptable for interior use where protected from freezing when wet.

NOTE 1—A surface coating may be applied to any class of brick of this standard when protected from freezing while wet. The function of the coating is to prevent penetration of dirt or liquids into the pores of the brick. Coatings should be applied only after complete drying of the paving.

3.1.2 *Traffic:*
3.1.2.1 *Type I*—Brick exposed to extensive abrasion, such as in driveways and entranceways to public or commercial buildings.

3.1.2.2 *Type II*—Brick exposed to intermediate traffic, such as floors in restaurants or stores and exterior walkways.

3.1.2.3 *Type III*—Brick exposed to low traffic, such as floors or patios in single-family homes.

4. Physical Requirements

4.1 *Durability*—The brick shall conform to the physical requirements for the class specified as prescribed in Table 1.

NOTE 2—The resistance of brick to weathering cannot be predicted with complete assurance at the present state of knowledge. There is no known test that can predict weathering resistance with complete accuracy.

Brick in general is superior in weathering resistance to other building materials. There are innumerable instances of satisfactory performance beyond 200 years and even into the thousands of years. Nevertheless, there are some brick that cannot survive a few winters of a severe freezing and thawing environment.

The durability requirements of the specification attempt to exclude such brick. This specification utilizes the best knowledge available at this time and is based on extensive research by several investigators. The durability requirements have an excellent correlation with in-use performance. Nevertheless, it is known that some brick that meet this specification may not be serviceable in severe climates. Furthermore, other brick that do not meet these specifications may show superior serviceability in the most severe climate. The best indication of brick durability is its service experience record.

4.2 *Performance Alternate*—If information on the performance of the units in a similar application of similar exposure and traffic is furnished by the manufacturer or his agent and is found acceptable by the specifier of the pavement material, or his agent, the physical requirements in 4.1 may be waived.

4.3 *Absorption Alternate*—If the average water absorption is less than 6 % after 24-h submersion in room-temperature water, the requirement for saturation coefficient shall be waived.

4.4 *Freezing and Thawing Test Alternate*—The require-

[1] This specification is under the jurisdiction of ASTM Committee C-15 on Manufactured Masonry Units, and is the direct responsibility of Subcommittee C15.02 on Clay Brick and Structural Clay Tile.
Current edition approved Nov. 10, 1995. Published January 1996. Originally published as C 902 – 79. Last previous edition C 902 – 93.
[2] *Annual Book of ASTM Standards*, Vol 04.05.
[3] *Annual Book of ASTM Standards*, Vol 04.02.

TABLE 1 Physical Requirements[A]

Designation	Compressive Strength, flatwise, gross area, min, psi (MPa)		Cold Water Absorption, max, %		Saturation Coefficient, max[B]	
	Average of 5 Brick	Individual	Average of 5 Brick	Individual	Average of 5 Brick	Individual
Class SX	8000 (55.2)	7000 (48.3)	8	11	0.78	0.80
Class MX	3000 (20.7)	2500 (17.2)	14	17	no limit	no limit
Class NX	3000 (20.7)	2500 (17.2)	no limit	no limit	no limit	no limit

[A] Minimum modulus of rupture values should be considered by the purchaser for uses of brick where support or loading may be severe.
[B] The saturation coefficient is the ratio of absorption by 24-h submersion in room temperature water to that after 5-h submersion in boiling water.

TABLE 2 Abrasion Requirements[A]

	(1) Abrasion Index,[B] max	(2) Volume Abrasion Loss,[C] max, cm³/cm²
Type I	0.11	1.7
Type II	0.25	2.7
Type III	0.50	4.0

[A] Select the sample according to the sampling procedure of Test Methods C 67. The brick shall meet the requirements of either column (1) or (2). The values listed shall not be exceeded by any individual unit within the sample.
[B] The abrasion index is calculated from the cold absorption in percent and the compressive strength in pounds per square inch as follows:

$$\text{Abrasion index} = \frac{100 \times \text{absorption}}{\text{compressive strength}}$$

Compressive strength values are influenced by specimen shape (particularly the height to width ratio of the test specimen). Therefore, a shape is specified which conforms to the data on which the abrasion index is based.[4]

The compressive strength shall be determined on specimens measuring 3⅞ by 3⅞ by 2¼ in. ± ¼ in. (98 by 98 by 57 mm ± 6 mm) for length, width, and height respectively. The brick shall be without core holes, other perforations or frogs. Other shaped specimens may be used provided that the producer submits evidence acceptable to the purchaser that the change in shape gives equivalent strength results to those of the specified shape.

The abrasion resistance should be determined according to Note 2 in those cases where the procedural requirements for compressive strength cannot be met.
[C] The volume abrasion loss shall be determined in accordance with Test Method C 418, with the following changes in procedure:

(1) The sand shall be a natural silica sand from Ottawa, IL, graded to pass a No. 50 (300-μm) sieve and retained on a No. 100 (150-μm) sieve.
(2) The test shall be run on dry brick.
(3) The duration of the test shall be 2 min.
(4) The rate of sand flow shall be 400 g/min.
(5) The volume loss shall be determined by filling the abraded depression with modeling clay, striking off level with the original surface of the brick, and removing and weighing the modeling clay. The volume loss shall be calculated from the bulk density of the modeling clay. The bulk density shall be determined on each lot of modeling clay.

An alternative method of determining the weight of clay used in filing the sand-blast cavity is to determine the weight of the modeling clay sample before and after filling the cavity.

ments for water absorption (24 h cold) and saturation coefficient specified in 4.1 shall be waived provided a sample of five brick, meeting all other requirements, passes the freezing and thawing test as described in the Rating section of the Freezing and Thawing procedures of Test Methods C 67.

4.4.1 No breakage and not greater than 0.5 % loss in dry weight of any individual unit.

NOTE 3—The 50 cycle freezing and thawing test is specified only as an alternative when brick do not conform to either Table 1 requirements for maximum water absorption and saturation coefficient, or to the restrictive absorption requirements in 4.3.

TABLE 3 Maximum Permissible Extent of Chippage from Edges and Corners

NOTE—The aggregate length of chips on a single unit shall not exceed 10 % of the perimeter of the exposed face of the brick.

Application	Chippage in Inches (Millimetres) in from	
	Edge	Corner
PS	5⁄16 (7.9)	½ (12.7)
PX	¼ (6.4)	⅜ (9.5)
PA	as specified by purchaser	

TABLE 4 Tolerances on Dimensions

Dimension, in. (mm)	Maximum Permissible Variation from Specified Dimension, plus or minus in. (mm)		
	Application PS	Application PX	Application PA
3 (76) and under	⅛ (3.2)	1⁄16 (1.6)	no limit
Over 3 to 4 (76 to 102) incl	3⁄16 (4.7)	3⁄32 (2.4)	no limit
Over 5 to 8 (127 to 203) incl	¼ (6.4)	⅛ (3.2)	no limit
Over 8 (203)	5⁄16 (7.9)	7⁄32 (5.6)	no limit

4.5 *Sulfate Soundness Test Alternate*—The requirements for water absorption (24 h cold) and saturation coefficient specified in 4.1 shall not be required if a sample of five brick survives 15 cycles of the sulfate soundness test in accordance with Sections 4, 5, and 8 of Test Method C 88 with no visible damage.

NOTE 4—The sulfate soundness test is an optional substitute test for the freezing-and-thawing test (4.4).

4.6 *Abrasion Resistance*—The brick shall meet the requirements of either column (1) or (2) of Table 2 for the applicable traffic use (see 3.1.2).

4.7 *Molded Brick (Soft Mud, Semi-Dry Pressed, and Dry Pressed Brick)*—The requirements listed in Table 1 shall be changed for molded brick to permit maximum absorption of 16 % average and 18 % individual, and minimum compressive strengths of 4000 psi (27.6 MPa) average and 3500 psi (24.1 MPa) individual for Class SX, provided that the requirements for saturation coefficient of Table 1 are met.

4.8 Unless otherwise specified by the purchaser, brick of Classes SX and MX shall be accepted instead of Class NX, and Class SX shall be accepted instead of Class MX. Surface coatings will not be required of Classes SX and MX when used instead of Class NX. Types I and II shall be accepted instead of Type III, and Type I shall be accepted instead of Type II.

NOTE 5—Skid/slip resistance should be considered by the purchaser for uses of brick where pedestrian traffic is anticipated. Methods of testing this characteristic are under study and it is hoped that a specification for this property can be added in future revisions of this standard when suitable test methods are developed.

---

[4] McBurney, J. W., Brink, R. H., Eberle, A. R., "Relation of Water Absorption and Strength of Brick to Abrasive Resistance," *Proceedings, ASTM*, Vol 40, 1940, pp. 1143–1151.

5. Efflorescence

5.1 When paving brick are tested in accordance with Test Methods C 67, the rating for efflorescence shall be: "not effloresced."

6. Size

6.1 The size of the brick shall be as specified by the purchaser or produced by the manufacturer as a stock item.

6.2 The tolerance on dimension shall depend on the bond pattern and method of installation of the units. Three different methods of applications are covered:

6.2.1 *Application PS*—Floor and patio brick intended for general use and installed with a mortar joint between individual units, or in an installation without mortar joints between units when they are laid in running or other bonds not requiring extremely close dimensional tolerances.

6.2.2 *Application PX*—Floor and patio brick intended for installation without mortar joints between the units, where exceptionally close dimensional tolerances are required as a result of special bond patterns or unusual construction requirements.

6.2.3 *Application PA*—Floor and patio units manufactured and selected to produce characteristic architectural effects resulting from nonuniformity in size, color, and texture of individual units. (The textures may exhibit inclusion of nonuniform nodules of mineral substances or purposely introduced cracks that enhance the appearance of the units.) The requirements on warpage as specified in 4.7 do not apply to this application.

6.3 When the application is not specified, the requirements for Application PS shall govern.

6.4 *Warpage*—The concave and convex warpage (distortion) of any face intended to be the exposed surface or edge of the paving shall not exceed the values of Table 5 when sampled and measured in accordance with Test Methods C 67.

TABLE 5 Tolerances on Distortion

Specified Dimension in. (mm)	Permissible Distortion, max in. (mm)	
	Type PX	Type PS
8 (203) and under	1/16 (1.6)	3/32 (2.4)
Over 8 (203) to 12 (305)	3/32 (2.4)	1/8 (3.2)
Over 12 (305) to 16 (406)	1/8 (3.2)	5/32 (4.0)

7. Visual Inspection

7.1 The brick shall be free of cracks or other imperfections detracting from the appearance of a designated sample when viewed from a distance of 15 ft (4.6 m) for Application PX and a distance of 20 ft (6 m) for Application PS.

7.2 The parts of the brick that will be exposed in place shall be free of chips that exceed the limits given in Table 3.

7.3 Unless otherwise agreed upon by the purchaser and the seller, a delivery of brick shall contain not more than 5 % brick that do not meet the combined requirements of Tables 2, 3, and 4, and including broken brick.

7.4 After brick are placed in usage the manufacturer or the manufacturer's agent shall not be held responsible for compliance of brick with the requirements of this specification for chippage and dimensional tolerances.

8. Keywords

8.1 brick; light traffic; masonry unit; paving; pedestrian traffic

Designation: C 1019 – 89a (Reapproved 1993)[ε1]

Standard Test Method for
Sampling and Testing Grout[1]

This standard is issued under the fixed designation C 1019; the number immediately following the designation indicates the year of original adoption or, in the case of revision, the year of last revision. A number in parentheses indicates the year of last reapproval. A superscript epsilon (ε) indicates an editorial change since the last revision or reapproval.

This standard has been approved for use by agencies of the Department of Defense. Consult the DoD Index of Specifications and Standards for the specific year of issue which has been adopted by the Department of Defense.

[ε1] NOTE—Section 11 was added editorially in June 1993.

1. Scope

1.1 This test method covers procedures for both field and laboratory sampling and compression testing of grout used in masonry construction.

1.2 The values stated in inch-pound units are to be regarded as the standard.

1.3 *This standard does not purport to address all of the safety problems, if any, associated with its use. It is the responsibility of the user of this standard to establish appropriate safety and health practices and determine the applicability of regulatory limitations prior to use.*

2. Referenced Documents

2.1 *ASTM Standards:*
C 39 Test Method for Compressive Strength of Cylindrical Concrete Specimens[2]
C 143 Test Method for Slump of Hydraulic Cement Concrete[2]
C 511 Specification for Moist Cabinets, Moist Rooms, and Water Storage Tanks Used in the Testing of Hydraulic Cements and Concretes[3]
C 617 Practice for Capping Cylindrical Concrete Specimens[2]

3. Significance and Use

3.1 Grout used in masonry is a fluid mixture of cementitious materials and aggregate with a high water content for ease of placement.

3.1.1 During construction, grout is placed within or between absorptive masonry units. Excess water must be removed from grout specimens in order to provide compressive strength test results more nearly indicative of the grout strength in the wall. In this test method, molds are made from masonry units having the same absorption and moisture content characteristics as those being used in the construction.

3.2 This test method can be used to either help select grout proportions by comparing test values or as a quality control test for uniformity of grout preparation during construction.

3.3 The physical exposure condition and curing of the grout are not exactly reproduced, but this test method does subject the grout specimens to absorption conditions similar to those experienced by grout in the wall. Test results of grout specimens taken from a wall should not be compared to test results obtained with this test method.

4. Apparatus

4.1 *Maximum-Minimum Thermometer.*

4.2 *Straightedge,* a steel straightedge not less than 6 in. (152.4 mm) long and not less than $\frac{1}{16}$ in. (1.6 mm) in thickness.

4.3 *Tamping Rod,* a nonabsorbent rod, either round or square in cross section nominally $\frac{5}{8}$ in. (15.9 mm) in dimension with ends rounded to hemispherical tips of the same diameter. The rod shall be a minimum length of 12 in. (304.8 mm).

4.4 *Wooden Blocks,* wooden squares with side dimensions equal to one half the desired grout specimen height, within a tolerance of 5 %, and of sufficient quantity or thickness to yield the desired grout specimen height, as shown in Figs. 1 and 2.

NOTE 1—Certain species of wood contain sugars which cause retardation of cement. In order to prevent this from occurring, new wooden blocks shall be soaked in limewater for 24 h, sealed with varnish or wax, or covered with an impermeable material prior to use.

5. Sampling

5.1 *Size of Sample*—Grout samples to be used for slump and compressive strength tests shall be a minimum of $\frac{1}{2}$ ft^3 (0.014 m^3).

5.2 *Field Sample*—Take grout samples as the grout is being placed into the wall. Field samples may be taken at any time except for the first and last 10 % of the batch volume.

NOTE 2—Frequency of sampling and age of test is to be determined by the specifier of this test method and is usually found in the contract documents.

6. Test Specimen and Sample

6.1 Each grout specimen shall have a square cross-section, nominally 3 in. (76.2 mm) or larger on the sides and twice as high as its width. Dimensional tolerances shall be within 5 % of the nominal width selected.

6.2 Three specimens shall constitute one sample to be tested at each age of test.

7. Procedure

7.1 Select a level location where the molds can remain undisturbed for 48 h.

7.2 *Mold Construction:*

---

[1] This method is under the jurisdiction of ASTM Committee C-12 on Mortars for Unit Masonry and is the direct responsibility of Subcommittee C12.02 on Research and Methods of Test.
Current edition approved March 31 and May 26, 1989. Published July 1989. Originally published as C 1019 – 84. Last previous edition C 1019 – 84.
[2] *Annual Book of ASTM Standards*, Vol 04.02.
[3] *Annual Book of ASTM Standards*, Vol 04.01.

NOTE—Front masonry unit stack not shown to allow view of specimen.

FIG. 1 Grout Mold (Units 6 in. (152.4 mm) or Less in Height, 2¼ in. (57.2 mm) High Brick Shown)

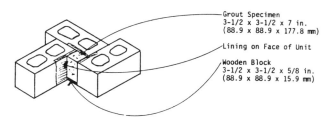

NOTE—Front masonry unit not shown to allow view of specimen.

FIG. 2 Grout Mold (Units Greater than 6 in. (152.4 mm) High, 8 in. (203.2 mm) High Concrete Masonry Unit Shown)

7.2.1 The mold space should simulate the grout location in the wall. If the grout is placed between two different types of masonry units, both types should be used to construct the mold.

7.2.2 Form a space with a square cross-section, nominally 3 in. (76.2 mm) or larger on each side and twice as high as its width, by stacking masonry units of the same type and moisture condition as those being used in the construction. Place wooden blocks, cut to proper size and of the proper thickness or quantity, at the bottom of the space to achieve the necessary height of specimen. Tolerance on space and specimen dimensions shall be within 5 % of the specimen width. See Figs. 1 and 2 and accompanying notes.

NOTE 3—Other methods of obtaining grout specimens and specimens of different geometry have been employed in grout testing, but are not covered by this test method. Since test results will vary with methods of forming and specimen geometry, comparison of test results should be confined to a single specimen shape and method of forming.

7.2.3 Line the masonry surfaces that will be in contact with the grout specimen with a permeable material, such as paper towel, to prevent bond to the masonry units.

NOTE 4—The lining is used to aid in stripping the grout specimen from the mold. Proper installation of the lining prevents irregularly sized specimens and varying test results.

7.3 Measure and record the slump of the grout in accordance with the requirements of Test Method C 143.

7.4 Fill the mold with grout in two layers. Rod each layer 15 times with the tamping rod. Rod the bottom layer through its depth. Slightly overfill the mold. Rod the second layer with the tamping rod penetrating ½ in. (12.7 mm) into the lower layer. Distribute the strokes uniformly over the cross section of the mold.

7.5 Level the top surface of the specimen with a straightedge and cover immediately with a damp absorbent material such as cloth or paper towel. Keep the top surface of the

sample damp by wetting the absorbent material and do not disturb the specimen for 48 h.

7.6 Protect the sample from freezing and variations in temperature. Store an indicating maximum-minimum thermometer with the sample and record the maximum and minimum temperatures experienced prior to the time the specimens are placed in the moist room.

7.7 Remove the masonry units after 48 h.

7.7.1 Transport field specimens to the laboratory, keeping the specimens damp and in a protective container.

7.8 Store in a moist room conforming to Specification C 511.

7.9 Cap the specimens in accordance with the applicable requirements of Practice C 617.

NOTE 5—Practice C 617 refers to capping cylindrical specimens; therefore, the alignment devices may need to be modified to ensure proper use with the rectangular prism specimens of this method. All other sections of Practice C 617 are applicable.

7.10 Measure and record the width of each face at mid-height. Measure and record the height of each face at mid-width. Measure and record the amount out of plumb at mid-width of each face.

7.11 Test the specimens in a damp condition in accordance with the applicable requirements of Test Method C 39.

8. Calculations

8.1 Determine the average cross-sectional area by measuring the width of each face at its mid-height, calculating the average width of opposite faces, and multiplying the averages.

8.2 Calculate the compressive strength by dividing the maximum load by the average cross-sectional area and express the result to the nearest 10 psi (69 kPa).

9. Report

9.1 The report shall include the following:

9.1.1 Mix design,

9.1.2 Slump of the grout,

9.1.3 Type and number of units used to form mold for specimens,

9.1.4 Description of the specimens—dimensions, amount out of plumb in percent,

9.1.5 Curing history, including maximum and minimum temperatures, and age of specimen when transported to laboratory and when tested,

9.1.6 Maximum load and compressive strength of each specimen and average compressive strength of the sample, and

9.1.7 Description of failure.

10. Precision and Bias

10.1 *General:*

10.1.1 The masonry units used to form the mold have different absorption rates and will remove slightly different amounts of water from each specimen. Thus the standard deviation for this test method is higher than that using a nonabsorbent mold.

10.1.2 The standard deviation from field samples of grout will be higher than that for laboratory samples. There is less control of grout ingredients, conditions of units for mold

FIGURE 1 - 10.1.2   ASTM C 1019   279

TABLE 1 Statistics of Laboratory-Prepared Samples

Number of Specimens	Mean, psi (MPa)	Standard Deviation, psi (MPa)	Coefficient of Variation, %
5	3784 (26.1)	306 (2.11)	8.1
5	2494 (17.2)	220 (1.52)	8.8
5	3178 (21.9)	634 (4.37)	20.0
6	5480 (37.8)	899 (6.2)	16.4
10	5350 (36.9)	826 (5.7)	15.4
12	3872 (26.7)	333 (2.30)	8.6

TABLE 2 Statistics of Field-Prepared Samples

Number of Specimens	Mean, psi (MPa)	Standard Deviation, psi (MPa)	Coefficient of Variation, %
3	3583 (24.7)	118 (0.81)	3.3
6	5455 (37.6)	324 (2.23)	5.9
6	3992 (27.5)	228 (1.57)	5.7

construction, and initial curing environment in field-prepared samples.

10.2 Limited test data are available for analysis at this time. A more detailed statement will be provided later. The following summary of available data is provided for review.

10.2.1 *Laboratory Samples*—The coefficients of variation for a series of laboratory samples of three specimens each ranged from 5.8 % with a mean value of 6452 psi (44.45 MPa) to 24.6 % with a mean value of 1373 psi (9.47 MPa) averaging 14.4 %. The standard deviation ranged from 211 psi (1.46 MPa) to 505 psi (3.48 MPa). Samples with a larger number of specimens had the characteristics found in Table 1.

10.2.2 *Field Samples*—Test reports from one project show the characteristics found in Table 2.

11. Keywords

11.1 cementitious; compressive strength; grout; masonry units

Standard Specification for
Thin Veneer Brick Units Made From Clay or Shale[1]

This standard is issued under the fixed designation C 1088; the number immediately following the designation indicates the year of original adoption or, in the case of revision, the year of last revision. A number in parentheses indicates the year of last reapproval. A superscript epsilon (ε) indicates an editorial change since the last revision or reapproval.

1. Scope

1.1 This specification covers thin veneer brick units made from clay, shale, fire clay, sand, or mixtures thereof, and fired to incipient fusion for use in adhered or fastened veneer applications. Three types of thin veneer brick units in each of two grades are covered. In this specification, the term thin veneer brick shall be understood to mean clay masonry unit with a maximum thickness of 1¾ in. (44.45 mm).

NOTE 1—Brick intended for paving should be specified under Specification C 902.

1.2 The property requirements of this standard apply at the time of purchase. The use of results from testing of brick extracted from masonry structures for determining conformance or non-conformance to the property requirements (Section 5) of this standard is beyond the scope of this standard.

1.3 Brick covered by this specification are manufactured from clay, shale, or similar naturally occurring substances and subjected to a heat treatment at elevated temperatures (firing). The heat treatment must develop sufficient fired bond between the particulate constituents to provide the strength and durability requirements of the specification. (See "firing" and "fired bond" in Terminology C 43.)

1.4 The values stated in inch-pound units are to be regarded as standard. The metric equivalents may be approximate.

1.5 The text of this standard references notes and footnotes which provide explanatory material. These notes and footnotes (excluding those in tables and figures) shall not be considered as requirements of the standard.

2. Referenced Documents

2.1 *ASTM Standards:*
C 43 Terminology of Structural Clay Products[2]
C 67 Test Methods for Sampling and Testing Brick and Structural Clay Tile[2]
C 902 Specification for Pedestrian and Light Traffic Paving Brick[2]
E 835/E 835M Guide for Modular Coordination of Clay and Concrete Masonry Units[3]

3. Grades

3.1 Two grades of thin veneer brick units are covered for exposure conditions to weather and are defined in Table 1 as Interior and Exterior.

4. Types

4.1 Three types of thin veneer brick units are covered as follows:

4.1.1 *Type TBS (Standard)*—Thin veneer brick for general use in masonry.

4.1.2 *Type TBX (Select)*—Thin veneer brick for general use in masonry where a higher degree of precision and lower permissible variation in size than permitted for Type TBS is required.

4.1.3 *Type TBA (Architectural)*—Thin veneer brick for general use in masonry selected to produce characteristic architectural effects resulting from nonuniformity in size and texture of the individual units.

4.2 When the type is not specified, the requirements for Type TBS will govern.

5. Physical Properties

5.1 *Durability*—The thin veneer brick shall conform to the physical requirements in Table 1 for the grade specified. When the grade is not specified, the requirements for Grade Exterior shall govern. If the average water absorption is less than 8.0 % after submersion in cold water for 24 h, the requirements for saturation coefficient shall be waived. If exterior or interior grade thin veneer brick are intended for use on interiors only, the requirements for water absorption (5-h boiling) and for saturation coefficient for interior grade in Table 1 shall govern.

5.2 *Freezing and Thawing*—The requirements specified in 5.1 for water absorption (5-h boiling) and saturation coefficient shall be waived provided a sample of 5 typical exterior grade thin veneer brick, meeting all other requirements, complies with the following requirements when subjected to 50 cycles of the freezing-and-thawing test:

5.2.1 *Grade Exterior*—No breakage and not greater than 0.5 % loss in dry weight of any individual brick.

6. Efflorescence

6.1 When the thin veneer brick are tested in accordance

---

[1] This specification is under the jurisdiction of ASTM Committee C-15 on Manufactured Masonry Units and is the direct responsibility of Subcommittee C15.02 on Clay Brick and Structural Clay Tile.
Current edition approved Dec. 10, 1996. Published January 1997. Originally published as C 1088 – 88. Last previous edition C 1088 – 94.
[2] *Annual Book of ASTM Standards*, Vol 04.05.
[3] *Annual Book of ASTM Standards*, Vol 04.07.

TABLE 1 Physical Requirements

Designation	Maximum Water Absorption by 5-h Boiling, %		Maximum Saturation Coefficient[A]	
	Average of 5 units	Individual	Average of 5 units	Individual
Grade Exterior	17.0	20.0	0.78	0.80
Grade Interior	22.0	25.0	0.88	0.90

[A] The saturation coefficient is the ratio of absorption by 24-h submersion in cold water to that after 5-h submersion in boiling water.

TABLE 2 Maximum Permissible Extent of Chippage from the Edges and Corners of Finished Face or Faces onto the Surface

Type	Percentage Allowed[A]	Chippage in in. (mm) in from		Percentage Allowed[A]	Chippage in in. (mm) in from	
		Edge	Corner		Edge	Corner
TBX	5 % or less	1/8 to 1/4 (3.2 to 6.4)	1/4 to 3/8 (6.4 to 9.5)	95 to 100 %	0 to 1/8 (0 to 3.2)	0 to 1/4 (0 to 6.4)
TBS (smooth)	10 % or less	1/4 to 5/16 (6.4 to 7.9)	3/8 to 1/2 (9.5 to 12.7)	90 to 100 %	0 to 1/4 (0 to 6.4)	0 to 3/8 (0 to 9.5)
TBS (rough)	15 % or less	5/16 to 7/16 (7.9 to 11.1)	1/2 to 3/4 (12.7 to 19.1)	85 to 100 %	0 to 5/16 (0 to 7.9)	0 to 1/2 (0 to 12.7)
TBA	as specified by the purchaser			as specified by the purchaser		

[A] Percentage of exposed brick allowed in the wall with chips measured the listed dimensions in from an edge or corner.

with Test Methods C 67, the rating for efflorescence shall be "not effloresced."

7. Material, Finish, and Manufacturer Limitation

7.1 Units shall not show surface defects and deficiencies, nor effects of surface treatments including coating in the manufacturing process, that interfere with installation of the brick or significantly impair the performance of the construction.

7.2 Colors and textures produced by application of inorganic coatings to the faces of the thin veneer brick are permitted if approved by the purchaser, provided that evidence is furnished of the durability of the coatings.

7.3 *Face or Faces:*

7.3.1 The face or faces that will be exposed in place shall be free of chips that exceed the limits given in Table 2. The aggregate length of chips shall not exceed 10 % of the perimeter of the face or faces of the thin veneer brick.

7.3.2 The face or faces shall not contain cracks or other imperfections that detract from the appearance of the designated sample when viewed from a distance of 15 ft (4.6 m) for Type TBX and a distance of 20 ft (6.1 m) for Types TBS and TBA.

7.4 Unless otherwise agreed upon between the purchaser and the seller, a delivery of thin veneer brick is permitted to contain not more than 5 % brick, including broken brick, that do not meet the requirements for chippage and tolerances.

7.5 After thin veneer brick are installed, the manufacturer or his agent shall not be held responsible for compliance of thin veneer brick with the requirements of this specification for chippage and dimensional tolerances.

8. Color and Texture

8.1 If brick having a particular color, color range, or texture are desired, these features shall be specified separately by the purchaser. The texture of the finished surfaces that will be exposed when in place shall conform to an approved sample consisting of not less than four typical stretcher thin veneer brick, each representing the desired texture. The color range shall be indicated by the approved sample.

9. Size

9.1 *Size*—The face size of thin veneer brick shall be as specified by the purchaser. In a representative sample of ten units selected to include the extreme range of color and dimensions of thin veneer brick to be supplied for each size and color combination in the purchase order, no thin brick shall depart from the specified size by more than the individual tolerance for the type specified as prescribed in

TABLE 3 Tolerances on Dimensions[A]

Specified Dimension, in. (mm)	Maximum Permissible Variation from Specified Dimension, ±in. (mm)	
	Type TBX	Type TBS
3 (76) and under	1/16 (1.6)	3/32 (2.4)
Over 3 to 4 (76 to 102) incl	3/32 (2.4)	1/8 (3.2)
Over 4 to 6 (102 to 152) incl	1/8 (3.2)	3/16 (4.7)
Over 6 to 8 (152 to 203) incl	5/32 (4.0)	1/4 (6.4)
Over 8 to 12 (203 to 305) incl	7/32 (5.6)	5/16 (7.9)
Over 12 to 16 (305 to 406) incl	9/32 (7.1)	3/8 (9.5)

[A] Tolerances for Type TBA shall be listed in purchase specification.

TABLE 4 Tolerances on Distortion[A]

Maximum Face Dimension, in. (mm)	Maximum Permissible Distortion, in. (mm)	
	Type TBX	Type TBS
8 (203) and under	1/16 (1.6)	3/32 (2.4)
Over 8 to 12 (203 to 305) incl	3/32 (2.4)	1/8 (3.2)
Over 12 to 16 (305 to 406) incl	1/8 (3.2)	5/32 (4.0)

[A] Tolerances for Type TBA shall be listed in purchase specification.

Table 3. Tolerances on dimensions for Type TBA shall be as specified by the purchaser.

NOTE 2—For a list of modular sizes see Guide E 835/E 835M. Sizes listed in this standard are not produced in all parts of the United States. Brick names denoting sizes may be regional and therefore may not be included in all reference books. Purchasers should ascertain the size of brick available in their locality and should specify accordingly, stating the desired dimensions (width by height by length).

9.2 *Warpage*—Tolerances for distortion or warpage of face or edges of individual units from a plane surface shall not exceed the maximum for the type specified as prescribed in Table 4. Tolerances on distortion for Type TBA shall be as specified by the purchaser.

10. Sampling and Testing

10.1 For purposes of tests, units that are representative of the commercial product shall be selected by a competent person appointed by the purchaser, the place or places of selection to be designated when the purchase order is placed. The sample or samples shall include specimens representative of ten thin veneer brick samples for each size and color combination in purchase order of the thin veneer brick supplied or to be supplied. The manufacturer or the seller shall furnish specimens for tests without charge.

10.2 The thin veneer brick shall be sampled and tested in accordance with Test Methods C 67. The provision of the brick Sampling Section in Test Methods C 67 shall govern the number of samples tested.

C 1088

Designation: C 1314 – 95$^{\epsilon 1}$

Standard Test Method for
Constructing and Testing Masonry Prisms Used to Determine Compliance with Specified Compressive Strength of Masonry[1]

This standard is issued under the fixed designation C 1314; the number immediately following the designation indicates the year of original adoption or, in the case of revision, the year of last revision. A number in parentheses indicates the year of last reapproval. A superscript epsilon (ϵ) indicates an editorial change since the last revision or reapproval.

$^{\epsilon 1}$ NOTE—Figure 1 was editorially corrected in April 1996.

1. Scope

1.1 This test method covers procedures for masonry prism construction and testing, and procedures for determining the compressive strength of masonry, f_{mt}, used to determine compliance with the specified compressive strength of masonry, f'_m.

NOTE 1—Although this test method is similar in many respects to Test Methods E 447, Test Methods E 447 requires collection of additional detailed information associated with research tests but not needed for compliance tests.

1.2 The values stated in inch-pound units are to be regarded as the standard. The values given in parentheses are for information only.

1.3 *This standard does not purport to address all of the safety concerns, if any, associated with its use. It is the responsibility of the user of this standard to establish appropriate safety and health practices and determine the applicability of regulatory limitations prior to use.*

2. Referenced Documents

2.1 *ASTM Standards:*
C 67 Test Methods of Sampling and Testing Brick and Structural Clay Tile[2]
C 140 Test Methods of Sampling and Testing Concrete Masonry Units[2]
E 447 Test Methods for Compressive Strength of Masonry Prisms[2]

3. Terminology

3.1 *Definition:*
3.1.1 *set*—a set consists of at least three prisms constructed of the same material and tested at the same age.
3.2 *Notations:*
3.2.1 f'_m—specified compressive strength of masonry.
3.2.2 f_{mt}—compressive strength of masonry.
3.2.3 h_p—prism height.
3.2.4 t_p—least actual lateral dimension of prism.

4. Significance and Use

4.1 This test method provides a means of verifying that masonry materials used in construction result in masonry that meets the specified compressive strength.

5. Masonry Prism Construction

5.1 Construct prisms of units representative of those used in the construction. If units have flutes or ribs that project ½ in. (12.5 mm) or more from the surface of the unit, remove those flutes or ribs by saw cutting flush with the surface of the unit at the base of the flute or rib.

5.2 Construct a set of prisms for each combination of materials and each test age at which the compressive strength of masonry is to be determined.

5.3 Build each prism in an opened, moisture-tight bag large enough to enclose and seal the completed prism. Construct prisms on a flat, level base. Construct prisms in a location where they will remain undisturbed until transported for testing.

5.4 Construct prisms as shown in Fig. 1 with units laid in stack bond in stretcher position. Orient units in the prism as in the corresponding construction. At the time of prism construction, the surfaces of the units shall be free of moisture. Where the corresponding construction is of multi-wythe masonry having wythes composed of different units or mortar, build prisms representative of each different wythe and test separately.

5.5 The length of masonry prisms may be reduced from that of an individual unit by saw cutting units prior to prism construction. Prisms composed of units that contain closed cells shall have at least one complete cell with one full-width cross web on either end. Prisms composed of units without closed cells shall have as symmetrical a cross section as possible. The minimum length of prisms shall be 4 in. (100 mm).

5.6 Build masonry prisms with full mortar beds (mortar all webs and face shells of hollow units). Use mortar representative of that used in the corresponding construction. Use mortar joint thickness and a method of positioning and aligning units, that are representative of the corresponding construction. Use mortar joints that are cut flush. For prisms to be grouted, remove mortar "fins" that protrude into the grout space.

5.7 Build prisms a minimum of two units high with a height-to-thickness ratio, h_p/t_p, between 1.3 and 5.0.

5.8 Immediately following the construction of the prism, seal the moisture-tight bag around the prism.

5.9 *Grouted Prisms.*
5.9.1 Where the corresponding construction is to be

---

^{1} This test method is under the jurisdiction of ASTM Committee C-15 on Manufactured Masonry Units and is the direct responsibility of Subcommittee C15.04 on Research.
Current edition approved Nov. 10, 1995. Published February 1996.
^{2} *Annual Book of ASTM Standards*, Vol 04.05.

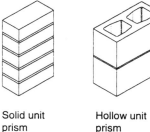
Solid unit
prism

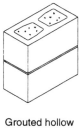
Hollow unit
prism

Grouted hollow
prism

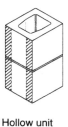
Hollow unit

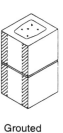
Grouted
hollow unit

Prisms reduced by saw cutting

FIG. 1 Masonry Prism Construction

solidly grouted, solidly grout the prisms not less than 24 h nor more than 48 h following the construction of the prisms. Use grout representative of that used in the corresponding construction. Before placing grout, remove mortar droppings from the grout space. Use grout consolidation and reconsolidation procedures representative of those used in the construction. Place additional grout into the prisms as necessary after each consolidation. Screed off excess and finish the grout so that it is level with the top of the prism and in contact with the units at the perimeter of the grout space. Grouted prisms shall contain no reinforcement.

5.9.2 Where the corresponding construction is to be partially grouted, construct two sets of prisms; grout one set solid as described in 5.9.1 and leave the other set ungrouted.

5.9.3 Where open-end units or prisms containing grout between similar wythes are to be grouted, use similar masonry units as forms to confine the grout during placement. Brace forms to prevent displacement during grouting. Grout as described in 5.9.1.

5.9.4 Immediately following the grouting operation, reseal the moisture-tight bag around the prism.

5.10 Keep all prisms from freezing. Do not disturb or move prisms for the first 48 h after construction and grouting. Keep prisms in the moisture-tight bags until 48 h prior to testing.

6. Transporting Masonry Prisms

6.1 Prior to transporting prisms, strap or clamp each prism to prevent damage during handling and transportation. Secure prisms to prevent jarring, bouncing, or tipping over during transporting.

7. Curing

7.1 After the initial 48 h of curing, maintain the bagged prisms in an area with a temperature of 75 ± 15°F (24 ± 8°C). Two days prior to testing, remove the moisture-tight bags and continue storing at a temperature of 75 ± 15°F (24 ± 8°C). Test prisms at an age of 28 days or at the designated test ages. Test a set of prisms at each age. Prism age shall be determined from the time of laying units for ungrouted prisms, and from the time of grouting for grouted prisms.

8. Preparation for Testing

8.1 *Measuring Prisms*—Measure the length and width at the edges of the top and bottom faces of the prisms to the nearest 0.05 in. (1.3 mm). Determine the length and width by averaging the four measurements of each dimension. Measure the height of the prism at the center of each face to

the nearest 0.05 in. (1.3 mm). Determine the height by averaging the four measurements.

8.2 *Capping Prisms*—Smooth irregularities of the prism's bearing surfaces using a method that will not reduce the integrity of the prism. Cap top and bottom of prisms prior to testing with sulfur-filled capping or with high-strength gypsum cement. Sulfur-filled capping material shall be 40 to 60 % sulfur by weight; the remainder may be ground fire clay or other suitable inert material passing a No. 100 (150 mm) sieve, with or without a plasticizer. Spread the capping material over a level surface that is plane within 0.003 in. (0.076 mm) in 16 in. (406 mm). Bring the surface to be capped into contact with the capping compound; firmly press down the specimen, holding the prism so that its axis is at right angles to the capping surfaces, and the capped ends are parallel within one degree. A spirit level (for example, bulls eye level, torpedo level) or equivalent method shall be used during the capping process to comply with these prism alignment criteria. The average thickness of the cap shall not exceed ⅛ in. (3 mm). Age the caps at least 2 h before testing the specimens.

9. Procedure

9.1 *Test Apparatus*—The test machine shall have an accuracy of plus or minus 1.0 % over the anticipated load range. The upper bearing shall be a spherically seated, hardened metal block firmly attached at the center of the upper head of the machine. The center of the sphere shall lie at the center of the surface held in its spherical seat but shall be free to turn in any direction, and its perimeter shall have at least ¼ in. (6.3 mm) clearance from the head to accommodate specimens whose bearing surfaces are not parallel. The diameter of the bearing surface shall be at least 6 in. (150 mm). A hardened metal bearing block may be used beneath the specimen to minimize wear of the lower platen of the machine. The bearing block surfaces intended for contact with the specimen shall have a hardness not less than HRC 60 (BHN 620). These surfaces shall not depart from plane surfaces by more than 0.001 in. (0.03 mm) in any 6-in. (150 mm) dimension. When the bearing area of the spherical bearing block is not sufficient to cover the area of the specimen, a single-thickness steel plate with surfaces machined plane within plus or minus 0.001 in. (0.03 mm) in any 6-in. (150 mm) dimension, and with a thickness equal to at least the distance from the edge of the spherical bearings to the most distant corner of the specimen; shall be placed between the spherical bearing block and the capped specimen. The length and width of the steel plate shall be at least

¼ in. (6 mm) greater than the length and width of the prisms.

9.2 *Installing the Prism in The Test Machine*—Wipe clean the bearing faces of the upper and lower platens or bearing blocks and of the test specimen and place the test specimen on the lower platen or bearing block. Align both centroidal axes of the specimen with the center of thrust of the test machine. As the spherically seated block is brought to bear on the specimen, rotate its movable portion gently by hand so that uniform seating is obtained.

9.3 *Loading*—Apply the load to the prism up to one-half of the expected total load at any convenient rate. Apply the remaining load at a uniform rate in not less than 1 nor more than 2 min.

9.4 *Observations*—Describe the mode of failure as fully as possible or illustrate, or both, crack patterns and spalling on a sketch or photograph. Note whether failure occurred on one side or one end of the prism prior to failure of the opposing side or end of the prism.

10. Calculation

10.1 Calculate test results as follows:

10.1.1 *Net Cross-Sectional Area*—Take the net cross-sectional area of ungrouted prisms as the net cross-sectional area of masonry units, which are cut as in the prism, determined by measurement or from a representative sample of units that are cut as in the prism, in accordance with Test Methods C 140 for concrete masonry and with Test Methods C 67 for clay masonry. Determine net cross-sectional area of fully grouted prisms by multiplying the length and width of the prism (see 8.1).

NOTE 2—Net area of concrete masonry units other than 100 % solid units is determined by Test Methods C 140. Net area determined by Test Methods C 140 is usually slightly different from the minimum net cross-sectional area.

10.1.1.1 Consider clay masonry units whose net cross-sectional area is at least 75 % of the gross cross-sectional area as 100 % solid.

10.1.2 *Masonry Prism Strength*—Calculate each masonry prism strength by dividing each prism's maximum compressive load sustained by the net cross-sectional area of that prism, and express the result to the nearest 10 psi (69 kPa).

10.1.2.1 Where sets of grouted and ungrouted prisms are tested, calculate the masonry prism strength separately for the grouted set and the ungrouted set.

10.1.2.2 Where a set of prisms is tested for each wythe of a multiwythe wall, calculate the masonry prism strength for each wythe.

10.1.3 *Compressive Strength of Masonry:*

TABLE 1   Height to Thickness Correction Factors for Masonry Prism Compressive Strength

h_p/t_p[A]	1.3	1.5	2.0	2.5	3.0	4.0	5.0
Correction Factor	0.75	0.86	1.0	1.04	1.07	1.15	1.22

[A] h_p/t_p—Ratio of prism height to least lateral dimension of prism.

10.1.3.1 Calculate the h_p/t_p ratio for each prism using the height and the least lateral dimension of that prism. Determine the correction factor from Table 1. If a prism's height to thickness ratio lies between the h_p/t_p values of Table 1, determine the corresponding correction factor by linear interpolation between the given values.

10.1.3.2 Multiply the masonry prism strength by the correction factor for the respective prism.

10.1.3.3 Calculate the compressive strength of masonry, f_{mt}, for each set of prisms by averaging the values obtained.

11. Report

11.1 Report the following information:

11.1.1 Name of testing laboratory,

11.1.2 Designation of each prism tested and description of prism including width, height, and length dimensions; h_p/t_p ratio; mortar type; and grout and masonry unit used in the construction,

11.1.3 Age of prism at time of test,

11.1.4 Maximum compressive load sustained by each prism in pounds force or newtons,

11.1.5 Net cross-sectional area of each prism in square inches or square millimetres, and method used to calculate area,

11.1.6 Test observations for each prism in accordance with 9.4,

11.1.7 Compressive strength of each prism calculated to the nearest 10 psi or 69 kPa (see 10.1.3.2), and

11.1.8 Compressive strength of masonry, f_{mt}, for each set of prisms calculated to the nearest 10 psi or 69 kPa (see 10.1.3.3).

12. Precision and Bias

12.1 Due to the variety of materials and combinations of materials involved, no statement is made concerning the precision or bias of this test method. Sufficient test data for all materials and combinations of materials are not available to permit the development of precision and bias statements.

13. Keywords

13.1 compressive strength of masonry; masonry prism; masonry prism strength; specified compressive strength of masonry

Standard Test Methods for
Compressive Strength of Masonry Prisms[1]

This standard is issued under the fixed designation E 447; the number immediately following the designation indicates the year of original adoption or, in the case of revision, the year of last revision. A number in parentheses indicates the year of last reapproval. A superscript epsilon (ϵ) indicates an editorial change since the last revision or reapproval.

This standard has been approved for use by agencies of the Department of Defense. Consult the DoD Index of Specifications and Standards for the specific year of issue which has been adopted by the Department of Defense.

1. Scope

1.1 These test methods cover two compression tests of masonry prisms:

1.1.1 *Method A*—For determining comparative data on the compressive strength of masonry built in the laboratory with either different masonry units or mortar types, or both.

1.1.2 *Method B*—For determining the compressive strength of masonry built at the job site with the same materials and workmanship to be used, or being used, in a particular structure.

NOTE 1—Method B is intended to establish the compressive strength of a particular set of materials by a preconstruction evaluation or to evaluate quality of materials and workmanship during construction.

1.2 Test specimens for both test methods are short compression prisms: the influence of slenderness ratio is taken into account either by proportions of the fabricated specimens or by the application of a correction factor.

1.3 The values stated in SI units are to be considered as the primary standard units.

1.4 *This standard does not purport to address all of the safety problems, if any, associated with its use. It is the responsibility of the user of this standard to establish appropriate safety and health practices and determine the applicability of regulatory limitations prior to use.*

2. Referenced Documents

2.1 *ASTM Standards:*

C 39 Test Method for Compressive Strength of Cylindrical Concrete Specimens[2]

C 67 Test Method of Sampling and Testing Brick and Structural Clay Tile[3]

C 140 Test Methods of Sampling and Testing Concrete Masonry Units[3]

C 143 Test Method for Slump of Hydraulic Cement Concrete[2]

C 144 Specification for Aggregate for Masonry Mortar[3]

C 270 Specification for Mortar for Unit Masonry[3]

C 476 Specification for Grout for Masonry[3]

C 780 Test Method for Preconstruction and Construction Evaluation of Mortars for Plain and Reinforced Unit Masonry[3]

C 1019 Test Method of Sampling and Testing Grout[3]

E 4 Practices for Force Verification of Testing Machines[4]

E 6 Terminology Relating to Methods of Mechanical Testing[4]

E 105 Practice for Probability Sampling of Materials[5]

E 111 Test Method for Young's Modulus, Tangent Modulus, and Chord Modulus[4]

E 575 Practice for Reporting Data from Structural Tests of Building Constructions, Elements, Connections, and Assemblies[6]

E 631 Terminology of Building Constructions[6]

3. Terminology

3.1 *Definitions*—For definitions and terminology used in these test methods, refer to Terminology E 6 and E 631.

4. Apparatus

4.1 The testing machine shall conform to the requirements given in Practices E 4.

4.2 The upper bearing shall be a spherically seated, hardened metal block firmly attached at the center of the upper head of the machine. The center of the sphere shall lie at the center of the surface held in its spherical seat, but shall be free to turn in any direction, and its perimeter shall have at least 6-mm (¼-in.) clearance from the head to allow for specimens whose bearing surfaces are not exactly parallel. The diameter of the bearing surface shall be at least 125 mm (5 in.). A hardened metal bearing block shall be used beneath the specimen to minimize wear of the lower platen of the machine. The bearing block surfaces intended for contact with the specimen should have a hardness not less than 60 HRC (620 HB). These surfaces shall not depart from plane surfaces by more than 0.003 mm (0.0001 in.) in any 150-mm (6-in.) dimension. When the bearing area of the spherical bearing block is not sufficient to cover the area of the specimen, a steel plate with surfaces machined to true planes within ±0.003 mm (0.0001 in.), and with a thickness equal to at least one half of the distance from the edge of the spherical bearing to the most distant corner shall be placed between the spherical bearing block and the capped specimen.

5. Test Specimen

5.1 *Sampling and Testing of Masonry Units:*

5.1.1 Select representative masonry units and test for

---

[1] These test methods are under the jurisdiction of ASTM Committee C-15 on Manufactured Masonry Units and are the direct responsibility of Subcommittee C 15.04 on Research.

Current edition approved Oct. 15, 1992. Published December 1992. Originally published as E 447 – 72. Last previous edition E 447 – 92a.

[2] *Annual Book of ASTM Standards*, Vol 04.02.

[3] *Annual Book of ASTM Standards*, Vol 04.05.

[4] *Annual Book of ASTM Standards*, Vol 03.01.

[5] *Annual Book of ASTM Standards*, Vol 14.02.

[6] *Annual Book of ASTM Standards*, Vol 04.07.

percent void area for hollow clay masonry and average net area for hollow concrete masonry and compressive stength in accordance with the following applicable ASTM methods:

Masonry Units	ASTM Designation
Building and face brick	C 67
Sandlime brick	C 140
Concrete brick	C 140
Concrete block	C 140
Structural clay tile	C 67

5.1.2 Select random samples of masonry unit test specimens in accordance with Recommended Practice E 105.

5.2 *Sampling and Testing of Mortar*—Use one of the types of mortar as specified in Specification C 270 unless otherwise required. Sample and test in accordance with Test Method C 780. Determine and report the physical properties by the standards indicated:

5.2.1 *Consistency*—Test Method C 780, Annex A1.

5.2.2 *Air Content*—Test Method C780, Annex A5.

5.2.3 *Compressive Strength*—Test Method C 780, Annex A6.

5.2.4 *Sieve Analysis*—Specification C 144.

5.3 *Sampling and Testing of Grout*—Use one of the types of grout as specified in Specification C 476 unless otherwise required. Sample and test in accordance with Method C 1019. Determine and report the physical properties by the test methods indicated:

5.3.1 *Slump*—Test Method C 143.

5.3.2 *Compressive Strength*—Method C 1019.

5.3.4 *Fine Grout*—Mold a minimum of three 50-mm (2-in.) cubes from a sample of the fine grout and test in the same manner as prescribed in 5.2, except moist cure cubes 48 h before releasing from molds.

5.3.5 *Coarse Grout*—Mold a minimum of three 75 by 150-mm (3 by 6-in.) cylinders from a sample of the coarse grout and moist cure for 48 h before releasing from molds. Cap and test them in compression in accordance with applicable provisions of Test Method C 39.

5.4 *Masonry Prisms*:

5.4.1 *Method A*—Build and test a minimum of three prism specimens for each combination of variables. Each test prism shall be a single-wythe specimen laid in stack bond, with a height-to-thickness ratio of not less than two nor more than five. The mortar joints shall be 10 mm (³⁄₈ in.) in thickness. Spread a full bed of mortar on each solid masonry unit and allow no furrowing of the mortar bed. Fully bed the face shells of hollow masonry units with mortar. Strike mortar joints flush with the face of the masonry without tooling. The length of the prism shall be greater than its thickness.

5.4.2 *Method B*—Build and test a minimum of three prisms using the materials and workmanship being used in the structure. In constructing the prisms, the mortar bedding, the thickness and tooling of the joints, the grouting, the condition of the units, and the bonding arrangement shall be in so far as possible the same as being used in the structure, except that no structural reinforcement shall be included in the test specimen; however, metal ties may be used. Prisms shall contain no grout unless all hollow cells and spaces in the actual wall construction are to be grouted.

5.4.2.1 The thickness of the prisms shall be the same as the thickness of the masonry part of the wall in the structure.

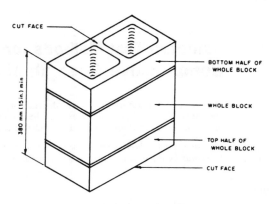

FIG. 1 Masonry Prism Construction for 200-mm (8-in.) High Units

The length of the prism shall be equal to or greater than the thickness of the prism. The height of the prism shall be at least twice the thickness, contain at least two mortar joints and be a minimum of 380 mm (15 in.). Where prisms composed of 200-mm (8-in.) high units are to be tested, and where the compression testing equipment will not accommodate three full units in height, the configuration shown in Fig. 1 may be used.

5.5 *Handling and Curing Conditions*:

5.5.1 *Method A*—Generally, cure all prisms for 28 days. However, lesser periods of time may be used provided the relation between that period and the 28-day strength of the masonry is established by a suitable statistical method (Practice E 105) (Note 2). Cure the prisms together with the corresponding mortar and grout specimens in laboratory air maintained at a temperature of 24 ± 8°C (75 ± 15°F), with a relative humidity between 30 and 70 %, and free of drafts. These environmental conditions will not require, generally, special air-conditioning equipment. Make a continuous graphical record of temperatures and humidity to detect unusual dryness or excessive moisture together with unusual fluctuations of temperature.

NOTE 2—A preferred method of determining the relation between the strengths of 28-day prisms and those cured for shorter periods of time is to build three additional prisms to be tested after curing the shorter time. Such a procedure is especially recommended for Method B and should be done in advance of start of construction if the prisms built and tested during construction are for the purpose of job control of materials and workmanship and the construction schedule requires a curing period of less than 28 days.

5.5.2 *Method B*—Prior to transporting prisms for testing, place a 20-mm (³⁄₄-in.) plywood sheet cut the same size as the prisms on the top and bottom and wire or strap tightly to prevent joints from being disturbed while the prisms are being transported to the laboratory for testing. Protect the prisms and transport them in such a manner as not to break the bond between mortar and units.

5.5.2.1 Remove prisms from moisture-tight bag one day before the scheduled compression test. Test at 28 days from fabrication. Where seven-day tests are used, the relationships between the 7-day and 28-day strengths of the masonry assemblage has been previously established for the materials used.

6. Procedure

6.1 *Capping Test Specimens*—Cap the ends of the prisms in the same manner as set forth for capping the units in either Methods C 67 or C 140.

6.2 *Specimen Measurements*—Determine the length and thickness of the prism to the nearest 0.3 mm (0.01 in.) by averaging three measurements taken at the center and quarter points of the height of the specimen. Measure the height of the specimen including caps to the nearest 3 mm (0.1 in.).

6.3 *Placing the Specimen*—Place the plain (lower) bearing block, with its hardened face up, on the table or platen of the testing machine directly under the spherically seated (upper) bearing block. Wipe clean the bearing faces of the upper and lower bearing blocks and of the test specimen and place the test specimen on the lower bearing block. Carefully align both centroid axes of the specimen with the center of thrust of the spherically seated block. As the spherically seated block is brought to bear on the specimen, rotate its movable portion gently by hand so that uniform seating is obtained.

6.4 *Rate of Loading*—Apply the load, up to one half of the expected maximum load, at any convenient rate, after which adjust the controls of the machine so that the remaining load is applied at a uniform rate in not less than 1 nor more than 2 min.

6.5 *Determination of Young's Modulus*—When stipulated, determine Young's modulus in accordance with Test Method E 111. Follow the designated method to the greatest extent possible.

6.6 *Observations*—During the course of the test, note the loading at which the first cracking sounds are heard. Note whether a stethoscope or other instrument is used for hearing. If observed, note the load corresponding to the appearance of the first crack either in the unit itself or spalling of the mortar joint, or both. Upon increasing the load to a maximum sustained by the prism, describe the mode of failure as fully as possible. In the cases of walls with two or more wythes, apply suitable sensitive gages transversely across the collar joint at the midheight of the specimen to detect the formation of cracks in the plane of the collar joint. (Bonded wire gages have been found to be suitable for these measurements.)

NOTE 3—The measurements of transverse strains across collar joints in double- or multi-wythe walls are essential to determine whether a plane of weakness exists along a continuous vertical joint such as a collar joint that may develop significant tensile strains indicating potential premature cracking.

7. Report

7.1 *Method A*—Prepare the report in conformance with Practice E 575 and include the following:

7.1.1 The compressive strength and other specified physical properties of masonry units, including the name and address of the manufacturer and his unit designation number or name.

7.1.2 The compressive strength and other specified physical properties of the mortar and grout, a sand-sieve analysis, and other specified physical properties of the constituent materials, including the name and address of the manufac-

turer of each constituent material, and his material designation number or name.

7.1.3 Brief description of specimen fabrication including method of bonding, joint thickness, and prism dimensions.

7.1.4 Age of prism at time of test.

7.1.5 Maximum load for each prism in pounds-force (or newtons).

7.1.6 Cross-sectional area of each specimen in square millimetres (or square inches). The gross area shall be used for solid masonry prisms and the net area for prisms built with hollow units not solidly grouted.

7.1.7 Compressive strength of each prism calculated to the nearest 70 kPa (10 psi).

7.1.8 Average compressive strength, standard deviation, and coefficient of variation of the sample.

7.1.9 When required, the stress-strain curve for each prism, determined in accordance with applicable provisions of Test Method E 111, shall be plotted. The secant modulus of elasticity shall be reported for suitable values of stress and strain. In the case of double-wythe walls, the compressive stress versus the transverse strain measured across the collar joint shall also be plotted.

7.1.10 A full description of the mode of failure. Whenever useful clarification will result, use photographs, and

7.1.11 Unusual defects in either specimens or caps shall be noted, including causes, when known, of unusual test results.

7.2 *Method B*—Prepare the report in conformance with Practice E 575 and include the following:

7.2.1 The compressive strength and other specified physical properties of masonry units, including the name and address of the manufacturer and his unit designation number or name.

7.2.2 Brief description of specimen fabrication including method of bonding, joint thickness, and prism dimensions.

7.2.3 Age of prism at time of test.

7.2.4 Maximum load for each prism in newtons (or pounds-force).

7.2.5 Cross-sectional area of each specimen in square millimeters (or square inches). The gross area shall be used for solid masonry prisms and the net area for prisms built with hollow units not solidly grouted.

7.2.6 Compressive strength of each prism calculated to the nearest 70 kPa (10 psi).

7.2.7 Average compressive strength, standard deviation, and coefficient of variation of the sample.

7.2.8 A full description of the mode of failure. Whenever useful clarification will result, use photographs.

7.2.9 Unusual defects in either specimens or caps shall be noted, including causes, when known, of unusual test results.

8. Precision and Bias

8.1 No statement is made either on the precision or on the bias of these test methods due to the variety of material and combinations of materials involved. Sufficient test data for all materials and combinations of materials are not presently available to permit the development of precision and bias statements.

9. Keywords

9.1 compressive strength; masonry; prisms

E 447

SELECTED

QUALITY CONTROL

STANDARDS

RATE OF ABSORPTION OF BRICK

The rate of absorption of brick has an important effect on the bond between brick and mortar.

For maximum bond strength, brick at the time of laying should have a rate of absorption not exceeding 0.035 ounces of water per sq. in. of surface after being placed in $\frac{1}{8}$ in. of water for one minute.

There is no consistent relationship between total absorption and rate of absorption. Some brick have high total absorption and low rate of absorption.

FIELD TEST FOR RATE OF ABSORPTION FOR BRICK

1.	A rough but effective test for rate of absorption is to place a 25-cent coin on a brick draw a circle around it with a pencil, then trace around the circle using a wax crayon.

2.	Using a medicine dropper, very quickly drop water within the circle completely filling it, taking care that the water does not flow over marked circle, and continue until 20 drops have been placed.

3.	Note that the time required for all of the water to be absorbed into the brick beginning with the time the circle is first filled.

4.	If the time exceeds $1\frac{1}{2}$ minutes, the brick need not be wetted; if less than $1\frac{1}{2}$ minutes, they must be pre-wetted.

VALUES FOR COMPRESSIVE STRENGTH

Following are minimum field strength values required by the California Office of the State Architect Title 24 and UBC Standard 21-16 for compressive strengths or mortar and grout test specimens

		28 Day Compressive Strength
MORTAR	1 part Portland Cement, $\frac{1}{2}$ part lime and $4\frac{1}{2}$ parts dry loose sand --	1500 psi
GROUT	1 part Portland Cement, 3 parts dry loose sand with from 1 to 2 parts pea gravel---------------------	2000 psi
CORES	Cut from wall ---	1500 psi

MASONRY PRISM TESTING

Prism testing of masonry is described in the enclosed Uniform Building Code, The Uniform Building Code Standards 21-17 and the American Society for Testing and Materials (ASTM C-1314-95) *Standard Test Methods for Constructing and Testing Masonry Prisms Used to Determine Compliance with Specified Compressive Strength of Masonry.*

Prism testing consists essentially of making sample assemblages of the proposed wall construction. Such testing is primarily to assure that the assemblage of masonry units, mortar and grout will provide an ultimate compressive strength that fulfills or exceeds the design requirements.

The user should be aware the strength of the masonry units and grout must be about one-third greater than the specified f'_m to obtain the assemblage strength required. Testing for final verification should be done early enough to allow time for adjustments.

GUIDE SPECIFICATIONS AND NOTES TO SPECIFIER

The following Guide Specifications were written as the framework for a specifications section. The Guide Specifications appear in the CSI three-part format on the left hand pages with explanatory information on the adjacent right hand pages titled *NOTES TO SPECIFIER.*

Each Guide Specification addresses a particular type or use of masonry construction. They should serve as a guide only and they must be revised and customized for each particular project.

When using any specification the Specifier must consider applicable building codes, local practices and the particular features of the project. The Specifier may find it useful to combine some of these specifications into one specification for smaller projects.

SECTION 04101
GUIDE SPECIFICATION FOR
MORTAR

PART 1 - GENERAL

1.01   SECTION INCLUDES:

   A. Mortar for Masonry.

   B. Repointing Mortar (Tuck Pointing).

1.02   RELATED SECTIONS:

   A. Quality Control, Testing and Laboratory Services.

   B. Concrete, Division 3.

   C. Clay Brick Masonry.

   D. Concrete Unit Masonry.

   E. Masonry Accessories.

   F. Masonry Restoration and Cleaning.

   G. Rough Carpentry, Division 6.

   H. Grout for Masonry.

   I. Masonry and Stone Veneer.

   J. Caulking, Sealants and Waterproofing, Division 7.

   K. Door Frames, Grouting Door Frames, Division 8.

1.03   PRODUCTS INSTALLED BUT NOT FURNISHED UNDER THIS SECTION:

   A. Joint Reinforcement.

   B. Metal Accessories.

   C. Masonry Units

   D. Flashing and Steel Metal.

   E. Veneer Ties.

1.02   Some of these broadscope items may not need to be included depending on the particular job. Likewise other broadscope or narrowscope items may need to be included for certain job conditions.

1.04   REFERENCES:

A. ASTM C 91-91, (UBC Standard 21-11), *Standard Specification for Masonry Cement.*

B. ASTM C 144-93, *Standard Specification for Aggregate for Masonry Mortar.*

C. ASTM C 150-89, (UBC Standard 19-1), *Standard Specification for Portland Cement.*

D. ASTM C 207-91, (UBC Standard No. 21-13), *Standard Specification for Hydrated Lime for Masonry Purposes.*

E. ASTM C 270-91a, (UBC Standard No. 21-15), *Standard Specification for Mortar for Unit Masonry.*

F. ASTM C 387-87, *Standard Specification for Packaged, Dry, Combined Materials for Mortar and Concrete.*

G. ASTM C 595-86, (UBC Standard No. 19-1), *Standard Specification for Blended Hydraulic Cements.*

H. ASTM C 780-91, *Standard Test Method for Preconstruction and Construction Evaluation of Mortars for Plain and Reinforced Unit Masonry.*

I. ASTM C 1142-92, *Standard Specification for Ready Mixed Mortar for Unit Masonry.*

J. ASTM C 1180-93a, *Standard Terminology of Mortar and Grout for Unit Masonry.*

1.05   SUBMITTALS:

A. Submit data indicating specifications used for mortar.

B. Submit test reports for mortar materials indicating conformance to ASTM C 270 (UBC Standard No, 21-15).

C. Submit test reports for field sampling and testing mortar in conformance to ASTM C 780.

D. Samples: Submit two ribbons of mortar for conformance with color.

1.06   DELIVERY, STORAGE AND HANDLING:

A. Deliver and store manufactured products in original, unopened containers.

B. Store materials in a clean, dry location protected from dampness and freezing.

1.04 Update the references, as applicable. UBC Standard numbers are shown, where applicable. Delete those standards which are not applicable.

1.05.A ASTM C 270 and UBC Standard 21-15 require that mortar be specified by proportion or property, not both.

1.05.B ASTM C 270 and UBC Standard 21-15 require comparison of laboratory prepared mortars to establish proportions for field mixed mortars when the property specifications are used.

1.05.D The actual mortar color may vary slightly from ribbon color samples.

1.06.A Cement, lime and any admixtures should be delivered in sealed containers and palletized so that these materials are not in direct contact with the ground.

C. Store cementitious ingredients in weather-tight enclosures and protect against contamination and warehouse set.

D. Stockpile and handle aggregates to prevent contamination from foreign materials.

E. Store admixtures to prevent contamination of damage from excessive temperature changes.

F. Keep water clean and free from harmful materials.

1.07   ENVIRONMENTAL REQUIREMENTS:

A. Heat sand and mixing water when the air temperature is below 40° F to provide mortar and grout temperatures between 40° F and 120° F when used.

B. Do not heat sand or water above 120° F.

PART 2 - PRODUCTS

2.01   MORTAR MATERIALS:

A. Cement:

   1. Portland Cement: ASTM C 150 (UBC Standard No. 19-1), Type I or II

   2. Mortar Cement: UBC Standard 21-19

   3. Masonry Cement: ASTM C 91 (UBC Standard 21-16)

   4. Plastic Cement: ASTM C 150 except in respect to limitations on insoluble residue, air entrainment, and additions subsequent to calcination. Approved types of plasticizing agents may be added to Types I and II portland cements, but not in excess of 12 percent of the total volume.

B. Hydrated Lime: ASTM C 207 (UBC Standard No. 21-13).

C. Sand: ASTM C 144.

D. Admixtures:

   1. The use of admixtures shall not be permitted except as specified by the Architect/Engineer and as approved by the Building Official.

   2. No air entraining admixtures or material containing air entraining admixtures may be used.

   3. No antifreeze compounds shall be added to mortar.

1.06.C Cement, lime and admixtures must be covered when there is any possibility of adverse conditions.

2.01.A.1 Type I or Type II Portland cement is generally used. California Division of the State Architect (DSA) allows types I, IA, II, IIA, III and IIIA.

2.01.A.3 Masonry cement is not permitted to be used as part of the structural system in seismic zone Nos. 2, 3 and 4. DSA does not permit masonry cement to be used.

2.01.A.4 Plastic cement is not permitted to be used as part of the structural system in seismic zone Nos. 2, 3 and 4. DSA does not permit plastic cement to be used.

2.01.C. Mortar made with sand not conforming to ASTM C 144 must meet the property specification requirements.

4. No admixtures containing chlorides shall be added to mortar.

E. Water:

1. Water shall be clean, potable and free from deleterious quantities of acids, alkalis and organic materials.

2. Water shall come from a domestic supply.

F. Mortar Pigment:

1. Mortar pigment shall not exceed 10% of the weight of portland cement.

2. Carbon black shall not exceed 3% of the weight of portland cement.

2.02 MORTAR MIXES:

A. Mortar: ASTM C 270 (UBC Standard No. 21-15 and UBC Table 21-A), Type M, S, N or O.

*** or ***

A. Ready- Mixed Mortar: ASTM C 1142, Type RM, RS, RN or RO.

PART 3 - EXECUTION

3.01 FIELD MIXING MORTAR:

A. All cementitious materials and aggregate shall be mixed between 3 and 10 minutes in a mechanical mixer with the amount of water to produce a spreadable, workable consistency. Dry mixes for mortar which have been preblended in a factory shall be mixed at the jobsite until workable, but not to exceed 10 minutes.

B. Control batching procedure to ensure proper proportions by measuring material by volume.

C. The consistency of mortar and grout may be adjusted to the satisfaction of the mason by retempering with water. Mortar may be retempered once within 2 ½ hours after initial mixing to compensate for water lost due to initial evaporation. Retempering shall be done by adding water into a formed basin within the mortar and then working the mortar into the water. Mortar shall not be retempered by splashing water over the surface.

D. Discard all mortar which has begun to harden. Also discard mortar if more than 2 ½ hours old.

3.02 APPLICATION OF MORTAR:

A. Ends of solid masonry units shall be buttered with sufficient mortar to fill head joints. Hollow unit masonry shall be mortared so that the head joint thickness is equal to the face shell thickness.

2.01.F   Limits on the amount of pigments should be halved when using masonry cement mortars.

2.02.A   Mortar mixes can be specified by ASTM C 270, UBC Standard No. 20-15 for job mix mortar or ASTM C 1142 for ready mixed mortars. Specify the mortar type: Type M, S or N for mortar specified by ASTM C 270 and UBC Standard No. 21-15 or Type RM, RS or RN for ready mixed mortar. Note: Types N or O mortar are not permitted to be used as part of the structural system in seismic zone Nos. 3 and 4. Type O mortar is not permitted in structural system in seismic zone 2.

3.01   ASTM C 270 can be referenced for field mixing mortar.

B. Mortar beds for solid units shall be slightly beveled towards the center of the wall so that the bed joints will be sufficiently filled when the masonry unit is brought into line. Furrowing of the joints is not permitted.

C. Closures shall be rocked into place with mortared head joints against two adjacent brick in place.

D. Corners and jambs may not be pounded into position to fit stretcher units.

E. Units which have been displaced after the mortar has begun to set shall be cleaned of all mortar and reset with fresh mortar.

F. Mortar fins and protrusions which protrude more than ½ inch into cells or spaces to grouted are to be avoided.

G. Mortar Joints shall be tooled as directed in Division 4.

3.03   REPOINTING MORTAR (TUCK POINTING):

A. Use Type __ mortar for repointing.

B. Prehydrate the repointing mortar as follows:

1. Mix dry ingredients thoroughly.

2. Add enough water to make damp, stiff mix which will retain its form when pressed into a ball.

3. After 1 to 2 hours, add sufficient water to bring it to the proper consistency.

END OF SECTION

3.02.B The beveled bed may leave some unfilled spaces at the back of the units but these will be filled by the fluid grout.

3.02.G Specific mortar joint type here or direct contractor to appropriate section on tooling.

3.03.A Specify the type of repointing mortar. Typically, repointing mortar should be Type N or O.

SECTION 04102
GUIDE SPECIFICATION FOR GROUT

PART 1 - GENERAL

1.01   SECTION INCLUDES:

A.   Grout for Masonry.

1.02   RELATED SECTIONS:

A.   Quality Control, Testing and Laboratory Services.

B.   Concrete, Division 3.

C.   Clay Brick Masonry.

D.   Concrete Unit Masonry.

E.   Masonry Accessories.

F.   Masonry and Stone Veneer.

G.   Masonry Restoration and Cleaning.

H.   Rough Carpentry, Division 6.

I.   Caulking, Sealants and Waterproofing, Division 7.

J.   Door Frames, Grouting Door Frames, Division 8.

K.   Mortar for Masonry.

1.03   PRODUCTS INSTALLED BUT NOT FURNISHED UNDER THIS SECTIONS:

A.   Reinforcing Steel.

B.   Metal Accessories.

C.   Masonry Units.

D.   Flashing and Steel Metal.

1.04   REFERENCES:

A.   ASTM C 150-89, Standard Specification for Portland Cement (UBC Standard No. 19-1).

1.02 Some of these broadscope items may not need to be included depending on the particular job. Likewise other broadscope or narrowscope items may need to be included for certain job conditions.

1.04 Update the references, as applicable. UBC Standard numbers are shown, where applicable. Delete those standards which are not applicable.

B. ASTM C 207-91, (UBC Standard No. 21-13), Standard Specification for Hydrated Lime for Masonry Purposes.

C. ASTM C 387-87, Standard Specification for Packaged, Dry, Combined Materials for Mortar and Concrete.

D. ASTM C 404-93, Standard Specification for Aggregates for Masonry Grout.

E. ASTM C 476-91, (UBC Standard No. 21-19), Standard Specification for Grout for Masonry.

F. ASTM C 595-86 Standard Specification for Blended Hydraulic Cements.

G. ASTM C 1019-89a, (UBC Standard No. 21-18), Standard Method of Sampling and Testing Grout.

1.05   SUBMITTALS:

A. Submit test reports for grout materials including conformance to ASTM C 476 (UBC Standard No. 21-19).

B. Submit test reports for field sampling and testing grout in conformance to ASTM C 1019 (UBC Standard No. 21-18).

1.06   DELIVERY, STORAGE AND HANDLING:

A. Grout may be plant-batched and shipped to project in ready mix trucks or grout may be mixed at project site.

B. Deliver and store manufactured products in original, unopened containers.

C. Store materials in a clean, dry location protected from dampness and freezing.

D. Store cementitious ingredients in weather-tight enclosures and protect against contamination and warehouse set.

E. Stockpile and handle aggregates to prevent contamination from foreign materials.

F. Store admixtures to prevent contamination of damage from excessive temperature changes.

G. Keep water clean and free from harmful materials.

1.07   ENVIRONMENTAL REQUIREMENTS:

A. Heat sand and mixing water when the air temperature is below 40° F to provide mortar and grout temperature between 40° F and 120° F when used.

1.05 A   A design mix can be submitted when grout is batched in a ready mix plant and delivered in transit mix truck

1.05.B   ASTM C 1019 and UBC Standard 21-18 are used to test uniformity of grout preparation during construction.

1.06.B   Applies to packaged cement, lime and dry grout mixes. Pre-blended dry bulk grout mix should be protected in enclosed bulk containers.

B. Do not heat sand or water above 120° F.

PART 2 - PRODUCTS

2.01 GROUT MATERIALS:

A. Portland Cement: ASTM C 150 (UBC Standard No. 19-1).

B. Hydrated Lime: ASTM C 207 (UBC Standard No. 21-13).

C. Aggregate: ASTM C 404.

D. Admixtures:

1. The use of admixtures shall not be permitted except as specified by the Architect/Engineer and as approved by the Building Official.

2. An admixture shall be used in high lift grouting to counteract water loss and volume reduction.

E. Water:

1. Water shall be clean, potable and free from deleterious quantities of acids, alkalis and organic materials.

2. Water shall come from a domestic supply.

2.02 GROUT MIXES:

A. Grout: ASTM C 476 (UBC Standard No. 21-19 and UBC Table 21-B).

1. Fine Grout (1 part portland cement; 2¼ to 3 parts sand).

2. Coarse Grout (1 part portland cement; 2¼ to 3 parts sand; 1 to 2 parts gravel).

3. Slump: 8 to 10 inches.

4. Minimum strength 2,000 psi.

2.01.A Fly ash may be used as a partial cement replacement for the cementitious material for plant-batched ready mix grout.

2.01.D Grout admixtures which decrease grout shrinkage and compensate for volume loss due to water absorption into the masonry unit are used when high lift grouting and may be used when low lift grouting.

2.02.A The use of fine or coarse grout is based on the width of the grout space and the height of the grout pour (See UBC Table No. 21-C).

2.02.A.2 In order to provide more volume of aggregate and reduce shrinkage, pea gravel is most commonly used.

2.02.A.3 The slump of grout should be between 8 and 10 inches so that the grout can flow and fill voids and cavities.

2.02.A.4 If strengths greater than the minimum compressive strength of 2,000 psi are required, it should be stated.

PART 3 - EXECUTION

3.01   FIELD MIXING GROUT:

A. All cementitious material and aggregate shall be mixed between 3 and 10 minutes in a mechanical mixer with the amount of water to produce a spreadable, workable consistency. Dry mixes for grout which have been preblended in a factory shall be mixed at the jobsite until workable, but not to exceed 10 minutes.

B. Control batching procedure to ensure proper proportions by measuring material by volume.

C. The consistency of grout may be adjusted to the satisfaction of the masonry by retempering with water.

D. Discard all grout which has begun to harden. Also discard grout which is more that 1½ hours old.

3.02   LOW LIFT GROUTING:

A. Grout pours 12 inches and less:

1.   If necessary, clean or roughen concrete foundation by sandblasting, chipping or other means to remove laitance.

2.   Lay one course of masonry making sure no mortar extends into grout spaces.

3.   Place all reinforcement which extends into grouted areas. Reinforcement shall be secured prior to grouting.

4.   Grout to below one-half of the top unit height and consolidate by puddling to eliminate voids in the grout.

5.   Lay an additional 12 inches of masonry units.

6.   Grout each 12 inches as the units are laid. Hold the top of each grout pour approximately 1½ inches below the top of the wall. Provide at least ½ inch of grout cover above horizontal reinforcing steel.

7.   At the completion of each wall, grout flush to the top of the units.

8.   Remove all grout droppings as the work progresses.

B. Grout pours more than 12 inches and up to 5 feet.

1.   Construct the masonry wall up to 5 feet above the foundation. Install all reinforcing steel, anchors and embedded items as masonry work progresses.

3.02.A.4 The grout must be made to flow into place by puddling with a stick or by vibrating. Puddling with a trowel blade is not permitted.

2. For two wythe walls, bond the wythes together with rectangular ties or joint reinforcing so that one cross wire secures approximately two square feet of wall.

3. For walls that are to be partially grouted, use expanded metal mesh or other material which will not interfere with bond to restrict the grout into only those cells which are to be grouted.

4. After the mortar joints have set, grout the wall to 1½ inches below the top of the wall. Where bond beams occur, stop grout pour a minimum of ½ inch below top of masonry.

5. Consolidate the grout using a mechanical vibrator and reconsolidate after the excess water is absorbed into the masonry units.

6. Continue to lay up masonry and reinforcing steel, up to 5 feet at a time. After the mortar has set, grout and consolidate.

7. At the completion of the wall, fill the grout space flush with the top of the units and consolidate.

3.03   HIGH LIFT GROUTING:

A. Construct the masonry wall up to a maximum of 24 feet above the foundation. Provide cleanout openings at the base of the wall at all vertical reinforcing bars but at a spacing no more than 32 inches on centers for solid grouted walls or a maximum of 48 inches on centers for partially grouted walls.

B. Install horizontal reinforcing steel, anchors and embedded items as masonry work progresses. Vertical reinforcing steel may be placed after the wall is constructed provided it is supported every 200 bar diameters with wire positioners or other devices to hold it in place. All reinforcement must be in place prior to grouting.

C. For two wythe walls, bond the wythes with rectangular ties or joint reinforcing so that one cross wire secures approximately two square feet of wall.

D. Install vertical grout dams at a maximum horizontal spacing of 30 feet to control the horizontal flow of grout.

For walls that are to be partially grouted, use expanded metal lath mesh or other material which will not interfere with bond to restrict the grout into only those cells which are to be grouted.

E. After the mortar joints have set, remove protruding mortar fins which excessively constrict the grout space. Remove all such droppings and debris through the cleanouts at the base of the wall.

F. After the cleanouts have been inspected, seal and brace the cleanouts.

G. Grout the walls in six foot lifts. All cells and spaces containing reinforcing steel shall be solidly grouted for partially grouted walls. For solid grouted walls, all cells shall be grouted.

3.02.B.5 It is important to consolidate the grout to remove any cracks or voids which form within the grout due to the loss of volume by the water being absorbed into the masonry. Reconsolidation of the grout further reduces the possibility of internal voids and cracks in the grout.

3.03.D Grout barriers and dams are required to control the flow of grout and to prevent any possible aggregate segregation.

3.03.G See Notes to Specifier for Sec. 3.02.B.5.

H. Consolidate the grout using a mechanical vibrator and reconsolidate after the excess water is absorbed into the masonry units.

I. Stop the grout 1½ inches below the top of the uppermost grouted unit if the grouting is to be stopped for more than one hour.

J. Continue to grout the wall in six foot lifts, consolidating and reconsolidating each lift.

K. Where additional masonry is to be laid above this point, stop the grout 1½ inches below the top of the masonry units. Otherwise, fill the grout space flush with the top of the units at the top of the wall and consolidate.

END OF SECTION

SECTION 04210
GUIDE SPECIFICATION
REINFORCED GROUTED BRICK MASONRY

PART 1–GENERAL

1.01   GENERAL REQUIREMENTS

Division 1 requirements, Drawing, General Conditions, Supplementary General Conditions, and Special Conditions apply to this section.

1.02   WORK INCLUDED:

A. Furnish and install clay brick units, mortar, grout and reinforcing steel in the masonry. Provide equipment necessary for their installation.

B. Install items furnished by others including:

    1.   Bolts, anchors and shelf angles: Section 05___.

    2.   Built-in nailing blocks: Section 06___.

1.03   RELATED WORK SPECIFIED ELSEWHERE:

A. Site Utilities: Section 02___.

B. Concrete: Section 03___.

C. Masonry Accessories: Section 04___.

D. Structural Steel: Section 05___.

E. Rough Carpentry: Section 06___.

F. Caulking, Sealants and Waterproofing: Section 07___.

G. Flashing and Sheet Metal Work: Section 07___.

H. Insulation: Section 07___.

1.04   QUALITY ASSURANCE:

A. Submittals:

    1.   Samples: Submit___ full size clay brick units of each type, including special shapes required to show range of colors, textures, finishes and dimensions.

    2.   Test Reports:

1.01 This includes all the work the masonry contractor is to accomplish.

1.02 and **1.03** "WORK INCLUDED" and "RELATED WORK" are listed here to define these facets, though they are sometimes not made part of the Division Specifications.

If these paragraphs are included, the Specifier must check that there are no portions overlapping or not covered elsewhere.

Many of these items are not part of the masonry section but may be required for the completed project such as bolts, dowels, and shoring and must be specified in the appropriate sections of the specification.

1.04.A Specify which units and quantity required for approval.

1.04.A.1 Insert number if brick required for approval and approved set of brick should be returned to the contractor

104.A.2 In lieu of test reports manufacturers certification may be accepted.

   a.   Provide Architect/Engineer with test reports for each type of brick.
   b.   Testing and reports are to be completed by an independent laboratory.
   c.   Test reports shall show:

      (1) Compressive strength.
      (2) 24-hr cold water absorption.
      (3) 5-hr boil absorption.
      (4) Saturation coefficient.
      (5) Initial rate of absorption (suction).

3.   Certification: Furnish manufacturer's certification that clay brick units provided meet or exceed the requirements of this specification.

B. Sample Panels:

1.   Construct ___ft x ___ft sample panel with at least one a 90 degree corner. This panel may be part of the actual masonry system.

2.   Show color range, texture range, bond, mortar color, joint tooling, critical design details and quality of workmanship.

3.   Masonry construction may not proceed until the Architect/Engineer approves sample panel mock-up.

4.   Sample panel shall remain on the project site for comparison to the actual masonry work. If the sample panel is not included in the actual masonry work, it shall be demolished and removed from the job site after the masonry work for the project has been accepted.

C. Testing:

1.   Cost for testing of units after delivery shall be borne by the purchaser, unless tests indicate that the units do not conform to the requirements of the specifications, in which case the cost shall be borne by the seller.

2.   Test brick in accordance with ASTM C 67 - ___, *Test Methods of Sampling and Testing Brick and Structural Clay Tile.*

3.   Grout shall be tested in accordance with ASTM C 1019-___, *Standard Method of Sampling and Testing Grout* (UBC Standard No. 21-18).

4.   Masonry prisms shall be tested in accordance with ASTM C 1314-95, *Standard Test Method for Constructing and Testing Masonry Prisms Used to Determine Compliance with Specified Compressive Strength of Masonry* (UBC Standard No. 21-17).

D. Inspection: Continuous inspection of the masonry construction shall be provided.

1.04.B   For normal brick masonry construction sample panels are not usually necessary. If a sample is desired, specify the desired size (often 4 feet by 4 feet) and any other special requirements.

The panel may be included as part of the project wall if approved by the architect.

1.04.C   Testing is often not performed unless required for certain code jurisdictions and/or for masonry utilizing higher than code allowed stresses.

When testing is necessary specify all tests that are to be conducted along with who shall pay for such tests.

1.04.C.4 Half design stresses are limited for f'_m = 2600 psi for clay masonry in seismic zones Nos 3 and 4 unless the f'_m is verified in accordance with UBC 2105.3.2, Item 2 (Masonry Prisms).

1.04.D   The requirement for continuous inspection may be deleted when the design stresses are within the limits set by the code for construction without continuous inspection. Periodic inspection requirements may also be inserted here if they are applicable.

1.05 PRODUCT DELIVERY, STORAGE, AND HANDLING

A. Store masonry units above ground to prevent contamination by mud, dust or other materials likely to cause staining or other defects.

B. Cover and protect masonry units from inclement weather to maintain quality control and physical requirements.

1.06 EXTREME WEATHER CONDITIONS

A. Cold Weather Conditions (40° F or below):

1. Wet or frozen masonry units shall not be used. Temperature of units when laid may not be less than 20° F.

2. Aggregates and mixing water shall be heated to produce mortar and grout temperatures between 40° F and 120° F.

3. Maintain mortar temperature on mortar boards above 32° F.

4. Remove any ice or snow which has inadvertently formed on the masonry bed by carefully applying heat to the surface until it is dry to the touch. Remove any frozen or damaged masonry.

5. When the air temperature is expected to fall below 25° F, provide salamanders or other heat sources on both sides of walls under construction. Additionally, employ wind breaks when the wind speed exceeds 15 miles per hour.

6. When the air temperature is expected to fall below 20° F, provide enclosures and heat sources to maintain the air temperature above 32° F.

7. Protect completed masonry and masonry not being worked on:

a. When the mean daily air temperature is between 40° F and 32° F, cover masonry with weather-resistive membrane for at least 24 hours.

b. When the mean daily air temperature is between 32° F and 25° F, cover masonry completely with weather resistive membrane for at least 24 hours.

c. When the mean daily air temperature is between 25° F and 20° F, cover masonry completely with insulating blankets or other equal protection for at least 24 hours.

d. When the mean daily air temperature is below 20° F, the masonry temperature shall be maintained above 32° F for at least 24 hours by enclosure and supplementary heat, electric heating blankets, infrared heat lamps or other acceptable methods.

1.06   These conditions can be deleted or selectively referred to for the applicable conditions.

B. Hot Weather Conditions:

1. Protect masonry construction from direct exposure to wind and sun when erected in an ambient air temperature of 99° F in the shade with a relative humidity less than 50 percent.

2. A very light fog spray may be applied to masonry surfaces for three days after construction in desert areas where the atmosphere is dry.

PART 2–PRODUCTS

2.01 CLAY BRICK UNITS:

A. Provide brick similar in texture and physical properties to those available for inspection at the Architect/Engineer's office.

B. Do not exceed variations in color and texture of samples accepted by the Architect/Engineer.

C. Facing Brick:

1. ASTM C 216-___, *Specifications for Facing Brick (Solid Units made from Clay or Shale)* (UBC Standard No. 21-1), Grade ___, Type ___.

2. Dimensions: ___ in. x ___ in. x ___ in.

3. Minimum compressive strength of units: ___ psi.

D. Building Brick:

1. ASTM C 62-___, *Specification for Facing Brick (Solid Units made from Clay or Shale)* (UBC Standard No. 21-1), Grade ___, Type ___.

2. Dimensions: ___ in. x ___ in. ___ x ___ in.

3. Minimum compressive strength of units: ___ psi.

2.01   Specify the grade, type and size of units along with any special requirements regarding the configuration, texture, color and strength of the units.

Where special units are required, they should be specified.

2.01.C.1 Specify Grade (MW or SW) and Type (FBA, FBS, or FBX). Grade SW facing brick are used where the masonry will be in contact with earth or where the weathering index is greater than 50. Use Grade MW elsewhere.

Bricks which are classified as Type FBS, or Face Brick Standard, have a wider color range and a greater variation in size and are therefore usually more economical. FBX brick (Face Brick Excellent) have a narrow color, range, minimum variation in size and a high degree of mechanical perfection for high quality face brick work. Brick classified as FBA (Face Brick Assorted) are non-uniform in size, color and texture and are used for special architectural blends.

2.01.C.2 Specify the sizes of the brick units based on local availability.

2.01.C.3 Where high strength masonry is required, specify the strength here (2000 psi to 14,000 psi).

2.01.D.1 Specify Grade (NW, MW, or SW). Grade SW brick are used where the masonry will be in contact with earth or where the weathering index is greater than 50. Grade NW brick are used for interior work and in back-up areas. Use Grade MW elsewhere.

2.01.D.2 Specify the sizes of the brick units based on local availability.

2.01.D.3 For special construction, note minimum compressive strength of units here.

E. Hollow Brick:

    1. ASTM C 652-___, *Specification for Hollow Brick (Hollow Units made from Clay or Shale)* (UBC Standard No. 21-1), Grade___, Type ___.

    2. Dimensions: ___ in. x ___ in. ___ x ___ in.

    3. Minimum compressive strength of units:___psi.

2.02 MORTAR

Mortar shall conform to Section 04101.

2.03 GROUT:

Grout shall conform to Section 04102.

2.04 REINFORCING STEEL:

A. Steel reinforcing shall conform to ASTM ____-___, Grade ___.

B. All horizontal steel, #5 or smaller, shall be furnished in twenty foot lengths plus a 15 percent allowance for lap and scrap loss. All horizontal steel, #6 or larger, shall be detailed, fabricated and supplied, as shown on the drawings. If the horizontal bars, #5 and smaller size, are to be detailed, fabricated and supplied in accordance with local practice, the supplier shall so specify in his bid.

C. All vertical steel in closed end concrete block construction shall be furnished in specified lengths plus laps.

D. Reinforcing bar hooks shall be fabricated in accordance with UBC Sec. 2107.2.2.5

E. Wire joint reinforcement shall conform to UBC Standard No. 21-10, *Joint Reinforcement for Masonry.*

F. Joint Reinforcing wire shall be fabricated as shown on the drawings. Fabrication shall be by electric welding.

G. Reinforcement shall be clean and free from loose rust, scale, dirt and any other coatings that may reduce bond.

H. Metal Ties and Anchors:

Metal ties and anchors shall meet the requirements of UBC Section 2102.2.7.

2.01.E.1 Specify Grade (MW, or SW) and Type (HBS, HBX, HBA or HBB). Grade SW facing brick are used where the masonry will be in contact with earth or where the weathering index is greater than 50. Use Grade MW elsewhere.

2.01.E.2 Specify the sizes of hollow brick units based on local availability.

2.01.E.3 Note minimum net compressive strength of units (2000 psi to 14,000 psi).

2.02 See Guide Specification for Mortar.

2.03 See Guide Specification for Grout.

2.04 Specify the ASTM specification that the reinforcing steel must meet (ASTM A 615, A 616, A 617, A 706, A 767 or A 775) along with the reinforcing grade. Typically ASTM A 615, Grade 60 reinforcing steel is used.

204.E One must be careful that this type wire is used instead of a soft annealed wire. ASTM A82 wire develops high strength and has dependable structural qualities which are required in structural work. The stiffness of this wire is a factor to aid in keeping it straight for proper placement in joints.

Horizontal steel in reinforced grouted masonry may be placed in the grout as the work progresses, and this is considered as adequate positioning for this steel, unless the designer has specified and considered in his design that the horizontal steel is the principal steel and hence must be held accurately in position.

Joint reinforcing wire shall be straight and true when placed in the joints.

I. Coatings. Reinforcement shall be clean at the time of installation, as hereinafter specified, and shall be uncoated. Galvanizing shall be in accordance with ASTM Standard A 153 or Federal Specification QQ-W-4611f, Finish 5.

PART 3–EXECUTION

3.01   PREPARATION

A. Materials:

1.   Masonry materials at the job site shall be stored off the ground to insure they are kept clean and protected from the elements.

2.   All masonry units shall be sound, free of cracks or other defects that would interfere with the proper placing of the unit or impair the strength of construction. Minor cracks incidental to the usual method of manufacture or minor chipping resulting from customary method of handling and shipping and delivery shall not be deemed grounds for rejection.
3.   Reinforcing bars shall be free of kinks or bends except for bends detailed on the drawings. Remove any loose rust, ice, oils and other deleterious coatings from the reinforcing steel.

B. Layout and Foundation:

1.   If site conditions or layout is in anyway improper, masonry work should not begin until cleared by governing authority.

2.   Foundation shall be level and at correct grade so that the initial bed joint shall not be less than ¼ inch nor more than 1 inch.

3   Surface of foundation shall be clean and free of laitance and other deleterious materials. Foundation surface shall be roughened to a full amplitude of $^1/_{16}$ inch.

4.   When a foundation dowel does not align with a vertical cell, it may be bent to a slope of not more than 1 inch horizontally to 6 inches vertically. Remove dowels which do not align properly and replace with new bars of equal capacity.

C. Wetting of Brick:

1.   When being laid, the brick shall have suction sufficient to hold the mortar and to absorb water from the mortar and grout. The brick shall be sufficiently damp so that the mortar will remain plastic enough to permit the brick to be leveled and plumbed immediately after being laid without destroying the bond.

2.   Wet brick with absorption rates in excess of 30 g/30 sq. in./min. as determined by ASTM C 67-___, so that the rate of absorption when laid does not exceed this amount.

2.04.I   When galvanizing is required, delete "and shall be uncoated" from the first sentence.   When there is no requirement for galvanizing, delete the entire second sentence.   In the event galvanizing is required for some reinforcement, it may be galvanized before or after fabrication.

The ASTM is for galvanizing after fabrication, and the Federal Specification is for galvanizing of wire units before fabrication.

3.01.A   Masonry materials must be stored properly to ensure they will remain clean and structurally sound.

3.01.B.2 The foundation must be brought into proper alignment either by chipping or by placing additional concrete or possibly by cutting the masonry units to fit.

3.01.B.4 Dowels may be replaced with approved inserts an properly installed or with new dowels dry packed or epoxied into place.

3. When wetting is necessary to obtain brick which are nearly saturated, surface dry at the time of laying, place a hose on the brick pile until water runs from the pile. Normally this should be done one day prior to the use of the brick but in warm, dry weather, the brick may need to be wetted only a few hours prior to laying.

3.02 INSTALLATION OF MASONRY UNITS:

A. General

1. All masonry units shall be laid true, level, plumb and in uniform coursing in accordance with the drawings. All corners and angles shall be square unless otherwise indicated on the drawings.

2. Unless otherwise specified or shown on the drawings, units shall be laid in running bond.

3. Use masonry saws to cut and fit masonry units.

4. Adjust masonry units into final position while mortar is soft and plastic. If units are displaced after mortar has stiffened, remove the mortar, clean the joints and units and relay the units with fresh mortar.

B. Protection of Work:

1. Protect masonry surfaces from mortar and grout droppings. Any mortar or grout which comes in contact with exposed masonry shall be cleaned immediately to prevent staining.

2. Cover the top of partially completed walls at all times when work is not in progress. Extend covers minimum of 2 feet down both sides of the wall and securely hold the cover in place. Provide additional protection of the masonry work as required in this specification for extreme weather conditions.

C. Mortar Joints:

1. The initial mortar joint at the foundation shall not be less than ¼ inch nor more than 1 inch in thickness. Provide full mortar coverage on the bed joint at the foundation except that mortar shall not project into cells to be grouted.

2. Nominal mortar joint thickness shall be ___ ± $\frac{1}{8}$ inch.

3. Mortar joints shall be straight, clean and uniform in thickness.

4. Head and bed joints of hollow units shall be filled with mortar for the thickness of the face shell. Solid units shall have full head and bed joints. Mortar for bed joints for solid units should be beveled and sloped towards the center of the wall in such a manner that the bed joints will be filled when the brick is finally brought into line. Furrowing of bed joints will not be permitted.

3.02.A.1 Specifier may include tolerance requirements for masonry work and positioning of reinforcing steel.

3.02.A.2 Masonry is typically laid in running bond. If a special pattern is required, specify it here.

3.02.B.2 Protection of masonry work is particularly important if it is subjected to inclement weather, either rain or snow.

3.02.C.2 Nominal mortar joints may be from $\frac{1}{4}$" to $\frac{5}{8}$" depending on the actual size of the unit and the layout module.

3.02.C.4 The bevelled bed may leave some space at the back unfilled with mortar, but that will be filled by the fluid grout.

5.   Avoid mortar fins which project into cells or spaces to be grouted. Fins which occur should be removed if they exceed ½" or the bed joint thickness in such a manner so as not to drop on any fresh grout below.

6.   For all exposed masonry below grade, provide tooled concave mortar joints.

7.   Unless otherwise specified or noted on the drawings, all mortar joints shall be tooled with a concave surface. Tooling shall be done when the mortar is partially set but still sufficiently plastic to bond. All tooling shall be done with a tool that compacts the mortar.

8.   If it is necessary to remove a unit after it has been set in place, the unit shall be removed from the wall, cleaned and set in fresh mortar.

9.   Expansion and control joints shall be placed and constructed as shown on the drawings. Keep these joints clean from mortar droppings and other debris.

3.03   BUILT-IN WORK:

A. Install bolts, anchors, nailing blocks, inserts, frames, vents, flashings, conduits and other built-in items as masonry work progresses.

B. Solid grout all spaces around built-in items unless otherwise noted on the drawings.

3.04   INSTALLATION OF REINFORCING STEEL:

A. Reinforcing shall be placed as detailed on the drawings and shall be secured against displacement prior to grouting at intervals not greater than 200 bar diameters. Horizontal bars may rest on the cross web of hollow units.

B. Tolerances for the placement of reinforcing steel in walls and flexural members shall be as follows:

1. $\pm$ ½ inch for d equal to 8 inches or less.

2. $\pm$ 1 inch for d equal to 24 inches or less but greater than 8 inches.

3. $\pm$ 1¼ inches for d equal to 24 inches or less.

4. $\pm$ 2 inches for longitudinal location of reinforcement.

3.02.C.5 It is better to leave small projecting fins in place rather than having them drop into the grout space.

3.02.C.6 Tooled concave joints provide the best type of joint to prevent water penetration. Specify if another joint type is desired.

3.02.C.9 Proper location and details for joints should be given on the drawings to insure allowance for movement due to moisture and temperature changes.

3.04.A Bracing of reinforcing steel is to prevent the bars from moving out of position.

The designer should be aware that the splices in the vertical reinforcing permitted by the guide specification are commonly used in the masonry industry for low lift grouted construction. If the designer does not want the bars spliced, he should specifically state this requirement and specify the use of open end units.

If high lift grouting construction is indicated on the drawings or in the specifications or if open end units are used, all vertical steel will be furnished full height or in accordance with the drawings. Vertical steel in multi-story construction should provide for a lap at each floor, or as shown on the drawings.

3.04.B These tolerances are in accordance with the requirements of the UBC Sec. 2104.5.

C. Clearance between reinforcing steel and the surface of the masonry shall not be less than ¼ inch for fine grout and ½ inch for coarse grout.

D. Horizontal joint reinforcement shall be fully embedded in mortar with a minimum $\frac{5}{8}$ inch between the joint reinforcement and the exposed surface. Mortar joints with wire reinforcement shall be at least as thick as twice the diameter of the wire.

E. Reinforcing bars and wire joint reinforcing shall be lapped as shown on the drawings.

F. Positioners. Reinforcement shall be held securely and properly in position. Steel in the grout spaces must be held in position by wiring or positioners spaced at not more than 200 times the bar diameter. Where these positioners are within ½ inches of the surface of the masonry they shall be galvanized according the ASTM Standard A 153.

3.05 GROUTING:

Masonry work shall be grouted as noted on the drawings and Section 04101.

3.06 POINTING AND CLEANING:

A. Point and tool any holes in mortar joints to produce a uniform, tight joint.

B. Cement, mortar and grout stains shall be removed immediately from all surfaces.

C. At the conclusion of masonry work, the mason contractor shall remove scaffolding and equipment used in the work along with all debris, refuse and surplus masonry materials from the premises.

3.07 SEALING:

Upon completion of the cleaning operations, and their approval by the architect, apply _____ water repellent coating manufactured by _____ in strict accordance with Section 07___ and the Manufacturer's installation instructions.

END OF SECTION

3.04.C   These clearances are in accordance with UBC Sec. 2106.3.3.

3.04.F   It is not necessary that the steel be held accurately in some position, e.g., temperature steel. In general, it should be near the center line of the wall.

Steel designed to resist load should be accurately positioned, either at the center of toward one face, depending on the design and as shown on the drawings. Steel with bends not shown on the drawing shall not be placed.

3.06.A   Specify the type of repointing mortar. Typically repointing mortar should be made to match the existing mortar.

SECTION 04220
GUIDE SPECIFICATION
REINFORCED CONCRETE UNIT MASONRY

PART 1–GENERAL

1.01   GENERAL REQUIREMENTS:

Division 1 requirements, Drawing, General Conditions, Supplementary General Conditions, and Special Conditions apply to this section.

1.02   WORK INCLUDED:

A. Furnish and install concrete masonry units, mortar, grout and reinforcing steel in the masonry. Provide equipment necessary for their installation.

B.   Install items furnished by others including:

    1. Bolts, anchors and shelf angles: Section 05___.

    2. Built-in nailing blocks: Section 06___.

1.03   RELATED WORK SPECIFIED ELSEWHERE:

A. Site Utilities: Section 02___.

B. Concrete: Section 03___.

C. Masonry Accessories: Section 04___.

D. Structural Steel: Section 05___.

E. Rough Carpentry: Section 06___.

F. Caulking, Sealants and Waterproofing: Section 07___.

G. Flashing and Sheet Metal Work: Section 07___.

H. Insulation: Section 07___.

1.04   QUALITY ASSURANCE:

A. Submittals:

1. Samples: Submit___ full size concrete masonry units of each type, including special shapes required to show range of colors, textures, finishes and dimensions.

1.01   This includes all the work the masonry contractor is to accomplish.

1.02 and **1.03**   "WORK INCLUDED" and "RELATED WORK" are listed here to define these facets, though they are sometimes not made part of the Division Specifications.

If these paragraphs are included, the Specifier must check that there are no portions overlapping or not covered elsewhere.

Many of these items are not part of the masonry section but may be required for the completed project such as bolts, dowels, and shoring and must be specified in the appropriate sections of the specification.

1.04.A   Specify which units and quantity required for approval.

2. Certification: Furnish manufacturer's certification that masonry units provided meet or exceed the requirements of this specification.

B. Sample Panels:

1. Construct ___ft x ___ft sample panel with at least one a 90 degree corner. This panel may be part of the actual masonry system.

2. Show color range, texture range, bond, mortar color, joint tooling, critical design details and quality of workmanship.

3. Masonry construction may not proceed until the Architect/Engineer approves sample panel mock-up.

4. Sample panel shall remain on the project site for comparison to the actual masonry work. If the sample panel is not included in the actual masonry work, it shall be demolished and removed from the job site after the masonry work for the project has been accepted.

C. Testing:

1. Cost for testing of units after delivery shall be borne by the purchaser, unless tests indicate that the units do not conform to the requirements of the specifications, in which case the cost shall be borne by the seller.

2. Concrete Masonry Units shall be tested in accordance with ASTM C 140 - ___, *Sampling and Testing Concrete Masonry Units* and ASTM C 426-___, *Standard Test Method for Drying Shrinkage of Concrete Block*

3. Grout shall be tested in accordance with ASTM C 1019-___, *Standard Method of Sampling and Testing Grout* (UBC Standard No. 21-18).

4. Masonry prisms shall be tested in accordance with ASTM C 1314-95, *Standard Test Method for Constructing and Testing Masonry Prisms Used to Determine Compliance with Specified Compressive Strength of Masonry* (UBC Standard No. 21-17).

D. Inspection: Continuous inspection of the masonry construction shall be provided.

1.05 PRODUCT DELIVERY, STORAGE, AND HANDLING:

A. Store masonry units above ground to prevent contamination by mud, dust or other materials likely to cause staining or other defects.

B. Cover and protect masonry units from inclement weather to maintain quality control and physical requirements. Prevent wetting of units prior to use.

1.04.B For normal block masonry construction sample panels are not usually necessary. If a sample is desired, specify the desired size (often 4 feet by 6 feet) and any other special requirements.

The panel may be included as part of the project wall if approved by the architect.

1.04.C Testing is often not performed unless required for certain code jurisdictions and/or for masonry utilizing higher than code allowed stresses.

When testing is necessary specify all tests that are to be conducted along with who shall pay for such tests.

1.04.C.4 Half stresses are limited to $f'_m = 1500$ for reinforced concrete unit masonry unless the f'_m is verified in accordance with UBC 2105.3.2, Item 2 (Masonry Prisms).

1.04.D The requirement for continuous inspection may be deleted when the design stresses are within the limits set by the code for construction without continuous inspection. Periodic inspection requirements may also be inserted here if they are applicable.

1.05.B It is important to cover the concrete masonry units if there is inclement weather, either rain or snow, to prevent them from becoming excessively wet.

1.06   EXTREME WEATHER CONDITIONS

A. Cold Weather Conditions (40° F or below):

1. Wet or frozen masonry units shall not be used. Temperature of units when laid may not be less than 20° F.

2. Aggregates and mixing water shall be heated to produce mortar and grout temperatures between 40° F and 120° F.

3. Maintain mortar temperature on mortar boards above 32° F.

4. Remove any ice or snow which has inadvertently formed on the masonry bed by carefully applying heat to the surface until it is dry to the touch. Remove any frozen or damaged masonry.

5. When the air temperature is expected to fall below 25° F, provide salamanders or other heat sources on both sides of walls under construction. Additionally, employ wind breaks when the wind speed exceeds 15 miles per hour.

6. When the air temperature is expected to fall below 20° F, provide enclosures and heat sources to maintain the air temperature above 32° F.

7. Protect completed masonry and masonry not being worked on:

a. When the mean daily air temperature is between 40° F and 32° F, cover masonry with weather-resistive membrane for at least 24 hours.

b. When the mean daily air temperature is between 32° F and 25° F, cover masonry completely with weather resistive membrane for at least 24 hours.

c. When the mean daily air temperature is between 25° F and 20° F, cover masonry completely with insulating blankets or other equal protection for at least 24 hours.

d. When the mean daily air temperature is below 20° F, the masonry temperature shall be maintained above 32° F for at least 24 hours by enclosure and supplementary heat, electric heating blankets, infrared heat lamps or other acceptable methods.

B. Hot Weather Conditions:

1. Protect masonry construction from direct exposure to wind and sun when erected in an ambient air temperature of 99° F in the shade with a relative humidity less than 50 percent.

2. A very light fog spray may be applied to masonry surfaces for three days after construction in desert areas where the atmosphere is dry.

1.06   These conditions can be deleted or selectively referred to for the applicable conditions.

PART 2–PRODUCTS

2.01   CONCRETE MASONRY UNITS:

A. Hollow and solid load-bearing concrete masonry units.

   1. ASTM C 90-___, *Specifications for Load-Bearing Concrete Masonry Units.* (UBC Standard No. 21-4), Grade ___, Type ___.

   2. Nominal face dimensions: ___ in. x ___ in.

   3. Net area compressive strength of the units: ___psi.

B. Building Brick:

   1. ASTM C 55-___, *Specification for Concrete Building Brick,* (UBC Standard No. 21-3), Type ___, Grade ___.

   2. Nominal face dimensions: ___ in. x ___ in.

   3. Net area compressive strength of units:___psi.

C. Hollow non-Load Bearing Masonry Units:

   1. ASTM C 129-___, *Specification for Non-Load-Bearing Concrete Masonry Units,* (UBC Standard No. 21-5), Type ___.

   2. Nominal face dimensions: ___ in. x ___ in.

D. Provide _____ weight units.

2.02   MORTAR:

2.01   Specify the grade, type and size of units along with any special requirements regarding the configuration, texture, color and strength of the units.

Note that split face block, slumped block and special architectural units comply with the strength requirements of UBC Standard No. 21-4 but many times do not comply with the dimensional requirements. Open end units do not comply with the UBC Standard on equivalent web thickness but are satisfactory since the open cells are grouted solid. Were special units are required, they should be specified.

2.01.A.1 Specify Type (I or II) and Grade (N or S). Note ASTM C 90-93 has deleted the Grade designations. In 1992 ASTM C 145 (solid load bearing concrete masonry units) was incorporated into ASTM C 90.

2.01.A.2 Specify the nominal sizes of the units.

2.01.A.3 Unless otherwise specified, concrete masonry units conforming to ASTM C 90 will have a minimum net compressive strength of 1900 psi. If stronger units are required, specify strength here.

2.01.B.1 Specify Type (I or II) and Grade (N or S).

2.01.B.2 Specify the nominal sizes of the units.

2.01.B.3 For special construction, note minimum net compressive strength of units.

2.01.C.1 Specify Type (I or II).

2.01.C.2 Specify the nominal sizes of the units.

2.01.D   Specify weight of units:

Lightweight units - less than 105 pounds per cubic foot
Medium weight units - 105 to less than 125 pounds per cubic foot.
Normal weight units - 125 pounds per cubic foot and more.

2.02   See Guide Specification for Mortar.

Mortar shall conform to Section 04101.

2.03 GROUT:

Grout shall conform to Section 04102.

2.04 REINFORCING STEEL:

A. Steel reinforcing shall conform to ASTM ____-___, Grade ___.

B. All horizontal steel, #5 or smaller, shall be furnished in twenty foot lengths plus a 15 percent allowance for lap and scrap loss. All horizontal steel, #6 or larger, shall be detailed, fabricated and supplied, as shown on the drawings. If the horizontal bars, #5 and smaller size, are to be detailed, fabricated and supplied in accordance with local practice, the supplier shall so specify in his bid.

C. All vertical steel in closed end concrete block construction shall be furnished in specified lengths plus laps.

D. Reinforcing bar hooks shall be fabricated in accordance with UBC Sec. 2107.2.2.5

E. Wire joint reinforcement shall conform to UBC Standard No. 21-10, *Joint Reinforcement for Masonry.*

F. Joint Reinforcing wire shall be fabricated as shown on the drawings. Fabrication shall be by electric welding.

G. Reinforcement shall be clean and free from loose rust, scale, dirt and any other coatings that may reduce bond.

H. Metal Ties and Anchors:

Metal ties and anchors shall meet the requirements of UBC Section 2102.2.7.

I. Coatings. Reinforcement shall be clean at the time of installation, as hereinafter specified, and shall be uncoated. Galvanizing shall be in accordance with ASTM Standard A 153 or Federal Specification QQ-W-4611f, Finish 5.

PART 3–EXECUTION

3.01 PREPARATION

A. Materials:

1. Masonry materials at the job site shall be stored off the ground to insure they are kept clean and protected from the elements.

2.03   See Guide Specification for Grout.

2.04   Specify the ASTM specification that the reinforcing steel must meet (ASTM A 615, A 616, A 617, A 706, A 767 or A 775) along with the reinforcing grade.

Typically ASTM A 615, Grade 60 reinforcing steel is used.

2.04.E   One must be careful that this type wire is used instead of a soft annealed wire. ASTM A 82 wire develops high strength and has dependable structural qualities which are required in structural work. The stiffness of this wire is a factor to aid in keeping it straight for proper placement in joints.

Horizontal steel in reinforced grouted masonry may be placed in the grout as the work progresses, and this is considered as adequate positioning for this steel, unless the designer has specified and considered in his design that the horizontal steel is the principal steel and hence must be held accurately in position.

Joint reinforcing wire shall be straight and true when placed in the joints.

2.04.I   When galvanizing is required, delete "and shall be uncoated" from the first sentence. When there is no requirement for galvanizing, delete the entire second sentence. In the event galvanizing is required for some reinforcement, it may be galvanized before or after fabrication.

The ASTM is for galvanizing after fabrication, and the Federal Specification is for galvanizing of wire units before fabrication.

3.01.A   Masonry materials must be stored properly to ensure they will remain clean and structurally sound.

2.   All masonry units shall be sound, free of cracks or other defects that would interfere with the proper placing of the unit or impair the strength of construction. Minor cracks incidental to the usual method of manufacture or minor chipping resulting from customary method of handling and shipping and delivery shall not be deemed grounds for rejection.

3.   Reinforcing bars shall be free of kinks or bends except for bends detailed on the drawings. Remove any loose rust, ice, oils and other   deleterious coatings from the reinforcing steel.

B.   Layout and Foundation:

1.   If site conditions or layout is in anyway improper, masonry work should not begin until cleared by governing authority.

2.   Foundation shall be level and at correct grade so that the initial bed joint shall not be less than ¼ inch nor more than 1 inch.

3.   Surface of foundation shall be clean and free of laitance and other deleterious materials. Foundation surface shall be roughened to a full amplitude of $\frac{1}{16}$ inch.

4.   When a foundation dowel does not align with a vertical cell, it may be bent to a slope of not more than 1 inch horizontally to 6 inches vertically. Remove dowels which do not align properly and replace with new bars of equal capacity.

3.02   INSTALLATION OF MASONRY UNITS:

A.   General

1.   All masonry units shall be laid true, level, plumb and in uniform coursing in accordance with the drawings. All corners and angles shall be square unless otherwise indicated on the drawings.

2.   Unless otherwise specified or shown on the drawings, units shall be laid in running bond.

3.   Use masonry saws to cut and fit masonry units.

4.   Lay only dry concrete masonry units. Concrete masonry units shall not be wetted unless otherwise approved.

5.   Adjust masonry units into final position while mortar is soft and plastic. If units are displaced after mortar has stiffened, remove the mortar, clean the joints and units and relay the units with fresh mortar.

B.   Protection of Work:

1.   Protect masonry surfaces from mortar and grout droppings. Any mortar or grout which comes in contact with exposed masonry shall be cleaned immediately to prevent staining.

3.01.B.2 The foundation must be brought into proper alignment either by chipping or by placing additional concrete or possibly by cutting the masonry units to fit.

3.01.B.4 Dowels may be replaced with approved inserts and properly installed or with new dowels dry packed or epoxied into place.

3.02.A.1 Specifier may include tolerance requirements for masonry work and positioning of reinforcing steel.

3.02.A.2 Masonry is typically laid in running bond. If a special pattern is required, specify it here.

2. Cover the top of partially completed walls at all times when work is not in progress. Extend covers minimum of 2 feet down both sides of the wall and securely hold the cover in place. Provide additional protection of the masonry work as required in this specification for extreme weather conditions.

C. Mortar Joints:

1. The initial mortar joint at the foundation shall not be less than ¼ inch nor more than 1 inch in thickness. Provide full mortar coverage on the bed joint at the foundation except that mortar shall not project into cells to be grouted.

2. Nominal mortar joint thickness shall be $^3/_8 \pm {}^1/_8$ inch for precision units and $^1/_2 \pm {}^1/_8$ inch for slumped units.

3. Mortar joints shall be straight, clean and uniform in thickness.

4. Head and bed joints of hollow units shall be filled with mortar for the thickness of the face shell. Solid units shall have full head and bed joints.

5. Provide tooled concave mortar joints for all exposed below grade masonry. Walls which are to be plastered shall have flush cut or sacked mortar joints.

6. Unless otherwise specified or noted on the drawings, all mortar joints shall be tooled with a concave surface. Tooling shall be done when the mortar is partially set but still sufficiently plastic to bond. All tooling shall be done with a tool that compacts the mortar.

7. If it is necessary to remove a unit after it has been set in place, the unit shall be removed from the wall, cleaned and set in fresh mortar.

8. Control joints shall be placed and constructed as shown on the drawing. Keep these joints clean from mortar drippings and other debris.

3.03 BUILT-IN WORK:

A. Install bolts, anchors, nailing blocks, inserts, frames, vents, flashings, conduits and other built-in items as masonry work progresses.

B. Solid grout all spaces around built-in items unless otherwise noted on the drawings.

3.04 INSTALLATION OF REINFORCING STEEL:

A. Reinforcing shall be placed as detailed on the drawings and shall be secured against displacement prior to grouting at intervals not greater than 200 bar diameters. Horizontal bars may rest on the cross web of hollow units.

3.02.B.2 Protection of masonry work is particularly important if it is subjected to inclement weather, either rain or snow.

3.02.C.5 and **C.6** Concave tooled joints provide the best weather resistance and strength and are therefore used widely. Specify if another joint type is desired.

3.02.C.8 Proper location and detailing of control joints should be given on the drawings to insure allowance for movement or shrinkage due to moisture and temperature changes.

3.04.A Bracing of reinforcing steel is to prevent the bars from moving our of position.

The designer should be aware that the splices in the vertical reinforcing permitted by the guide specification are commonly used in the masonry industry for low lift grouted construction. If the designer

B. Tolerances for the placement of reinforcing steel in walls and flexural members shall be as follows:

1. $\pm$ ½ inch for d equal to 8 inches or less.

2. $\pm$ 1 inch for d equal to 24 inches or less but greater than 8 inches.

3. $\pm$ 1¼ inches for d equal to 24 inches or less.

4. $\pm$ 2 inches for longitudinal location of reinforcement.

C. Clearance between reinforcing steel and the surface of the masonry shall not be less than ¼ inch for fine grout and ½ inch for coarse grout.

D. Horizontal joint reinforcement shall be fully embedded in mortar with a minimum ⅝ inch between the joint reinforcement and the exposed surface. Mortar joints with wire reinforcement shall be at least as thick as twice the diameter of the wire.

E. Reinforcing bars and wire joint reinforcing shall be lapped as shown on the drawings.

F. Positioners. Reinforcement shall be held securely and properly in position. Steel in the grout spaces must be held in position by wiring or positioners spaced at not more than 200 times the bar diameters. Where these positioners are within ½ inches of the surface of the masonry they shall be galvanized according to ASTM Standard A 153.

3.05 GROUTING:

Masonry work shall be grouted as noted on the drawings and Section 04102.

3.06 POINTING AND CLEANING:

A. Point and tool any holes in mortar joints to produce a uniform, tight joint.

B. Cement, mortar and grout stains shall be removed immediately from all surfaces.

does not want the bars spliced, he should specifically state this requirement and specify the use of open end units.

If high lift grouting construction is indicated on the drawings or in the specifications or if open end units are used, all vertical steel will be furnished full height or in accordance with the drawings. Vertical steel in multi-story construction should provide for a lap at each floor, or as shown on the drawings. Steel with bends not shown on the drawings shall not be placed.

3.04.B   These tolerances are in accordance with the requirements of the UBC Sec. 2104.5.

3.04.F   It is not necessary that the steel be held accurately in some position, e.g., temperature steel. In general, it should be near the center line of the wall.

Steel designed to resist load should be accurately positioned, either at the center of toward one face, depending on the design and as shown on the drawings.

3.05   See Guide Specification for Grout

3.06.A   Proper pointing of bee holes and lie pin holes reduces the possibility of water penetration.

3.06.B   Care should be taken to keep the work clean during construction and this will reduce the clean up work at the end of the masonry construction.

C. At the conclusion of masonry work, the mason contractor shall remove scaffolding and equipment used in the work along with all debris, refuse and surplus masonry materials from the premises.

3.07 WATERPROOFING:

Upon completion of the cleaning operations, and their approval by the architect, apply _____ water repellent coating manufactured by _____ in strict accordance with Section 07___ and the Manufacturer's installation instructions.

END OF SECTION

The Specifier should specify the preferred method of cleaning the block surface (such as sandblasting, acid washing, etc.). Note that specific instructions on the application of these methods is required as sandblasting can damage some surfaces.

3.07   Sealing may be specified here or in a separate section or may be deleted if the block does not require sealing.

SECTION 04240
GUIDE SPECIFICATION
FOR ANCHORED VENEER

PART 1 - GENERAL

1.01   GENERAL REQUIREMENTS:

Division 1 requirements, Drawing, General Conditions, Supplementary General Conditions, and Special Conditions apply to this section.

1.02   WORK INCLUDED:

A. This section includes labor, material, equipment, tools and appliances for the complete installation of anchored veneer in the areas shown on the Drawings.

B. Installing, pointing and cleaning of the veneer.

C. Attaching of anchorage to supports provided by others.

D. Placing of anchorages in the veneer as detailed.

E. Setting into the veneer all bolts, anchors, metal attachments and inserts as furnished and located by others.

F. Removal of surplus veneer material and waste after completion of the veneer work.

1.03   RELATED WORK SPECIFIED ELSEWHERE:

A. Reinforcing Steel: Section 04 ___.

B. Metal Accessories: Section 04 ___.

C. Mortar: Section 04___.

D. Rough Carpentry: Section 06 ___.

E. Caulking, Sealants and Waterproofing: Section 07___.

1.04   QUALITY ASSURANCE:

A. Submittals:

    1. Samples: Submit _____veneer units of each type, including special shapes required to show range of colors, textures, finishes and dimensions.

    2. Certifications: Furnish Manufacturer's certification that the veneer units provided meet or exceed the requirements of this Specification.

1.01   Division 1, General Requirements, is that portion of the CSI Format for Building Specifications which contains all provisions pertaining to the job as a whole, including the "Special Conditions."

1.02 and **1.03**   "WORK INCLUDED" and "RELATED WORK" are listed here to define these facets, though they are sometimes not made part of the Division Specifications.

If these paragraphs are included, the Specifier must check that there are no portions overlapping or not covered elsewhere.

1.04   Specify any special submittals or tests required to assure the materials and construction procedure will be of the quality desired.

1.04.A.1 Specify the number of veneer units required for approval.   Coordinate submittal requirements with your standard procedures.

B. Sample Panels:

1. Construct a sample panel of the work of this Section at the location on the side indicated by the Architect.

2. Provide on sample panel for each combination of veneer patterns, bond pattern, and mortar color.

3. Sample panels shall be ___ft. (__m) x___ft. (__ m)

4. Show color range, texture range, mortar color, joint tooling, bond pattern, cleaning, quality of workmanship and other aspects of the work relating to this Section.

5. Make necessary adjustments in the sample panels and secure the Architects approval. Construction may not proceed until the Architect approves sample panel.

6. Sample panels shall remain on the project site to serve as a datum for comparison with the remainder of the work of this Section for the purpose of acceptance or rejection.

7. If the sample panel is not included in the actual masonry work, it shall be demolished after the masonry work for the project has been accepted.

1.05   PRODUCT DELIVERY, STORAGE AND HANDLING:

A. Store veneer units, cement, lime and other materials of this section off ground to prevent contamination with mud, water or other harmful materials.

B. Veneer units shall be handled and stored in such a manner so as to prevent chipping, breaking and cracking.

C. Cover and protect materials from inclement weather to maintain quality control and physical requirements.

PART 2 - PRODUCTS

2.01   VENEER UNITS:

A. The anchored veneer shall be _____as produced by _____.

B. Reference Specification: ASTM C ___-___.

C. Dimensions: ___ in. (__ mm) wide by ___ in. (.. mm) high by ___ in. (__ mm) long.

D. Provide all units with straight cut edges and square corners unless otherwise shown or specified.

1.04.B Sample panels and mock-ups are a good way to assure that workmen know what is expected of them. The sample panels may be part of the finished project.

1.04.B.3 Specify the size of sample panel required along with any special requirements on its construction. A 4' x 4' (1.2 m x 1.2 m) sample panel is common unless special conditions require otherwise.

1.05 The intent of the Section is to keep materials clean, dry and in an acceptable condition during and after job delivery.

2.01.A The Specifier must call out the type of veneer and desired supplier.

This will have been decided by preliminary samples and type of work previously determined by the Designer and the Supplier.

E. Do not permit veneer units to vary more than $^3/_8$" (9.5 mm) in any dimension from those specified.
Provide units similar in texture and color to those approved by the Architect.

2.02   MORTAR

Mortar shall comply with the requirements of Section 04101

2.03   GROUT:

A. Grout, if used between the veneer units and the backing, shall be of pourable consistency.

B. Grout shall be fine grout.

    1.   Fine grout shall be 1 part portland cement and 3 parts sand.

2.04   OTHER MATERIALS:

A. All ties and anchors shall be of non-corrodible material or corrosion protected material in compliance with local Building Code requirements.

B. "Paper-backed mesh" shall consist of waterproof paper or a membrane equivalent to asphalt saturated felt free from holes and breaks and weighing not less than 14 pounds per 100 square feet (.68 kg/m^2). Felt shall be applied weatherboard

fashion, lapped not less than __ inches (__ mm) at horizontal joints and not less than ___ inches (__ mm) at vertical joints. Over this apply wire mesh not less than 16 gauge (1.575 mm) at ___ inches (__ mm) o.c. with provisions for furring the mesh from the membrane, for embedment of wire.

PART 3 - EXECUTION

3.01   WORKMANSHIP:

A. Examine the areas and conditions under which work of this Section will be performed. Correct conditions that are detrimental to timely and proper completion of the work. Do not proceed until unsatisfactory conditions are corrected.

B. Veneer shall be laid in a pattern as shown on Drawing number ___.

C. Veneer shall be clean and free of dust and shall be laid with not less than one inch (25 mm) of air space, mortar or grout between the veneer and the backing wall. The space behind the veneer shall be solidly filled with mortar or grout or left void with spot bedding as required.

D. All joints shall be filled solidly with mortar. Provide tooled concave mortar joints unless otherwise specified or noted on the Drawings. Tooling shall be done when the mortar is partially set but still sufficiently plastic to bond.

2.01.E   ASTM C 126, Table 4 lists tolerances on dimensions.

2.02   Mortar is often specified here. The editors choose to make a separate mortar specification for simplicity.

2.04   Specify any other materials pertinent to the particular job.

2.04.B   Specify type of paper-backed mesh along with lapping requirements and any additional requirements.

3.01.C   The space between the veneer and back up may be solid grouted or left void as a cavity with provisions for weep holes.

3.01.D   Tooled concave joints provide the best types of joint to prevent water penetration. Specify if another joint type is desired.

E. No veneer unit shall exceed ten inches (254 mm) in thickness at any point measured from back to face as laid. When veneer is placed in its bed joint, it shall remain at rest before the remaining mortar is applied around and behind.

There shall be no outside support other than the veneer beneath it and the anchor ties to the backing wall. In no case shall the overall distance from the backing wall to the exposed face exceed 11 inches (279 mm).

F. Veneer ties shall have at least $^5/_8$ inch (16 mm) of mortar coverage from exterior surface to prevent corrosion.

G. Wire mesh shall be attached to studs as shown on Drawings.

H. Anchors shall be installed to provide not less than the support shown on the details and in such as manner as to eliminate looseness or lack of fit of dovetails in slots, improper bends on kinks in wire, etc.

3.02   CLEANING:

A. Mortar stains shall be removed with clear water as work progresses. Upon completion, all exposed surfaces shall be cleaned with a 10% solution of commercial muriatic acid, removing all stains with fiber brushes and then washing with clean water.

B. In the event ordinary cleaning is not adequate, the Contractor shall use special cleaning methods such as sandblasting, acid washing, chipping, etc., as approved by the Architect.

C. At the conclusion of masonry work, the mason contractor shall remove scaffolding and equipment used in the work along with all debris, refuse and surplus masonry materials from the premises.

3.03   SEALING:

Upon completion of the cleaning operation, and their approval by the Architect, apply _____ water repellent coating manufactured by _____-in strict accordance with Section 07 ___ and the Manufacturer's installation instructions.

END OF SECTION

3.01.E Depth of veneer, including mortar and grout, is recommended to be a minimum of three inches (76 mm) to a maximum of approximately 11 inches (279 mm).

In some instances, thicker light weight veneer may be used. The limits so prescribed have been selected merely because of customary use in local codes.

3.01.H Details of veneer anchors and other veneer requirements should be shown on the Drawings.

The designer should detail the type of anchorage of the veneer. These should show clearly the number, type, method of attachment, etc. The basic requirements for ties is that they be permanently adequate, of a material proven to be corrosion resistant.

Ordinary reinforcing bars or ordinary steel wire in joints, when protected by the cement mortar and grout have generally proven to be satisfactory.

Anchors that will be exposed to a corrosive atmosphere must be galvanized, coated or of stainless steel, etc.

3.02.A Specify the desired cleaning method. Note that a careful selection is necessary as acid washing can discolor some brick and stone and sandblasting can harm the surface of many materials.

3.03 Sealing may be specified here or in a separate Section.

SECTION O4255
GUIDE SPECIFICATION FOR
ADHERED THIN CLAY BRICK
OR STONE SLAB MASONRY VENEER

PART 1 - GENERAL

1.01   GENERAL REQUIREMENTS:

Division 1 Requirements, Drawings, General and Supplementary General Conditions, and Special Conditions apply to this section.

1.02   WORK INCLUDED:

Furnish and install brick or stone slab veneer units and mortar. Provide equipment necessary for their installation.

1.03   RELATED WORK SPECIFIED ELSEWHERE:

   A.   Plaster undercoats: Section 09__.

   B.   Rough Carpentry: Section 06__.

   C.   Caulking, Sealants and Waterproofing: Section 07___.

1.04   QUALITY ASSURANCE:

   A.   Submittals:

   1.   Sample: Submit ___ brick or stone slab veneer units of each type, including special shape required to show range of colors, textures, finished and dimensions.

   2.   Certification: Furnish Manufacturer's certification that the brick units provided meet or exceed the requirements of this specification.

   B.   Sample Panels:

   1.   Construct a sample panel of the work of the Section at the location on the site indicated by the Architect.

   2.   Provide one sample panel for each combination of brick veneer patterns, bond pattern, and mortar color.

   3.   Sample panels shall be ____ft (__m) x ____ft (__m).

   4.   Show color range, texture range, mortar color, joint tooling, bond pattern, cleaning, quality of workmanship and other aspects of the work relating to this Section.

1.01   This includes all the work the masonry contractor is to accomplished.

1.02 and **1.03** "WORK INCLUDED" and "RELATED WORK" are listed here to define these facets, though they are sometimes not made part of the Division Specifications.

   If these paragraphs are included, the Specifier must check that there are not portions overlapping or not covered elsewhere.

1.03.A   Plaster undercoats may be part of "lath and plaster" or part of this Section.

1.04   Specify any special submittal or tests required to assure the materials and construction procedure will be of the quality desired.

1.04.A   Specify the number of veneer units required for approval. Coordinate submittal requirements with your standard procedures.

1.04.B   Sample panels and mock-ups are a good way to assure that workmen know what is expected for them to provide. Sample panels can be made as part of the finished project.

   Sample panels may not be necessary for small jobs.

1.04.B.3 Specify the size of sample panel required along with any special requirements on its construction.

5. Make necessary adjustments in the sample panels and secure the Architect's approval. Construction may not proceed until the Architect approves sample panel mock-up.

6. Sample panels shall remain on the project site to serve as a datum for comparison with the remainder of work of this Section for the purpose of acceptance or rejection.

7. If the sample panel is not included in the actual masonry work, it shall be demolished after the masonry work for the project has been accepted.

1.05 PRODUCT DELIVERY, STORAGE AND HANDLING:

A. Store brick or stone slab veneer, cement, lime and other materials of this section off the ground to prevent contamination with mud, water or other harmful materials. Veneer units shall be handled and stored in such a manner so as to prevent chipping, breaking and cracking.

B. Cover and protect materials from inclement weather to maintain quality control and physical requirements.

PART 2 - PRODUCT

2.01 BRICK VENEER UNITS:

A. Units shall have _____face texture and _____color as manufactured by _____.

B. Reference Specification: ASTM C 1088-___, *Standard Specification for Thin Veneer Brick Units Made from Clay or Shale*, Type ____, Grade _____.

C. Dimensions:___in.(___mm) wide by ___in.(___mm) long by ___in. (___mm) thick.

D. Provide all units with straight cut edges and square corners unless noted otherwise.

E. Do not permit brick veneer units to vary more than $^3/_8$" (9.5 mm) in any dimension from those specified. Provide units similar in texture and color to those approved by the Architect.

2.02 STONE SLAB UNITS:

A. Limestone: [Cut Indiana Oolitic limestone.]_____.

B. Marble: ASTM C503, Classification [I__Calcite;] [II__Dolomite;] [III__Serpentine;] [IV__Travertine;] [with [filled] [unfilled] surface.]

C. Granite: ASTM C615, [sawed.] [cut.] [split.]_____.

D. Slate: ASTM C629, Classification [I__Exterior;] [II__Interior;] [sawed.] [cut.] [split.]_____.

1.05 The intent of this section is to keep materials clean, dry and in an acceptable condition during and after job delivery.

2.01 Numerous different types, sizes, styles and colors of veneer units are available. Verify that the color, texture and dimensions you select are available from the Manufacturer named.

For thin brick units, specify the grade (exterior or interior) and type (TBX, TBS or TBA). Type TBS (Standard) thin brick is for general use in exposed exterior and interior masonry walls. For a higher degree of precision and a lower variation in size, specify Type TBX (Select) units. Type TBA (Architectural) brick veneer units have intentional nonuniformities in size, color and texture.

2.01.E ASTM C 1088, Table 3 lists tolerances on dimensions.

2.02 Specify the types, sizes, grades and colors of the desired veneer units. Add any special requirements as necessary.

If a travertine stone is selected, indicate whether it is filled or unfilled.

Surface finishes vary considerably between different stone materials; specify accordingly. When two or more surface finishes are required, consider using a schedule at the end of this Section.

E. Sandstone: ASTM C616 Classification [I___Sandstone.] [II__Quartzitic Sandstone.] [III__Quartzite].

F. Color:_____.

G. Surface Finish:_____.

H. Grade:_____; free of defect.

I. Thickness:_____inch (____mm) nominal.

J. Face Size:___x___inch (__x__mm) nominal.

2.03   MORTAR:

A. Mortar shall comply with the requirements of Section 04101 with the following exceptions:

1. Thin set mortar (if used): Provide a commercial grade mortar specifically formulated for the purpose and composed of latex mixed with Portland cement, or provide such other thin set mortar as is specifically approved in advance by the Architect.

2. Pointing mortar: Provide a mix of one part Portland cement and ___part lime to ___parts sand by volume. Color shall be _____.

PART 3 - EXECUTION

3.01   SURFACE PREPARATION:

A. Examine the areas and conditions under which work of this Section will be performed. Correct conditions detrimental to timely and proper completion of the work. Do not proceed until unsatisfactory conditions are corrected.

B. Thoroughly clean the substrata of debris, dust, and finishes which will not bond with the mortar.

1. Remove sealers, bond breakers, and other applied finishes by use of a light sandblast or other means approved by Architect.

2.03 Mortar materials may be specified in a separate specification or in this specification. The editor's have selected to provide a separate mortar specification for simplicity in this publication.

See Guide Specification for Mortar (Section 04101)

2.03.A.1 Commercially prepared latex mortars are available and have been used with great success. Consult their manufacturers for detailed information and help in selection.

2.03.A.2 Pointing mortar, if not incorporated with the setting process, should be specified and should state color desired. Typically pointing mortar may be type N mortar with 1 part portland cement, ½ to 1 ¼ parts lime and 2 ¼ to 3 parts sand.

3.01.A It is crucial that the substrata be prepared properly so that the veneer construction will perform properly. The supporting surface should be relatively rigid so that it may support the thin veneer adequately. The presence of cracks or yielding of the surface should be checked as they may indicate an unsatisfactory substrata surface.

3.01.B.1 Proper bond depends upon having suitable substrata. The substrata must be compatible with portland cement and must be clean. Oil, dust residues, sealers, bond breakers, form releases and other substances can significantly reduce the bond adhesion between the mortar and the substrata and therefore must be completely removed. Sandblasting is one of the easiest and best methods to remove these substances although chipping, scrapping and wire brushing have also been used successfully.

2. Provide substrata with the ability to receive and bond with the mortar. Dampen the backing surface with clean water by fogging. Surface shall be damp, but not wet or saturated, at the time of laying.

3.02 INSTALLATION:

A. General

1. Position all veneer units level, plumb and true and in uniform coursing in accordance with the Drawings. All corners and angles shall be square unless otherwise indicated on the drawings.

2. Unless otherwise indicated on the Drawings, place units in running bond. Install control and expansion joints as shown on the drawings and at all control joints in the backing surface.

3. Lay out each wall or panel in a manner to minimize cutting of brick veneer units. Use masonry saws to cut and fit masonry units.

B. Spread the approved mortar on the surface of the setting bed and the back of the veneer units using a notched trowel. Provide a uniform setting bed of $^1/_8$" minimum thickness.

C. Slide veneer units into place and press firmly to achieve full contact of the mortar. Veneer units (brick) shall be lifted and removed occasionally to verify that full bedding has been achieved.

D. Provide _____ in.(___ mm) wide mortar joints.

3.03 POINTING:

A. Allow mortar to set up fully prior to starting the filling of the mortar joints. Take special care so as not to move the installed brick units.

B. Point all mortar with the approved mortar.

C. Tool all mortar joints to a slight concave unless otherwise called for on the Drawings.

3.04 CLEANING:

A. Promptly clean the brick veneer units as the work progresses to minimize the need for final cleaning. Remove all cement and mortar stains from the face of the brick veneer units.

3.01.B.2 The area to be faced is often fog sprayed about 2 hours prior to veneer mortar installation. This procedure keeps the water in the mortar from being sucked into the substrata too quickly which can reduce the bond adhesion. The actual time and amount of wetting depends on the absorption of the backing materials, the temperature and the humidity. Note that extremely absorptive veneer units are sometimes wetted prior to laying as well.

3.02.A.2 State here the bond pattern you desire. Place expansion and control joints as required by the materials and as directed by the building code. Joints in the veneer should always be located at joints in the substrata to allow adequate expansion, shrinkage and temperature movements and to avoid cracking of the veneer.

3.02.B In some instances it may be advantageous to permit the mason to only apply the mortar to the back of the veneer surface. By fully covering the back surface of the veneer only, the mason can lay the units much more rapidly.

3.02.D Specify the desired joint width (usually $^3/_8$ in. $\pm$ $^1/_8$ in.(10 mm $\pm$ 3 mm)).

3.03 Mortar joints must be filled fully to prevent moisture penetration. Tooled concave joints provide the best weather resistance and are therefore recommended. Specify if another joint type is desired.

3.03.B Pointing mortar should be Type N (1 part portland cement, 1 part hydrated lime and 6 parts sand) or Type O (1 part portland cement, 2 parts hydrated lime and 9 parts sand).

3.04 Care should be taken to keep the work clean during construction. This will significantly reduce the amount of cleaning required at the end of construction.

The Specifier should specify the preferred method of cleaning the veneer surface (such as sandblasting, acid washing, etc.). Note that specific instructions on the application of these methods is required as sandblasting can damage some surfaces and acid washing can cause green staining to occur on some bricks.

B. At the conclusion of masonry work, the mason contractor shall remove scaffolding and equipment used in the work along with all debris, refuse and surplus masonry materials from the premises.

3.05   SEALING:

Upon completion of the cleaning operations, and their approval by the Architect, apply _____ water repellent coating manufactured by _____ in strict accordance with Section 07___and the Manufacturer's installation instructions.

3.06   SCHEDULE:

Provide a schedule when stone surface finishes, colors or textures vary for different locations.

END OF SECTION

3.05   Sealing may be specified here or in a separate section.

3.06   Schedule may include special instructions; i.e. variable width head and bed mortar joints or other particular features.

SECTION 04270
GUIDE SPECIFICATION FOR
GLASS UNIT MASONRY

PART 1 - GENERAL

1.01   DESCRIPTION:

This section includes materials and installation of :

A.   Glass Unit Masonry.

B.   Mortar for Glass Unit Masonry.

C.   Reinforcement for Glass Unit Masonry.

D.   Expansion Strips.

E.   Asphalt Emulsion.

F.   Packing.

G.   Panel Anchors

H.   General Terms and Conditions apply.

I.   UBC Section 2110, Glass Block Masonry.

1.02   RELATED WORK SPECIFIED ELSEWHERE:

A.   Concrete Reinforcement.

B.   Concrete.

C.   Concrete Unit Masonry.

D.   Structural Steel, Aluminum, and Miscellaneous Metal.

E.   Caulking and Sealants.

F.   Framed openings to receive Glass Masonry

1.03   SUBMITTALS:

A.   Submit two full-size glass masonry units of each type and manufacturer's catalog data.

1.01 and **1.02** "DESCRIPTION" and "RELATED WORK SPECIFIED ELSEWHERE" are list here to define these facets, though they are sometimes not made part of the Division Specifications.

If these paragraphs are includes, the specifier must check that there are not portions overlapping or not covered elsewhere.

Many of these items are not part of the masonry section, but may be required for the completed project such as bolts, parts of anchorage system, and protection and must be specified in the appropriate sections of the specification.

1.03.A Glass block may be solid or hollow units and may contain inserts.

B. Furnish manufacturer's certificate and test results to confirm that glass units comply with the cited ASTM specification.

C. Provide statement from the glass block supplier giving results of ASTM tests when requested by Owner's Representative.

D. Submit ____ copies of a report from a testing laboratory verifying that aggregate material is asbestos-free and conforms to the specified gradations or characteristics when requested by Owner's representative.

E. Prepare a minimum 2 foot long by 2 foot high panel of each type and pattern of glass unit masonry, including special features. Conform installed masonry to the approved panels. Approved panels may be part of the permanent construction if so approved and conforming to all other requirements indicated and specified.

1.04   QUALITY CONTROL:

Knowledge of the following information is essential for proper use and installation of glass units.

A. Glass block panels are non-load bearing. Adequate provisions must be made for support of construction above panels.

B. Provisions for expansion and movement must be made at jambs and heads of all panels. Mortar must not bridge expansion spaces.

C. Sills of all panels must be painted with a heavy coat of asphalt emulsion and must dry for two hours before first mortar bed is placed.

D. Mortar should be mixed and applied in accordance with manufacturers recommendations.

E. All exterior panels must be well caulked to prevent penetration of moisture.

F. Size of structural members supporting panels should be determined by structural analysis to avoid excessive or harmful deflection.

G. Glass units can become stained or etched by substances released from concrete surfaces or weathering steel exposed to rainwater. Head details should be designed to keep drip water away from Glass units.

1.05   MEASUREMENT AND PAYMENT:

Payment for the work in this section shall be included as part of the lump-sum bid amount stated in the Proposal.

1.04.C   In accordance with manufacturers recommendations.

1.04.E   Related work specified elsewhere.

PART 2 - MATERIALS

2.01   GLASS BLOCK UNITS:

Glass Block Units shall conform to UBC Section 2110, solid or hollow, and may contain inserts, and all surfaces that contact mortar shall be treated to ensure adhesion between the mortar and glass block.

2.02   MORTAR AND GROUT

Mortar shall be type N, in accordance with Table 21-A, Mortar Proportions for Unit Masonry, and ASTM C 207. Type S mortar is permitted.

An optimum mortar mix recommendation is 1 part Portland Cement, 1 part lime and 6 parts sand.

Mortar shall consist of the following:

A. Portland Cement - Shall be Type 1 conforming to the Standard Specifications for Portland Cement, ASTM Designation C 150. Use only one brand.

B. Lime - Shall be hydrated, conforming to ASTM C 207, Type S.

C. Sand - Shall conform with the Standard Specification for Aggregate for Masonry Mortar, ASTM Designation C 144 for thin joints.

D. Integral Type Waterproofer - Metallic-stearate type or Latex type.

2.03   EXPANSION STRIPS:

Where shown or required, shall be either glass fiber or polyethylene as furnished or recommended by glass block manufacturer. Expansion joints shall not be less than 3/8 inch.

2.04   PANEL REINFORCEMENT:

Panel Reinforcing shall be steel double-wire mesh formed of 2 parallel 9-gauge minimum wires either 1-5/8" or 2" on center with electrically welded cross wires 9 gauge minimum at regular intervals, hot dipped galvanized after fabrication.

2.05   PANEL ANCHORS:

Where shown on drawings shall be No. 20-gauge perforated steel strips 24" long by 1-3/4" wide galvanized after perforating. Embed on approximately 24" centers in standard width Glass Block panels and 16" centers in narrow width Glass Block panels.

2.06   ASPHALT EMULSION:

Where shown or required. shall be a water-based asphalt emulsion of correct consistency for application.

2.07   PACKING (BACKER RODS):

Where indicated on drawings or required as a lateral cushioning for Glass Block panels at jambs, heads, and intermediate supports, shall be polyethylene foam, neoprene, non staining, dry rot treated oakum or equal filler approved by the sealant manufacturer.

PART 3 - EXECUTION

3.01   PRODUCT DELIVERY, STORAGE, AND HANDLING:

Glass block units and/or pre-fabricated panels shall be stored in a clean, dry, secure location prior to installation. Cracked, chipped, and otherwise defective glass block and/or panels will be rejected.

3.02   PREPARATION:

Verify that steel frames to receive glass units are correctly installed plumb and level. Install expansion strips as shown and as recommended by the manufacturer. If applicable, mechanically connect panel anchors to steel frames, accurately aligned with the horizontal mortar joints to receive joint reinforcing. Clean surfaces to receive mortar of dust, dirt, oil, grease, and other deleterious substances before placing units. Clean unit edges prior to setting. Coat all sill surfaces to receive mortar with a heavy coat of asphalt emulsion and allow to dry for at least two hours before mortar is placed.

3.03   MIXING AND HANDLING MORTAR:

Place half of the required water and sand in an operating machine mixer; then add portland cement, remainder of sand and water, and then hydrated lime. Machine mix not less than 10 minutes after all ingredients are charged. Mix mortar to a consistency as stiff as will permit good workability using the minimum water feasible. Mortar shall be drier than mortar for ordinary masonry. Retempering the mortar after it has taken its initial set shall not be permitted.

3.04   PLACEMENT OF REINFORCEMENT:

A. Install Panel Reinforcement in horizontal joints., Reinforcement shall be installed on 16" centers and in joints immediately above and below all openings within panels. Reinforcing steel shall run continuously from end to end of panels and shall be lapped not less than 6" whenever it is necessary to use more than one length. Do not bridge expansion joints with reinforcing. Fully embed all joint reinforcement in mortar.

B. Place full mortar bed for joints not requiring panel reinforcement - do not furrow.

3.01   Glass block units shall not be reused after being removed from an existing panel.

3.04.B   All mortar joints are to be completely filled.

C. All vertical and horizontal mortar joints shall be not less than $^1/_4$" and not more than $^3/_8$" thick and shall be completely filled.

3.05 LAYING GLASS UNIT MASONRY:

A. Sill area to be covered by mortar shall have a heavy coat of Asphalt Emulsion. Allow emulsion to dry before placing mortar.

B. Erect glass units plumb and with true plumb and horizontal joints throughout. Joints shall be of uniform size on each surface. Keep the glass blocks clean and free of mortar on exposed surfaces as the installation progresses. Match approved sample panels.

C. Adhere Expansion Strips to jambs and head with gobs of asphalt emulsion. Make certain expansion strip extends to sill.

D. Set full mortar bed joint, applied to sill. Set lower course of block. All mortar joints must be full and not furrowed. Steel tools must not be used to tap blocks into position. For solid glass block it may be necessary to use wedges in the mortar joints of the lower courses to prevent the mortar from being "squeezed" out.

E. Where indicated for lateral support and cushioning of glass units, pack the space between the glass units and frames with the approved packing material rammed tight and firm at sill, jambs, and heads of glass unit panels, and space where required for sealant caulking.

F. Exterior panels shall not exceed 144 square feet of unsupported area, nor 15 feet in any dimension, and interior panels shall not exceed 250 square feet unsupported area, nor 25 feet in any dimension.

G. Do not install glass block units when temperature is 40° F and falling.

3.06 PROTECTION OF WORK:

Protect sills, ledges, and offsets from mortar drippings and other damage during construction. Protect face materials against staining by removing misplaced mortar or grout immediately, and by brushing the masonry surface with a non-metallic stiff bristled brush, at the end of each day's work.

3.07 JOINTS:

A. Strike joints smooth while mortar is still plastic and before final set. At this time rake out all spaces requiring sealant to a depth equal to the width of the spaces. Remove surplus mortar from faces of glass blocks and wipe dry. Tool joints smooth and concave, before mortar takes final set. (Remove wedges from lower courses of solid blocks and point the voids with mortar).

B. After final mortar set install packing tightly between glass block panel and jamb and head construction. Leave space for sealing.

C. Apply sealant evenly to the full depth of recesses as indicated on the drawings and in accordance with the manufacturer's application manual and instructions.

3.08   CLEANING:

A. Surplus mortar shall be removed and the faces of the blocks wiped dry at the time joints are tooled. Cleaning is facilitated by the use of an ordinary household scrub brush having stiff bristles. Final cleaning shall be done by others after mortar has attained final set but before becoming dry on the block surfaces. Do not use abrasive cleaners (steel wool, wire brush) in conjunction with removing mortar of dirt from the faces of glass block. Remove mortar with damp cloth before final set occurs.

B. Glass Block units should be cleaned at least 3-4 times a year so that materials such as metal ions, alkali rundown from concrete, stucco, etc., are not permitted prolonged contact with the block's coated surface. Edge drips must be designed to prevent moisture rundown from these surfaces.

END OF SECTION

3.07.C Related work specified elsewhere.

SECTION 07175
GUIDE SPECIFICATION FOR
BRICK WATER REPELLENT COATING

This section is intended to include water or solvent based liquid, usually applied to clay or brick surfaces as a water repellent coating to retard moisture absorption into above grade porous surfaces, and for stain or soil resistance to interior surfaces. Be aware that some masonry associations do not recommend use of such coating.

PART 1-GENERAL

1.01   WORK INCLUDED

A.   Water repellent coating to exterior and interior clay or brick surfaces.

B.   Schedule of surfaces as indicated on drawings.

1.02   RELATED WORK SPECIFIED ELSEWHERE

A.   Concrete Surfaces

B.   Other Masonry Surfaces

C.   Joint Sealers

D.   Stucco Surfaces

E.   Elastomeric Coatings

1.03   REFERENCES

Include only reference standards that are to be indicated within the text of this Section.

A.   ASTM C 62 Standard Specification for Building Brick

B.   ASTM C 216 Standard Specification for Facing Brick

C.   ASTM E 514 Standard Test Method for Water Penetration and Leakage through Masonry

D.   ASTM G 53 Standard Practice for Operating Light-Exposure and Water-Exposure Apparatus for Exposure of Nonmetallic Materials.

1.04   PERFORMANCE

A.   Apply coating to exhibit ability to allow ___ percent maximum moisture absorption in material being treated.

1.01.B Schedule of surfaces may be included in specifications.

1.02 Limit to clarifying coordination and to prevent duplication of bids.

1.03.D. ASTM G 53 describes the test apparatus and procedure for determining weathering and ultraviolet stability of materials such as clear water repellent treatments.

1.04 In according with manufacturers recommendations.

B. Weathering and UV Stability-Provide treatment materials tested on clay or brick masonry in accordance with ASTM G 53.

1.05 QUALITY ASSURANCE

A. Manufacturer: Company specializing in manufacture of water repellent coatings with three years minimum experience.

B. Applicator: Acceptable to manufacturer with a minimum of 2 years experience in water repellent coating application.

C. Mock up

1. Apply sample application of water repellent according to manufacturer's recommendations to serve as standard of acceptable quality for the project. Apply clear water repellent treatment to left side of mock-up and allow to cure prior to application of treatment to right side of mock-up.

2. Verify that substrate is coated with sufficient water repellent to effectively repel moisture off the surface.

3. Verify that application of water repellent material to substrate produces no surface discoloration.

1.06 SUBMITTALS

A. Submit product data under provisions of General Requirements (Division 1).

B. Include details of product description, tests performed, limitations to coating, cautionary procedures required during application and chemical properties, including percentage of solids.

When manufacturer's instructions for specific installation requirements are utilized, carefully edit PART 3 EXECUTION requirements to avoid conflict with those instructions.

C. Submit manufacturer's installation instructions under provisions of General Requirements.

D. Submit manufacturer's certificate under provisions of General Requirements that coating meets or exceeds specified requirements.

1.07 PRODUCT DELIVERY, STORAGE AND HANDLING

A. Deliver materials in original sealed containers, clearly marked with the manufacturer's name, brand name and type of material.

B. Store materials in area where temperatures are not less than 40° F (4.5° C)

1.04.B The procedure in ASTM C 53 uses UV lamps to intentionally degrade the water repellent coating for the purpose of correlating laboratory testing to natural weathering.

1.08   ENVIRONMENTAL REQUIREMENTS

A. Do not apply coating when surface temperature is lower than 50° F (10° C) or higher than 100° F (38° C).

B. Do not apply water repellent in rainy conditions.

C. Do not apply water repellent coating to damp surfaces.

1.09   WARRANTY

A. Special warranty: Provide for correcting failure of water repellent treatment to resist penetration of water.

1.   Warranty period: five years.

PART 2-PRODUCTS

2.01   ACCEPTABLE MANUFACTURERS

A. Acceptable manufacturers and product for the water repellent treatment are:

B. Substitutions acceptable under the provisions of the General Requirements.

2.02   MATERIALS

A. Water repellent shall be a penetrating VOC (Volatile Organic Compound) compliant emulsion.

B. Water repellent, when dry, shall not stain or discolor the clay or brick surface.

PART 3-EXECUTION

3.01   INSPECTION

A. Verify that joint sealers are installed and cured.

B. Verify surfaces to be coated are dry, clean, and free of efflorescence, oil, or other matter detrimental to application of coating.

C. Verify that masonry joints found unsound, hollow, or otherwise defective, have been raked out to a depth of $\frac{1}{2}$ inch (13 mm) and repointed with mortar.

D. Verify that cracks which exceed $\frac{1}{64}$ inch (0.4 mm) wide have been filled with pointing mortar or caulking material.

1.09.A Noted is project warranty, not the standard manufacturer material warranty. Modify where longer term is considered necessary or where repair of materials damaged by leaks is to be included.

2.01.A State name of manufacturer and product acceptable for application.

3.01.A Note that some water repellent treatments should be applied to damp surfaces in accordance with manufacturer's recommendations.

3.01.D The crack dimension is a variable that will change depending on the characteristics of the product. Check with manufacturers product specifications for crack bridging dimension.

E. Verify that all flashing and caulking materials have been properly installed.

F. Commencement of installation means acceptance of substrate.

3.02   PREPARATION

A. Clean surfaces to remove dust, dirt, oil, wax, efflorescence and other foreign materials that may interfere with uniform penetration.

B. If necessary, remove oil or foreign substance with a chemical solvent which will not affect coating.

C. Curing agents and/or form release agents must be removed and the surface cleaned before sealer can be applied.

D. Scrub and rinse surfaces with water and let dry.

Some coatings can easily attack and etch aluminum and glass surfaces and kill vegetation. Specify special protection requirements if deemed appropriate.

E. Protect adjacent surfaces not scheduled to receive coating.

F. If applied on unscheduled surfaces, remove immediately, by approved methods.

G. Protect landscaping, property and vehicles.

3.03   APPLICATION

A. Delay work until the clay or brick substrate is cured in accordance with manufacturer's recommendations.

B. Apply coating in accordance with manufacturer's instructions, at the rate specified by the manufacturer.

C. Apply in uniformly in accordance with the coating procedure recommended by the manufacturer.

D. Start application at the top of wall working downward, keeping a flooding wet edge at all times. Avoid run downs longer than approximately 12 inches (300 mm) unless applicator can pick up run downs before they dry. Strike off or roll off any obvious drips or puddles of sealer remaining on overhangs or horizontal surfaces after 15-30 minutes of application.

E. On light colored surfaces it is recommended that the application start at the bottom of the wall and work upward. This will reduce the possibility of long run-down streaks picking up concentrations of foreign material ahead of the application and being permanently sealed into the surface.

3.02.E   Check with manufacturer for items requiring protection.

F. When spraying, hold spray tip 12-18 inches (300-450 mm) from the wall using overlapping spray pattern and fanning motion at the end of each horizontal pass.

3.04   SCHEDULE

Include a schedule of surfaces on the Drawings to clearly identify locations for application.

END OF SECTION

SECTION 07180
CLEAR WATER REPELLENTS
FOR CONCRETE UNIT MASONRY

PART 1 - GENERAL

1.01   SUMMARY:

A. Section includes: Provide clear water repellent treatment for exposed exterior surfaces of special color and texture concrete masonry construction.

B. Related Sections:

   1.   Section 04220 - Concrete Unit Masonry: Mock-ups and testing requirements for systems with clear water repellent treatment.

   2.   Section 07900 - Joint Sealers: Coordination for compatibility.

   3.   Section 09830 - Elastomeric Coatings: Opaque elastomeric waterproof coatings.

1.02   REFERENCES:

A. ASTM C 90 Standard Specification for Load-Bearing Concrete Masonry Units.

B. ASTM C 140 - Methods for Sampling and Testing Concrete Masonry Units.

C. ASTM E 514 Standard Test Method for Water Penetration and Leakage Through Masonry

D. ASTM G 53 - Standard Practice for Operating Light- and Water-Exposure Apparatus for Exposure of Nonmetallic Materials.

1.03   SYSTEM DESCRIPTION:

A. Performance Requirements:

   1.   Absorption: Provide treatment materials which have been treated on concrete masonry to indicate degree of change of absorption of concrete masonry units based on ASTM C 140.

      a.   Effectiveness: Minimum 85% over control units.

   2.   Water Vapor Transmission: Provide treatment materials which have been tested on concrete masonry to indicate degree of change of water vapor transmission through concrete masonry units based on ASTM E 96.

      a.   Change: No significant change in water vapor transmission.

   3.   Weathering and UV Stability: Provide treatment materials tested on concrete masonry in accordance with ASTM G 53.

1.01.B   Limit to clarifying coordination and to prevent duplication of bids.

1.02.A   The finished concrete masonry wall should not contain any cracks or beeholes what would inhibit the effectiveness of the water repellent.

1.02.D   ASTM G 53 describes the test apparatus and procedure for determining weathering and ultraviolet stability of materials such as clear water repellent treatments.

1.03.A.3 The procedure in ASTM G 53 uses UV lamps to intentionally degrade the water repellent coating for the purpose of correlating laboratory testing to natural weathering.

a. Change: No significant change in unit color or absorption.

1.04   SUBMITTALS:

A. Product Data: Submit manufacturer's literature for clear water repellent treatment.

B. Samples: Submit samples of concrete masonry units approved for use in Project with water repellent treatment applied to half of each sample face; indicate which half has been coated.

C. Quality Assurance Submittals:

1. Test Reports: Submit test reports indicating compliance with performance requirements for absorption, water vapor transmission, and weathering and UV stability.

2. Certificates:

a. Submit installer qualification certificates.
b. Submit manufacturer's installer approved certificates.
c. Submit certification indicating materials comply with applicable Volatile Organic Compound (VOC) limitations.

3. Manufacturer's Instructions: Provide copies of manufacturer's instructions to field office.

4. Manufacturer's Field Reports: Submit report of manufacturer's represent-atives indicating materials have been installed in accordance with manufacturer's instructions and recommendations.

1.05   QUALITY ASSURANCE:

A. Qualifications:

1. Qualification of Manufacturer: Firm with minimum five years record of successful in-service experience of clear water repellent treatments manufactured for concrete masonry unit application.

2. Qualification of Installers: Applicator with minimum five years successful experience in projects of similar scope using specified or similar treatment materials and approved by treatment manufacturer.

B. Regulatory Requirements: Provide materials with not more than the maximum VOC as required by applicable authorities.

C. Mock-Up: Prior to commencing work, including bulk purchase and delivery of material, apply clear water repellent treatment to concrete masonry mock-up indicated in Contract Documents.

1.03.A.3.a   Where gloss is intended revise wording to "No significant change in unit gloss, color or absorption".

1.05.B   Where known, VOC limitations should be included.   Where unknown, or subject to regular revision, use general statement

1. Testing: Provide field testing of concrete masonry system mock-up, including clear water repellent treatment; test to be observed by Owner's representative and water repellent treatment manufacturer's representative.

    a. Application: Apply clear water repellent treatment to left side of mock-up and allow to cure prior to application of treatment to right side of mock-up.

    b. Test: Twenty days after completion of application of treatment, test mock-up with 5/8 inch garden hose with spray nozzle located ten feet from wall and aimed upward so water strikes wall at 45 degree downward angle.

        1) Run water continuously for two hours.

        2) Observe back side of mock-up for water penetration and leakage; where leakage is detected make changes as needed to correct and retest.

    c. Results: Cooperate with testing procedures and modify Project treatment application as required to pass mock-up tests for water penetration and leakage resistance.

2. Approval: Proceed with clear water repellent treatment work only after successful completion of field test application and approval of mock-up.

D. Pre-Installation Meeting:

1. Attend pre-installation meeting required prior to commencement of concrete masonry installation.

2. Review procedures and coordination required between concrete masonry and clear water repellent treatment work and between treatment work and work which could be affected by or affect treatment.

3. Convene additional pre-installation meeting prior to water repellent treatment application for coordination with work not previously coordinated including joint sealers as needed.

1.06 DELIVERY, STORAGE AND HANDLING:

A. Delivery: Deliver water repellent treatment products to job site in manufacturer's container with label intact and listing product identification, batch number and date of manufacture.

B. Storage: Keep materials in original, unopened containers; prevent contamination by foreign materials.

1.07 SITE CONDITIONS:

A. Environmental Requirements: Comply with manufacturer's recommendations; do not apply clear water repellent treatments under the following conditions:

1.05.C.1 ASTM E 514 is a laboratory test to determine water penetration through masonry described is a field test similar to the Navy hose stream test. This test requires access to the opposite side of the concrete masonry wall.

1. During inclement weather, when air temperature is below 50 degrees Fahrenheit or above 100 degrees Fahrenheit.

2. When rain or temperatures below 40 degrees Fahrenheit are predicted for a period of 24 hours before or after treatment application.

3. Earlier than 3 days after surfaces became wet.

4. When substrates might be frozen.

5. When surface temperature is less than 40 degrees Fahrenheit.

1.08   WARRANTY:

A. Special Warranty: Provide for correcting failure of water repellent treatment to resist penetration of water.

1. Warranty Period: Five Years.

PART 2 - PRODUCTS

2.01   MATERIALS:

A. Clear Water Repellent Treatment:

1. Appearance: Clear, non-yellowing water repellent treatment shall not alter appearance, color, or texture of substrate under any lighting conditions.

B. Compatibility: Provide products which are recommended by manufacturer to be fully compatible with indicated substrates and joint sealers which are in contact with water repellent treatment.

PART 3 - EXECUTION

3.01   EXAMINATION:

A. Verification of Conditions: Examine substrates; do not apply treatment to damp, dirty, dusty, or otherwise unsuitable surfaces; comply with manufacturer's recommendations.

1. Do not begin application of clear water repellent treatment until voids and beeholes visible from 5 feet, and cracks greater than 0.02 inches wide in masonry substrates have been repaired.

1.08.A Noted is the project warranty, not the standard manufacturer material warranty. Modify where longer term is considered necessary or where repair of materials damaged by leaks is to be included.

2.01.A Provide information regarding acceptable manufacturers and products. or sufficient description to allow competitive bidding.

2.01.A.1 Modify as applicable. Requirements for type (penetrating or surface seal), chemical composition, gloss, percentage of solids, and other attributes can be included as applicable--verify.

3.01.A Note that some water repellent treatments should be applied to damp surfaces.

3.02   PREPARATION:

A. Protection: Provide masking or covering for materials which could be damaged by application of clear water repellent treatment.

    1. Sealant Coordination: Assure treatment compatibility with each type of joint sealer within or adjacent to surfaces receiving clear water repellent treatment.

        a. Coordinate treatment application with joint sealers; where recommended by joint sealant manufacturer, apply treatment after application and cure of joint sealers.

        b. Mask surfaces indicated to receive joint sealers which would be adversely affected by clear water repellent treatment where treatment must be applied prior to application of joint sealers.

    2. Protect glass, glazed products and prefinished products from contact with water repellent treatment.

    3. Protect landscape materials with breathing type drop cloths; plastic covers are not acceptable.

B. Surface preparation: Prepare substrates in accordance with water repellent treatment manufacturer's recommendations.

    1. Clean surfaces of dust, dirt, and foreign matter detrimental to proper application of water repellent treatment.

3.03   APPLICATION:

A. General: Apply treatment in accordance with clear water repellent treatment manufacturer's instructions and applicable recommendations, including number of coats, maximum allowable coverage, and equipment.

    1. Review procedures used for application of treatment to mock-up and recommendations for changes needed based on water penetration tests conducted on mock-up.

    2. Consult with manufacturer's representative for site inspection, for proper application techniques not fully covered in manufacturer's instructions, and for applicable recommendations.

3.04   REPAIR:

A. Repair or replace materials damaged by application of water repellent treatment.

3.02.A.1 Verify requirements based on compatibility of treatment and sealants; where compatibility is unknown use wording given. Very application of sealant on mock-up panel. Check for staining.

3.02.A.2 Check manufacturer for items requiring protection.

3.03.A Review manufacturer's literature. Avoid unnecessary repetition. Where items noted as recommendations are given, specifications can list them as requirements where desired and when appropriate.

3.05 FIELD QUALITY CONTROL:

A. Site Inspections: Manufacturer's representative shall inspect application of water repellent treatment in progress to verify compliance with manufacturer instructions and recommendations.

3.06 CLEANING:

A. Clean water repellent treatment from surfaces not indicated to be treated immediately; comply with recommendations of materials manufacturers for proper cleaning technique to prevent damage.

END OF SECTION

BIA Technical Notes: SUBJECT HEADING INDEX

Copies of BIA Technical Notes can be purchased directly from Brick Institute of America, 11490 Commerce Park Drive, Reston, Virginia 20191-1525, Tel: (703) 620-0010, Fax: (703) 620-3928 or through http://www.bia.org.

TEK MANUAL FOR CONCRETE MASONRY DESIGN AND CONSTRUCTION

Copies of NCMA Technical Notes can be purchased directly from National Concrete Masonry Association, 2302 Horse Pen Road, Herdon, Virginia 20117-3499, Tel: (703) 713-1900, Fax: (703) 713-1910 or through http://www.ncma.org.

Table of Contents

* These TEK also available in Spanish

CROSS REFERENCE UBC/ACI - ACI/UBC

The following cross reference directory is meant to assist the user in understanding the relationship between the 1997 edition of the Uniform Building Code and the 1995 edition of ACI 530/ASCE 5/TMS 402, Building Code Requirements for Masonry Structures and ACI 530.1/ASCE 6/TMS 602, Specification for Masonry Structures (commonly referred as ACI 530.1).

The Uniform Building Code (UBC) is the governing code in the western half of the United States. The Northeast is regulated by the Building Officials & Code Administrators (BOCA), while the Southeast follows the Southern Standard Building Code (SBC). Both BOCA and SBC reference ACI 530 and ACI 530.1 for their masonry building code.

Three prominent national organizations, The Masonry Society, American Concrete Institute and American Society of Civil Engineers sponsor the Building Code Requirements for Masonry Structures (ACI 530) and the Specifications for Masonry Structures (ACI 530.1), as reported by the Masonry Standards Joint Committe.

Titles that are printed in italics do not contain any Code text, therefore, may not have a cross reference correlation. The notation "Not covered" is not intended to mean that the complementary code specifically excludes the code text. There are circumstances wherein a cross reference would be misleading, for example, a requirement contained in the Uniform Building Code Strength Design section intentionally would not be linked to the ACI 530/ASCE 5/TMS 402 since the latter does not contain the Strength Design method.

It should also be noted that certain criteria that governs code may not be referenced in this list. For example, the Uniform Building Code relies upon Chapter 31 for the construction of chimneys and fireplaces, whereas the other codes make reference to the provisions contained in the One and Two Family Dwelling Code.

The following pages cross-reference UBC to ACI 530 and ACI 530 to UBC.

1997 UNIFORM BUILDING CODE		ACI 530-95/ASCE 5-95/TMS 402-95	
Section	**Title**	**Section**	**Title**
2101	*General*	*530-95/Part 1*	*General*
2101.1	Scope	530-95/1.1	Scope
2101.2	Design methods	530-95/5.1.1	Design scope
2101.2.1	Working stress design	530-95/Part 3	Analysis and design
2101.2.2	Strength design		Not covered
2101.2.3	Empirical design	530-95/Chapter 9	Empirical design of masonry
2101.2.4	Glass masonry	530-95/Chapter 11	Glass unit masonry
2101.3	Definitions	530-95/2.2	Definitions
2101.4	Notations	530-95/2.1	Notations
2102	*Material standards*	*530.1-95/1.3*	*References*
2102.1	Quality	530.1-95/1.3	References
2102.2	Standards of quality	530.1-95/1.3	References
2103	*Mortar and grout*	*530.1-95/2.1*	*Mortar materials*
2103	*Mortar and grout*	*530.1-95/2.2*	*Grout materials*
2103.1	General	ASTM C 270	Mortar
2103.1	General	ASTM C 476	Grout
2102.2	Materials	ASTM C 270	Mortar
2102.2	Materials	ASTM C 476	Grout
2103.3	*Mortar*	*ASTM C 270*	*Mortar for unit masonry*
2103.3.1	General	ASTM C 270	Mortar for unit masonry
2103.3.2	Selecting proportions	ASTM C 270	Mortar for unit masonry
2103.4	*Grout*	*530.1-95/2.2*	*Grout materials*
2103.4.1	General	ASTM C 476	Grout for masonry
2103.4.2	Selecting proportions	ASTM C 476	Grout for masonry
2103.5	*Additives and admixtures*	*530.1-95/2.6.A*	*Mortar*
2103.5	*Additives and admixtures*	*530.1-95/2.6.B*	*Grout*
2103.5.1	General	ASTM C 270	Mortar
2103.5.1	General	ASTM C 476	Grout
2103.5.2	Antifreeze compounds	ASTM C 270/4.1.4	Antifreeze compounds
2103.5.2	Antifreeze compounds	ASTM C 476/3.1.7	Antifreeze compounds
2103.5.3	Air entrainment	ASTM C 270/4.1.4	Air entrainment
2103.5.3	Air entrainment	ASTM C 476/3.1.2	Air-entraining admixtures
2103.5.4	Colors	530.1-95/2.6.A.2	Maximum mortar color
2104	*Construction*	*530.1-95*	*Specifications for masonry structures*
2104.1	General	530.1-95/1.1	Summary
2104.2	Materials: handling, storage & prep.	530.1-95/1.7	Delivery, storage & handling
2104.3	*Cold weather construction*	*530.1-95/1.8.C*	*Cold weather construction*
2104.3.1	General	530.1-95/1.8.C	Cold weather construction.
2104.3.2	Preparation		Not covered
2104.3.3	Construction	530.1-95/1.8.C	Cold weather construction.
2104.3.4	Protection	530.1-95/1.8.B & C	Masonry protection
2104.3.5	Placing grout and protection		Not covered

1997 UNIFORM BUILDING CODE		ACI 530-95/ASCE 5-95/TMS 402-95	
Section	Title	Section	Title
2104.4	Placing masonry units	530.1-95/3.3	Masonry erection
2104.4.1	Mortar	530.1-95/3.3.B	Placing mortar and units
2104.4.2	Surfaces	530.1-95/3.2.B	Clean surfaces
2104.4.3	Solid masonry units	530.1-95/3.3.B.4	Solid units
2104.4.4	Hollow-masonry units	530.1-95/3.3.B.3	Hollow units
2104.5	Reinforcement placing	530.1-95/3.4.E	Reinforcement
2104.6	Grouted masonry	530.1-95/3.5	Grout placement
2104.6.1	General conditions	530.1-95/3.5	Grout placement
2104.6.1	General conditions	530.1-95/3.2.D	Debris
2104.6.1	General conditions	530.1-95/3.3.B.1	Mortar fins
2104.6.2	Construction requirements	530.1-95/3.2.E	Reinforcement
2104.6.2	Construction requirements	530.1-95/3.2.F	Cleanouts
2104.6.2	Construction requirements	530.1-95/3.3.B.1	Mortar fins
2104.7	Aluminum equipment	530.1-95/3.3.D.9	Aluminum
2104.8	Joint reinforcement	530.1-95/3.4.C.2	Joint reinforcement
2105	Quality assurance	530-95/Part 2	Quality assurance and construction requirements
2105	Quality assurance	530.1-95/1.6	Quality assurance
2105.1	General	530-95/3.1.1	Material labor and const
2105.2	Scope	530-95/3.1.1	Material labor and const.
2105.3	Compliance with f'_m	530-95/3.2	Acceptance relative to strength requirements
2105.3.1	General	530-95/3.2	Acceptance relative to strength requirements
2105.3.2	Masonry prism testing	530.1-95/1.4.B.3	Prism test method
2105.3.2	Masonry prism testing	530.1-95/3.6.A	Prism testing
2105.3.3	Masonry prism test record		Not covered
2105.3.4	Unit strength method	530.1-95/1.4.B.2	Unit strength method
2105.3.5	Testing prisms from constructed masonry		Not covered
2105.4	Mortar testing	530.1-95/3.6.B	Mortar testing
2105.5	Grout testing	530.1-95/3.6.C	Grout testing
2106	General design requirements	530-95/Part 3	Analysis and design
2106.1	General	530-95/5.1	Scope
2106.1.1	Scope	530-95/5.1.1	Scope
2106.1.2	Plans	530-95/1.2.1	Project drawings
2106.1.3	Design loads	530-95/5.2.2	Loading
2106.1.4	Stack bond		Not covered
2106.1.5	Multiwythe walls	530-95/5.8	Multiwythe walls
2106.1.5.1	General	530-95/5.8.2.1	Multiwythe walls
2106.1.5.2	Wall ties in cavity wall construction	530-95/5.8.2.5	Wall ties
2106.1.5.2	Wall ties in cavity wall construction	530-95/5.8.3.2	Wall ties

1997 UNIFORM BUILDING CODE		ACI 530-95/ASCE 5-95/TMS 402-95	
Section	Title	Section	Title
2106.1.5.3	Wall ties for grouted multiwythe construction	530-95/5.8.2.5	Wall ties
2106.1.5.4	Joint reinforcement	530-95/8.4.2	Joint reinforcement
2106.1.6	Vertical support	530.1-95/3.3.B.1	Bed and head joints
2106.1.7	Lateral support	530-95/5.7	Lateral load distribution
2106.1.8	Protection of ties & joint reinforcement	530.1-95/3.4.C.4	Protection for reinforcement
2106.1.9	Embedded pipes and conduits	530-95/4.1.1.1	Design-displaced masonry
2106.1.10	Load tests		Not covered
2106.1.11	Reuse of masonry units		Not covered
2106.1.12	*Special seismic risk provisions*	*530-95/Chapter 10*	*Seismic design req"ments*
2106.1.12.1	General	530-95/10.2.1	General
2106.1.12.2	Special provisions for Seismic Zones 0 and 1	530-95/10.3	Seismic Performance Category A
2106.1.12.2	Special provisions for Seismic Zones 0 and 1	530-95/10.4	Seismic Performance Category B
2106.1.12.3	Special provisions for Seismic Zone 2	530-95/10.5	Seismic Performance Category C
2106.1.12.4	Special provisions for Seismic Zones 3 and 4	530-95/10.6	Seismic Performance Category D
2106.1.12.4	Special provisions for Seismic Zones 3 and 4	530-95/10.7	Seismic Performance Category E
2106.2	*Working stress design and strength design requirements for unreinforced and reinforced masonry*		
2106.2.1	General	530-95/7.1.2	General requirements
2106.2.2	Specified comp. strength of masonry	530-95/5.4	Design strength
2106.2.3	*Effective thickness*		
2106.2.3.1	Single-wythe walls		Not covered
2106.2.3.2	Multiwythe walls		Not covered
2106.2.3.3	Cavity walls		Not covered
2106.2.3.4	Columns		Not covered
2106.2.4	Effective height	530-95/2.2	Effective height (definition)
2106.2.5	Effective area		Not covered
2106.2.6	Effective width of intersecting walls	530-95/5.13.4.2.c	Wall intersection
2106.2.7	Distribution of concentrated vertical loads in walls	530-95/5.12.1	Concentrated loads
2106.2.8	Loads on non-bearing walls	530-95/5.15	Framed construction
2106.2.9	Vertical deflection	530-95/5.6	Deflection of beams & lintels
2106.2.10	Structural continuity	530-95/5.11.1	Load transfer at horizontal connections
2106.2.11	Walls intersecting with floors and roofs	530-95/4.2	Masonry attachment to structural frames
2106.2.12	*Modulus of elasticity of materials*		
2106.2.12.1	Modulus of elasticity of clay masonry	530-95/5.5.2.2	Clay masonry
2106.2.12.1	Modulus of elasticity of concrete	530-95/5.5.2.3	Concrete masonry

1997 UNIFORM BUILDING CODE		ACI 530-95/ASCE 5-95/TMS 402-95	
Section	Title	Section	Title
	masonry		
2106.2.12.2	Modulus of elasticity of steel	530-95/5.5.2.1	Steel reinforcement
2106.2.13	Shear modulus of masonry	530-95/5.5.2	Elastic moduli
2106.2.14	Placement of embedded anchor bolts		
2106.2.14.1	General		Not covered
2106.2.14.2	Minimum edge distance	530-95/5.14.2.3	Allowable load in shear, l_{be}
2106.2.14.3	Minimum embedment depth	530-95/5.14.2.1	Minimum embedment length
2106.2.14.4	Minimum spacing between bolts.		Not covered
2106.2.15	Flexural resistance of cavity walls	530-95/5.8.3.1.d	Loads acting transverse
2106.3	Working stress design and strength design requirements for reinforced masonry		
2106.3.1	General	530-95/7.1.2	Scope
2106.3.2	Plain bars	530.1-95/2.4.A	Reinforcing steel
2106.3.3	Spacing of longitudinal reinforcement	530-95/8.3.1	Clear distance between parallel bars
2106.3.3	Spacing of longitudinal reinforcement	530-95/8.3.5	Reinforcement embedded in grout
2106.3.4	Anchorage of flexural reinforcement	530-95/8.5.1	Development of reinforcement
2106.3.4	Anchorage of flexural reinforcement	530-95/8.5.3	Embedment of flexural reinforcement
2106.3.5	Anchorage of shear reinforcement	530-95/8.5.6.1	Bar and wire reinforcement
2106.3.6	Lateral ties	530-95/5.9.1.6	Lateral ties
2106.3.7	Column anchor bolt ties		Not covered
2106.3.8	Effective width b of compression area	530-95/7.3.3.1	Width of the compression area
2107	Working stress design of masonry		
2107.1	General	530-95/Part 3	Analysis and design
2107.1.1	Scope	530-95/5.1.1	Scope
2107.1.2	Allowable masonry stresses		Not covered
2107.1.3	Minimum dimensions for masonry structures located in Seismic Zones 3 and 4		Not covered
2107.1.3.1	Bearing walls		Not covered
2107.1.3.2	Columns	530-95/5.9.1.1	Minimum column dimension
2107.1.4	Design assumptions		Not covered
2107.1.5	Embedded anchor bolts	530-95/5.14	Anchors bolts in masonry
2107.1.5.1	General	530-95/5.14.1	Test design requirements
2107.1.5.2	Tension	530-95/5.14.2.2	Allowable load in tension
2107.1.5.3	Shear	530-95/5.14.2.3	Allowable load in shear
2107.1.5.4	Combined shear and tension	530-95/5.14.2.4	Combined shear and tension
2107.1.6	Compression in walls and columns		
2107.1.6.1	Walls, axial loads	530-95/6.3.1	Axial compression and flexure

1997 UNIFORM BUILDING CODE		ACI 530-95/ASCE 5-95/TMS 402-95	
Section	Title	Section	Title
2107.1.6.2	Columns, axial loads	530-95/5.9.1.5	Stresses
2107.1.6.3	Columns, bending or combined bending and axial loads	530-95/5.9.1.5	Stresses
2107.1.7	Shear walls, design loads		Not covered
2107.1.8	Design, composite construction		
2107.1.8.1	General	530-95/5.13.1.2	Composite action design
2107.1.8.2	Determination of moduli of elasticity	530-95/5.5	Determination of material properties
2107.1.8.3	Structural continuity		
2107.1.8.3.1	Bonding of wythes	530-95/5.8.5	Wythes of walls
2107.1.8.3.2	Material properties	530-95/5.5	Material properties
2107.1.8.4	Design procedure, transformed sections	530-95/5.13.1.2	Composite stress computations
2107.1.9	Reuse of masonry units		Not covered
2107.2	Design of reinforced masonry	530-95/Chapter 7	Reinforced masonry
2107.2.1	Scope	530-95/7.1	Scope
2107.2.2	Reinforcement	530-95/7.3	Axial compression & flexure
2107.2.2.1	Maximum reinforcement size	530-95/8.2.1	Max. reinforcement size
2107.2.2.2	Cover	530-95/8.4.1	Reinforcing bars min. cover
2107.2.2.3	Development length	530-95/8.5.2	Embedment of bars & wires
2107.2.2.3	Development length	530-95/8.5.7.1	Lap splices
2107.2.2.4	Reinforcement bond stress		Not covered
2107.2.2.5	Hooks	530-95/8.5.4	Standard hooks
2107.2.2.6	Splices	530-95/8.5.7	Splices of reinforcement
2107.2.3	Design assumptions	530-95/7.1.1	Design of structures
2107.2.4	Nonrectangular flexural elements		Not covered
2107.2.5	Allowable flexural compressive stress and force	530-95/7.3.2.1	Axial compressive stress
2107.2.6	Allowable flexural compressive stress	530-95/7.3.2.2	Flexural compressive stress
2107.2.7	Combined compressive stresses, unity formula	530-95/6.3.1	Axial compression and flexure
2107.2.8	Allowable shear stress in flexural members	530-95/7.5.2.2.a	Allowable flexural stresses
2107.2.8	Allowable shear stress in flexural members	530-95/7.5.2.3.a	Allowable flexural stresses
2107.2.9	Allowable shear stress in shear walls	530-95/7.5.2.2.b	Allowable shear stresses
2107.2.9	Allowable shear stress in shear walls	530-95/7.5.2.3.b	Allowable shear stresses
2107.2.10	Allowable bearing stress	530-95/5.12.3	Bearing stress limitation
2107.2.11	Allowable stresses in reinforcement	530-95/7.2.1	Allowable stresses
2107.2.12	Lap splice increases		Not covered
2107.2.13	Reinforcement for columns	530-95/5.9.1	Design of columns
2107.2.13.1	Vertical reinforcement	530-95/5.9.1.4	Vertical column reinforcement.
2107.2.14	Compression in walls and columns	530-95/7.3.2	Allowable forces and stresses

1997 UNIFORM BUILDING CODE		ACI 530-95/ASCE 5-95/TMS 402-95	
Section	Title	Section	Title
2107.2.14.1	General	530-95/7.3.2.2	Compressive stress in reinforced masonry
2107.2.14.2	Walls, bending or combined bending and axial loads	530-95/7.3.1	Members subjected to combined loads
2107.2.15	Flexural design, rectangular flexural elements		Not covered
2107.2.16	Bond of flexural reinforcement		Not covered
2107.2.17	Shear in flexural members and shear walls	530-95/7.5.2.1	Calculated shear stress
2107.3	*Design of unreinforced masonry*	*530-95/Chapter 6*	*Unreinforced masonry*
2107.3.1	General	530-95/6.1	Scope
2107.3.2	Allowable axial compressive stress	530-95/6.3.1	Axial compression and flexure
2107.3.3	Allowable flexural compressive stress	530-95/6.3.1.c	Axial compression and flexure
2107.3.4	Combined compressive stresses, Unity formula	530-95/6.3.1	Axial compression and flexure
2107.3.5	Allowable tensile stress	530-95/6.3.1.1	Allowable tensile stresses
2107.3.6	Allowable shear stress in flexural members	530-95/6.5.2	In-plane shear stresses
2107.3.7	Allowable shear stress in shear walls	530-95/6.5.2	In-plane shear stresses
2107.3.8	Allowable bearing stress	530-95/5.12.3	Bearing stress limitation
2107.3.9	Combined bending and axial loads, compressive	530-95/6.3.1	Axial compression and flexure
2107.3.10	Compression in walls and columns	530-95/6.3.1	Axial compression and flexure
2107.3.11	Flexural design		Not covered
2107.3.12	Shear in flexural members and shear walls	530-95/6.5.1	Shear stresses
2107.3.13	Corbels	530-95/9.9.4	Corbelling
2107.3.14	Stack bond	530-95/5.16	Stack bond masonry
2108	*Strength design of masonry*		
2108-1	*General*		*Not covered*
2108.1.1	General provisions		Not covered
2108.1.2	Quality assurance provisions		Not covered
2108.1.3	Required strength		Not covered
2108.1.4	Design strength		Not covered
2108.1.4.1	*Beams, piers and columns*		*Not covered*
2108.1.4.1.1	Flexure		Not covered
2108.1.4.1.2	Shear		Not covered
2108.1.4.2	Wall design for out-of-plane loads		Not covered
2108.1.4.2.1	Walls with factored axial load of 0.04 f'_m or less		Not covered
2108.1.4.2.2	Walls with factored axial load greater than 0.04 f'_m		Not covered

1997 UNIFORM BUILDING CODE		ACI 530-95/ASCE 5-95/TMS 402-95	
Section	Title	Section	Title
2108.1.4.3	Wall design for in-plane loads		Not covered
2108.1.4.3.1	Axial load		Not covered
2108.1.4.3.2	Shear		Not covered
2108.1.4.4	Moment-resisting wall frames		Not covered
2108.1.4.4.1	Flexure with or without axial load		Not covered
2108.1.4.4.2	Shear		Not covered
2108.1.4.5	Anchor		Not covered
2108.1.4.6	Reinforcement		Not covered
2108.1.4.6.1	Development		Not covered
2108.1.4.6.2	Splices		Not covered
2108.1.5	Anchor bolts		Not covered
2108.1.5.1	Required strength		Not covered
2108.1.5.2	Nominal anchor bolt strength		Not covered
2108.1.5.3	Anchor bolt placement		Not covered
2108.2	Reinforced masonry		Not covered
2108.2.1	General		Not covered
2108.2.1.1	Scope		Not covered
2108.2.1.2	Design assumptions		Not covered
2108.2.2	Reinforcement requirements and details		Not covered
2108.2.2.1	Maximum reinforcement		Not covered
2108.2.2.2	Placement		Not covered
2108.2.2.3	Cover		Not covered
2108.2.2.4	Standard hooks		Not covered
2108.2.2.5	Minimum bend diameter for reinforcing bars		Not covered
2108.2.2.6	Development		Not covered
2108.2.2.7	Splices		Not covered
2108.2.3	Design of beams, piers and columns		Not covered
2108.2.3.1	General		Not covered
2108.2.3.2	Design assumptions		Not covered
2108.2.3.3	Balanced reinforcement ratio for compression limit state		Not covered
2108.2.3.4	Required strength		Not covered
2108.2.3.5	Design strength		Not covered
2108.2.3.6	Nominal strength		Not covered
2108.2.3.6.1	Nominal axial and flexural strength		Not covered
2108.2.3.6.2	Nominal shear strength		Not covered
2108.2.3.7	Reinforcement		Not covered
2108.2.3.8	Seismic design provisions		Not covered
2108.2.3.9	Dimensional limits		Not covered
2108.2.3.10	Beams		Not covered
2108.2.3.10.1	Scope		Not covered

1997 UNIFORM BUILDING CODE		ACI 530-95/ASCE 5-95/TMS 402-95	
Section	Title	Section	Title
2108.2.3.10.2	Longitudinal reinforcement		Not covered
2108.2.3.10.3	Transverse reinforcement		Not covered
2108.2.3.10.4	Construction		Not covered
2108.2.3.11	Piers		Not covered
2108.2.3.11.1	Scope		Not covered
2108.2.3.11.2	Longitudinal reinforcement		Not covered
2108.2.3.11.3	Transverse reinforcement		Not covered
2108.2.3.12	Columns		Not covered
2108.2.3.12.1	Scope		Not covered
2108.2.3.12.2	Longitudinal reinforcement		Not covered
2108.2.3.12.3	Lateral ties		Not covered
2108.2.3.13.4	Construction		Not covered
2108.2.4	Wall design for out-of-plane loads		Not covered
2108.2.4.1	General		Not covered
2108.2.4.2	Maximum reinforcement		Not covered
2108.2.4.3	Moment and deflection calculations		Not covered
2108.2.4.4	Walls w/axial load of 0.04 f'_m or less		Not covered
2108.2.4.5	Walls w/axial load greater than 0.04 f'_m		Not covered
2108.2.4.6	Deflection design		Not covered
2108.2.5	Wall design for in-plane loads		Not covered
2108.2.5.1	General		Not covered
2108.2.5.2	Reinforcement		Not covered
2108.2.5.3	Design strength		Not covered
2108.2.5.4	Axial strength		Not covered
2108.2.5.5	Shear strength		Not covered
2108.2.5.6	Boundary members		Not covered
2108.2.6	Design of moment-resisting wall frames		Not covered
2108.2.6.1	General requirements		Not covered
2108.2.6.1.1	Scope		Not covered
2108.2.6.1.2	Dimensional limits		Not covered
2108.2.6.1.3	Analysis		Not covered
2108.2.6.2	Design procedure		Not covered
2108.2.6.2.1	Required strength		Not covered
2108.2.6.2.2	Design strength		Not covered
2108.2.6.2.3	Design assumptions for nominal strength		Not covered
2108.2.6.2.4	Reinforcement		Not covered
2108.2.6.2.5	Flexural members (beams)		Not covered
2108.2.6.2.6	Members subjected to axial force and flexure		Not covered
2108.2.6.2.7	Pier design forces		Not covered
2108.2.6.2.8	Shear design		Not covered

1997 UNIFORM BUILDING CODE		ACI 530-95/ASCE 5-95/TMS 402-95	
Section	Title	Section	Title
2108.2.6.2.9	Joints		Not covered
2109	*Empirical design of masonry*	*530-95/CHAPTER 9*	*Empirical design of masonry*
2109.1	General	530-95/9.1	Scope
2109.2	Height	530-95/9.2	Height
2109.3	Lateral stability	530-95/9.3	Lateral stability
2109.4	*Compressive stresses*	*530-95/9.4*	*Compressive stress requirements*
2109.4.1	General	530-95/9.4.1	Compressive stresses in masonry
2109.4.2	Allowable stresses	530-95/9.4.2	Compressive stresses
2109.4.3	Stress calculations	530-95/9.4.2.1	Calculated compressive stresses
2109.4.4	Anchor bolts	530-95/4.2.1	Anchorage of masonry
2109.5	Lateral support	530-95/9.5	Lateral support
2109.6	*Minimum thickness*	*530-95/9.6*	*Thickness of masonry*
2109.6.1	General	530-95/9.6.2.1	Nominal masonry thickness
2109.6.2	Variation in thickness	530-95/9.6.2.3	Change in thickness
2109.6.3	Decrease in thickness	530-95/9.6.2.3	Change in thickness
2109.6.4	Parapets	530-95/9.6.4	Parapet walls
2109.6.5	Foundation walls	530-95/9.6.3	Foundation walls
2109.7	*Bond*	*530-95/9.7*	*Bond*
2109.7.1	General	530-95/9.7.1	General
2109.7.2	Masonry headers	530-95/9.7.2	Bonding w/masonry headers
2109.7.3	Wall ties	530-95/9.7.3	Bonding with wall ties
2109.7.4	Longitudinal bond	530-95/9.7.5	Longitudinal bond
2109.8	*Anchorage*	*530-95/9.8*	*Anchorage*
2109.8.1	Intersecting walls	530-95/9.8.2	Intersecting walls
2109.8.2	Floor and roof anchorage	530-95/9.8.3	Floor and roof anchorage
2109.8.3	Walls adjoining structural framing	530-95/9.8.4	Walls adjoining structural framing
2109.9	*Unburned clay masonry*		*Not covered*
2109.9.1	General		Not covered
2109.9.2	Bolts		Not covered
2109.10	*Stone masonry*	*530-95/9.7.4*	*Natural or cast stone*
2109.10.1	General	530-95/9.7.4	Natural or cast stone
2109.10.2	Construction	530-95/9.7.4.1	Ashlar masonry
2109.10.2	Construction	530-95/9.7.4.2	Rubble stone masonry
2109.10.3	Minimum thickness	530-95/9.6.2.2	Minimum thickness
2110	*Glass masonry*	*530-95/Chapter 11*	*Glass unit masonry*
2110.1	General	530-95/11.1	Scope
2110.2	Mortar joints		Not covered
2110.3	Lateral support	530-95/11.4.3	Lateral
2110.4	Reinforcement	530-95/11.7	Reinforcement
2110.5	Size of panels	530-95/11.3	Panel size

1997 UNIFORM BUILDING CODE		ACI 530-95/ASCE 5-95/TMS 402-95	
Section	Title	Section	Title
2110.6	Expansion joints	530-95/11.5	Expansion joints
2110.7	Reuse of units		Not covered
2111	*Chimneys, fireplaces and barbecues*		*Not covered*
Table 21-A	Mortar proportions for units masonry	ASTM C 270	Table 1 proportion specification req.
Table 21-B	Grout proportions by volume	ASTM C 476	Table 1 grout proportions by volume
Table 21-C	Grouting limitations	530-95/Table 3.1.2	Grout space req.
Table 21-C	Grouting limitations	530.1-95/Table 5	Grout space req.
Table 21-D	Specified compressive strength of masonry f'_m (psi) based on specifying the compressive strength of masonry units	530.1-95/Table 1	Compressive strength of masonry based on the compressive strength of clay masonry units and type of mortar used in construction.
Table 21-D	Specified compressive strength of masonry f'_m (psi) based on specifying the compressive strength of masonry units	530.1-95/Table 2	Compressive strength of masonry based on the compressive strength of concrete masonry units and type of mortar used in construction.
Table 21-E-1	Allowable tensions, B_t for embedded anchor bolts for clay and concrete masonry, pounds		Not covered
Table 21-E-2	Allowable tensions, B_t for embedded anchor bolts for clay and concrete masonry, pounds		Not covered
Table 21-F	Allowable shear, B_v for embedded anchor bolts for clay and concrete masonry, pounds		Not covered
Table 21-G	Minimum diameters of bend	530-95/Table 8.5.5.1	Minimum diameters of bend
Table 21-H-1	Radius of gyration for concrete masonry units	530-95/5.13.3	Radius of gyration
Table 21-H-2	Radius of gyration for clay masonry unit length,16 inches	530-95/5.13.3	Radius of gyration
Table 21-H-3	Radius of gyration for clay masonry unit length, 12 inches	530-95/5.13.3	Radius of gyration
Table 21-I	Allowable flexural tensions (psi)	530-95/Table 6.3.1.1	Allowable flexural tension
Table 21-J	Maximum nominal shear strength values		Not covered
Table 21 K	Nominal shear strength coefficient		Not covered
Table 21-L	Shear wall spacing req. for empirical design of masonry	530-95/Table 9.3.1	Shear wall spacing requirements
Table 21-M	Allowable compressive stresses for empirical design of masonry	530-95/Table 9.4.2	Allowable compr. stresses for empirical design of masonry

1997 UNIFORM BUILDING CODE		ACI 530-95/ASCE 5-95/TMS 402-95	
Section	Title	Section	Title
Table 21-N	Allowable shear on bolts for empirically designed masonry except unburned clay units		Not covered
Table 21-O	Wall lateral support requirements for empirical design of masonry	530-95/Table 9.5.1	Wall lateral support requirements
Table 21-P	Thickness of foundation walls for empirical design of masonry	530-95/Table 9.6.3.1	Foundation wall construction
Table 21-Q	Allowable shear on bolts for masonry of unburned clay units		Not covered
1403	*Veneer*	*530-95/CHAPTER 12*	*Veneers*
1403.1	Scope	530-95/12.1	Scope
1403.1.1	General	530-95/12.1.1	Scope
1403.1.2	Limitations	530-95/12.5.1.2	Height limit
1403.2	Definitions	530-95/2.2	Definitions
1403.3	Materials	530.1-95/1.3	References
1403.4	Design	530-95/12.2	General design requirements
1403.4.1	General	530-95/12.2.1	Design requirements
1403.4.2	Adhered veneer		Not covered
1403.4.3	Anchored veneer		Not covered
1403.5	Adhered veneer		Not covered
1403.5.1	Permitted backing		Not covered
1403.5.2	Area limitations		Not covered
1403.5.3	Unit size limitations		Not covered
1403.5.4	Application		Not covered
1403.5.5	Ceramic tile		Not covered
1403.6	Anchored veneer		
1403.6.1	Permitted backing	530-95/12.2.1	Backing system of exterior veneer
1403.6.2	Height and support limitations	530-95/12.5.1	Weight of anchored veneer
1403.6.3	Area limitations		Not covered
1403.6.4	Application		
1403.6.4.1	General		Not covered
1403.6.4.2	Masonry and stone units [5 inches (127 mm) maximum in thickness]	530-95/12.7	Anchor requirements
1403.6.4.2	Masonry and stone units [5 inches (127 mm) maximum in thickness]	530-95/12.8	Masonry veneer anchored to wood backing
1403.6.4.3	Stone units [10 in. (254 mm) max. in thickness]	530-95/12.9	Masonry veneer anchored to steel backing
1403.6.4.3	Stone units [10 in. (254 mm) max. in thickness]	530-95/12.10	Masonry veneer anchored to masonry or concrete backing
1403.6.4.4	Slab-type units [2 inches (51 mm) maximum in thickness]		Not covered
1403.6.4.5	Terra cotta or ceramic units.		Not covered
Table 14-A	Ceramic tile setting mortars		Not covered

ACI 530-95ASCE 5-95/TMS 402-95		1997 UNIFORM BUILDING CODE	
Section	Title	Section	Title
530-95/Part 1	General	2101	General
530-95/Chapter 1	General Requirements		
530-95/1.1	Scope	2101.1	Scope
530-95/1.2	Contract documents		
530-95/1.2.1	Project drawings	2106.1.2	Plans
530-95/1.2.2	Drawing items	106.3.3	Information on plans & specifications
530-95/1.2.3	Document coordination	106.3.5	Inspection and observation program
530-95/1.2.4	Design calculations	106.3.1	Application for permit
530-95/1.3	Approval of special systems		
530-95/1.3.1	Alternate design and construction	104.2.8	Alternate methods, designs
530-95/1.4	Standards cited in this Code		
530-95/1.4.1	Standards		Not covered
530-95/Chapter 2	Notations and Definitions		
530-95/2.1	Notations	2101.4	Notations
530-95/2.2	Definitions	2101.3	Definitions
530-95/2.2	Effective height (Definition)	2106.2.4	Effective height
530-95/2.2	Definitions	1403.2	Definitions
530-95/Part 2	Quality Assurance and Construction Requirements	2105	Quality Assurance
530-95/Chapter 3	General		
530-95/3.1.1	Material labor and construction	2105.1	General
530-95/3.1.1	Material labor and construction.	2105.2	Scope
530-95/3.1.2	Grouting, minimum spaces	2104.6.1	General conditions
530-95/Table 3.1.2	Grout space requirements	Table 21-C	Grouting limitations
530-95/3.2	Acceptance relative to strength requirements	2105.3	Compliance with f_m
530-95/3.2	Acceptance relative to strength requirements	2105.3.1	General
530-95/Chapter 4	Embedded Items-Anchorage of Masonry to Framing and Existing Construction		
530-95/4.1	Embedded conduits, pipes	2106.1.9	Pipes and conduits embedded in masonry
530-95/4.2	Masonry attachment to structural frames	2106.2.11	Walls intersecting with floors and roofs
530-95/4.2.1	Anchorage of masonry	2109.4.4	Anchor bolts
530-95/4.3	Connectors		
530-95/4.3.1	Location of connectors		Not covered
530-95/Part 3	Analysis and Design	2106	General Design Requirements
530-95/Chapter 5	General Analysis and Design Requirements		
530-95/5.1	Scope	2106.1	General
530-95/5.1.1	Design scope	2101.2	Design methods
530-95/5.1.1	Scope	2106.1.1	Scope
530-95/5.1.1	Scope	2107.1.1	Scope

ACI 530-95ASCE 5-95/TMS 402-95		1997 UNIFORM BUILDING CODE	
Section	Title	Section	Title
530-95/5.1.2	Non-lateral force members		Not covered
530-95/5.1.3	Empirical design	2109.1	General
530-95/5.1.4	Glass unit masonry design		Not covered
530-95/5.1.5	Masonry veneer design	1403.4.1	General design
530-95/5.2	*Loading*		
530-95/5.2.1	Loading assumptions	2106.1.3	Design loads
530-95/5.2.2	Loading	2106.1.3	Design loads
530-95/5.2.3	Load resistance	2106.1.3	Design loads
530-95/5.2.4	Load considerations		Not covered
530-95/5.3	*Load combinations*		
530-95/5.3.1	Load combinations		Not covered
530-95/5.3.2	One third load increase		
530-95/5.4	Design strength	2106.2.2	Specified compressive strength of masonry
530-95/5.5	Determination of material properties	2107.1.8.2	Determination of moduli of elasticity
530-95/5.5	Material properties	2107.1.8.3.2	Material properties
530-95/5.5.1	General	2106.2.12.1	Modulus of elasticity of masonry
530-95/5.5.2	Elastic moduli	2106.2.13	Shear modulus of masonry
530-95/5.5.2.1	Steel reinforcement	2106.2.12.2	Modulus of elasticity of steel
530-95/5.5.2.2	Clay masonry	2106.2.12.1	Modulus of elasticity of clay masonry
530-95/Table 5.5.2.2	Clay masonry moduli		Not covered
530-95/5.5.2.3	Concrete masonry	2106.2.12.1	Modulus of elasticity of concrete masonry
530-95/Table 5.5.2.3	Concrete masonry moduli		Not covered
530-95/5.5.2.4	Grout		Not covered
530-95/5.5.3	*Thermal expansion coefficients*		
530-95/5.5.3.1	Clay masonry		Not covered
530-95/5.5.3.2	Concrete masonry		Not covered
530-95/5.5.4	Moisture expansion coefficient		Not covered
530-95/5.5.5	*Shrinkage coefficients*		
530-95/5.5.5.1	Moisture controlled units		Not covered
530-95/5.5.5.2	Non moisture controlled units		Not covered
530-95/5.5.6	*Creep coefficients*		
530-95/5.5.6.1	Clay masonry		Not covered
530-95/5.5.6.2	Concrete masonry		Not covered
530-95/5.6	Deflection of beams and lintels	2106.2.9	Vertical deflection
530-95/5.7	Lateral load distribution	2106.1.7	Lateral support
530-95/5.8	*Multiwythe walls*	*2106.1.5*	*Multiwythe walls*
530-95/5.8.1	General	2106.1.5.1	General
530-95/5.8.2	Wythes of walls	2107.1.8.3.1	Bonding of wythes
530-95/5.8.2.1	Multiwythe walls	2106.1.5.1	General
530-95/5.8.2.2.	Shear stresses		Not covered
530-95/5.8.2.3	Stress determination		Not covered

ACI 530-95ASCE 5-95/TMS 402-95		1997 UNIFORM BUILDING CODE	
Section	Title	Section	Title
530-95/5.8.2.4	Header requirements		Not covered
530-95/5.8.2.5	Wall ties	2106.1.5.2	Wall ties in cavity wall construction
530-95/5.8.2.5	Wall ties	2106.1.5.3	Wall ties for grouted multiwythe construction
530-95/5.8.3	Noncomposite action		Not covered
530-95/5.9	Columns		
530-95/5.9.1	Design of columns	2107.2.13	Reinforcement for columns
530-95/5.9.1.1	Minimum column dimension	2107.1.3.2	Columns
530-95/5.9.1.2	Height ratio		Not covered
530-95/5.9.1.3	Eccentricity resistance		Not covered
530-95/5.9.1.4	Vertical column reinforcement	2107.2.13.1	Vertical reinforcement
530-95/5.9.1.5	Stresses	2107.1.6.2	Columns, axial loads
530-95/5.9.1.5	Stresses	2107.1.6.3	Columns, bending or combined bending and axial loads
530-95/5.9.1.6	Lateral ties	2106.3.6	Lateral ties
530-95/5.10	Pilasters		Not covered
530-95/5.11	Load transfer at horiz. connection		
530-95/5.11.1	Load transfer at horiz. connection	2106.2.10	Structural continuity
530-95/5.11.2	Design of devices		Not covered
530-95/5.12	Concentrated loads		
530-95/5.12.1	Concentrated loads	2106.2.7	Distribution of concentrated vertical loads in walls
530-95/5.12.2	Bearing stresses	2107.2.10	Allowable bearing stress
530-95/5.12.3	Bearing stress limitation	2107.2.10	Allowable bearing stress
530-95/5.12.3	Bearing stress limitation	2107.3.8	Allowable bearing stress
530-95/5.13	Section properties		
530-95/5.13.1	Stress computations		
530-95/5.13.1.1	Stress computations		Not covered
530-95/5.13.1.2	Composite action design	2107.1.8.1	General
530-95/5.13.1.2	Composite stress computations	2107.1.8.4	Design procedure, transformed sections
530-95/5.13.2	Stiffness		Not covered
530-95/5.13.3	Radius of gyration	Table 21-H-1	Radius of gyration for concrete masonry units
530-95/5.13.3	Radius of gyration	Table 21-H-2	Radius of gyration for clay masonry unit length, 16 Inches
530-95/5.13.3	Radius of gyration	Table 21-H-3	Radius of gyration for clay masonry unit length, 12 Inches
530-95/5.13.4	Intersecting walls		
530-95/5.13.4.1	Wall intersections		Not covered
530-95/5.13.4.2	Design of wall intersections		Not covered
530-95/5.13.4.2.a	Running bond		Not covered
530-95/5.13.4.2.b	Flanges		Not covered
530-95/5.13.4.2.c	Width of flange	2106.2.6	Effective width of intersecting walls
530-95/5.13.4.2.d	Shear design		Not covered

ACI 530-95ASCE 5-95/TMS 402-95		1997 UNIFORM BUILDING CODE	
Section	Title	Section	Title
530-95/5.13.4.2.e	Intersecting wall connections		Not covered
530-95/5.14	*Anchors bolts in masonry*	*2107.1.5*	*Embedded anchor bolts*
530-95/5.14.1	Test design requirements	2107.1.5.1	General
530-95/5.14.1.1	Anchor bolt testing		Not covered
530-95/5.14.1.2	Anchor bolt load limitations	2107.1.5	Embedded anchor bolts
530-95/5.14.2	Plate, headed and bent bar bolts	2107.1.5	Embedded anchor bolts
530-95/5.14.2.1	Minimum embedment length	2106.2.14.3	Minimum embedment depth
530-95/5.14.2.2	Allowable load in tension	2107.1.5.2	Tension
530-95/5.14.2.3	Allowable load in shear, l_{be}	2106.2.14.2	Minimum edge distance
530-95/5.14.2.3	Allowable load in shear	2107.1.5.3	Shear
530-95/5.14.2.4	Combined shear and tension	2107.1.5.4	Combined shear and tension
530-95/5.15	Framed construction	2106.2.8	Loads on non-bearing walls
530-95/5.16	Stack bond masonry	2107.3.14	Stack bond
530-95/Chapter 6	*Unreinforced Masonry*	*2107.3*	*Design of Unreinforced Masonry*
530-95/6.1	Scope	2107.3.1	General
530-95/6.2	Stresses in reinforcement		Not covered
530-95/6.3	*Axial compression and flexure*		
530-95/6.3.1	Axial compression and flexure	2107.1.6.1	Walls, axial loads
530-95/6.3.1	Axial compression and flexure	2107.2.7	Combined compressive stresses, unity formula
530-95/6.3.1	Axial compression and flexure	2107.3.2	Allowable axial compressive stress
530-95/6.3.1	Axial compression and flexure	2107.3.4	Combined compressive stresses, unity formula
530-95/6.3.1	Axial compression and flexure	2107.3.9	Combined bending and axial loads, compressive
530-95/6.3.1	Axial compression and flexure	2107.3.10	Compression in walls and columns
530-95/6.3.1.c	Axial compression and flexure	2107.3.3	Allowable flexural compressive stress
530-95/6.3.1.1	Allowable tensile stresses	2107.3.5	Allowable tensile stress
530-95/Table 6.3.1.1	Allowable flexural tension	Table 21-I	Allowable flexural tensions (psi)
530-95/6.4	Axial tension		Not covered
530-95/6.5	*Shear*		
530-95/6.5.1	Shear stresses	2107.3.12	Shear in flexural members and shear walls
530-95/6.5.2	In-plane shear stresses	2107.3.6	Allowable shear stress in flexural members
530-95/6.5.2	In-plane shear stresses	2107.3.7	Allowable shear stress in shear walls
530-95/6.5.3	Shear stress limitations		Not covered
530-95/Chapter 7	*Reinforced Masonry*	*2107.2*	*Design of Reinforced Masonry*
530-95/7.1	*Scope*	*2107.2.1*	*Scope*
530-95/7.1.1	Design of structures	2107.2.3	Design assumptions
530-95/7.1.2	General requirements	2106.2.1	General
530-95/7.1.2	Scope	2106.3.1	General
530-95/7.2	*Steel reinforcement*		
530-95/7.2.1	Allowable stresses	2107.2.11	Allowable stresses in reinforcement

ACI 530-95ASCE 5-95/TMS 402-95		1997 UNIFORM BUILDING CODE	
Section	Title	Section	Title
530-95/7.3	*Axial compression and flexure*	*2107.2.2*	*Reinforcement*
530-95/7.3.1	Members subjected to combined loads	2107.2.14.2	Walls, bending or combined bending and axial loads
530-95/7.3.2	Allowable forces and stresses	2107.2.14	Compression in walls and columns
530-95/7.3.2.1	Axial compressive stress	2107.2.5	Allowable flexural compressive stress and force
530-95/7.3.2.2	Flexural compressive stress	2107.2.6	Allowable flexural compressive stress
530-95/7.3.2.2	Compressive stress in reinforced masonry	2107.2.14.1	General
530-95/7.3.3	*Effective compressive width*		
530-95/7.3.3.1	Width of the compression area	2106.3.8	Effective width *b* of compression area
530-95/7.3.3.2	Effective compressive width	2106.3.8	Effective width *b* of compression area
530-95/7.3.4	Beams		Not covered
530-95/7.4	Axial tension	2107.2.3	Design assumptions
530-95/7.5	*Shear*		
530-95/7.5.1	Flexural member non-tension design		Not covered
530-95/7.5.1.1	Reinforcement		Not covered
530-95/7.5.1.2	Calculated shear stress		Not covered
530-95/7.5.2	Flexural member tension design		Not covered
530-95/7.5.2.1	Calculated shear stress	2107.2.17	Shear in flexural members
530-95/7.5.2.2.a	Allowable flexural stresses	2107.2.8	Allowable shear stress in flexural members
530-95/7.5.2.2.b	Allowable shear stresses	2107.2.9	Allowable shear stress in shear walls
530-95/7.5.2.3.a	Allowable flexural stresses	2107.2.8	Allowable shear stress in flexural members
530-95/7.5.2.3.b	Allowable shear stresses	2107.2.9	Allowable shear stress in shear walls
530-95/7.5.2.4	M/Vd ratio		Not covered
530-95/7.5.3	Minimum area of shear reinf.	2107.2.17	Shear in flexural members
530-95/7.5.3.1	Spacing of shear reinforcement	2106.1.12.4.2	Shear walls-reinforcement
530-95/7.5.3.2	Spacing of shear reinforcement	2106.1.12.4.2	Shear walls-wall reinforcement
530-95/7.5.4	Shear stresses in collar joints		Not covered
530-95/7.5.5	Cantilever beams		Not covered
530-95/Chapter 8	*Details of Reinforcement*		
530-95/8.1	Scope		Not covered
530-95/8.2	*Size of reinforcement*		
530-95/8.2.1	Maximum reinforcement size	2107.2.2.1	Maximum reinforcement size
530-95/8.2.2	Maximum reinforcement diameter		Not covered
530-95/8.2.3	Joint reinforcement	2106.1.5.4	Joint reinforcement
530-95/8.3	*Placement limits for reinforcement*		
530-95/8.3.1	Clear distance between parallel bars	2106.3.3	Spacing of longitudinal reinforcement
530-95/8.3.2	Column clear bar distance	2107.2.13.1	Vertical reinforcement
530-95/8.3.3	Contact lap splice	2107.2.2.6	Splices
530-95/8.3.4	Bundled bars		Not covered

ACI 530-95ASCE 5-95/TMS 402-95		1997 UNIFORM BUILDING CODE	
Section	Title	Section	Title
530-95/8.3.5	Reinforcement embedded in grout	2106.3.3	Spacing of longitudinal reinforcement
530-95/8.4	*Protection for reinforcement*		
530-95/8.4.1	Reinforcing bars minimum cover	2107.2.2.2	Cover
530-95/8.4.2	Joint reinforcement	2106.1.5.4	Joint reinforcement
530-95/8.4.3	Wall tie protection	2106.1.8	Protection of ties and joint reinforcement
530-95/8.5	*Development of reinforcement*		
530-95/8.5.1	Development of reinforcement	2106.3.4	Anchorage of flexural reinforcement
530-95/8.5.2	Embedment of bars and wires in tension	2107.2.2.3	Development length
530-95/8.5.3	Embedment of flexural reinforcement	2106.3.4	Anchorage of flexural reinforcement
530-95/8.5.4	Standard hooks	2107.2.2.5	Hooks
530-95/8.5.5	*Minimum bend diameter*		
530-95/8.5.5.1	Diameter of bend	2107.2.2.5	Hooks
530-95/Table 8.5.5.1	Minimum diameters of bend	Table 21-G	Minimum diameters of bend
530-95/8.5.5.2	Standard hook development		Not covered
530-95/8.5.5.3	Hook compression	2107.2.2.5.6	Hook compression
530-95/8.5.6	*Development of shear reinforcement*		
530-95/8.5.6.1	Bar and wire reinforcement	2106.3.5	Anchorage of shear reinforcement
530-95/8.5.6.2	Welded wire fabric		Not covered
530-95/8.5.7	Splices of reinforcement	2107.2.2.6	Splices
530-95/8.5.7.1	Lap splices	2107.2.2.3	Development length
530-95/8.5.7.2	Welded splices	2107.2.2.6	Splices
530-95/8.5.7.3	Mechanical connections	2107.2.2.6	Splices
530-95/8.5.7.4	End bearing splices		Not covered
530-95/Chapter 9	*Empirical Design of Masonry*	*2109*	*Empirical Design of Masonry*
530-95/9.1	Scope	2109.1	General
530-95/9.2	Height	2109.2	Height
530-95/9.3	Lateral stability	2109.3	Lateral stability
530-95/Table 9.3.1	Shear wall spacing requirements	Table 21-L	Shear wall spacing requirements for Empirical design of masonry
530-95/9.4	*Compressive stress requirements*	*2109.4*	*Compressive stresses*
530-95/9.4.1	Compressive stresses in masonry	2109.4.1	General
530-95/9.4.2	Compressive stresses (Allowable)	2109.4.2	Allowable stresses
530-95/Table 9.4.2	Allowable compressive stresses for Empirical design of masonry	Table 21-M	Allowable compressive stresses for Empirical design of masonry
530-95/9.4.2.1	Calculated compressive stresses	2109.4.3	Stress calculations
530-95/9.4.2.2	Multiwythe walls	2109.4.2	Allowable stresses
530-95/9.5	Lateral support	2109.5	Lateral support
530-95/Table 9.5.1	Wall lateral support requirements	Table 21-O	Wall lateral support requirements for Empirical design of masonry

ACI 530-95ASCE 5-95/TMS 402-95		1997 UNIFORM BUILDING CODE	
Section	Title	Section	Title
530-95/9.6	*Thickness of masonry*	*2109.6*	*Minimum thickness*
530-95/9.6.1	General	2109.6.1	General
530-95/9.6.2	*Thickness of walls*		
530-95/9.6.2.1	Nominal thickness of masonry	2109.6.1	General
530-95/9.6.2.2	Minimum thickness	2109.10.3	Minimum thickness
530-95/9.6.2.3	Change in thickness	2109.6.2	Variation in thickness
530-95/9.6.2.3	Change in thickness	2109.6.3	Decrease in thickness
530-95/9.6.3	Foundation walls	2109.6.5	Foundation walls
530-95/Table 9.6.3.1	Foundation wall construction	Table 21-P	Thickness of foundation walls for Empirical design of masonry
530-95/9.6.4	Parapet walls	2109.6.4	Parapets
530-95/9.7	*Bond*	*2109.7*	*Bond*
530-95/9.7.1	General	2109.7.1	General
530-95/9.7.2	Bonding with masonry headers	2109.7.2	Masonry headers
530-95/9.7.3	Bonding with wall ties	2109.7.3	Wall ties
530-95/9.7.4	Natural or cast stone	2109.10	Stone masonry
530-95/9.7.4	Natural or cast stone	2109.10.1	General
530-95/9.7.4.1	Ashlar masonry	2109.10.2	Construction
530-95/9.7.4.2	Rubble stone masonry	2109.10.2	Construction
530-95/9.7.5	Longitudinal bond	2109.7.4	Longitudinal bond
530-95/9.8	*Anchorage*	*2109.8*	*Anchorage*
530-95/9.8.1	General		Not covered
530-95/9.8.2	Intersecting walls	2109.8.1	Intersecting walls
530-95/9.8.3	Floor and roof anchorage	2109.8.2	Floor and roof anchorage
530-95/9.8.4	Walls adjoining structural framing	2109.8.3	Walls adjoining structural framing
530-95/9.9	Miscellaneous requirements		Not covered
530-95/9.9.1	Chases and recesses		Not covered
530-95/9.9.2	Lintels		Not covered
530-95/9.9.3	Support on wood		Not covered
530-95/9.9.4	Corbelling	2107.3.13	Corbels
530-95/Chapter 10	*Seismic Design Requirements*	*2106.1.12*	*Special Provisions in Areas of Seismic Risk*
530-95/10.1	Scope	2106.1.12.1	General
530-95/10.2	*General*		
530-95/10.2.1	General	2106.1.12.1	General
530-95/10.2.2	Strength requirements		Not covered
530-95/10.3	Seismic Performance Category A	2106.1.12.2	Special provisions for Seismic Zones 0 and 1
530-95/10.4	Seismic Performance Category B	2106.1.12.2	Special provisions for Seismic Zones 0 and 1
530-95/10.5	Seismic Performance Category C	2106.1.12.3	Special provisions for Seismic Zone 2
530-95/10.6	Seismic Performance Category D	2106.1.12.4	Special provisions for Seismic Zones 3 and 4

ACI 530-95ASCE 5-95/TMS 402-95		1997 UNIFORM BUILDING CODE	
Section	Title	Section	Title
530-95/10.7	Seismic Performance Category E	2106.1.12.4	Special provisions for Seismic Zones 3 and 4
530-95/Chapter 11	*Glass Unit Masonry*	*2110*	*Glass Masonry*
530-95/11.1	Scope	2110.1	General
530-95/11.2	Units		Not covered
530-95/11.3	Panel size	2110.5	Size of panels
530-95/11.4	*Support*		
530-95/11.4.1	Isolation		Not covered
530-95/11.4.2	Vertical		Not covered
530-95/11.4.3	Lateral	2110.3	Lateral support
530-95/11.5	Expansion joints	2110.6	Expansion joints
530-95/11.6	Mortar		Not covered
530-95/11.7	Reinforcement	2110.4	Reinforcement
530-95/Chapter 12	*Veneers*	*1403*	*Veneer*
530-95/12.1	Scope	1403.1	Scope
530-95/12.1.1	Scope	1403.1.1	General
530-95/12.2	General design requirements	1403.4	Design
530-95/12.2.1	Design requirements	1403.4.1	General
530-95/12.2.1	Backing system of exterior veneer	1403.6.1	Permitted backing
530-95/12.3	Alternative design of anchored masonry veneer		Not covered
530-95/12.4	Prescriptive requirements for anchored masonry veneer		Not covered
530-95/12.5	Vertical support of anchored masonry veneer		Not covered
530-95/12.5.1	Weight of anchored veneer	1403.6.2	Height and support limitations
530-95/12.5.1.1	Wood support		Not covered
530-95/12.5.1.2	Height limit	1403.1.2	Limitations
530-95/12.5.1.3	Story support	1403.6.2	Height and support limitations
530-95/12.5.1.4	Veneer on wood framing	1403.1.2	Limitations
530-95/12.6	Masonry units		Not covered
530-95/12.7	Anchor requirements	1403.6.4.2	Masonry and stone units [5 inches (127 mm) maximum in thickness]
530-95/12.8	Masonry veneer anchored to wood backing	1403.6.4.2	Masonry and stone units [5 inches (127 mm) maximum in thickness]
530-95/12.9	Masonry veneer anchored to steel backing	1403.6.4.3	Stone units [10 in. (254 mm) max. in thickness]
530-95/12.10	Masonry veneer anchored to masonry or concrete backing	1403.6.4.3	Stone units [10 in. (254 mm) max. in thickness]
530-95/12.11	Veneer laid on other than running bond		Not covered
530-95/12.12	Requirements in seismic areas	1403.6.4.2	Masonry and stone units [5 inches (127 mm) maximum in thickness]

ACI 530-95ASCE 5-95/TMS 402-95		1997 UNIFORM BUILDING CODE	
Section	Title	Section	Title
530.1-95	*Specifications for Masonry Structures*	*2104*	*Construction*
530.1-95/Part 1	*General*		
530.1-95/1.1	Summary	2104.1	General
530.1-95/1.2	Definitions	2101.3	Definitions
530.1-95/1.3	*References*	*2102*	*Material standards*
530.1-95/1.3	References	2102.1	Quality
530.1-95/1.3	References	2102.2	Standards of quality
530.1-95/1.3	References	1403.3	Materials
530.1-95/1.4	*System description*		
530.1-95/1.4.A	Compressive strength requirements		Not covered
530.1-95/1.4.B.1	Alternatives for determination of compressive strength	2105.3.1	Compliance with f_m'/General
530.1-95/1.4.B.2	Unit strength method	2105.3.4	Unit strength method
530.1-95/1.4.B.3	Prism Test Method	2105.3.2	Masonry prism testing
530.1-95/Table 1	Compressive strength of masonry based on the compressive strength of clay masonry units and type of mortar used in construction.	Table 21-D	Specified compressive strength of masonry f_m' (psi) based on specifying the compressive strength of masonry units
530.1-95/Table 2	Compressive strength of masonry based on the compressive strength of concrete masonry units and type of mortar used in construction.	Table 21-D	Specified compressive strength of masonry f_m' (psi) based on specifying the compressive strength of masonry units
530.1-95/Table 3	Correction factors for clay masonry prism strength	Table 21-17-A	UBC Standard/Correction factors
530.1-95/Table 4	Correction factors for concrete masonry prism strength	Table 21-17-A	UBC Standard/Correction factors
530.1-95/1.5	Submittals		Not covered
530.1-95/1.6	Quality assurance	2105	Quality assurance
530.1-95/1.7	Delivery, storage and handling	2104.2	Materials: handling, storage and preparation
530.1-95/1.8	Project conditions		Not covered
530.1-95/1.8.A	Construction loads		Not covered
530.1-95/1.8.B	Masonry protection	2104.3.4	Protection
530.1-95/1.8.C	Cold weather construction	2104.3	Cold weather construction
530.1-95/1.8.C	Cold weather construction	2104.3.1	General
530.1-95/1.8.C	Cold weather construction.	2104.3.3	Construction
530.1/95/1.8.D	Hot weather construction		Not covered
530.1-95/Part 2	*Products*		
530.1-95/2.1	Mortar materials	2103	Mortar and grout
530.1-95/2.2	Grout materials	2103	Mortar and grout

ACI 530-95ASCE 5-95/TMS 402-95		1997 UNIFORM BUILDING CODE	
Section	Title	Section	Title
530.1-95/2.2	Grout materials	2103.4	Grout
530.1-95/2.3	Masonry materials	2102.1	Quality
530.1-95/2.4	Reinforcement and metal accessories		
530.1-95/2.4.A	Reinforcing steel	2102.2.10	Reinforcement
530.1-95/2.4.B	Joint reinforcement	2102.2.10	Reinforcement
530.1-95/2.4.C	Anchors, ties and accessories	2102.2.7	Connectors
530.1-95/2.4.D	Stainless steel		Not covered
530.1-95/2.4.E	Coatings for corrosion protection	2102.2.7	Connectors
530.1-95/2.5	Accessories		Not covered
530.1-95/2.6	Mixing		
530.1-95/2.6.A	Mortar	2103.5	Additives and admixtures
530.1-95/2.6.A.1	Mixing time	2104.2.7	Mixing time
530.1-95/2.6.A.2	Maximum mortar color	2103.5.4	Colors
530.1-95/2.6.A.3	Chlorides	2103.5.2	Antifreeze compounds
530.1-95/2.6.A.4	Glass unit masonry		Not covered
530.1-95/2.6.B	Grout	2103.5	Additives and admixtures
530.1-95/2.7	Fabrication		Not covered
530.1-95/2.8	Source quality	2102.2	Standards of quality
530.1-95/Part 3	Execution		
530.1-95/3.1	Inspection		Not covered
530.1-95/3.2	Preparation		
530.1-95/3.2.A	Cleaning of reinforcement		
530.1-95/3.2.B	Clean surfaces	2104.4.2	Surfaces
530.1-95/3.2.C	Wetting masonry units	2104.2.2	Materials/Reinforcement
530.1-95/3.2.D	Debris	2104.6.1	General conditions
530.1-95/3.2.E	Reinforcement	2104.6.2	Construction requirements
530.1-95/3.2.F	Cleanouts	2104.6.2	Construction requirements
530.1-95/3.3	Masonry erection	2104.4	Placing masonry units
530.1-95/3.3.A	Bond pattern		Not covered
530.1-95/3.3.B	Placing mortar and units	2104.4.1	Mortar
530.1-95/3.3.B.1	Mortar fins	2104.6.1	General conditions
530.1-95/3.3.B.1	Mortar fins	2104.6.2	Construction requirements
530.1-95/3.3.B.1	Bed and head joints	2106.1.6	Vertical support
530.1-95/3.3.B.2	Collar joints		Not covered
530.1-95/3.3.B.3	Hollow units	2104.4.4	Hollow-masonry units
530.1-95/3.3.B.4	Solid units	2104.4.3	Solid masonry units
530.1-95/3.3.B.5	Glass units	2110.2	Mortar joints
530.1-95/3.3.B.6	All units		Not covered
530.1-95/3.3.C	Prefabricated concrete and masonry items		Not covered
530.1-95/3.3.D	Embedded items and accessories	2104.7	Aluminum euipment
530.1-95/3.3.E	Bracing of masonry		Not covered

ACI 530-95ASCE 5-95/TMS 402-95		1997 UNIFORM BUILDING CODE	
Section	Title	Section	Title
530.1-95/3.3.F	Site tolerances		Not covered
530.1-95/3.4	*Reinforcement installation*		
530.1-95/3.4.A	Basic requirements	2104.5	Reinforcement placing
530.1-95/3.4.B	Securing reinforcement	2104.5	Reinforcement placing
530.1-95/3.4.C	*Details of reinforcement*		
530.1-95/3.4.C.1	Clear distances	2106.3.3	Spacing of longitudinal reinforcement
530.1-95/3.4.C.2	Joint reinforcement	2104.8	Joint reinforcement
530.1-95/3.4.C.3	Bending of reinforcement		Not covered
530.1-95/3.4.C.4	Protection for reinforcement	2106.1.8	Protection of ties and joint reinforcement
530.1-95/3.4.D	Wall ties	2106.1.5.2	Wall ties in cavity wall construction
530.1-95/3.4.E	Reinforcement	2104.5	Reinforcement placing
530.1-95/3.4.F	Glass unit masonry panel anchors	2110.3	Lateral support
530.1-95/3.5	*Grout placement*	*2104.6*	*Grouted masonry*
530.1-95/3.5	Grout placement	2104.6.1	General conditions
530.1-95/Table 5	Grout space requirements	Table 21-C	Grouting limitations
530.1-95/3.6	*Field quality control*		
530.1-95/3.6.A	*Prism testing*	*2105.3.2*	*Masonry prism testing*
530.1-95/3.6.B	Mortar testing	2105.4	Mortar testing
530.1-95/3.6.C	Grout testing	2105.5	Grout testing
530.1-95/3.7	Cleaning		Not covered

INDEX

A

B

INDEX

F

G

INDEX

H

INDEX

INDEX

INDEX

INDEX

W

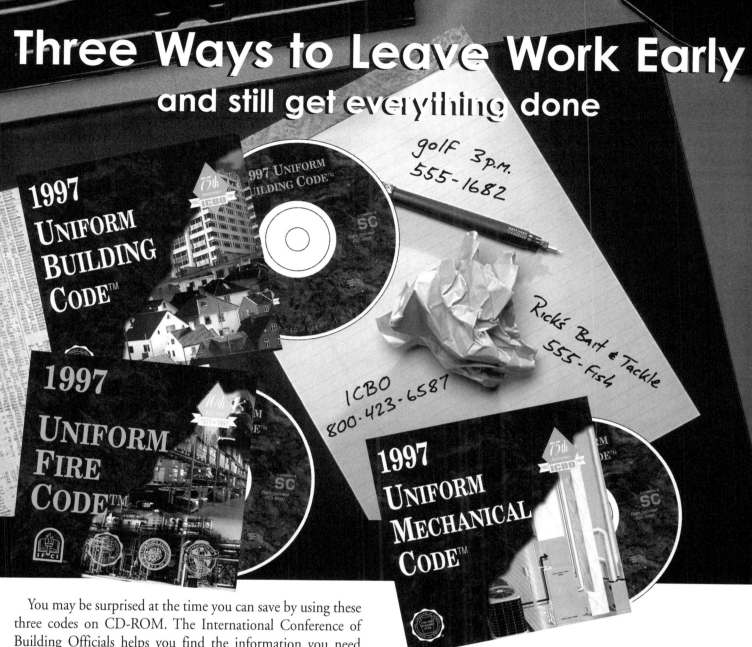